Lecture Notes in Mathematics 1573

Editors:
A. Dold, Heidelberg
B. Eckmann, Zürich
F. Takens, Groningen

V. P. Havin N. K. Nikolski (Eds.)

Linear and Complex Analysis Problem Book 3

Part I

Springer-Verlag
Berlin Heidelberg New York
London Paris Tokyo
Hong Kong Barcelona
Budapest

Editors

Victor P. Havin
Department of Mathematics and Mechanics
St. Petersburg State University
Staryi Peterhof
St. Petersburg, 198904, Russia

Nikolai K. Nikolski
UFR de Mathématiques
Université Bordeaux-I
351, cours de la Libération
33405 Talence CEDEX, France

Mathematics Subject Classification (1991): 30B, 30C, 30D, 30E, 30H, 31, 32, 46B, 46D, 46H, 46J, 46K, 46L, 42A, 42B, 45, 47A, 47B, 93B

ISBN 3-540-57870-6 Springer-Verlag Berlin Heidelberg New York
ISBN 0-387-57870-6 Springer-Verlag New York Berlin Heidelberg

CIP-data applied for

Printed in Germany

SPIN: 10078827 46/3140-543210 - Printed on acid-free paper

CONTENTS

VOLUME 1

VOLUME 2

PREFACE

In 1978 we published a book entitled "99 unsolved problems of Linear and Complex Analysis" (Volume 81 of "Zapiski nauchnyh seminarov LOMI"; English translation in Journal of Soviet Mathematics, **26** (1984), No. 5). It consisted of short problem articles sent by mathematicians of many countries in response to our invitation headed by the following lines:

'Which problems of Linear and Complex Analysis would you propose to your numerous colleagues if you had a possibility to address them all simultaneously?
The editorial board of "Investigations in Linear Operators and Function Theory" edited by the Leningrad Branch of the V. A. Steklov Mathematical Institute of the Academy of Sciences of the USSR (LOMI) has decided to put this question to a hundred specialists joined in an invisible collective working on the common circle of problems and to publish their answers as "Collection of unsolved Problems of Linear and Complex Analysis". Such "Collection ..." may be useful not only to its authors but to their colleagues including the beginner analysts.'

It seems we were right. In 1984 the second edition appeared. Instead of 99, its title mentioned 199 problems.* Both editions have interested many colleagues. A big part of problems is now solved, but time has brought with it more new problems and questions. That is why one more (yet again enlarged) publication of the Collection seemed desirable. Its third version reproduces a large part of the second with addition of new problems and of information concerning the old ones. Our motives, the style and the general direction of the book were described in detail in the preface to the 2nd edition. Excerpts from that preface are reproduced below. There is not much to add. We only make several remarks on some *new* moments.

The first is the increase of the size. Instead of 13, the number of chapters is now 20; the total number of problems is 341 (they were 199 in 1983). The book consists of two volumes (both preceding editions were one-volume books). This growth can be explained by the abundance of new results and ideas in Spectral Operator-and-Function Theory. Our purely operator-theoretic chapters are now six (namely, Chapters 4–9); they were only two in the second edition. This fact bears witness to the intense activity of operator theorists gaining new areas and discovering new connections. So much for Operator Theory, a key subject underlying and unifying the whole book; the word "Linear" in the title refers mainly to this theme. As to the word "Complex", this part of the book is also enriched by the inclusion of new chapters 13, 18, 19, not to mention new problems gathered under the "old" titles.

The second moment is the new technique of preparation of the text. Both preceding versions of the book (as a whole) were prepared by its editors (though assisted by a collective of collaborators). This time every chapter had its own editor (or editors). The

---

Linear and complex Analysis Problem Book. 199 Research Problems.* Lect. Notes Math. **1043, Springer–Verlag, 1984

initiative and organization and coordination problems were ours, a difficult task, to say the least (see also the explanations in *Acknowledgements* below). Almost all chapters are provided with introductions by the chapter editors. In these introductions they try to help the reader to grasp the general direction of the chapter, to record additional bibliography, and sometimes also to explain their point of view on the subject or to make historical comments.

Chapters are divided into sections. They total 341 (in 1984 and 1978 there were 199 and 99 respectively). We treat the words "section" and "problem" as synonymous for the purposes of classification (though a section may contain more than one problem). "Problem 1.25" means the 25-th section of the first chapter; "Problem 1.26 old" ("Problem 1.26 v. old") mean that Problem 1.26 is reproduced from the 1984 edition (1978 edition, respectively) and has not been completely solved (as far as we know); "Problem **S**.1.27" means the 27-th section of Chapter 1 representing a solution of a problem from the previous edition. Some notation (used sometimes without further explanations) is indicated at the end of the book. A subject index and an author index are provided. We took the liberty to modify the section titles in "Contents" to make it shorter.

And **the third moment** in which this edition differs from its predecessors is the unfavorable situation in former Soviet mathematics caused by the well-known events that interfered brutally with our project just when it was started and could not be stopped. As we already mentioned, both preceding versions were prepared by "an informal editorial board" consisting for the most part of the members of the (then) Leningrad Seminar of the Spectral Function-and-Operator Theory. It was a numerous and energetic group of enthusiasts whose participation ensured the success of the undertaking. In 1990, proposing the project of the 3rd edition to our colleagues throughout the world, we hoped that we still could rely upon the same group. We also reckoned with the technical group of LOMI (now POMI), the Leningrad Branch of the Steklov Institute, remembering our experience of 1978 and 1983. But when our project was really started the situation changed dramatically. Our group melted away and soon became unable to achieve a joint effort, and POMI couldn't support us anymore (such things as, say, keyboarding, paper and so on, are now a big problem in Russia). In fact, the project turned out to be a purely private enterprise of the editors.

But now, after all, thanks to generous help of our friends and colleagues (see *Acknowledgements* below) this book lies before its reader. We hope that it will serve "the invisible community" of analysts working in Linear and Complex Analysis and will help them in solving and discovering many new and exciting problems.

From the preface to the previous edition

This volume offers a collection of problems concerning analytic functions, linear function spaces and linear operators.

The most exciting challenge to a mathematician is usually not what he understands, but what still eludes him. This book reports what eluded a rather large group of analysts in 1983 whose interests have a large overlap with those of our Seminar.* Consequently,

*i.e., the Seminar on Spectral Theory and Complex Analysis consisting principally of mathematicians working in the Leningrad Branch of the V. A. Steklov Mathematical Institute (LOMI) and in Leningrad University.

therefore, the materials contained herein are chosen for some sort of mild homogeneity, and are not at all encyclopaedic. Thus, this volume differs markedly from some well-known publications which aim at universality. We confine ourselves to the (not very wide) area of Analysis in which we work, and try—within this framework—to make our collection as representative as possible. However, we confess to obeying the Bradford law (the exponential increase of difficulties in obtaining complete information). One of our purposes is to publish these problems promptly, before they lose the flavour of topicality or are solved by their proposers or other colleagues.

This Problem Book evolved from the earlier version published as volume 81 of "Zapiski Nauchnyh Seminarov LOMI" in 1978 (by the way, much of the work arising from the above mentioned Seminar is regularly published in this journal). It is now twice the size, reflecting the current interests of a far wider circle of mathematicians. For five years now the field of interests of the "invisible community" of analysts we belong to has enlarged and these interests have drifted towards a more intense mixing of Spectral Theory with Function Theory. And the volume as a whole is rather accurate reflection of this process.

We are pleased that almost a half of the problems recorded in the first edition, 50 of 99, have been solved, partly or completely. The problems of 1978 (we call them "old" problems) are sometimes accompanied with commentary reporting what progress towards their solution has come to our attention. Moreover, those "old" problems which have been almost completely solved are assembled under the title "SOLUTIONS" at the end of each chapter (including information as to how and by whom they have been solved).

When we decided to prepare this new edition, we solicited the cooperation of many colleagues throughout the world. Some two hundred responded with ample and helpful materials, doubling the number of collaborators of the first edition. Their contributions ranged from carefully composed articles (not always short) to brief remarks. This flow it was our task to organize and to compress into the confines of a single volume. To effectuate this we saw no alternative to making extensive revisions (more exactly, abbreviations) in the texts supplied. We hope that we have succeeded in preserving the essential features of all contributions and have done no injustice to any.

At first sight the problems may appear very heterogeneous. But they display a certain intrinsic unity, and their approximate classification (i.e. division into chapters) did not give us much trouble. We say "approximate" because every real manifestation of life resists systematization. Some problems did not fit into our initial outline and so some very interesting ones are collected under the title "Miscellaneous Problems" ...

EDITORS

Acknowledgements

The publication of these volumes would have been impossible without the generous and self-denying help of our colleagues. To explain this, we start by describing some of the obstacles we had to overcome.

Our instructions, sent to all chapter editors, were very thorough and detailed. They contained a lot of technical explanations, and TEX-macros* (prepared by A. V. SUDAKOV). Unfortunately, they were largely ignored or neglected (except by those editors who worked in LOMI). We got a huge collection of texts in disorder; they required enormous work to coordinate and unify them (which could have been dispensed with almost completely if our instructions had been followed). We had to typeset hundreds of pages anew; many solved problems had to be detected and separated from the unsolved ones. Innumerable instances "to appear" from the preceding edition had to be replaced by correct bibliographical data (actually, some 1000 new references have been added!). Dozens of new commentaries had to be written (some chapter editors practically didn't revise "old" problems). And we had no technical staff necessary to turn a motley set of chapters into a book.

Meanwhile, as a result of the deteriorating situation and the decay of all structures in the former USSR (a malignant process whose rapidity we underestimated starting the project), the collective we could rely upon had disappeared and its members dispersed throughout the world. The e-mail became the only way of communication between them (including the authors of these lines). The situation looked desperate and the project could not have been rescued without assistance of our colleagues. This assistance was really invaluable to us. These skilled mathematicians in their most active years put their research aside and did a huge amount of purely technical work, making it possible for this book, to see the light of day.

The job has been done by three consecutive "technical teams". The first was headed by A. A. BORICHEV, the second by V. V. KAPUSTIN, the third by V. I. VASYUNIN, whose contribution to the project was especially great. The teams spent much time and energy retyping the text, tidying it up, hunting out inconsistencies and omissions to make the book a well-organized and handy tool for the user. We are not sure the debugging process has been completed and all defects have been discovered and removed: we apologize to the reader for remaining flaws. Our possible excuse is the fact that collectives capable of doing gratuitous work of such proportions existed only in the USSR, the country where the project was conceived, but which disappeared just at the final (and the hardest) stage of editing.

We are happy to thank the Mathematical Department of the University Bordeaux-I for its financial support. Its Graduate School (then headed by J.-L. JOLY and P. FABRIE) made possible the work of V. Vasyunin, putting at his disposal all necessary technical facilities.

We hope that this introduction is sufficient to explain why our gratitude to all who contributed to the book is especially deep and sincere. It is our duty and pleasure to name the following colleagues.

*This book was typeset using AMS-TEX macro package.

General Technical Directors:

Aleksander BORICHEV
Vladimir KAPUSTIN
Vasily VASYUNIN

Proof-reading and English Editing:

Serguei KISLIAKOV

Proof-reading:

Maria GAMAL

Keyboarding Advisor:

Andrei SUDAKOV

Keyboarding:

Cathy ANTONOVSKAYA
Yuri YAKUBOVICH

Checking References and Addresses:

Evgueni ABAKUMOV

Indices Compilers:

Maria GAMAL
Andrei GROMOV

Technical Advice and Various Help:

Ludmila DOVBYSH
Dmitri YAKUBOVICH

Episodical Advice:

William BADE
Philip CURTIS
Anton SERGEYEV
Rouslan SIBILEV
Elisabeth STROUSE
Yuri VYMENETS

We are indebted very much to all of them as well as to the Mathematical Editorial Board of Springer–Verlag for patience and support of our project.

EDITORS

LIST OF PARTICIPANTS

Adams D. R. 12.31
Adamyan V. M. 4.2
Ahern P. 10.19
Aizenberg L. A. ch.17, 1.18, 17.4
Aleksandrov A. B. 10.10, S.10.25, 11.15
Alexander H. 17.12
Anderson J. 3.7, S.10.24
Arov D. Z. 4.2, 4.3, 4.7
van Assche W. 13.6
Atzmon A. 5.4
Axler S. S.7.22
Azarin V. 16.8, 16.9
Azizov T. Ya. 5.9, 20.3
Bade W. 2.8
Baernstein A. 10.15, 18.10, 18.16
Bagby T. 12.14
Belitskiĭ G. 16.13
Belyi V. I. 12.7, 18.9
Ben-Artzi M. 6.7, 6.8
Berenstein C. A. 11.8, 11.9, 17.15, 17.16
Berg Ch. 12.5
Bielefeld B. ch.19, 19.1
Birman M. Sh. ch.6, 5.15, 6.5, 6.6
Bishop C. J. 18.1
Boivin A. 12.8
Bollobás B. S.2.33
Böttcher A. 7.18
Bourgain J. 1.2
de Branges L. 2.31, 6.1, 14.10
Brennan J. E. ch.12, ch.14, 12.9, 12.10
Brown G. 2.22
Brudnyi Yu. A. 15.8
Bruna J. 11.26, 15.5, 15.7
Carleson L. 19.5
Casazza P. G. 10.21
Chang S.-Y. A. 10.17, 10.18
Clark D. N. 5.11
Coburn L. 7.14
Conway J. B. ch.8, 8.7
Curtis P. C. Jr. 2.3
Dales H. G. ch.2, 2.6, 2.7
Davidson K. R. 5.3
Davis Ch. S.5.18
Devaney R. 19.9
Devinatz A. 6.7, 6.8, S.14.20
Dijksma A. 4.8
Djrbashyan M. M. 14.1
Domar Y. 11.17
Douglas R. 7.8
Duren P. L. ch.18, 18.11, 18.15
Dyakonov K. M. 10.23
Dym H. 12.4
Dyn'kin E. M. ch.10, 11.19, 14.6
Ecalle J. 14.9
Eiermann M. 13.1
Eremenko A. E. 16.8, 16.15, S.16.19, S.16.20, 19.10
Esterle J. 17.18
Faddeev L. D. 4.5, S.6.11
Fan Q. 7.6
Fel'dman I. A. 2.20, 5.14
Forelli F. 10.22, S.11.27, 17.13
Frankfurt R. 11.12
Fritzsche B. 4.7
Fuglede B. V. 16.15
Gamelin T. W. 2.14
Gaposhkin V. F. 3.4
Garnett J. 10.14
Gauthier P. 12.8
Gay R. 17.15, 17.16, 17.17
Ginzburg Yu. 4.4
Gohberg I. ch.4, 4.9
Goldberg A. A. ch.16, 16.2, 16.6, 16.7, S.16.19, S.16.20
Gonchar A. A. 12.17
Goodman A. W. 18.5
Gorin E. A. 2.9, 2.12, 20.4
Gorkin P. 2.17
Grimmett G. 14.14
Grishin A. 16.8
Grinshpan A. Z. 18.7
Gulisashvili A. B. 20.7
Gubreev G. 15.4
Guivarc'h Y. 3.8
Gurariĭ V. P. 11.16, 11.18
Haslinger F. 1.17
Hasumi M. 10.20
Havin V. P. ch.12, ch.14, S.10.25, 14.3, 14.4
Havinson S. Ya. 16.14
Hayman W. 12.24, 16.16
Hedenmalm H. 12.12, 12.13, 17.11
Hedberg L. I. S.12.33
Heinonen J. 18.14
Helemskiĭ A. Ya. ch.2, 2.1
Helson H. 11.23
Henkin G. 12.22, 17.1

LIST OF PARTICIPANTS

Herrero D. A. 5.6
Holbrook J. A. R. 5.1
Hruščëv S. V. 3.3, 14.3, 14.12
Ibragimov I. A. 3.2
Igari S. 2.21
Iohvidov I. S. 5.9
Iserles A. 13.2
Ivanov L. D. 12.26
Ivanov O. V. 2.24
Janas J. 7.15
Janson S. 7.3
Jarnicki M. 17.3
Jones P. W. 1.6, 10.2, 10.9, 12.2, 12.32
Jöricke B. 14.4, 17.6
Kaashoek M. A. ch.4, 4.9
Kadec M. I. 16.11
Kahane J.-P. ch.3, 3.9, 11.20
Kapustin V. V. 9.3
Kargaev P. P. 10.10
Karlovich Yu. 7.12
Kaufman R. 14.7, 16.18
Kérchy L. 9.4
Khavinson D. 12.20
Khurumov Yu. V. 17.5
Kirstein B. 4.7
Kisliakov S. V. ch.1, ch.10, 10.12
Kitover A. K. 2.9, 5.13
Komarchev I. A. 1.10
Koosis P. 14.5
Korenblum B. I. 11.13, 18.2
Král J. 10.3, 12.27, 16.17
Krasichkov-Ternovskiĭ I. F. 11.4, 11.6
Krein M. G. 4.2, 7.16, 15.11
Krein S. G. 20.2
Kriete T. 12.3, 14.2
Krupnik N. Ya. 2.20, 7.11, 10.6
Krushkal S. L. 17.14
Krzyż J. G. 10.4, 18.6
Kurina G. A. 20.2
Langer H. 4.8
Langley J. 16.12
Latushkin Yu. D. 7.20
Leiterer J. 4.6
Leont'ev A. F. 15.2
Levin B. ch.16, 11.22, 16.1, S.16.21, S.16.22
Lewis J. 14.16, 14.17
Lin V. Ya. 20.5, 20.6
Littlejohn L. L. 13.2
Litvinchuk G. S. 7.20
Lubinsky D. S. 13.3
Lykova Z. A. 2.2
Lyubarskiĭ Yu. I. 1.8
Lyubich M. Yu. ch.19, 19.7, 19.10
Lyubich Yu. I. S.5.19
MacCluer B. 14.2
Magnus A. P. 13.4
Makarov B. M. 1.10, 1.11
Makarov N. G. S.6.12, 8.11, 14.4
Manfredi J. 14.15
Markus A. 2.20, 5.14
Marshall D. 12.23
Matsaev V. I. 14.8
Maz'ya V. 12.30
McGuire P. J. 8.1
McKean H. P. 3.1, 6.3
McMullen C. 19.2
Meise R. 11.7
Mel'nikov M. S. 12.25
Méril A. 17.17
Milin I. M. 18.13
Milman V. 1.1
Milnor J. 19.3, 19.6
Momm S. 11.7
Moran W. 2.22
Mortini R. 2.26
Muckenhoupt B. 10.5
Müller P. 1.4, 10.11
Murai T. 12.28
Naboko S. N. 9.2
Napalkov V. 14.13
Nevai P. ch.13
Nikolski N. K. ch.9, ch.11, ch.15, 8.11, 9.1, S.9.5, 11.11, 15.3
Novikov R. G. 17.1
O'Farrell A. 12.15
Olin R. F. 8.2
Ostrovskii I. V. ch.16, 16.2–16.7, S.16.20, S.16.21
Ovcharenko I. 15.12
Palamodov V. 11.2
Pavlov B. S. S.6.11, S.9.5
Pedersen H. 12.5
Peetre J. ch.7
Pełczyński A. 1.9
Peller V. V. 3.3, 5.12, 7.2, 7.7
Peng L. 7.6
Perez-Gonzalez F. 12.11
Pflug P. 17.3
Power S. C. 7.1
Prössdorf S. 7.19
Przytycki F. 19.8
Pták V. 2.28
Putinar M. 4.10, 8.5, 8.6
Putnam C. R. 8.3, 8.9, 8.10
Radjavi H. 5.2
Ransford T. J. 12.21
Rees M. 19.4
Reshetihin N. Yu. 4.5
Rochberg R. 2.19, 7.4, 10.8
Rodman L. 4.9, 5.5
Ronkin L. I. 17.9

Rosenthal P. ch.5
Rovnyak J. 18.8
Rubel L. A. 12.19, 16.12, 17.2
Sakhnovich L. A. 6.9, 7.13, 15.13
Saitoh S. 1.7, 18.4
Samokhin M. 2.25
Sarason D. 2.16, 7.9, 10.16
Semënov E. M. 1.5, 1.12, 1.13
Semënov-Tian-Shansky M. A. 7.21
Semiguk O. S. 1.15
Semmes S. 10.7
Sergeev A. G. 17.7, 17.8
Shamoyan F. A. 11.24
Shields A. L. 11.14
Shirokov N. A. 15.6
Shishkin S. 20.1
Shul'man V. S. 2.13
Shwartsman P. A. 15.8
Siddiqi J. A. 14.11
Silbermann B. 7.18, S.7.23
Simon B. 5.17
Sjögren P. 10.13
Skiba N. I. 1.15
de Snoo H. S. V. 4.8
Sodin M. L. 16.4, 16.5, 16.10, 16.15
Solev V. N. 3.2
Solomyak M. Z. 5.15, 5.16
Spitkovskiĭ I. M. 5.5, 7.12, 7.16
Stahl H. B. 13.1
Stephenson K. 18.12
Stray A. 12.6
Struppa D. C. 11.8
Sudakov V. N. 3.6
Sundberg C. 7.10
Sutherland S. 19.11
Szőkefalvi-Nagy B. S.4.11
Tarkhanov N. 12.1
Taylor B. 15.1
Teodorescu R. S.9.6
Thomas M. P. 11.21
Tkachenko V. A. 11.5, 16.13
Tolokonnikov V. A. 2.27
Totik V. 13.5
Trutnev V. M. 1.19, 11.3
Tumarkin G. C. 10.1
Vasyunin V. I. ch.9, S.9.5, S.9.6
Verbitskiĭ I. È. 7.11, 10.6
Verdera J. 12.16
Vershik A. M. 3.5, 5.10
Villamor E. 14.15
Vinogradov S. A. 15.9
Vitushkin A. G. 12.25
Vladimirov V. 7.17
Voiculescu D. 2.18, 6.2, 8.8
Volberg A. L. ch.12, ch.14, S.14.19
Volovich I. V. 7.17
Waelbroeck L. 11.1
Wallin H. 12.18
Wermer J. 2.15, 2.32
Widom H. 6.4
Williams D. 11.25
Willis G. 2.5
Wodzicki M. 1.14, 2.4
Wojtaszczyk P. 1.3
Wolff T. 11.10, 14.18
Wolniewicz T. M. 1.4
Wu J.-M. 18.14
Wu P. Y. 5.7
Xia D. 8.4
Yafaev D. R. 6.10
Yakubovich D. V. 5.8
Yakubovich V. A. 20.1
Yger A. 11.9, 17.15–17.17
Young N. J. 2.29, 4.1
Zafran M. 15.10
Zaidenberg M. G. 20.6
Zakhariuta V. P. 1.15, 1.16
Żelazko W. 2.11, 2.23
Zemánek J. 2.10, 2.30, 12.29
Zhu K. 7.5, 18.3
Znamenskiĭ S. V. 17.10

Chapter 1

BANACH SPACES

Edited by

S. V. Kisliakov
St. Petersburg branch of
Steklov Mathematical Institute
Fontanka 27
St. Petersburg, 191011
Russia

INTRODUCTION

The title of this chapter does not reflect quite properly its contents, because the last 5 problems (1.15–1.19) are devoted to *non*-Banach spaces (mainly various locally convex spaces of analytic functions). The fact that all these 5 problems are "old" probably may serve as an excuse.

The proper Banach space theoretic part of the chapter can be divided into 4 sections. This division (but not the section lengths!) reflects in fact the present situation in the field.

The first section consists of Problem 1.1 only and is devoted to *finite dimensional geometry*, the area in which there were several genuine breakthroughs in the last decade. The second section contains Problems 1.2–1.8. All of them are about *concrete function spaces*, mainly Hardy spaces and their relatives. The next group is formed by problems of *operator theoretic nature* (1.9–1.12). Finally, problems 1.13 and 1.14 pertain to pure *infinite dimensional geometry* of (more or less) general spaces.

I have excluded two problems that appeared in the corresponding chapter of 1984 edition and were solved "almost completely" even at that time. They are Wojtaszczyk's "Finite dimensional operators on spaces of analytic functions" and Casazza's "Complemented subspaces of A, H^1 and H^∞" (1.4 old and 1.5 old in the enumeration of 1984 edition). The only remaining open question in the first of them was as follows. Does there exist a constant q, $q > 1$ such that every basis for the disc algebra has the basis constant $> q$? This seems to be still unknown. As to the second problem, the question of primarity of H^1 was answered in the positive by P. Müller (Müller P.F.X., *On projections in H^1 and BMO*, Studia Math. **89** no. 2 (1988), 145–158). So the only remaining open question in Casazza's problem is that of primarity of the disc algebra.

1.1

PROPORTIONAL QUOTIENTS OF FINITE DIMENSIONAL NORMED SPACES

V. D. MILMAN

I would like to recall and to discuss one of the remaining unsolved bastions of the Local Theory of Normed Spaces. I will start with the strongest conjecture which was first passed at the Oberwolfach meeting in 1981.

First, a few standard notations (see [8] or [11] for any unexplained notion). Let $X = (\mathbb{R}^n, \|\cdot\|)$ be a normed space and $K(X)$ denote its unit ball $K(X) = \{x \in \mathbb{R}^n, \|x\| \leqslant 1\}$. Denote by $d_X = d(X, \ell_2^n)$ the Banach-Mazur distance of X to ℓ_2^n. By the volume ratio of X we mean the quantity

$$\text{v.r.}\, X = \inf\left\{ \left(\frac{\operatorname{vol} K(X)}{\operatorname{vol} \mathcal{E}}\right)^{1/n} \,\middle|\, \mathcal{E} \text{ is the maximum volume ellipsoid inscribed into } K(X) \right\}.$$

It was proved in [12] that $\text{v.r.}\, X < C$ implies that for any λ, $0 < \lambda < 1$, there is a subspace sX of X, $\dim sX \geqslant \lambda N$, such that $d_{sX} \leqslant f(\lambda; C)$. Also, by [2], if cotype 2 constant of X is $C_2(X)$ then there is a subspace sX with $\dim sX \geqslant c\varepsilon^2(\dim X)/C_2(X)^2$, and $d_{sX} \leqslant 1+\varepsilon$. Having these facts in mind let us ask our first problem. We will denote by fX a quotient space of X (from a "factor-space") instead of more standard qX keeping q for the cotype of X.

PROBLEM 1. *Is there an absolute constant C such that every $X = (\mathbb{R}^n, \|\cdot\|)$ contains a subspace E, $\dim E \geqslant n/2$, such that $C_2(E^*) \leqslant C$? (In other form, does any normed space X of dimension n have a quotient space $Y = fX$ such that $\dim Y \geqslant n/2$ and Y is of cotype 2 with $C_2(Y) \leqslant C$?).*

Many open problems of Local Theory would be solved if this Problem would be solved in the positive. The Quotient of Subspace Theorem [4] which states that any n-dimensional normed space X has a quotient of a subspace Y, $\dim Y \geqslant n/2$, which is (uniformly) isomorphic to a Euclidean space (i.e. $d_Y \leqslant C$, C is a universal constant) would be, of course, an immediate consequence. In fact, the first thinking about possibility for such a theorem came as a result of this conjecture. However, the proof of the Q.-S.-Theorem was originally realized through volume considerations. Also a version of the Problem involving volume ratio instead of cotype 2 condition was solved.

THEOREM. *For any λ, $0 < \lambda < 1$, there is $C(\lambda) < \infty$ such that for any n-dimensional normed space $X = (\mathbb{R}^n, \|\cdot\|)$ there is a quotient space $Y = fX$ of X, $\dim Y \geqslant \lambda n$, with volume ratio $\text{v.r.}\, Y \leqslant C(\lambda)$.*

The theorem follows from [5] and [6] (see also [7]). I am not sure any more that such a strong fact, as Problem 1 suggests, can be true even though I don't have in mind any potential counter-example. So I will pass to another, weaker conjecture which, being solved in positive, would be sufficient for most consequences of Problem 1.

PROBLEM 2. *For any* $\varepsilon > 0$ *is there* $C(\varepsilon)$ *such that for any integer* n *and any normed space* $X = (\mathbb{R}^n, \|\cdot\|)$ *there is a quotient* fX *of* X *of* $\dim fX \geqslant n/2$ *and of cotype* $(2+\varepsilon)$ *constant* $C_{2+\varepsilon}(fX) \leqslant C(\varepsilon)$?

Of course, if such a statement would be true we would expect, similarly, that also for any $\lambda < 1$ one may find a quotient fX of $\dim fX \geqslant \lambda n$ and $C_{2+\varepsilon}(fX) \leqslant C(\varepsilon; \lambda)$.

List some consequences of the positive solution of Problem 2.

PROBLEM 3. *Does, for any* $0 < \lambda < 1$ *and* $X = (\mathbb{R}^n, \|\cdot\|)$ *of finite cotype* $q < \infty, X$ *contain a subspace* sX, $\dim sX \geqslant \lambda n$, *of type 2? Exactly, does a function* $f(\lambda; q; A)$ *exist such that the type 2 constant*

$$T_2(sX) \leqslant f(\lambda; q, C_q(x))?$$

The positive solution of Problem 3 follows from the positive answer to Problem 1. However, slightly weaker version of Problem 3 with $T_2(sX)$ replaced by $T_{2-\varepsilon}(sX)$ (for any $\varepsilon > 0$ there is a subspace sX, $\dim sX \geqslant \lambda n$, and $T_{2-\varepsilon}(sX) \leqslant f(\varepsilon; \lambda; q; C_q(x))$ would follow from the positive solution of Problem 2. To show this, we use the following theorem of Pisier [10]:

Let $C_q(Y)$ and $C_{q^*}(Y^*)$ be the cotype q (respectably q^*) constants of a normed space Y and its dual Y^*; if $\frac{1}{q} + \frac{1}{q^*} > \frac{1}{2}$ then Y is a K-convex space, i.e. the Rademacher Projection of Y ($\equiv$ K-convexity constant) K_Y is bounded by some function $f(q, C_q(Y); q^*, C_{q^*}(Y^*))$ (or, in equivalent language, for some $p > 1$ the type p constant $T_p(Y)$ is bounded by a function of $q, q^*, C_q(Y)$ and $C_{q^*}(Y^*)$).

Apply Problem 2 to X^* and choose $\varepsilon > 0$ such that $\frac{1}{q} + \frac{1}{2+\varepsilon} > \frac{1}{2}$. Then apply Pisier's theorem to $Y = fX^*$ from Problem 2. Clearly, cotype q_{Y^*} of Y^* is at most the cotype q of X, because Y^* is a subspace of X. The positive answer in Problem 2 would imply that we could choose Y being of cotype $(2+\varepsilon)$. Therefore, Y is K-convex; then, by Maurey-Pisier [3] there is the duality between cotype of Y and type p_{Y^*} of Y^*: $\frac{1}{p_{Y^*}} + \frac{1}{2+\varepsilon} \leqslant 1$; so Y^* has type $(2-\varepsilon)$.

The following question of Pisier would follow from Problem 3 (also taken in its weaker form):

PROBLEM 4 (Pisier; around 80). *Let a space* $X = (\mathbb{R}^n, \|\cdot\|)$ *be 2-isomorphic to a subspace of* ℓ_∞^N *and let for* $q < \infty$ *the cotype* q *constant of* X *be* $C_q(X)$. *Is it true that* $N \geqslant \exp(cn)$ *for* $c = c(q; C_q(x)) > 0$?

(It would mean that every subspace of ℓ_∞^N of dimension much higher than $\log N$ does not have a finite cotype, i.e. it contains a smaller subspace $(1+\varepsilon)$-isometric to ℓ_∞^k for some large k; such fact would be a far reaching continuation of results of Figiel-Johnson [1]).

To derive a solution of Problem 4 from the positive solution of Problem 3 (or its weaker version) just use [9].

Let us note, in the end, that many other open problems of Local Theory are connected to the Problems 1 and 2.

References

1. Figiel T., Johnson W. B., *Large subspaces of ℓ_∞^n and estimates of the Gordon–Lewis constants*, Israel J. Math. **37** (1980), 92–112.
2. Figiel T., Lindenstrauss J., Milman V. D., *The dimension of almost spherical sections of convex bodies*, Acta Math. **129** (1977), 53–94.
3. Maurey B., Pisier G., *Séries de variables aleatoires vectorielles indépendantes et propriétés géometriques des espaces de Banach*, Studia Math. **58** (1976), 45–90.
4. Milman V. D., *Almost Euclidean quotients spaces of subspaces of finite dimensional normed spaces*, Proc. Amer. Math. Soc. **94** (1985), 445–449.
5. Milman V. D., *Geometrical inequalities and mixed volumes in Local Theory of Banach Spaces*, Asterisque **131** (1985), 373–400.
6. Milman V. D., *Inegalité de Brunn-Minkowski inverse et applications à la théorie locale des espaces normés*, C.R. Acad. Sci. Paris **302** (1986), 25–28.
7. Milman V. D., *Some applications of duality relations*, Springer LNM **1469** (1991), 13–40.
8. Milman V. D., Schechtman G., *Asymptotic theory of finite-dimensional normed spaces*, Springer LNM **1200** (1986).
9. Pisier G., *Remarques sur un résultat non publié de B. Maurey*, Sem. d'Anal. Fonctionnelle (1981), Ecole Polytechnique, Paris.
10. Pisier G., *Holomorphic semi-groups and the geometry of Banach spaces*, Ann. of Math. **115** (1982), 375–392.
11. Pisier G., *The volume of convex bodies and Banach space geometry*, Cambridge Tracts in Mathematics, vol. 94, 1989.
12. Szarek S., Tomczak-Jaegermann N., *On nearly Euclidean decompositions for some classes of Banach spaces*, Compositio Math. **40** (1980), 367–385.

Department of Mathematics
Tel Aviv University
Tel Aviv 69978
Israel

1.2
old

SOME QUESTIONS ON THE STRUCTURE OF THE SPACE OF UNIFORMLY CONVERGENT FOURIER SERIES

J. Bourgain

$A = A(\mathbb{D})$ denotes the disc algebra and

$$U = U(\mathbb{T}) = \{ f \in A \colon f \text{ is uniform limit of } f * D_n \},$$
$$D_n = \sum_{0 \leqslant k \leqslant n} e^{ik\Theta}.$$

The norm on U is given by $\|f\|_U = \sup_n \|f * D_n\|_\infty$.

Various analogies between the spaces A and U are known now. It was shown by D. Oberlin [3] that measure-zero compact subsets of $\mathbb{T}$ are peak-interpolation sets for U, an improvement of the Rudin–Carleson theorem. Using related techniques, I obtained the following

Proposition 1. *Let K be a compact subset of $\mathbb{T}$ and $\varepsilon > 0$. Then there exists $f \in U$ satisfying*

(i) $\|f\|_U \leqslant C$
(ii) $|f| < \varepsilon$ *on* K
(iii) $\|1 - f\|_2 < C(\log \frac{1}{\varepsilon})|K|^{1/2}$,

where C is a constant.

Fixing a finite sequence $g_1, \dots, g_r$ in U, $\|g_s\|_U \leqslant M$ $(1 \leqslant s \leqslant r)$, condition (i) can be strengthened by requiring in addition

$$\|f \cdot g_s\|_U \leqslant C(r, M) \qquad (1 \leqslant s \leqslant r).$$

Here are some corollaries of Prop. 1 for the Banach space theory of U (see [2]).

Proposition 2. 1. *The dual space U^* of U is weakly complete. In fact, bounded sequences in U have either a w^*-complemented ℓ^1-subsequence or a weakly convergent subsequence.*

2. *Reflexive subspaces of U^* are isomorphic to subspaces of L^1. In particular, they are of cotype 2.*

Recall that a normed space X has cotype $q \geqslant 2$ provided following inequality holds for all finite sequences $\{x_i\}$ in X:

$$\int \left\| \sum \varepsilon_i x_i \right\| d\varepsilon \geqslant \delta \left(\sum \|x_i\|^q \right)^{1/q},$$

where δ is a fixed constant, and (ε_i) is the usual Rademacher sequence.

No results seem to be known as far as the finite dimensional properties of U, U^* are concerned. In particular, the following problems can be posed.

PROBLEM 1. *Does U^* have any cotype $q < \infty$? Is U^* of cotype 2?*

PROBLEM 2. *Assume E a λ-complemented subspace of U, of dimension n. Is it true that E contains ℓ^∞_m-subspace for $m \sim n$? How well can E be embedded as a complemented subspace of ℓ^∞?*

These questions are solved for the disc algebra A (see [1]). Their solution for the space U probably requires different techniques.

REFERENCES

1. Bourgain J., *New Banach space properties of the disc algebra and H^∞*, Acta Math. **152** (1984), no. 1–2, 1–48.
2. Bourgain J., *Quelques propriétés linéaires topologiques de l'espace des séries de Fourier uniformément convergentes*, C.R.A.S. Paris **295** (1982), Sér.1, 623–625.
3. Oberlin D. M., *A Rudin–Carleson theorem for uniformly convergent Taylor series*, Michigan Math. J. **27** (1980), no. 3, 309–314.

INSTITUTE OF MATHEMATICS, IHES,
91440 BÛRES-SUR-YVETTE,
FRANCE

1.3
old

BASES IN H^p SPACES ON THE BALL

P. WOJTASZCZYK

By $H^p(B)$ we will mean the natural Hardy space of analytic functions on the unit ball of $\mathbb{C}^n$. $A(B)$ denotes the ball algebra of all functions continuous in $\overline{B}$ and analytic in B. We are interested in construction and existence problems for Schauder bases in these spaces.

Let me recall that a sequence of elements $(f_n)_{n=0}^{\infty}$ in a Banach space X is a Schauder basis for X if for every x in X there exists a unique sequence of scalars $(a_n)_{n=0}^{\infty}$ such that the series $\sum_{n=0}^{\infty} a_n f_n$ converges to x in the norm of X. The basis is called unconditional if for every x in X the corresponding series $\sum_{n=0}^{\infty} a_n f_n$ is unconditionally convergent.

For the ball algebra $A(B)$ the question of the existence of a Schauder basis is a well known open problem (cf. [1]). This seems to be the most concrete separable Banach space for which this question is still open today. It is known that $A(B)$ does not have an unconditional basis.

For $1 \leqslant p < \infty$ the situation is a little more intriguing. It is a relatively easy task to check that for $1 < p < \infty$ the monomials in a correct order form a Schauder basis for $H^p(B)$. However this basis is not unconditional for $p \neq 2$. It was proved in [6] that for $1 \leqslant p < \infty$ $H^p(B)$ is isomorphic as a Banach space to $H^p(\mathcal{D})$, the classical Hardy space on the unit disc. Since unconditional bases for $H^p(\mathcal{D})$ are well known, cf. [2], [4], [5], we get the existence of unconditional bases in $H^p(B)$ for $1 \leqslant p < \infty$. This argument however has one drawback, it is non-constructive, so we pose the following

PROBLEM. *Construct an unconditional basis in the space $H^p(B)$, $1 \leqslant p \neq 2 < \infty$.*

The most interesting case is $p = 1$. There is also an auxiliary question related to this:

Does there exist an orthonormal unconditional basis in $H^1(B)$?

The case $p < 1$ is less clear. In $H^p(\mathcal{D})$ we have unconditional bases, cf. [3], [5]. However in general variables the very existence of an unconditional basis in $H^p(B)$, $p < 1$ is still open. The proof of isomorphism between $H^1(B)$ and $H^1(\mathcal{D})$ given in [6] can be extended to $p < 1$ (after some technical modifications) provided the following question has positive answer.

QUESTION. *Is $H^p(\mathcal{D})$, $p < 1$ isomorphic to a complemented subspace of $H^p(B)$?*

REFERENCES

1. Pełczyński A., *Banach Spaces of Analytic Functions and Absolutely Summing Operators*, CBMS regional conference series, vol. 30.

2. Bočkarev S. V., *Existence of a basis in the space of functions analytic in the disk, and some properties of Franklin's system*, Mat. Sbornik **95 (137)** (1974), no. 1, 3–18 (Russian); English transl. in Math. USSR Sbornik **24** (1974), no. 1.
3. Sjölin P., Stromberg J.-O., *Basis properties of Hardy spaces*, Stockholms Universitet preprint (1981), no. 19.
4. Wojtaszczyk P., *The Franklin system is an unconditional basis in H^1*, Arkiv för Mat. **20** (1982), no. 2, 293–300.
5. Wojtaszczyk P., *H^p-spaces, $p < 1$ and spline systems*, Studia Math. **77** (1984), 289–320.
6. Wojtaszczyk P., *Hardy spaces on the complex ball are isomorphic to Hardy spaces on the disc,* $1 \leqslant p < \infty$, Annals of Math. **118** (1983), 21–34.

MATH.INST.POLISH ACAD.SCI.
00-950 WARSZAWA,
ŚNIADECKICH, 8
POLAND

CHAPTER EDITOR'S NOTE

Bourgain [7] has proved that the ball algebra possesses a basis. In connection with [6], one must also consult [8].

REFERENCES

7. Bourgain J., *Homogeneous polynomials on the ball algebra and polynomial bases*, Israel J. Math. **68** (1989), no. 3, 327–347.
8. Wolniewicz T. M., *On isomorphisms between Hardy spaces on complex balls*, Ark. för Mat. **27** (1989), 155–168.

1.4

ISOMORPHISMS BETWEEN H^1 SPACES

P. F. X. MÜLLER, T. M. WOLNIEWICZ

Let Ω be a strictly pseudoconvex domain in $\mathbb{C}^n$. By U we will denote the unit disk in $\mathbb{C}$. The Hardy space $H^p(\Omega)$ can be defined in many equivalent ways, for instance as the space consisting of all analytic functions f in Ω which admit a harmonic majorant for $|f|^p$.

It is known that for $1 < p < \infty$, $H^p(\Omega)$ is isomorphic to $L^p([0,1])$. In the case when $\Omega = B_n$, the unit ball of $\mathbb{C}^n$, it was shown in [Wol2] that $H^1(B_n)$ is isomorphic to $H^1(U)$. The method used in [Wol2], modeled after [M] seems to suggest a possible extension of the proof to all pseudoconvex domains. Thus we ask:

PROBLEM 1. *Is $H^1(\Omega)$ isomorphic to $H^1(U)$ for every strictly pseudoconvex domain in $\mathbb{C}^n$?*

On the negative side it must be said that H^1 on polydisks in dimensions greater than 1 is not isomorphic to $H^1(U)$ (cf. [B]) and [Wol1] contains an example of a nonsmooth domain $\mathcal{D}$, not equivalent to a product domain for which this isomorphism does not exist either.

It follows from the implementation of Maurey's method described in [Wol2] that Problem 1 can be reduced to a more specific question. To state it we must introduce some additional notation. Let ds^2 be a preferred metric of Stein in Ω (cf. [St]), Δ the corresponding Laplace–Beltrami operator and h_z the harmonic measure for Δ corresponding to $z \in \Omega$. Let, for $\xi \in \partial\Omega$, ν_ξ denote the normal vector to $\partial\Omega$ at ξ. On $\partial\Omega$ one introduces nonisotropic 'balls' $B(\xi, r)$ which have diameter of order r in the direction $i\nu_\xi$ and $\sqrt{r}$ in the perpendicular directions. Finally let π be the normal projection onto $\partial\Omega$ defined in a neighborhood V of $\partial\Omega$.

PROBLEM 2. *Is it true that*

$$\inf_{z \in \Omega \cap V} h_z\Big(B(\pi(z), \delta(z))\Big) > 0?$$

A positive answer to Problem 2 leads to a positive solution to Problem 1.

For a large class of uniformly elliptic operators the associated harmonic measure satisfies inequalities of the type considered here (cf. [CFMS]). The difficulty in our case comes from the fact that the Laplace–Beltrami operator we have to consider is not uniformly elliptic.

These non isotropic balls determine a quasi-metric on $\partial\Omega \times \partial\Omega$

$$d(x,y) = \inf\Big\{\mu(B(x,r)) : y \in B(x,r)\Big\} + \inf\Big\{\mu(B(y,s)) : x \in B(y,s)\Big\}$$

where μ is the surface measure on $\partial\Omega$ induced by the Lebesgue measure. As shown by E. M. Stein in [St] the triple $(\partial\Omega, d, \mu)$ forms a space of homogeneous type in the sense of Coifman and Weiss [C-W]. We denote by $H^1(\partial\Omega, d, \mu)$ the atomic Hardy space associated with the space of homogeneous type $(\partial\Omega, d, \mu)$ as defined by Coifman and Weiss in [C-W] p. 591, 592.

N. Th. Varopoulos in [V] proved that $H^1(\Omega)$ (the Hardy space of analytic functions) is isomorphic to a complemented subspace of $H^1(\partial\Omega, d, \mu)$.

Using this result we could give a positive solution to Problem 1 if we could solve

PROBLEM 3. *Is it true that $H^1(\partial\Omega, d, \mu)$ is isomorphic to a complemented subspace of $H^1(U)$?*

Problem 3 is a real variable problem concerning atomic H^1 spaces, which is a special case of the following

PROBLEM 4 (posed by P. Wojtaszczyk). *Under which conditions on a space of homogeneous type (X, d, μ) is the atomic Hardy space $H^1(X, d, \mu)$ isomorphic to $H^1(U)$?*

A first step to an answer of question 4 is contained in [Mü] where a sufficient condition (on (X, d, μ)) was given for $H^1(X, d, \mu)$ to contain a complemented subspace isomorphic to $H^1(U)$.

REFERENCES

[B] Bourgain J., *The non-isomorphisms of H^1-spaces in a different number of variables*, Bull. Soc. Math. Belg. Sér. B. **35** (1983), 127–136.

[CFMS] Caffarelli L., Fabes E., Mortola S., Salsea S., *Boundary behaviour of nonnegative solutions of elliptic operators in divergence form*, Indiana Univ. Math. J. **30** (1981), 621–640.

[C-W] Coiffman R. R., Weiss G., *Extensions of Hardy spaces and their use in analysis*, Bull. Amer. Math. Soc. **83** (1977), 569–645.

[M] Maurey B., *Isomorphismes entre espaces H^1*, Acta Math. **145** (1980), 79–120.

[Mü] Müller P., *On linear topological properties of H^1 on spaces of homogeneous type*, Trans. AMS **317** (1990), 463–484.

[St] Stein E. M., *Boundary behaviour of holomorphic functions of several complex variables*, Princeton University Press, Princeton NJ (1972).

[V] Varopoulos N. Th., *BMO functions and the $\bar{\partial}$-equation*, Pacific J. Math. **71** (1977), 221–273.

[Wol1] Wolniewicz T., *Independent inner functions in the classical domains*, Glasgow Math. J. **29** (1987), 229–236.

[Wol2] Wolniewicz T., *On isomorphisms between Hardy spaces on complex balls*, Ark. Math. **27** (1989), 155–168.

INSTITUT FÜR MATHEMATIK
J. KEPLER UNIVERSITÄT
A-4040 LINZ
AUSTRIA

INSTYTUT MATEMATYKI
UNIWERSYTET NIKOŁAJA KOPERNIKA
UL. CHOPINA 12/18
87-100 TORUŃ
POLAND

CHAPTER EDITOR'S NOTE (JULY 1993)

H. Arai [A] has recently resolved Problems 1, 2, and 3 in the positive in the case where Ω is a strongly pseudoconvex domain with smooth boundary.

REFERENCE

[A] Arai Hitoshi, *On isomorphisms between Hardy spaces on strongly pseudoconvex domains* (to appear).

1.5
v.old

SPACES OF HARDY TYPE

E. M. SEMËNOV

A Banach space E of measurable functions on $[0, 2\pi]$ is called a symmetric (or rearrangement invariant) space iff the norm of E is monotone and any two equimeasurable functions have equal norms ([1], chapter 2). The L^p-spaces ($1 \leqslant p \leqslant \infty$), the Orlicz spaces, and the Lorentz spaces can serve as examples. Remind that if the function Ψ is non-decreasing and concave on $[0, 2\pi]$, $\Psi(0) = 0$, then the Lorentz space $\Lambda(\Psi)$ consists of functions x such that

$$\|x\|_{\Lambda(\Psi)} = \int_0^1 x^*(t)\, d\Psi(t) < +\infty,$$

where x^* is the function non-increasing on $[0, 2\pi]$ and equimeasurable with x.

A symmetric space E gives rise to a space of complex functions on $\mathbb{T}$ consisting of functions with moduli from E. This space is also denoted by E. By $H(E)$ we denote the set of all functions f analytic in the unit circle $\mathbb{D}$ and satisfying $\|f\|_{H(E)} < \infty$,

$$\|f\|_{H(E)} \stackrel{\text{def}}{=} \sup_{o<r<1} \|f_r\|_E, \quad f_r(\xi) \stackrel{\text{def}}{=} f(r\xi).$$

Though the classical Hardy spaces $H^p = H(L^p)$ have been studied rather well, the theory of general spaces $H(E)$ is only fragmentary.

The set of extreme points of the unit ball of $\Lambda(\Psi)$ is contained in the set of functions $\dfrac{\chi_e \cdot \varepsilon}{\Psi(me)}$, where $|\varepsilon(t)| \equiv 1$ and χ_e is the characteristic function of a measurable set e, $e \subset [0, 2\pi]$. In the case where Ψ is strongly concave these two sets coincide. The following PROBLEM arises naturally: *describe the set of extreme points of the unit ball of $H(\Lambda(\Psi))$*. Some partial results are contained in [2]. The space $H(\Lambda(\Psi))$ is nothing but H^1 if $\Psi(t) = t$, and coincides with H^∞ if $\Psi(t) = \operatorname{sign} t$. In these two cases the set of extreme points of the unit ball is well known, see [3], part 9.

We believe that the solution of the above-mentioned problem will possibly be useful for describing all isometric operators on $H(\Lambda(\Psi))$. Some interesting results on isometric operators on a symmetric space are contained in [4].

REFERENCES

1. Krein S. G., Petunin Yu. I., Semenov E. M., *Interpolation of linear operators*, Nauka, Moscow, 1978 (Russian); English transl. 1982, AMS Providence.
2. Bryskin I. B., Sedaev A. A., *The geometric properties of the unit ball in spaces of the type of the Hardy classes*, Zapiski nauchn. sem. LOMI **39** (1974), 7–16 (Russian); English transl. in J. Soviet Math. **8** (1977), no. 1, 1–9.
3. Hoffman K., *Banach Spaces of Analytic Functions*, Prentice–Hall, Englewood Cliffs, New Jersey, 1962.

4. Zaidenberg M. G., *On isometric classification of symmetric spaces*, Doklady Akad. Nauk SSSR **234** (1977), no. 2, 283–286 (Russian); English transl. in Soviet Math. Dokl. **18** (1977), no. 3, 636–640.

DEPT. OF MATHEMATICS
VORONEZH STATE UNIVERSITY
UNIVERSITETSKAYA PL. 1.
VORONEZH, 394693
RUSSIA

EDITORS' NOTE

Here are some more articles connected with isometries of L^p-spaces of analytic and harmonic functions:

REFERENCES

5. Plotkin A. I., *Continuation of L^p-isometries*, Zapiski nauchn. sem. LOMI **22** (1971), 103–129 (Russian); English transl. in J. Soviet Math. **2** (1974), no. 2.
6. Plotkin A. I., *Isometric operators in L^p-spaces of analytic and harmonic functions*, Zapiski nauchn. sem. LOMI **30** (1972), 130–145 (Russian); English transl. in J. Soviet Math. **4** (1975).
7. Plotkin A. I., *An algebra that is generated by translation operators, and L^p-norms*, Functional Analysis, no.6, Theory of operators in linear spaces, Uljanovsk, 1976, pp. 112–121. (Russian)
8. Plotkin A. I., *Isometric operators in spaces of summable analytic and harmonic functions*, Doklady Akad. Nauk SSSR **185** (1969), no. 5, 995–997 (Russian); English transl. in Soviet Math. Dokl. **10** (1969).

1.6
old

SPACES WITH THE APPROXIMATION PROPERTY?

PETER W. JONES

Recall that a Banach space X has the approximation property (a.p.) if for all compact $E \subset X$ and for all $\varepsilon > 0$ there is a bounded linear operator $T\colon X \to X$ such that $\|x - Tx\| < \varepsilon$ when $x \in E$, and such that T has finite rank. Not every Banach space has the a.p. [1].

Does H^∞ have the a.p.?

Some mild evidence that this might be true comes from the recent result that L^∞/H^∞ (i.e. BMO) has the a.p. [2]. Another interesting space for which the a.p. is unknown is $W^{k,\infty}(\mathbb{R}^n)$, $n \geqslant 2$. (When $n = 1$ the answer is easy and positive.) Here

$$W^{k,\infty}(\mathbb{R}^n) = \{\, f\colon \mathcal{D}^\alpha f \in L^\infty(\mathbb{R}^n), \quad 0 \leqslant |\alpha| \leqslant k \,\}.$$

REFERENCES

1. Enflo P., *A counter-example to the approximation problem in Banach spaces*, Acta Math. **130** (1973), 309–317.
2. Jones P. W., *BMO and the Banach space approximation problem*, Institut Mittag–Leffler report (1983), no. 2.

DEPARTMENT OF MATHEMATICS
YALE UNIVERSITY
BOX 2155 YALE STATION
NEW HAVEN, CT 06520
USA

1.7

REPRESENTATIONS OF THE NORMS IN BERGMAN – SELBERG SPACES ON SECTORS

SABUROU SAITOH

Let $\Delta(\alpha)$ be the sector $\{ z \mid |\arg z| < \alpha \}$ on the complex z plane, and $K_{\Delta(\alpha)}(z, \overline{u})$ be the Bergman kernel on $\Delta(\alpha)$ for the Bergman space $H_{K_{\Delta(\alpha)}}$ on $\Delta(\alpha)$ composed of all analytic functions $f(z)$ with finite norms

$$\|f\|^2_{H_{K_{\Delta(\alpha)}}} = \iint_{\Delta(\alpha)} |f(z)|^2 \, dx \, dy \qquad (z = x + iy).$$

For any $q > 0$, we set

$$K_q(z, \overline{u}) = \Gamma(2q)\pi^q K_{\Delta(\alpha)}(z, \overline{u})^q.$$

Then, for $q > \frac{1}{2}$, $K_q(z, \overline{u})$ is the reproducing kernel for the Bergman-Selberg space $H_{K_q}(\Delta(\alpha))$ on $\Delta(\alpha)$ composed of all analytic functions $f(z)$ on $\Delta(\alpha)$ with finite norms

$$\|f\|^2_{H_{K_q}(\Delta(\alpha))} = \frac{1}{\Gamma(2q-1)\pi^q} \iint_{\Delta(\alpha)} |f(z)|^2 K_{\Delta(\alpha)}(z, \overline{z})^{1-q} \, dx \, dy.$$

For $q = \frac{1}{2}$, $K_{\frac{1}{2}}(z, \overline{u})$ is the Szegö kernel for the Szegö space $H_{K_{\frac{1}{2}}}(\Delta(\alpha))$ on $\Delta(\alpha)$ composed of all analytic functions $f(z)$ on $\Delta(\alpha)$ such that

$$\sup_{|\theta| < \alpha} \int_0^\infty |f(re^{i\theta})|^2 \, dr < \infty,$$

and the norm in $H_{K_{\frac{1}{2}}}(\Delta(\alpha))$ is given by

$$\|f\|^2_{H_{K_{\frac{1}{2}}}(\Delta(\alpha))} = \frac{1}{2\pi} \int_{\partial\Delta(\alpha)} |f(z)|^2 \, |dz|$$

where $f(z)$ means the Fatou's nontangential boundary values of f on ∂D. When $0 < q < \frac{1}{2}$, for the integral expressions of $H_{K_{\frac{1}{2}}}(\Delta(\alpha))$ norms, see (3.14) in [4].

As generalizations of the result in [1], we obtain the following two theorems which are applied to the investigations of analyticity of the solutions of nonlinear partial differential equations in [3].

THEOREM 1. *For $0 < \alpha < \frac{\pi}{2}$ and for $f \in H_{K_1}(\Delta(\alpha))$, we have the expression*

$$\|f\|^2_{H_{K_1}(\Delta(\alpha))} = \sin(2\alpha) \sum_{n=0}^{\infty} \frac{(2 \sin \alpha)^{2n}}{(2n+1)!} \int_0^\infty x^{2n+1} |\partial_x^n f(x)|^2 \, dx. \tag{1}$$

Conversely, any C^∞ function $f(x)$ on the positive real line $(0, \infty)$ with convergent summation in (1) can be extended analytically to $\Delta(\alpha)$ and the extension $f(z)$ satisfies the identity (1).

THEOREM 2. *For $0 < \alpha < \frac{\pi}{2}$ and for $f \in H_{K_{\frac{1}{2}}}(\Delta(\alpha))$, we have the identity*

$$\|f\|^2_{H_{K_{\frac{1}{2}}}(\Delta(\alpha))} = \frac{\cos\alpha}{\pi} \sum_{n=0}^{\infty} \frac{(2\sin\alpha)^{2n}}{(2n)!} \int_0^{\infty} x^{2n} |\partial_x^n f(x)|^2 \, dx. \tag{2}$$

Conversely, any C^∞ function $f(x)$ on the positive real line $(0, \infty)$ with convergent summation in (2) can be extended analytically to $\Delta(\alpha)$ and the extension $f(z)$ satisfies the identity (2).

In the proofs in Theorems 1 and 2, we use the identities

$$\Gamma(1+it)\Gamma(1-it) = \frac{\pi t}{\sinh \pi t},$$

$$\Gamma(\frac{1}{2}+it)\Gamma(\frac{1}{2}-it) = \frac{\pi}{\cosh \pi t},$$

and the Parseval-Plancherel's theorem in the Mellin transform. By using the Gauss formula

$$F(a,b;c;1) = \frac{\Gamma(c)\Gamma(c-a-b)}{\Gamma(c-a)\Gamma(c-b)} \qquad c \neq 0, -1, -2, \ldots, \qquad Re(c-a-b) > 0$$

with

$$F(a,b;c;1) = \frac{\Gamma(c)}{\Gamma(a)\Gamma(b)} \sum_{n=0}^{\infty} \frac{\Gamma(a+n)\Gamma(b+n)}{n!\,\Gamma(c+n)},$$

we obtain in [4]

THEOREM 3. *For any $q > 0$ and $f \in H_{K_q}(\Delta(\frac{\pi}{2}))$, we have the identity*

$$\|f\|^2_{H_{K_q}(\Delta(\frac{\pi}{2}))} = \sum_{n=0}^{\infty} \frac{1}{n!\,\Gamma(n+2q+1)} \int_0^{\infty} \left|\partial_x^n (xf'(x))\right|^2 x^{2n+2q-1} \, dx. \tag{3}$$

Conversely, any C^∞ function $f(x)$ on the positive real line $(0, \infty)$ with convergent summation in (3) can be extended analytically to $\Delta(\frac{\pi}{2})$. The analytic extension $f(z)$ satisfying $\lim_{x\to\infty} f(x) = 0$ belongs to $H_{K_q}(\Delta(\frac{\pi}{2}))$ and the identity (3) is valid.

In Theorem 3, on the general sector $\Delta(\alpha)$ we have two representations

$$\|f\|^2_{H_{K_q}(\Delta(\alpha))} = \left(\frac{2\alpha}{\pi}\right)^{2q-1} \sum_{n=0}^{\infty} \left(\frac{\pi}{2\alpha}\right)^{2(n+1)}$$

$$\cdot \int_0^{\infty} \left| \left(\frac{\partial}{\partial x} x^{1-\frac{\pi}{2\alpha}}\right)^{(n-1)} \left(x(f(x)x^{(1-\frac{\pi}{2\alpha})q})'\right)' \right|^2 x^{\frac{\pi}{\alpha}(n+q-1)} \, dx$$

$$= \left(\frac{2\alpha}{\pi}\right)^{2q} \sum_{n=0}^{\infty} \frac{1}{n!\,\Gamma(n+2q+1)}$$

$$\cdot \int_0^{\infty} \left|\partial_x^n \left(x(f(x^{\frac{2\alpha}{\pi}})x^{(\frac{2\alpha}{\pi}-1)q})'\right)\right|^2 x^{2n+2q-1} \, dx.$$

Here, $(\frac{\partial}{\partial x} x^{1-\frac{\pi}{2\alpha}})^{(n-1)}$ denotes the iterated derivative. The second expression coincides with that obtained by using the conformal mapping $w = z^{\frac{\pi}{2\alpha}}$, Theorem 3 and the conformally invariant property of H_{K_q} norms. We now have the

PROBLEM 1.7

QUESTION. *Does there exist some simple representations of* $\|f\|_{H_{K_q}(\Delta(\alpha))}$ *as in Theorems 1–3 on the general sector* $\Delta(\alpha)$ *and for* $q > 0$*?*

REFERENCES

1. Aikawa H., Hayashi N., Saitoh S., *The Bergman space on a sector and the heat equation*, Complex Variables **15** (1990), 27–36.
2. Aikawa H., Hayashi N., Saitoh S., *Isometrical identities for the Bergman and the Szegö spaces on a sector*, J. Math. Soc. Japan **43** (1991), 195–201.
3. Hayashi N., *Solutions of the (generalized) Korteweg–de Vries equation in the Bergman and the Szegö spaces on a sector*, Duke Math. J. **62** (1991), 575–591.
4. Saitoh S., *Representations of the norms in Bergman–Selberg spaces on strips and half planes*, Complex Variables (to appear).

DEPARTMENT OF MATHEMATICS
FACULTY OF ENGINEERING
GUNMA UNIVERSITY
KIRYU 376
JAPAN

1.8

SPACES OF ANALYTIC FUNCTIONS IN THE UNIT DISK WHICH ARE GENERATED BY A MEASURE NONINVARIANT WITH RESPECT TO ROTATIONS

YU. I. LYUBARSKII

Let a finite measure μ in the closed unit disk $\overline{\mathbb{D}} = \{ z \in \mathbb{C} : |z| \leqslant 1 \}$ be given such that $(\operatorname{supp} \mu) \cap \{ z \in \mathbb{C} : |z| = 1 \} \neq \emptyset$ and such that the sequence of functions $\{z \mapsto z^n\}_{n=0}^{\infty}$ is minimal in the space $L^2(\mu)$ of all μ-square summable functions. Consider the set $A(\mu) \subset L^2(\mu)$ admitting the representation

$$f(z) = \sum_{n=0}^{\infty} a_n z^n \tag{1}$$

where the series converges in $L^2(\mu)$. As can be easily seen, every function $f \in L^2(\mu)$ (it is defined on $\operatorname{supp} \mu$) may be continued analytically on the whole open disk $\mathbb{D}$.

PROBLEM 1. *What is an independent description of the set $A(\mu)$ as a set of functions analytic in $\mathbb{D}$?*

The linear space $A(\mu)$ admits introducing of the natural Banach structure. For any function $f \in A(\mu)$ represented in the form (1) let

$$\|f\|_{A(\mu)} := \sup_{n \geqslant 0} \left\{ \left\| \sum_{k=0}^{n} a_k z^k \right\|_{L^2(\mu)} \right\}$$

PROBLEM 2. *What Banach spaces may be obtained (up to isomorphism) in such a way?*

INSTITUTE FOR LOW TEMPERATURE
PHYSICS & ENGINEERING
47, LENIN AVE., KHARKOV, 310164,
UKRAINE

1.9
old

COMPACTNESS OF ABSOLUTELY SUMMING OPERATORS

A. PEŁCZYŃSKI

Every concrete absolutely summing surjection with an infinite dimensional range allows to prove the classical Grothendieck theorem on absolutely summing operators from ℓ^1 into ℓ^2. On the other hand, given a Banach space X so that every absolutely summing operator from every ultrapower of X to ℓ^2 is compact, one can consider new local characteristics of operators on X (e.g. in spirit of Problem 3′ below).

Our knowledge of what concerns the existence of non-compact absolutely summing operators with a given domain space is however less than satisfactory.

PROBLEM 1. *Let X be a Banach space. Are the following conditions equivalent:*

(a) *there is an absolutely summing non-compact operator from X into a Hilbert space ℓ^2;*

(b) *there is an absolutely summing surjection from X onto ℓ^2?*

Observe that if one replaces in (a) and (b) "absolutely summing" by "2-absolutely summing" then the "new (a)" is equivalent to the "new (b)" and is equivalent to the fact that X contains an isomorph of ℓ^1 (cf. [1], [2]).

An obvious example of a space satisfying (a) and (b) is any $\mathcal{L}^1$-space of infinite dimension (by the Grothendieck theorem). Another example is the disc algebra A. A well-known example of a 1-absolutely summing surjection from A onto ℓ^2 is the so-called "Paley projection"

$$f \mapsto \left(\frac{1}{2\pi}\int_0^{2\pi} e^{-i2^n t} f(t)\,dt\right)_{1\leqslant n\leqslant\infty},$$

cf. [3].

PROBLEM 2. *Are the following conditions equivalent:*

(a) *every absolutely summing operator from X to ℓ^2 factors through a Hilbert–Schmidt operator;*

(b) *X^{**} is isomorphic to a quotient of a $C(K)$-space?*

It is a well-known consequence of the classical result of Grothendieck that (b) implies (a).

Observe that if ℓ^2 in (a) is replaced by ℓ^1, then the modified property (a) is equivalent to (b), cf. J. Bourgain and A. Pełczyński (in preparation).

Every $\mathcal{L}^\infty$-space satisfies (a). The $\mathcal{L}^\infty$-space constructed by Bourgain [4] which does not contain c_0 is not isomorphic to any quotient of a $C(K)$-space. So "X^{**}" in (b) cannot be replaced by "X".

Let $L(\ell^2)$ (respectively $K(\ell^2)$) stand for the spaces of all bounded operators (respectively all compact operators) from ℓ^2 into itself.

PROBLEM 3. *Is every absolutely summing operator from $L(\ell^2)$ into a Hilbert space compact?*

Obviously every absolutely summing operator from $K(\ell^2)$ into ℓ^2 is compact because the dual of $K(\ell^2)$ is separable. However Problem 3 has a local counterpart for $K(\ell^2)$.

PROBLEM 3'. *Does there exist a "modulus of capacity" $\varepsilon \mapsto N(\varepsilon)$ such that if $\pi_1(u\colon K(\ell^2) \to \ell^2) \leqslant 1$, then the ε-capacity of $u(B_{K(\ell^2)})$ does not exceed $N(\varepsilon)$ (here $B_{K(\ell^2)} = \{T \in K(\ell^2)\colon \|T\| \leqslant 1\}$)?*

The positive answer to Problem 3 will follow if one could establish the following structural property of $L(\ell^2)$:

let X be a subspace of $L(\ell^2)$ isomorphic to ℓ^1; then there exists a subspace Y of $L(\ell^2)$ isomorphic to a $C(K)$-space such that $Y \cap X$ is infinite-dimensional.

Our last problem concerns spaces of smooth functions.

PROBLEM 4. *Is every absolutely summing operator from $C^k(\mathbb{T}^n)$ into a Hilbert space compact?*

We do not know whether there exists a "Paley phenomenon" for $C^1(\mathbb{T}^n)$, i.e. whether there is an absolutely summing surjection from $C^1(\mathbb{T}^n)$ onto ℓ^2 (cf. comments to Problem 1). It seems to be unlikely that there exists an *invariant* absolutely summing surjection, as in the case of the disc algebra A.

The author would like to thank Prof. S. V. Kisliakov for a valuable discussion.

REFERENCES

1. Ovsepian R. I., Pełczyński A., *On the existence of a fundamental total and bounded biorthogonal sequence in every separable Banach space, and related constructions of uniformly bounded orthonormal systems in L^2*, Studia Math. **54** (1975), 149–159.
2. Weis L., *On strictly singular and strictly cosingular operators*, Studia Math. **54** (1975), 285–290.
3. Pełczyński A., *Banach spaces of analytic functions and absolutely summing operators*, Regional conference series in mathematics, vol. 30, AMS, Providence, 1977.
4. Bourgain J., Delbaen F., *A class of special $\mathcal{L}_\infty$-spaces*, Acta Math. **145** (1980), no. 3–4, 155–176.

INSTITUTE OF MATHEMATICS
POLISH ACADEMY OF SCIENCES
ŚNIADECKICH 8,
00-950 WARSAW
POLAND

CHAPTER EDITOR'S NOTE

M. Wojciechowski [5] has proved that there is no translation invariant absolutely summing surjection from $C^1(\mathbb{T}^n)$ onto ℓ^2. See also [6] for more information.

REFERENCES

5. Wojciechowski M., *Translation invariant projections on anisotropic Sobolev spaces on tori in L^1- and uniform norms*, Studia Math. **100** (1991), no. 2, 149–161.
6. Pełczyński A., Wojciechowski M., *Paley projections on anisotropic Sobolev spaces on tori*, Proc. London Math. Soc. **65** (1992), no. 2, 405–422.

1.10
v.old

WHEN IS $\Pi_2(X,\ell^2) = L(X,\ell^2)$?

I. A. KOMARCHEV, B. M. MAKAROV

Let X and Y be two infinite dimensional Banach spaces and let $L(X,Y)$ denote the space of all continuous linear operators from X to Y. An operator $T \in L(X,Y)$ is said to be *p-absolutely summing* if there exists a positive constant C such that

$$\sum_{k=1}^{n} \|Tx_k\|^p \leqslant C^p \cdot \sup\left\{\sum_{k=1}^{n} |\langle x_k, x'\rangle|^p : x' \in X^*, \|x'\| \leqslant 1\right\}$$

for each n in $\mathbb{N}$ and $x_1, x_2, \ldots, x_n \in X$. The set of all p-absolutely summing operators from X to Y is denoted by $\Pi_p(X,Y)$. The conditions for $\Pi_p(X,Y)$ to coincide with $\Pi_r(X,Y)$ or with $L(X,Y)$ have been the subject of a great number of publications (see [1]–[4]). The results obtained not only are of their own interest but also are widely used in problems connected with the isomorphic classification of Banach spaces.

It is easy to see that $\Pi_p(X,Y) \subset \Pi_r(X,Y)$ for $p < r$. The Dvoretzky theorem on almost Euclidean sections of convex bodies shows that the equality $\Pi_p(X,Y) = L(X,Y)$ has the highest chance to hold if Y is isomorphic to a Hilbert space, i.e. that $\Pi_p(X,\ell^2) = L(X,\ell^2)$ provided $\Pi_p(X,Y) = L(X,Y)$ at least for one infinite dimensional space Y. Besides, it is well-known that $\Pi_p(X,\ell^2) = \Pi_2(X,\ell^2)$ for $p \geqslant 2$. Thus the investigation of the problem whether $\Pi_p(X,Y)$ coincides with $L(X,Y)$ leads immediately to the question of conditions ensuring the equality

$$\Pi_2(X,\ell^2) = L(X,\ell^2). \tag{1}$$

A space X satisfying (1) will be called *2-trivial* (cf.[5]). Obviously a space X and its dual X^* are 2-trivial (or not) simultaneously.

A GENERAL QUESTION we want to raise is *to find out conditions* (in particular the conditions of geometrical nature) *under which a Banach space X is* (or is not) *2-trivial.*

It is known [6] that (1) is impossible for the space X not containing ℓ^1_n uniformly (for example if X is uniformly convex). On the other hand it is easy to verify that this condition is not sufficient for the 2-triviality. Indeed, the sequential Lorentz space $\Lambda(C)$ not only contains ℓ^1_n uniformly but is even saturated by subspaces isomorphic to ℓ^1 (to wit every infinite dimensional subspace of $\Lambda(C)$ contains a subspace isomorphic to ℓ^1). Nevertheless $\Lambda(C)$ fails to be 2-trivial and moreover it is a space of type $(\mathcal{E})$ (see the definition below).

It can be proved that X is not 2-trivial provided X satisfies the following condition: there exist two sequences $\{A_n\}_{n\geqslant 1}$ and $\{B_n\}_{n\geqslant 1}$ of operators such that

$$A_n\colon \ell^2_n \to X, \quad B_n\colon X \to \ell^2_n, \quad B_n \circ A_n = \mathrm{id}_{\ell^2_n}$$

and $\underline{\lim}_n n^{-1/2}|A_n|\cdot|B_n| = 0$. A space satisfying these conditions is said to be of type $(\mathcal{E})$.

It has been essentially proved in [6] that a space not containing ℓ_n^1 uniformly is a space of type $(\mathcal{E})$. However the condition of being of type $(\mathcal{E})$ is also not necessary for the non-2-triviality. As S. V. Kisliakov has pointed out, the reflexive "non sufficiently Euclidean" space built in [7] fails to be of type $(\mathcal{E})$ and simultaneously it can be proved that this space fails to be 2-trivial.

What has been said above indicates that the class of all 2-trivial spaces cannot be too large. The following conjecture looks therefore rather plausible.

CONJECTURE 1. *No infinite dimensional reflexive Banach space is 2-trivial.*

An equivalent statement: *there exists no infinite dimensional reflexive Banach space X such that each operator from $L(\ell^2, \ell^2)$, which can be factored through X, is a Hilbert–Schmidt operator.* We note that a positive solution to Conjecture 1 would obviously imply the solution (in the class of reflexive Banach spaces) to the Grothendieck Problem on the coincidence of the spaces of nuclear and compact operators.

The following QUESTIONS arise naturally.

1. *Under what conditions does 2-triviality of a space X imply the equality $\Pi_2(X, \ell^1) = L(X, \ell^1)$?*

2. *Which of the spaces of analytic or smooth functions are of type $(\mathcal{E})$?*

3. *Is it true that in any space X of type $(\mathcal{E})$ there exists a sequence of subspaces $\{X_n\}$ ($\dim X_n = n$) with one of the following two properties:*
   (a) $\sup_n \lambda(X_n, X) < \infty$, $d(X_n, \ell_n^2) = o(\sqrt{n})$ *or*
   (b) $\sup_n d(X, \ell_n^2) < \infty$, $\lambda(X_n, X) = o(\sqrt{n})$*?*

(Here $\lambda(X_n, X)$ is the relative projection constant).

The assumption that a 2-trivial space has an unconditional basis apparently rather drastically diminishes the class of such spaces. For example, each reflexive Banach space with an unconditional basis is not 2-trivial [8]. On the other hand, as it is shown in [9], the space $(\sum c_0)_{\ell^1}$ also fails to be 2-trivial (more precisely it is of type $(\mathcal{E})$). These results give some ground to the following

CONJECTURE 2. *If a 2-trivial infinite dimensional Banach space X has an unconditional basis, then X is isomorphic to either c_0 or ℓ^1 or $c_0 \oplus \ell^1$.*

To illustrate conjecture 2 we mention a result which follows from Theorem 1 in [8]: If X has an unconditional basis, if Y is not isomorphic to a Hilbert space, and if $\Pi_2(X, Y) = L(X, Y)$, then X is isomorphic to c_0.

REFERENCES

1. Lindenstrauss J., Pełczyński A., *Absolutely summing operators in $\mathcal{L}_p$-spaces and their applications*, Studia Math. **29** (1968), 275–326.
2. Kwapień S., *On a theorem of L. Schwartz and its applications to absolutely summing operators*, Studia Math. **38** (1970), 193–201.
3. Dubinsky E., Pełczyński A., Rosenthal H., *On Banach spaces X for which $\Pi_2(\mathcal{L}_\infty, X) = B(\mathcal{L}_\infty, X)$*, Studia Math. **44** (1972), 617–648.

4. Maurey B., *Théorèmes de factorisation pour les opérateurs linéaires à valeurs dans les espaces L^p*, Astérisque **11** (1974), 1–163.
5. Morrell J. S., Retherford J. R., *p-trivial Banach spaces*, Studia Math. **43** (1972), 1–25.
6. Davis W. J., Johnson W. B., *Compact nonnuclear operators*, Studia Math. **51** (1974), 81–85.
7. Johnson W. B., *A reflexive Banach space which is not sufficiently Euclidean*, Studia Math. **55** (1976), 201–205.
8. Komarchev I. A., *On 2-absolutely summing operators in Banach lattices*, Vestnik LGU, ser. mat., mekh., astr. (1980), no. 19, 97–98. (Russian)
9. Figiel T., Lindenstrauss J., Milman V., *The dimension of almost spherical sections of convex bodies*, Acta Math. **133** (1977), 53–94.

St. Petersburg Technical University
St. Petersburg
Russia

Dept. of Mathematics and Mechanics
St. Petersburg State University
Bibliotechnaya pl. 2.
Staryi Peterhof, St. Petersburg, 198904
Russia

Commentary by I. A. Komarchev

M. Rudel'son proved that Conjecture 2 was true. Moreover, he established the following theorem.

Theorem 1. (Cf. [10]). *Let X be a 2-trivial Banach space with normalized unconditional basis $(e_i)_{i\in\mathbb{N}}$. Then $\mathbb{N}$ splits into the union of two disjoint subsets A and B such that $\{e_i\}_{i\in A}$ is equivalent to the unit vector basis of $\ell^1_{|A|}$ and $\{e_i\}_{i\in B}$ is equivalent to the unit vector basis of $c_0^{|B|}$.*

Subsequently I succeeded to simplify the most difficult part of Rudel'son's proof. The argument is sketched below.

For $I \subset \mathbb{N}$, we denote by X_I the closed linear hull of $\{e_i\}_{i\in I}$. If $|I| < \infty$, we denote by $D_p(I)$ $(p = 1, \infty)$ the smallest number C such that the basis $\{e_i\}_{i\in I}$ is C-equivalent to the unit vector basis of $\ell^p_{|I|}$. For $s \in \mathbb{R}$ we set $\lambda_p(I, s) = \max\{\,|J| : J \subset I, D_p(J) \leqslant s\,\}$.

Since X is 2-trivial, there exists $Q > 0$ such that $\pi_2(T) \leqslant Q\|T\|$ for every operator $T\colon X \to \ell^2$.

The core of Theorem 1 is the following statement (once this statement is proved, the rest is relatively easy, see [10]).

Proposition 1. *There exists $\alpha > 0$ such that for every finite subset I of $\mathbb{N}$ we have either $\lambda_1(I, 2Q^2) \geqslant \alpha|I|$ or $\lambda_\infty(I, 2Q^2) \geqslant \alpha|I|$.*

To establish this proposition we need three lemmas.

Lemma 1. *There is $\sigma > 0$ such that for every finite subset I of $\mathbb{N}$,*

$$\lambda_1(I, 2Q^2)\lambda_\infty(I, 2Q^2) \geqslant \sigma|I|.$$

This is Lemma 3 of [10].

For technical reasons we assume in what follows that $\sigma < 1$.

LEMMA 2. *Let X be an n-dimensional Banach space with normalized 1-unconditional basis $\{e_i\}_{1\leqslant i\leqslant n}$. Let $\{T_k\}_{1\leqslant k\leqslant l}$ and $\{S_j\}_{1\leqslant j\leqslant m}$ be two partitions of $\{1,\dots,n\}$ satisfying the following property: $D_1(T_k)\leqslant C$, $D_\infty(S_j)\leqslant C$ for all k and j, $1\leqslant k\leqslant l$, $1\leqslant j\leqslant m$. Then there exist constants C_1 and C_2 (depending on C only) and a subspace Y of X such that*

(a) $\dim Y\geqslant C_1\min(n^2/ml^2, n^2/m^2l)$

(b) $d(Y,\ell^2_{\dim Y})\leqslant 2$

(c) *there is a projection P from X onto Y with $\|P\|\leqslant C_2 mln^{-1}\log n$.*

Proof. Consider the natural Euclidean norm $|\cdot|$ on X: $\left|\sum\alpha_ie_i\right|=\left(\sum|\alpha_i|^2\right)^{1/2}$. Clearly,

$$cl^{-1/2}\left|\sum\alpha_ie_i\right|\leqslant\left\|\sum\alpha_ie_i\right\|\leqslant cm^{1/2}\left|\sum\alpha_ie_i\right|$$

for every vector $\sum_{1\leqslant i\leqslant n}\alpha_ie_i\in X$. Let M and M^* be the medians of the functions $\|\cdot\|$ and $\|\cdot\|^*$ (the norm in the dual space X^*) with respect to the normalized surface measure mes on the unit sphere S^{n-1} of $|\cdot|$. It has been proved in [9] that there exists a subspace Y of X satisfying (b) and the relation

$$\dim Y\geqslant \text{const}\, n\min\left((Mm^{-1/2})^2,(M^*l^{-1/2})^2\right),$$

accompanied by a projection P from X onto Y with $\|P\|\leqslant(16/9)MM^*$.

Let us estimate the medians M and M^*. It follows from the well-known estimate of the median of the ℓ^1-norm (cf.[9]) that there exists a set $A\subset S^{n-1}$ with mes $A\geqslant 1/2$ such that $\sum_{1\leqslant i\leqslant n}|\alpha_i|\geqslant a\sqrt{n}$ for every $\sum_{1\leqslant i\leqslant n}\alpha_ie_i\in A$, a being a universal constant. Since by the hypotheses

$$\sum_{1\leqslant i\leqslant n}|\alpha_i|\leqslant\sum_{1\leqslant k\leqslant l}\sum_{i\in T_k}|\alpha_i|\leqslant l\max_k\sum_{i\in T_k}|\alpha_i|\leqslant Cl\left\|\sum_{1\leqslant i\leqslant n}\alpha_ie_i\right\|,$$

we obtain that

$$\left\|\sum_{1\leqslant i\leqslant n}\alpha_ie_i\right\|\geqslant ac^{-1}l^{-1}\sqrt{n}\qquad\text{for }\sum\alpha_ie_i\in A,$$

whence

$$M\geqslant ac^{-1}l^{-1}\sqrt{n}.$$

Now, using the well-known estimate of the median of the ℓ^∞-norm, we can find a subset B of S^{n-1}, mes $B\geqslant 1/2$, such that $\max_i|\alpha_i|\leqslant b\sqrt{n^{-1}\log n}$ whenever $\sum\alpha_ie_i\in B$. Since by the hypotheses $D_\infty(S_i)\leqslant C$ $(i=1,2,\dots,n)$, we obtain

$$\left\|\sum_{1\leqslant j\leqslant n}\alpha_je_j\right\|\leqslant\sum_{1\leqslant i\leqslant m}\left\|\sum_{j\in S_i}\alpha_je_j\right\|\leqslant C\sum_{1\leqslant i\leqslant m}\max_{j\in S_i}|\alpha_j|\leqslant Cm\max_{1\leqslant j\leqslant n}|\alpha_j|.$$

Hence

$$\left\|\sum\alpha_je_j\right\|\leqslant Cbm\sqrt{n^{-1}\log n}\qquad\text{whenever }\sum\alpha_je_j\in B,$$

whence

$$M \leqslant Cbm\sqrt{n^{-1}\log n}.$$

By symmetry, we also have

$$aC^{-1}\sqrt{n}m^{-1} \leqslant M^* \leqslant Cbl\sqrt{n^{-1}\log n},$$

and (a) and (c) follow. □

LEMMA 3. *Let* $0 \leqslant t_n \leqslant t_{n-1} \leqslant \ldots \leqslant t_1$, $r \in \mathbb{N}$, $\sigma > 0$. *Suppose that* $r\sigma^{-1}t_1 \leqslant \sum_{1\leqslant i\leqslant n} t_i$. *Then one can find* m *and* k *with* $m \leqslant k \leqslant n$ *and* $k - m \leqslant r - 1$ *satisfying* $\sum_{m\leqslant i\leqslant k} t_i \geqslant r(1+\sigma)^{-1}t_m$.

Proof. Suppose not. Write $n = lr + p$ with $l, p \in \mathbb{N}$ and $p < r$. We then have

$$\sum_{ir+1\leqslant j\leqslant (i+1)r} t_j < r(1+\sigma)^{-1}t_{ir+1} \qquad \text{for } 0 \leqslant i \leqslant l-1$$

and

$$\sum_{lr+1\leqslant j\leqslant n} t_j < r(1+\sigma)^{-1}t_{lr+1}.$$

Summing all these inequalities we obtain

$$\sum_{1\leqslant j\leqslant n} t_j < r(1+\sigma)^{-1}\sum_{0\leqslant i\leqslant l} t_{ir+1}.$$

Since $rt_{ir+1} \leqslant t_{(i-1)r+1} + \cdots + t_{ir}$ $(i = 1, 2, \ldots, l)$, it follows that

$$\sum_{1\leqslant j\leqslant n} t_j < r(1+\sigma)^{-1}t_1 + (1+\sigma)^{-1}\sum_{1\leqslant j\leqslant n} t_j,$$

or

$$\sigma(1+\sigma)^{-1}\sum_{1\leqslant j\leqslant n} t_j < r(1+\sigma)^{-1}t_1.$$

This contradicts the hypotheses. □

Now we are ready to prove Proposition 1. Suppose its conclusion fails. Then for every $r \in \mathbb{N}$ there is $I \subset \mathbb{N}$, $|I| < \infty$, with

$$\begin{aligned} \lambda_1(I, 2Q^2) &\leqslant (\sigma/2r)|I|, \\ \lambda_\infty(I, 2Q^2) &\leqslant (\sigma/2r)|I|. \end{aligned} \tag{1}$$

We construct by induction a finite collection of disjoint subsets $J_1, \ldots, J_n$ of I. It follows from (1) and Lemma 1 that $\lambda_1(I, 2Q^2) \geqslant 2r$. Pick a subset $J_1 \subset I$ of maximal cardinality such that $D_\infty(J_1) \leqslant 2Q^2$. Suppose that $J_1, J_2, \ldots, J_k$ have already been constructed. We stop if $\lambda_1(I \setminus \bigcup_{1\leqslant l\leqslant k} J_l, 2Q^2) < r$, otherwise we take for J_{k+1} a subset of maximal cardinality in $I \setminus \bigcup_{1\leqslant l\leqslant k} J_l$ for which $D_\infty(J_{k+1}) \leqslant 2Q^2$.

Clearly $J_1, J_2, \dots, J_n$ are mutually disjoint and $|J_n| \leqslant |J_{n-1}| \leqslant \dots \leqslant |J_1|$. It can easily be seen that $\left|\bigcup_{1\leqslant l\leqslant n} J_l\right| \geqslant |I|/2$ (otherwise, by (1) and Lemma 1, the construction could be continued).

Since

$$|J_1| = \lambda_1(I, 2Q^2) \leqslant (\sigma/2r)|I| \leqslant (\sigma/r)\left|\bigcup_{1\leqslant l\leqslant n} J_l\right| = (\sigma/r)\sum_{1\leqslant l\leqslant n} |J_l|,$$

we can find (by Lemma 3) positive integers m and k with $m \leqslant k$, $k - m \leqslant r - 1$ such that

$$r(\sigma+1)^{-1}|J_m| \leqslant \sum_{m\leqslant i\leqslant k} |J_i|. \tag{2}$$

We claim that if r is large enough, we can find r disjoint subsets $S_1, S_2, \dots, S_r$ of $\bigcup_{m+1\leqslant i\leqslant k} J_i \overset{\text{def}}{=} J'$, of cardinality $[r\sigma/2(\sigma+1)]$ each, such that $D_\infty(S_i) \leqslant 2Q^2$.

Indeed, suppose that for some $t \geqslant 1$ the sets S_l with $l < t$ are already constructed (this assumption is void for $t = 1$). Since $|J_m| = \lambda_1\big(I \setminus \bigcup_{i<m} J_i\big)$, we clearly have also $|J_m| = \lambda_1\big(J_m \bigcup (J' \setminus \bigcup_{l<t} S_l)\big)$. In view of Lemma 1

$$\lambda_\infty\Big(J_m \bigcup (J' \setminus \bigcup_{l<t} S_l)\Big) \geqslant \sigma\frac{|J_m| + |J'| - (t-1)\big[r\sigma/2(\sigma+1)\big]}{|J_m|}$$

$$\geqslant r\frac{\sigma}{\sigma+1}\,\frac{|J_m| + |J'| - (t-1)\big[r\sigma/2(\sigma+1)\big]}{|J_m| + |J'|}, \quad \text{by (2)}.$$

Again applying (2) and the fact that $|J_m| \geqslant r$, we obtain $|J_m| + |J'| \geqslant r(\sigma+1)^{-1}$, whence

$$\lambda_\infty\Big(J_m \bigcup (J' \setminus \bigcup_{l<t} S_l)\Big) \geqslant \frac{r\sigma}{\sigma+1}\left(1 - \frac{(t-1)\sigma}{2r}\right).$$

If $t < r$ then the factor in brackets on the right is greater than $1/2$ (we recall that σ has been supposed to be < 1), and the claim follows.

Now let $S = \bigcup_{1\leqslant i\leqslant r} S_i$ and $T_j = S \bigcap J_j$. Then we are under the hypotheses of Lemma 2. Since $|T| = r\big[r\sigma/(2(\sigma+1))\big]$ and the number of nonempty T_j's is at most r, we obtain by Lemma 2 that there is a subspace Y of X_T such that $d(Y, \ell^2_{\dim Y}) \leqslant 2$, $\dim Y \geqslant \text{const}\, r$ and there exists a projection P from X_T onto Y with $\|P\| \leqslant \text{const} \log r$. Since X is 2-trivial, we have $\pi_2(P) \leqslant 2Q\|P\| \leqslant \text{const}\, Q \log r$. On the other hand, $\pi_2(P) \geqslant (\dim Y)^{1/2} \geqslant \text{const}\sqrt{r}$. This contradiction proves Proposition 1. □

Reference

10. Rudel'son M. V., *A characterization of 2-trivial Banach spaces with unconditional basis*, Zapiski Nauchn. Semin. LOMI **157** (1987), 76–87. (Russian)

1.11

STABLY REGULAR OPERATORS. LATTICES OF OPERATORS

B. M. MAKAROV

Let $\mathcal{L}(L^p, L^q)$ denote the set of all (continuous linear) operators from L^p to L^q. An operator is called regular if it maps order-bounded sets to order-bounded sets.

1. DEFINITION (cf.[1], [2]). An operator U in $\mathcal{L}(L^p, L^q)$ is called stably regular from the left (from the right) if for every $W \in \mathcal{L}(L^q, L^q)$ (respectively, every operator $V \in \mathcal{L}(L^p, L^p)$) the operator WU (respectively, UV) is regular. If for every V, W as above the operator WUV is regular, then U is called (two-sided) stably regular.

The set of all two-sided stably regular operators in $\mathcal{L}(L^p, L^q)$ will be denoted by $\mathcal{L}^{\sim}_{\mathrm{st}}(L^p, L^q)$.

For $\min(p, q') \leqslant 2$ (where q' is defined by $1/q + 1/q' = 1$), a description of operators in $\mathcal{L}^{\sim}_{\mathrm{st}}(L^p, L^q)$ has been obtained in [1], [2], [3].

QUESTION. *Describe the operators in $\mathcal{L}^{\sim}_{\mathrm{st}}(L^p, L^q)$ for $1 \leqslant q < 2 < p \leqslant \infty$.*

2. It is known that for $1 \leqslant p \leqslant 2 \leqslant q \leqslant \infty$ the one-sided (left or right) stable regularity of an operator is equivalent to the two-sided stable regularity, and for $1 < p \leqslant 2$, $1 \leqslant q < 2$ the stable regularity from the right is equivalent to the two-sided stable regularity, but it is not implied by the stable regularity from the left ([1], [2], [3]).

QUESTION. *Describe the operators mapping L^p to L^q for $1 < p \leqslant 2$, $1 \leqslant q < 2$, which are stably regular from the left.*

3. QUESTION. *Is it true that for $1 \leqslant q < 2 < p \leqslant \infty$ the stable regularity from one side (either specified or not) implies the two-sided stable regularity?*

4. If $\min(p, q') \leqslant 2$ then the set $\mathcal{L}^{\sim}_{\mathrm{st}}(L^p, L^q)$ is a sublattice of the lattice of all regular operators (cf.[2], [3]).

QUESTION. *Is it true that the set $\mathcal{L}^{\sim}_{\mathrm{st}}(L^p, L^q)$ is a sublattice of the lattice of all regular operators for $1 \leqslant q < 2 < p \leqslant \infty$?*

5. DEFINITION. Let $0 < r < \infty$. An operator U in $\mathcal{L}(L^p, L^q)$ is called r-absolutely summing if there exists a number C such that for any natural number n and any vectors $x_1, \ldots, x_n$ in L^p the inequality

$$\left(\sum_{k=1}^{n} \|Ux_k\|^r\right)^{1/r} \leqslant C \sup\left\{\left(\sum_{k=1}^{n} |\langle x_k, f\rangle|^r\right)^{1/r} : f \in L^{p'}, \|f\| \leqslant 1\right\}$$

holds. The smallest possible constant C is denoted by $\pi_r(U)$, and the set of all r-absolutely summing operators in $\mathcal{L}(L^p, L^q)$ is denoted by $\Pi_r(L^p, L^q)$.

For $1 < p \leqslant \infty$ the inclusion $\Pi_{p'}(L^p, L^{p'}) \subset \mathcal{L}^{\sim}_{\text{st}}(L^p, L^{p'})$ is valid. This inclusion is proper if $1 < p < 2$, and it turns into equality if $p = 2$ or $p = \infty$. One has

PROPOSITION 1. *Let $2 < p < \infty$. Then $\mathcal{L}^{\sim}_{\text{st}}(L^p, L^{p'}) \subset \Pi_{p'+\varepsilon}(L^p, L^{p'})$ for all $\varepsilon > 0$.*

QUESTION. *Is it true that $\mathcal{L}^{\sim}_{\text{st}}(L^p, L^{p'}) = \Pi_{p'}(L^p, L^{p'})$ for $2 < p < \infty$?*

6. For $1 \leqslant r \leqslant 2$ the question as to whether the set $\Pi_r(L^p, L^q)$ is a sublattice of the lattice of all regular operators, has been solved for all p, q, r, except for the case $r = p' > q$ ([4], [5], [6]).

QUESTION. *Is it true that $\Pi_{p'}(L^p, L^q)$ is a sublattice of the lattice of all regular operators for $1 \leqslant q < p' < 2$?*

7. Let $2 < r < \infty$, $1 \leqslant p \leqslant 2 \leqslant q \leqslant \infty$. It is known that, for $q > r$, the set $\Pi_r(L^p, L^q)$ is not isomorphic to any Banach lattice (cf.[5], [6]). On the other hand, the following statement is true:

PROPOSITION 2 (V. G. Samarskij). *Let $1 \leqslant p \leqslant r' < 2 \leqslant q < r$. Then the space $\Pi_r(L^p, L^q)$ is a sublattice of the lattice of all regular operators and $\pi_r(V) \leqslant C\pi_r(U)$ for all operators U, V in $\Pi_r(L^p, L^q)$ satisfying the condition $|V| \leqslant |U|$ (here C is a constant that does not depend on U and V).*

One can prove that Proposition 2 is also valid for $p = 2$, $2 \leqslant q < r$.

QUESTION (V. G. Samarskij). *Is the statement of Proposition 2 valid for $r' < p < 2 < q < r$ and for $1 \leqslant p \leqslant 2$, $q = r$?*

8. P. Dodds and D. Fremlin have proved in [7] that if $1 < p, q < \infty$, then any regular operator from L^p to L^q with modulus having a compact majorant, is itself compact. One may also ask whether the same is true for r-absolutely summing operators. In the cases in which $\Pi_r(L^p, L^q)$ is a sublattice of the lattice of regular operators the answer is affirmative. The case where $\Pi_r(L^p, L^q)$ is not a sublattice of this lattice has not been completely studied. As it is shown by M. D. Ulymzhiev [3], in the cases $1 < r \leqslant 2 < q < \infty$ or $2 < r < q < \infty$ the answer to this question is negative, i.e. the analogue of Dodds–Fremlin theorem is not true in these cases.

QUESTION. *Does the analogue of Dodds–Fremlin theorem for the set $\Pi_r(L^p, L^q)$ hold in the cases $1 \leqslant p' < r \leqslant 2$, $1 \leqslant q \leqslant 2$ or $2 < r < \infty$, $2 < p < \infty$, $1 \leqslant q \leqslant r$, and also for the values of p, q, r mentioned in the question of Section 7?*

REFERENCES

1. Makarov B. M., *Stably regular operators and the uniqueness of operator ideals with local unconditional structure*, Sibirskij Matem. Zh. **28** (1987), no. 1, 157–162 (Russian); English transl. in Siberian Math. J. **28** (1987), no. 1, 120–124.
2. Makarov B. M., *p-absolutely summing operators and some applications*, Algebra i Analiz **3** (1991), no. 2, 1–76 (Russian); English transl. in St.Petersburg Math. J. **3** (1992), no. 2, 227–298.
3. Makarov B. M., Ulymzhiev M. D., *Lattices of stably regular operators*, Vestnik Leningr. Univ., Ser.1 (1991), no. 3, 44–50. (Russian)

4. Makarov B. M., Samarskij V. G., *A vector lattice structure of the spaces of absolutely summing operators*, Matem. Zametki **43** (1988), no. 4, 498–508 (Russian); English transl. in Math. Notes **43** (1988), no. 4, 287–292.
5. Samarskij V. G., *Absence of local unconditional structure in some operator spaces*, Zapiski nauchn. semin. LOMI **92** (1979), 300-306. (Russian)
6. Schütt C., *Unconditionality in tensor products*, Israel J. Math. **31** (1978), 209–216.
7. Dodds P., Fremlin D. H., *Compact operators in Banach lattices*, Israel J. Math. **34** (1979), 287–320.

DEPT. OF MATHEMATICS AND MECHANICS
ST. PETERSBURG STATE UNIVERSITY
BIBLIOTECHNAYA PL. 2.
STARYI PETERHOF, ST. PETERSBURG, 198904
RUSSIA

1.12
old

OPERATOR BLOCKS IN BANACH LATTICES

E. M. SEMËNOV

The operator Q_e of multiplication by the characteristic function of a measurable subset $e \subset [0,1]$ has the unit norm in every functional Banach lattice E on $[0,1]$ (see [1] for the definition).

Associate with every continuous linear operator $T\colon E \to E$ the number

$$\sigma(T,E) = \inf\{\, \|Q_e T Q_f\|_E\colon\ me \cdot mf > 0\,\}$$

and let

$$\mathcal{D}(E) = \{\, T \in \mathcal{L}(E,E)\colon\ \sigma(T,E) > 0\,\}.$$

PROBLEM. *Under what conditions on E the set $\mathcal{D}(E)$ is empty, i.e. $\sigma(T,E) = 0$ for every linear operator?*

This question arose for the first time in [2] (for concrete spaces) in connection with the contractibility problem of linear groups in Banach spaces. In particular, an isometry T of L^2 satisfying $\sigma(T,L^2) > 0$ was constructed there. On the other hand, it has been proved [3] that $\mathcal{D}(L^1) = \mathcal{D}(L^\infty) = \emptyset$ and that $\mathcal{D}(L^p) \neq \emptyset$ for $1 < p < \infty$ [4].

Recall now the definition of the Lorentz space $L^{p,q}$ (see [5] for their properties). For a measurable function x on $[0,1]$ let x^* denote the non-increasing rearrangement of $|x|$. Then

$$\|x\|_{L^{p,q}} = \left(\int_0^1 (x^*(t)\cdot t^{1/p})^q \,\frac{dt}{t}\right)^{\frac{1}{q}}, \qquad 1 < p < \infty,\ 1 \leqslant q \leqslant \infty.$$

CONJECTURE. *$\mathcal{D}(L^{p,q}) \neq \emptyset$ iff either $1 \leqslant q \leqslant p \leqslant 2$ or $2 \leqslant p \leqslant q \leqslant \infty$.*

It is well-known that $(L^{p,q})^* = L^{p',q'}$, where $p' = \frac{p}{p-1}$, $q' = \frac{q}{q-1}$. Therefore without loss of generality it can be assumed that $p \leqslant 2$. It is also known that $\mathcal{D}(L^{p,\infty}) = \emptyset$, $1 < p < 2$ and that $\bigcap_{q:1\leqslant q\leqslant p} \mathcal{D}(L^{p,q}) \neq \emptyset$, $1 \leqslant p \leqslant \infty$ [6].

For the case $p < q < +\infty$ the situation remains unclear.

Nothing is known about the set $\mathcal{D}(L^{2,q})$ except $q = 2$ when $\mathcal{D}(L^{2,2}) = \mathcal{D}(L^2) \neq \emptyset$.

The problem of non-emptiness of $\mathcal{D}(E)$ remains open for the Orlicz spaces.

The operators $T = T(p) \in \mathcal{D}(L^p)$ constructed in [4] depend on p and this is not a mere occasion. The set $\mathcal{D}(L^{p_1}) \cap \mathcal{D}(L^{p_2})$ is not empty (let $1 < p_1 < p_2 < \infty$ for the definiteness) iff $p_1 \leqslant 2 \leqslant p_2$ [6]. However it is not clear what conditions provide $\mathcal{D}(E_1) \cap \mathcal{D}(E_2) = \emptyset$ in the general case.

REFERENCES

1. Lindenstrauss J., Tzafriri L., *Classical Banach Spaces* II, Springer-Verlag, Berlin, 1979.
2. Mitjagin B. S., *The homotopy structure of a linear group of a Banach space*, Uspekhi Mat. Nauk **25** (1970), no. 5(155), 63-106 (Russian); English transl. in Russian Math. Surv. **25** (1970), no. 5, 59–104.
3. Edelstein I., Mityagin B., Semenov E., *The linear groups of C and $\mathcal{L}_1$ are contractible*, Bull. Acad. Polon. Sci., Ser. Math. **18** (1970), no. 1, 27–33.
4. Semenov E. M.,Tsirelson B. S., *The problem of smallness of operator blocks in L_p spaces*, Zeit. Anal. Anwendungen **2** (1983), no. 4, 367–373.
5. Krein S. G., Petunin Ju. I., Semenov E. M., *Interpolation of Linear Operators*, AMS Providence, 1982.
6. Semenov E. M., Shteinberg A. M., *Operator blocks in $L_{p,\theta}$-spaces*, Dokl. Akad. Nauk SSSR **272** (1983), no. 1, 38–40 (Russian); English transl. in Sov. Math. Dokl. **28** (1983), 333–335.

DEPT. OF MATHEMATICS
VORONEZH STATE UNIVERSITY
UNIVERSITETSKAYA PL. 1.
VORONEZH, 394693
RUSSIA

1.13

ORLICZ PROPERTY

E. M. SEMËNOV

A Banach space E is said to be of cotype q for some $q \geqslant 2$ if there exists a constant $C < \infty$ such that, for every finite set of vectors $\{x_k\}_{k=1}^n$ in E, we have

$$\frac{1}{C}\left(\sum_{k=1}^{n} \|x_k\|_E^q\right)^{1/q} \leqslant \int_0^1 \left\|\sum_{k=1}^{n} r_k(t)x_k\right\|_E dt.$$

Here $r_k(t)$ are Rademacher functions. Replacing the right hand side of this inequality by

$$\max_{0\leqslant t\leqslant 1}\left\|\sum_{k=1}^{n} r_k(t)x_k\right\|_E,$$

we obtain the definition of q-Orlicz property. The 2-Orlicz property is named the Orlicz property. It means that $\sum \|x_k\|_E^2 < \infty$ for each unconditionally convergent series $\sum x_k$.

It is evident that cotype 2 implies the Orlicz property. Many years ago B. Maurey stated the problem whether these concepts are equivalent. Recently M. Talagrand has given a negative answer to this problem [3]. However, Maurey's problem has a positive solution for some classes of Banach spaces.

A Banach lattice E on $[0,1]$ or $[0,\infty)$ is called rearrangement invariant if equimeasurable functions have equal norms. E. M. Semenov and A. M. Shteinberg proved that q-Orlicz property and cotype q are equivalent for rearrangement invariant spaces on $[0,1]$ or $[0,\infty)$ with Lebesgue measure [2]. G. Pisier gave an alternative proof of this statement.

These concepts are connected with factorization theorems. G. Pisier proved that a $(q,1)$-summing operator acting from $C(K)$ to a Banach space E factors through the Lorentz space $L_{q,1}(\mu)$ for some probability μ on a compact K [1]. The injection C into a rearrangement invariant space E is of cotype 2 iff it factors through some Lorentz space [1].

The space constructed by M. Talagrand in [3] has an unconditional basis. However it has no symmetric basis. On the other hand, the technique of [2] does not apply to rearrangement invariant spaces with discrete measure.

PROBLEM. *Are q-Orlicz property and cotype q equivalent for spaces with symmetric basis ($q \geqslant 2$)?*

There are some open problems on Orlicz property in [3].

REFERENCES

1. Pisier G., *Factorization of operators through $L_{p,\infty}$ or $L_{p,1}$ and non-commutative generalization*, Math. Ann. **276** (1986), 105–196.
2. Semenov E. M., Shteinberg A. M., *Orlicz property of symmetric spaces*, Doklady AN SSSR **314** (1990), no. 6, 1941–1944 (Russian); English transl. in Soviet Math. Dokl. **42** (1991), no. 2, 679–682.
3. Talagrand M., *Cotype of operators from $C(K)$*, Invent.-Math. **107** (1992), no. 1, 1–40.

DEPT. OF MATHEMATICS
VORONEZH STATE UNIVERSITY
UNIVERSITETSKAYA PL. 1.
VORONEZH, 394693
RUSSIA

1.14

HOMOLOGICAL DIMENSIONS OF BANACH SPACES

Mariusz Wodzicki

Various homological dimensions provide important invariants in the structural theory of modules over a given ring. It has been recently realized [6] that the category of Banach spaces has a rich homological structure. Banach space analogues of the module-theoretic dimensions can be defined, and the question about their eventual values arises naturally.

1. An exact sequence of Banach spaces and continuous linear mappings

$$0 \leftarrow X \leftarrow P_0 \leftarrow P_1 \leftarrow \cdots \leftarrow P_n \leftarrow 0 \tag{1}$$

will be called:

(**a**) a *projective* resolution of X if $P_i = l^1(S_i)$ for certain sets S_i $(i = 0, \ldots, n)$;
(**b**) a *flat* resolution of X if all P_i's are $\mathcal{L}_1$-spaces [5].

The natural number n is called the length of (1). The length of a shortest projective resolution of a given Banach space X is called the *projective dimension* of X and is denoted by $\operatorname{pd} X$. If X admits no projective resolution of finite length, we will write $\operatorname{pd} X = \infty$. The flat dimension $\operatorname{fd} X$ is defined similarly. If $\operatorname{pd} X = n < \infty$, then in any exact sequence (1) with $P_i = l^1(S_i)$ $(0 \leqslant i \leqslant n-1)$, the last term P_n is always isomorphic to $l^1(S_n)$ for some set S_n as follows from generalized Schanuel's Lemma [3, Thm 189] combined with a theorem of Köthe [4, 3.(6)]. Similarly, if $\operatorname{fd} X = n < \infty$ then in any exact sequence (1) with P_i $(0 \leqslant i \leqslant n-1)$ being $\mathcal{L}_1$-spaces, the last term P_n is an $\mathcal{L}_1$-space [6].

2. An exact sequence of Banach spaces and continuous linear mappings

$$0 \rightarrow X \rightarrow Q_0 \rightarrow Q_1 \rightarrow \cdots \rightarrow Q_n \rightarrow 0 \tag{2}$$

will be called:

(**a**) an *injective* resolution of X if every $Q_i$is an injective Banach space (cf., e.g., [2, Def. 1.31]),
(**b**) an *absolutely-pure* resolution of X if every Q_i is an $\mathcal{L}_\infty$ space [5],
(**c**) a *pure-injective* resolution if the chain complex

$$0 \leftarrow X^* \leftarrow Q_0^* \leftarrow Q_1^* \leftarrow \cdots \leftarrow Q_n^* \leftarrow 0$$

which is conjugate to (2) is continuously chain homotopic to zero and if each Q_i is complemented in Q_i^{**}.

We define the injective dimension $\operatorname{id} X$ to be the length of a shortest injective resolution of X. If no injective resolution of finite length exists, we write $\operatorname{id} X = \infty$. The *absolutely-pure* dimension $\operatorname{apd} X$ and the *pure-injective* dimension $\operatorname{pid} X$ are defined similarly. If $\operatorname{id} X = n < \infty$, then in any exact sequence (2) with Q_i $(0 \leqslant i \leqslant n-1)$ being injective, the last term Q_n is injective. The definitions of the absolutely-pure and of the pure-injective dimensions are similarly independent of the particular choice of corresponding resolutions.

3. The above five definitions are related to each other in several ways. Here are some of these relations [6]:

(a) $\operatorname{pd} X \geqslant \operatorname{fd} X = \operatorname{apd} X^* = \operatorname{fd} X^{**}$;
(b) $\operatorname{id} X \geqslant \operatorname{apd} X = \operatorname{fd} X^* = \operatorname{apd} X^{**}$;
(c) $\operatorname{id} X = \operatorname{apd} X$ if X is *pure-injective*, i.e., $\operatorname{pid} X = 0$;
(d) $\operatorname{id} X = \operatorname{pid} X$ if X is an $\mathcal{L}_\infty$-space, i.e., $\operatorname{apd} X = 0$;
(e) $\operatorname{pd} X = \operatorname{fd} X$ or $\operatorname{pd} X = \infty$.

PROBLEM. *Determine which values can be taken by any one of the above five dimensions in the class of all Banach spaces.*

PROBLEM. *Determine which values can be taken by the flat dimension in the class of Banach spaces of infinite projective dimension.*

It would be very interesting to find a characterization (say in terms of the asymptotic geometry of finite-dimensional subspaces) of the class of Banach spaces with finite projective (respectively, flat, injective, absolutely-pure, or pure-injective) dimension.

Comments.

(a) It is known at present that the projective dimension can take at least the following three values: 0, 1 and ∞. It follows from **3** (e) that any $\mathcal{L}_1$-space which is not isomorphic to $l^1(S)$, for some set S, has infinite projective dimension. Jean Bourgain constructed [1] an uncomplemented subspace V of L^1 which is isomorphic to l^1. Thus $W = L^1/V$ satisfies $\operatorname{fd} W = 1$ and $\operatorname{pd} W = \infty$ as well as $\operatorname{id} W^* = 1$. In his book [2, pp.107-8], Bourgain constructs a series of finite dimensional Banach spaces with badly complemented subspaces. (I am indebted for this information to G. Pisier). His construction leads easily to an example of an uncomplemented subspace Y of l^1 which is isomorphic to l^1. Thus $Z = l^1/Y$ satisfies $\operatorname{pd} Z = \operatorname{fd} Z = \operatorname{id} Z^* = \operatorname{apd} Z^* = 1$. I know of no examples of Banach spaces with the value of any of the above dimensions equal to 2.

(b) For the most of the classical Banach spaces (excluding obvious exceptions) the flat and absolutely-pure dimensions are expected to be equal to ∞.

(c) Any successful approach to the problems formulated above is likely to be based on clever constructions of finite-dimensional Banach spaces.

REFERENCES

1. Bourgain J., *A counterexample to a complementation problem*, Comp. math. **43** (1981), 133-144.
2. Bourgain J., *New classes of $\mathcal{L}^p$-spaces*, Springer Lecture Notes in Mathematics **889** (1981).
3. Kaplansky I., *Commutative rings*, Allyn and Bacon, Boston, 1970.
4. Köthe G., *Hebbare Lokalkonvexe Räume*, Math. Ann. **165** (1966), 181–195.
5. Lindenstrauss J., Pełczyński A., *Absolutely summing operators in $\mathcal{L}_1$-spaces and their applications*, Studia math. **29** (1968), 275–326.
6. Wodzicki M., unpublished notes (OSU, Columbus, February 1991).

DEPARTMENT OF MATHEMATICS
UNIVERSITY OF CALIFORNIA
BERKELEY, CALIFORNIA
CA 94720
U.S.A.

1.15
v.old

SPACES OF ANALYTIC FUNCTIONS (ISOMORPHISMS, BASES)

V. P. Zakharyuta, O. S. Semiguk, N. I. Skiba

$\mathcal{O}(\mathcal{D})$ will denote the space of all functions analytic in the domain $\mathcal{D}$, $\mathcal{D} \subset \widehat{\mathbb{C}}$. A domain $\mathcal{D}$ is called standard if $\mathcal{O}(\mathcal{D})$ is isomorphic (as a linear topological space) to one of three (mutually non-isomorphic) spaces

$$\mathcal{O}_1 \stackrel{\text{def}}{=} \mathcal{O}(\mathbb{D}), \qquad \mathcal{O}_\infty \stackrel{\text{def}}{=} \mathcal{O}(\mathbb{C}), \qquad \mathcal{O}_2 \stackrel{\text{def}}{=} \mathcal{O}_1 \times \mathcal{O}_\infty.$$

In [1] the class R of all standard domains was completely described. Moreover in [1] the properties of $\mathcal{D} \in R$ were found out, determining to which particular one of these three spaces the space $\mathcal{O}(\mathcal{D})$ is isomorphic. These properties involve the structure of the set of all irregular points of the boundary $\partial\mathcal{D}$ (see [2] for notions from the potential theory). The isomorphic classification of spaces $\mathcal{O}(\mathcal{D})$ for $\mathcal{D}$'s not in R remains unknown.

Any domain $\mathcal{D}(q,r) \stackrel{\text{def}}{=} \widehat{\mathbb{C}} \setminus (\bigcup_{j=1}^{\infty} \operatorname{clos} \mathbb{D}(q^j, r_j) \cup \{0\})$, where $\mathbb{D}(a,r) = \{\zeta : |\zeta - a| < r\}$, $q \in (0,1)$, $r = (r_j)_{j \geqslant 1}$ is a monotone sequence of positive numbers with

$$\sum_{j=1}^{\infty} \left(\log \frac{1}{r_j} \right)^{-1} < \infty, \tag{1}$$

does not belong to R (and $\mathcal{D}(q,r) \in R$ whenever the series in (1) diverges).

Conjecture. *There exists a continuum of mutually non-isomorphic spaces $\mathcal{O}(\mathcal{D}(q,r))$.*

This conjecture is stated also in [6] (problem 63).

Let us mention in connection with the open question on the existence of a basis in the space $\mathcal{O}(K)$ of all functions analytic on a compact set K, $K \subset \mathbb{C}$ that this question is open for $K = \mathbb{C} \setminus \mathcal{D}(q,r)$ as well (under condition (1)), though it was proved [7] for such K that $\mathcal{O}(K)$ has no basis in common with $\mathcal{O}(\mathcal{D})$, $\mathcal{D}$ being any regular (in the potential-theoretical sense) neighbourhood of K. From this fact it follows that $\mathcal{O}(K)$ has no basis of the form $\{ \sum_{j=0}^{n} P_{jn} z^j \}_{n=0}^{\infty}$.

Let Ω be 1-dimensional open Riemann surface.

(a) We say that Ω is regular iff there exists the Green function $G(\zeta, z)$ with $\lim_{n\to\infty} G(\zeta, z_n) = 0$, $\zeta \in \Omega$, for any sequence (z_n) with no limit point in Ω. Under additional restrictions (for example if Ω is a relatively compact subdomain of another Riemann surface Ω_1) it has been proved that $\mathcal{O}(\Omega)$ and $\mathcal{O}_1$ are isomorphic if Ω is regular (cf. [8] and references therein). *Is this true in the general case?* The necessity of this condition follows from general results for Stein manifolds ([3] and references therein).

(b) Let Ω be a Riemann surface with the ideal boundary of capacity zero. *Is then* $\mathcal{O}(\Omega)$ *isomorphic to* $\mathcal{O}_\infty$? The condition is necessary even in the multidimensional case (unpublished).

(c) The QUESTION about the existence of a basis in $\mathcal{O}(\Omega)$ is solved only under some additional restrictions (even for surfaces satisfying (a) and (b) above [8], [9]).

Clearly $\mathcal{O}(G)$ and $\mathcal{O}(K)$ are non-isomorphic whenever $\mathcal{O}$ is open and K is a compact set ($\mathcal{O}, K \subset \mathbb{C}$).

QUESTION. *Which other differences in topological properties of sets* E_1, E_2 ($\subset \mathbb{C}$) *imply that* $\mathcal{O}(E_1)$ *and* $\mathcal{O}(E_2)$ *are non-isomorphic?*

Here $\mathcal{O}(E)$ denotes the inductive limit of the net $\{\mathcal{O}(V)\}$ of countably normed spaces $\mathcal{O}(V)$, V running through the set of all open neighbourhoods of E. V. P. Erofeev proved (unpublished) that $\mathcal{O}(\mathbb{D} \cup \alpha) \not\approx \mathcal{O}(\mathbb{D} \cup \beta)$, α and β being an open and a closed subarcs of the unit circle $\partial\mathbb{D}$. It is not known whether $\mathcal{O}(\mathcal{D} \cup \{1\})$ and $\mathcal{O}(\mathcal{D} \cup \beta)$ are isomorphic if β is a closed non-degenerate arc of $\partial\mathbb{D}$.

In [4] a method was proposed to construct common bases for $\mathcal{O}(\mathcal{D})$ and $\mathcal{O}(K)$, $K \subset \mathcal{D}$. This method uses a special orthogonal basis common for a pair of Hilbert spaces H_0, H_1 and for the Hilbert scale* H^α generated by H_0 and H_1 essentially generalizing well-known results of V. P. Erokhin about common bases (see, e.g., [4], [3], [7], [8], [9]).

THEOREM ([4], [12], [8]). *Let* $K \subset \mathcal{D}$, $K = \{\, t \in \mathcal{D} : |f(t)| \leqslant \sup_K |f|, \forall f \in \mathcal{O}(\mathcal{D}) \,\}$ *and suppose* $\mathcal{D} \setminus K$ *is a regular domain in* $\widehat{\mathbb{C}}$ *(or a relatively compact domain on a Riemann surface). Then there exist Hilbert spaces* H_0, H_1 *with*

$$H_1 \hookrightarrow \mathcal{O}(\mathcal{D}) \hookrightarrow \mathcal{O}(K) \hookrightarrow H_0 \tag{2}$$

and for all spaces H^α *of the corresponding scale*

$$\mathcal{O}(\operatorname{clos} \mathcal{D}_\alpha) \hookrightarrow H^\alpha \hookrightarrow \mathcal{O}(\mathcal{D}_\alpha), \tag{3}$$

where $\mathcal{D}_\alpha = \{\, z \in \mathcal{D} : \omega(\mathcal{D}, K, z) < \alpha \,\} \cup K$, $\omega(\mathcal{D}, K, z)$ is the harmonic measure of $\partial\mathcal{D}$ with respect to $\mathcal{D} \setminus K$ ([3], p. 299). All embeddings in (2) and (3) are continuous. The common orthogonal basis $(e_n)_{n \geqslant 0}$ of the spaces H_0, H_1 is a common basis in $\mathcal{O}(\mathcal{D})$ and $\mathcal{O}(K)$.

A QUESTION arises: *how "far" is it possible to "move apart" the spaces* H_0, H_1 *satisfying* (2) *without breaking* (3)?

Let $H^\infty(\mathcal{D})$ be a Banach space of all bounded functions analytic in $\mathcal{D}$. We consider Hilbert spaces H_1 with

$$H^\infty(\mathcal{D}) \hookrightarrow H_1 \hookrightarrow \mathcal{O}(\mathcal{D}). \tag{4}$$

---

*The notion of a Hilbert scale introduced by S. G. Krein has a number of important applications to problems of the isomorphic classification of linear spaces and to the theory of bases. We refer to the paper by V. S. Mityagin and G. M. Henkin "*Linear problems of complex analysis*", Uspekhi Mat. Nauk, **26** no.4 (1971), 93–152 (Engl. transl. in Russian Math. Surv., **26** no.4 (1971), 99–164) containing many results concerning spaces of analytic functions, a list of unsolved problems and an extensive bibliography.—Ed.

The well-known Kolmogorov's problem about the validity of the asymptotic relation

$$\log d_n(A_K^{\mathcal{D}}) \sim -\frac{n}{\tau(K,\mathcal{D})}$$

for the n widths $d_n(A_K^{\mathcal{D}})$ of the compact set $A_K^{\mathcal{D}} \stackrel{\text{def}}{=} \{ f \in H^\infty(\mathcal{D}) : \max_K |f| \leqslant 1 \}$ ($\tau(K,\mathcal{D})$ is the Green's capacity of the compact set K with respect to $\mathcal{D}$) can be reduced to the following

PROBLEM. *Describe all domains $\mathcal{D}$ with* (4) $\Longrightarrow$ (3) (*for a suitable* H_0) ([8], see also [11]).

REFERENCES

1. Zaharjuta V. P., *Spaces of functions of a single variable that are analytic in open sets and on compacta*, Mat. Sb. **82** (1970), no. 1, 84–98. (Russian)
2. Landkof N. S., *Fundamentals of modern potential theory*, Izdat. "Nauka", Moscow, 1966 (Russian); English transl. 1972, Springer-Verlag, New York–Heidelberg.
3. Zaharjuta V. P., *Extremal plurisubharmonic functions, Hilbert scales, and the isomorphism of spaces of analytic functions of several variables.* I,II, Teor. Funkts. Funktsional. Anal. i Prilozhen. **19** (1974), 133–157; **21** (1974), Kharkov, 65–83. (Russian)
4. Zaharjuta V. P., *Continuable bases in spaces of analytic functions of one and several variables*, Sib. Matem. Zhurn. **8** (1967), no. 2, 277–292. (Russian)
5. Dragilev M. M., Zaharjuta V. P., Khaplanov M.G., *On certain problems concerning bases of analytic functions*, Actual problems of science, Rostov-on-Don, 1967, pp. 91-102. (Russian)
6. Unsolved Problems, *Proceedings of the International Colloquium on Nuclear Spaces and Ideals in Operator Algebras, Warsaw, 1969*, Warszawa–Wrocław, 1970, pp. 467–483.
7. Zakharyuta V.P., Kadampatta S.N., *Existence of continuable bases in spaces of functions that are analytic on compacta*, Mat. Zametki **27** (1980), no. 5, 701–713 (Russian); English transl. in Math. Notes **27** (1980), no. 5, 1334–1340.
8. Zaharjuta V. P., Skiba N. I., *Estimates of the n-widths of certain classes of functions that are analytic on Riemann surfaces*, Mat. Zametki **19** (1976), no. 6, 899–911 (Russian); English transl. in Math. Notes **19** (1976), no. 6, 525–532.
9. Semiguk O. S., *On existence of general bases in the space of analytic function on a compact Riemann surface*, Manuscript deposed in VINITI, no 620-77, Univ. of Rostov-on-Don, Rostov-on-Don, 1977. (Russian)
10. Widom H., *Rational approximation and n-dimensional diameter*, J. Approximation Theory **5** (1972), no. 2, 343–361.
11. Skiba N. I., *On an upper estimation of n-diameters of one class of holomorphic functions*, Collection of papers of young researches of the department of mathematics, Deposed in VINITI, no 1593-78, RIMI, Rostov-on-Don, 1978. (Russian)
12. Nguen Thanh Van, *Bases de Schauder dans certains espaces de fonctions holomorphes*, Ann. Inst. Fourier (Grenoble) **22** (1972), no. 2, 169–253.

ROSTOV STATE UNIVERSITY
ROSTOV-ON-DON, 344711
RUSSIA

AND

MATEMATIK BÖLÜMÜ
ORTA DOGY TEKNIK UNIV.
06531 ANKARA, TURKEY

ROSTOV STATE UNIVERSITY
ROSTOV-ON-DON, 344711
RUSSIA

ROSTOV INSTITUTE OF ENGINEERING
ROSTOV-ON-DON
RUSSIA

Commentary by the Authors

The problems (a), (b), (c) (in a more general situation, namely for Stein manifolds) have been solved in [13] by a synthesis of results on Hilbert scales of spaces of analytic functions [3] and last results on characterization of power series spaces of finite or infinite type [14], [15].

We formulate one of these results as an example.

Let Ω be a connected Stein manifold. Ω is said to be P-regular if there exists a plurisubharmonic function $u(z)$ such that $u(z) < 0$ in Ω and $u(z_n) \to 0$ for any sequence (z_n) without limit points in Ω.

THEOREM. *$A(\Omega) \simeq A(\mathbb{D}^n)$ if and only if Ω is P-regular.*

References

13. Zakharyuta V. P., *Isomorphism of spaces of analytic functions*, Doklady Akad. Nauk SSSR **255** (1980), no. 1, 11–14 (Russian); English transl in Soviet Math. Dokl. **22** (1980), no. 3, 631-634.
14. Vogt D., *Eine Charakterisierung der Potenzreihenräume vom endlichen Typ und ihre Folgerungen*, Manuscr. Math. **37** (1982), 269-301.
15. Vogt D., Wagner M. J., *Charakterisierung der Unterräume und Quotientenräumeder nuklearen stabilen Potenzreihenräume vom unendlichen Typ*, Studia Math. **70** (1981), no. 1, 63–80.

1.16
old

ON ISOMORPHIC CLASSIFICATION OF F-SPACES

V. L. ZAKHARYUTA

1. For a given family of positive sequences $\{a_{ip}\}$, let $K(a_{ip})$ be a Köthe space, i.e., the F-space of all sequences $x = \{x_n\}_{n\geqslant 1}$ satisfying

$$(1)\qquad |x|_p \stackrel{\text{def}}{=} \sum_{n\geqslant 1} |x_n| a_{np} < +\infty, \quad p = 1, 2, \dots .$$

The space $K(a_{ip})$ is endowed with the topology defined by the family of semi-norms (1). It is called *a power (Köthe) space* if $a_{ip} = h_p(i) a_i$, where $-\infty < h_p(i) \leqslant h_{p+1}(i)$, $h_p(i) - h_1(i) \leqslant C(p) < +\infty$, $i, p \in \mathbb{N}$. For example the so-called power series spaces

$$(2)\qquad E_\alpha(a) \stackrel{\text{def}}{=} K(\exp \alpha_p a_i), \quad \alpha_p \uparrow \alpha, \ -\infty < \alpha \leqslant +\infty, \ a = (a_i),$$

are power spaces in our sense. $E_\alpha(a)$ is said to be of finite (infinite) type if $\alpha < +\infty$ ($\alpha = +\infty$).

Consider two classes of power spaces:

(1) the class $\mathcal{E}$ of power spaces of the first kind [1], [2]:

$$E(\lambda, a) \stackrel{\text{def}}{=} K\left(\exp\left(-\frac{1}{p} + \lambda_i p\right) a_i\right),$$

(2) the class $\mathcal{F}$ of power spaces of the second kind:

$$F(\lambda, a) \stackrel{\text{def}}{=} K(\exp \varphi_p(\lambda_i) a_i),$$

where $\varphi_p(t) \stackrel{\text{def}}{=} -\frac{1}{p} + \min\{\frac{1}{t}, p\}$, $p \in \mathbb{N}$. Here $a = (a_i)$, $a_i > 0$; $\lambda = (\lambda_i)$, $0 \leqslant \lambda_i \leqslant 1$ in both cases.

If we consider isomorphic spaces as identical, then $\mathcal{E} \cap \mathcal{F}$ consists of spaces (2) and also of their cartesian products; $\mathcal{E}$ contains spaces $E_0(a) \widehat{\otimes} E_\infty(b)$ [2] and $\mathcal{F}$ contains spaces of all analytic functions on unbounded n-circular domain in $\mathbb{C}^n$ [3].

PROBLEM 1. *Give criteria of isomorphisms:*

$$(3)\qquad E(\lambda, a) \simeq E(\mu, b)$$

$$(4)\qquad F(\lambda, a) \simeq F(\mu, b)$$

in terms of (λ, a), (μ, b).

Articles [1], [2] contain a criterion of isomorphism (3) but under an additional requirement on $E(\lambda, a)$ (note that in AMS translation of [1] in Lemma 4 the important chain of quantifiers "$\forall p' \exists p \forall q \exists q' \forall r' \exists r \forall s \exists s' \exists c \forall t, \tau, \sigma > 0$" has been omitted).

Let us formulate one result on isomorphism (4), which somewhat generalizes the result of [3]. Denote by Λ the set of all sequences $\lambda = (\lambda_i)$, $0 < \lambda_i \leqslant 1$ such that there exist limits

$$\varphi_\lambda(\Theta) = \lim_{n \to \infty} \frac{|\{i : \lambda_i \leqslant \Theta,\ i \leqslant n\}|}{n}, \quad \Theta \in (0, 1],$$

and $\varphi_\lambda(\Theta)$ is strongly increasing in Θ.

THEOREM 1. *Let $\lambda, \mu \in \Lambda$ and $a_{2i} \asymp a_i$. Then* (4) *implies*

(1) $a_i \asymp b_i$,

(2) $\exists c\colon \frac{1}{c}\varphi_\mu(\frac{\Theta}{c}) \leqslant \varphi_\lambda(\Theta) \leqslant c\varphi_\mu(c\Theta)$, $0 < \Theta \leqslant 1$.

Note for a comparison that isomorphism (3) takes place for arbitrary $\lambda, \mu \in \Lambda$ whenever the condition $a_i \asymp b_i$ is fulfilled.

Class $\mathcal{F}$ is also of a great interest because the following conjecture seems plausible.

CONJECTURE 1. *There exists a nuclear power space of the second kind without the bases quasiequivalence property*.*

2. Let X be an F-space with the topology defined by a system of semi-norms $\{|\cdot|_p, p \in \mathbb{N}\}$ and $\varphi(t)$ be a convex increasing function on $[1, \infty)$. Denote by $\mathcal{D}_\varphi$ the class of all spaces X such that $\exists p \forall q \exists m, r, c$:

$$|f|_q \leqslant \big(\varphi(t)\big)^m |f|_p + \frac{c}{t}|f|_r, \quad t \geqslant 1,\ f \in X.$$

Classes $\mathcal{D}_\varphi$, being invariant with respect to isomorphisms, are a modified generalization of Dragilev's class $\mathcal{D}_1$ [4] (see also [5]); similar dual classes Ω_φ were considered in Vogt–Wagner [6].

Classes $\mathcal{D}_\varphi$ have been used in [7] to give a positive answer to a question of Zerner. Consider the family $\mathcal{O}$ of all domains $\mathcal{G}$ with a single cusp:

$$\mathcal{G}_\psi = \{\,(x, y) \in \mathbb{R}^2\colon |y| < \psi(x),\ 0 < x < 1\,\},$$

ψ being a non-decreasing C^1-function on $[0, 1]$, $\psi(0) = 0$. Then there exists a continuum of mutually non-isomorphic spaces $C^\infty(\overline{\mathcal{G}})$ with $\mathcal{G} \in \mathcal{O}$. The following theorem clarifies the role of classes $\mathcal{D}_\varphi$ in this problem.

THEOREM 2. *$C^\infty(\overline{\mathcal{G}}_\psi) \in \mathcal{D}_\varphi$ iff there exist $\mu, \nu > 0$ satisfying $1/\psi(x) \leqslant \big(\varphi(\frac{1}{x^\nu})\big)^\mu$ for $0 < x < x_0$.*

PROBLEM 2. *Are all spaces $C^\infty(\overline{\mathcal{G}})$ from the same class $\mathcal{D}_\varphi$ isomorphic or there exists a more subtle (than in Theorem 2) classification of these spaces?*

CONJECTURE 2. *There exists a modification (apparently very essential one) of Vogt–Wagner's classes Ω_φ which allows one to prove the conjecture on the existence of a continuum of mutually non-isomorphic spaces $\mathcal{O}\big(\mathcal{D}(q, r)\big)$* (see this Collection, Problem 1.15).

REFERENCES

1. Zaharjuta V. P., *On the isomorphism and quasiequivalence of bases for Köthe power spaces*, Dokl. Akad. Nauk SSSR **221** (1975), no. 4, 772-774 (Russian); English transl. in Soviet Math. Dokl. **16** (1975).

---

*The definition see e.g. in the paper by Mityagin B. S. *"Approximate dimension and bases in nuclear spaces"*, Uspekhi Mat. Nauk, **16** no.4 (1961), 63–132 (Engl. transl. in Russian Math. Surv., **16** no.4 (1961), 59–128)

2. Zaharjuta V. P., *The isomorphism and quasiequivalence of bases for Köthe echelon spaces*, Theory of operators in linear spaces, Proc. Seventh Winter School, Drogobych, 1974, Moscow, pp. 101–126. (Russian)
3. Zaharjuta V. P., *Generalized Mityagin invariants and a continuum of pairwise nonisomorphic spaces of analytic functions*, Funkts. Analiz i Prilozh. **11** (1977), no. 3, 24–30 (Russian); English transl. in Funct. Anal. and Appl. **11** (1977), no. 3, 182–188.
4. Zaharjuta V. P., *Some linear topological invariants, and the isomorphism of the tensor products of the centers of scales*, Izv. Severo–Kavkaz. Nauchn. Centra Vyss. Skoly (1974), no. 4, 62–64. (Russian)
5. Vogt D., *Charakterisierung der Unterräume von S*, Math.Z. **155** (1977), 109–117.
6. Vogt D., Wagner M. J., *Charakterisierung der Quotientenräume von S und eine Vermutung von Martineau*, Studia Math. **67** (1980), 225–240.
7. Goncharov A. P., Zakharyuta V. P., *Linear topological invariants and the spaces of infinitely differentiable functions*, Math. analysis and its appl., Interuniv. Work Collect. (1985), Rostov, 18–27. (Russian)

Rostov State University
Rostov-on-Don, 344711
Russia

and

Matematik Bölümü
Orta Dogy Teknik Univ.
06531 Ankara, Turkey

1.17
old

WEIGHTED SPACES OF ENTIRE FUNCTIONS

F. HASLINGER

Let $p\colon \mathbb{C} \to \mathbb{R}$ be a continuous function and define

$$\mathcal{F}_R = \{ f \text{ entire} : \|f\|_r \overset{\text{def}}{=} \sup_{z\in\mathbb{C}} \big| f(z) \exp\big(-p(rz)\big)\big| < \infty,\ \forall r > R \},$$

where $R \in \mathbb{R}_+$. We suppose that $\|\cdot\|_r \leqslant \|\cdot\|_s$ for $r > s > R$ and that $\mathcal{F}_R$ is not trivial. It is easily seen that $\mathcal{F}_R$ is a Fréchet space, the topology of which is strictly stronger than the topology of uniform convergence on the compact subsets of $\mathbb{C}$. With the help of the Riesz representation theorem the dual space of $\mathcal{F}_R$ can be identified with the space of all complex valued measures μ on $\mathbb{C}$ such that

$$\int_{\mathbb{C}} \exp\big(rp(z)\big)\, d|\mu|(z) < \infty$$

for an $r > R$ (see [5]).

As an example consider $p(z) = |z|^\alpha$, for $\alpha > 0$, then $\mathcal{F}_R$ is the space of all entire functions of order α and type R^α (see [7]); in this case the monomials $\{z^n\}_{n\geqslant 0}$ constitute a Schauder basis in $\mathcal{F}_R$ and $\mathcal{F}_R$ is topologically isomorphic to the space $\mathcal{H}$ of all holomorphic functions on the disc $\mathbb{D}$ if $R > 0$ and to the space $\mathcal{H}(\mathbb{C})$ of all entire functions if $R = 0$, both spaces $\mathcal{H}(\mathbb{D})$ and $\mathcal{H}(\mathbb{C})$ endowed with the topology of compact convergence. Here, $\mathcal{F}_R$ is also a nuclear Fréchet space (see [7]) and the dual space can be identified with a space of germs of holomorphic functions (Köthe duality [4]). All this can be used to find a solution for interpolation problems such as for instance $f^{(n)}(\lambda_n) = a_n$, $n = 0, 1, 2, \ldots$ in the space $\mathcal{F}_R$ by means of methods from functional analysis (see [3], [1]).

If we take $p(z) = |e^z| = e^x$, $z = x + iy$ (here p is not a function of $|z|$), then the corresponding spaces $\mathcal{F}_R$ do not contain the polynomials and properties similar to the above example are not known. Another example of interest is due to Gel'fand and Shilov [2]:

$$S^{\beta,B}_{\alpha,A} = \{ f \in C^\infty(\mathbb{R})\colon \sup_{x\in\mathbb{R}} |x^k f^{(q)}(x)| / (\tilde{A}^k \tilde{B}^q k^{k\alpha} q^{q\beta}) < \infty,\ \tilde{A} > A,\ \tilde{B} > B \}$$

$$k, q = 0, 1, 2, \ldots, \text{ where } \alpha, \beta, A, B > 0 \text{ and } \alpha + \beta \geqslant 1.$$

In fact, each function $f \in S^{\beta,B}_{\alpha,A}$ can be extended to $\mathbb{C}$ and $S^{\beta,B}_{\alpha,A}$ coincides with the space of all entire functions such that

$$\sup_{z\in\mathbb{C}} \big| f(z) \exp\big(a'|x|^{1/\alpha} - b'|y|^{1/(1-\beta)}\big)\big| < \infty,$$

where $0 < a' < a$, $b' > b > 0$ (see [2]).

The following problems are of special interest if the weight p is not a function of $|z|$.

PROBLEM 1. *Is it possible to find a representation of the dual space $\mathcal{F}'_R$ of $\mathcal{F}_R$ as a space of certain holomorphic functions or germs of holomorphic functions, analogous to the so called "Köthe-duality" [4] for the space $\mathcal{H}(\mathbb{D})$ or $\mathcal{H}(\mathbb{C})$?*

PROBLEM 2. *For which weights p is the space $\mathcal{F}_R$ nuclear?*

Mityagin [6] proved the nuclearity of the spaces $S^{\beta,B}_{\alpha,A}$.

PROBLEM 3. *Existence of Schauder bases in $\mathcal{F}_R$.*

This problem seems to be quite difficult. If $\mathcal{F}_R$ is nuclear and has a Schauder basis, then $\mathcal{F}_R$ can be identified with a Köthe sequence space (see [7]). If the monomials $\{z^n\}_{n\geqslant 0}$ constitute a Schauder basis in $\mathcal{F}_R$, as in the first example, then $\mathcal{F}_R$ is a so called power-series-space (see [7]). Let $h \in \mathcal{F}_R$ be an entire function which is not of the form $az+b$, $a, b \in \mathbb{C}$, then $\operatorname{span}(h^n : n \geqslant 1) \neq \mathcal{F}_R$: by our assumption on h, there exist two points $z_1, z_2 \in \mathbb{C}$, $z_1 \neq z_2$ with $h(z_1) = h(z_2)$, now set $\mu = \delta_{z_1} - \delta_{z_2}$, where δ_{z_i} denotes the Dirac measures $(i = 1, 2)$; then $\mu \in \mathcal{F}'_R$ and $<h^n, \mu> = 0$ $\forall n \in \mathbb{N}$, therefore, by the Hahn–Banach theorem, $\operatorname{span}(h^n : n \geqslant 1) \neq \mathcal{F}_R$. So, if $\mathcal{F}_R$ does not contain the monomials, $\mathcal{F}_R$ cannot have a Schauder basis of the form $\{h^n\}_{n\geqslant 1}$. B. A. Taylor [8] constructed an example of a weighted space of entire functions containing the polynomials and the function $\exp(z)$, but where $\exp(z)$ cannot be approximated by polynomials.

REFERENCES

1. Berenstein C. A., Taylor B. A., *A new look at interpolation theory for entire functions of one variable*, Adv. of Math. **33** (1979), 109–143.
2. Gel'fand I. M., Shilov G. E., *Verallgemeinerte Funktionen* II, III, VEB Deutscher Verlag der Wissenschaften, Berlin, 1962.
3. Haslinger F., Meyer M., *Abel–Gončarov approximation and interpolation*, preprint.
4. Köthe G., *Topologische lineare Räume*, Springer Verlag, Berlin, Heidelberg, New York, 1966.
5. Martineau A., *Equations différentielles d'ordre infini*, Bull. Soc. Math. de France **95** (1967), 109–154.
6. Mityagin B. S., *Nuclearity and other properties of spaces of type S*, Trudy Mosk. Mat. Ob-va **9** (1960), 317–328 (Russian); English transl. in Amer. Math. Soc. Transl. **93** (1970), 45–60.
7. Rolewicz S., *Metric Linear Spaces*, Monographie Matematyczne, vol. 56, Warsaw, 1972.
8. Taylor B. A., *On weighted polynomial approximation of entire functions*, Pacif. J. Math. **36** (1971), 523–539.

INSTITUT FÜR MATHEMATIK
UNIVERSITÄT WIEN
STRUDLHOFGASSE 4
A-1090 WIEN, AUSTRIA

CHAPTER EDITOR'S NOTE

The author informed the editors about his paper [9] where it was proved that for certain convex weight functions the dual space of the corresponding Fréchet space of entire functions could be identified with a DF-space of entire functions, and that under certain assumptions these spaces possessed bases. See also the subsequent paper [10].

REFERENCES

9. Haslinger F., *Weighted spaces of entire functions*, Indiana Univ. Math. J. **35** (1986), no. 1, 193–208.
10. Haslinger F., Smejkal M., *Representation and duality in weighted Fréchet spaces of entire functions*, Lecture Notes Math. **1275** (1987), 168–196.

1.18
v.old

LINEAR FUNCTIONALS ON SPACES OF ANALYTIC FUNCTIONS AND THE LINEAR CONVEXITY IN $\mathbb{C}^n$

L. A. Aizenberg

A domain $\mathcal{D}$ in $\mathbb{C}^n$ is called *linearly convex* (l.c.) if for each point ζ of its boundary $\partial\mathcal{D}$ there exists an analytic plane $\{ z \in \mathbb{C}^n : a_1 z_1 + \cdots + a_n z_n + b = 0 \}$ passing through ζ and not intersecting $\mathcal{D}$. A set E is said to be *approximable from inside (from outside)* by a sequence of domains $\mathcal{D}_k$, $k = 1, 2, \ldots$ if $\operatorname{Clos}\mathcal{D}_k \subset \mathcal{D}_{k+1}$ (resp., $\operatorname{Clos}\mathcal{D}_{k+1} \subset \mathcal{D}_k$) and $E = \bigcup_k \mathcal{D}_k$ (resp., $E = \bigcap_k \mathcal{D}_k$). A compact set M is called *linearly convex* (l.c.) if there exists a sequence of l.c. domains approximating M from the outside. Applications of these notions to a number of problems of Complex Analysis, similar concepts introduced by A. Martineau and references may be found in [1]–[5].

If $\mathcal{D}$ is a bounded l.c. domain with C^2-boundary, then every function continuous in $\operatorname{Clos}\mathcal{D}$ and holomorphic in $\mathcal{D}$ has a simple integral representation in terms of its boundary values. The representation follows from the Cauchy–Fantappié formula [7] and is written explicitly in [8], [1], [2]. It leads to a description of the conjugate space of the space $O(\mathcal{D})$ (resp. $O(M)$) of all functions holomorphic in a l.c. domain $\mathcal{D}$ (resp. on a compact set M) which can be approximated from inside (from the outside) by bounded l.c. domains with C^2-boundary (see [9] for convex domains and compact sets and [2] for linear convex sets; the additional condition of approximating domains imposed in [2] can be removed). Such an approximation is not always possible [6]. This description of the conjugate space is a generalization of well-known results by G. Köthe, A. Grotendieck, Sebastião e Silva, C. L. da Silva Dias and H. G. Tillman for the case $n \geqslant 1$. Let $0 \in E$. Then $\widetilde{E} = \{ w \in \mathbb{C}^n : w_1 z_1 + \cdots + w_n z_n \neq 1 \text{ for every } z \in E \}$ is called the *conjugate set* and plays the role of "the exterior" in this description. Let

$$\mathcal{D}_m = \{ z : \Phi_m(z) < 0 \}, \qquad 0 \in \mathcal{D}_m, \quad m = 1, 2, \ldots$$

be the approximating domains specified above, $\Phi_m \in C^2$, $\operatorname{grad}\Phi_m \neq 0$ on $\partial\mathcal{D}_m$. Consider a differential form

$$\omega(u, z) = \frac{(n-1)!}{(2\pi i)^n \langle u, z\rangle^n} \sum_{k=1}^{n} (-1)^{k-1} u_k \, du_1 \wedge \cdots \wedge du_{k-1} \wedge du_{k+1} \wedge \cdots \wedge du_n \wedge dz_1 \wedge \cdots \wedge dz_n,$$

where $\langle u, z\rangle = u_1 z_1 + \cdots + u_n z_n$. Let $\tau(\Phi) = (\tau_1(\Phi), \ldots, \tau_n(\Phi))$, where $\tau_k(\Phi) = \Phi'_{z_k} \langle \operatorname{grad}\Phi(z), z\rangle^{-1}$. Every linear continuous functional F on $O(\mathcal{D})$ (on $O(M)$) has a representation

$$F(f) = \int_{\partial\mathcal{D}_m} f(z)\varphi(\tau(\Phi_m))\omega(\operatorname{grad}\Phi_m(z), z), \tag{1}$$

where $\varphi \in O(\widetilde{\mathcal{D}})$ (respectively, $\varphi \in O(\widetilde{M})$), m depends only on φ. Formula (1) establishes an isomorphism between the linear topological spaces $O'(\mathcal{D})$ and $O(\widetilde{\mathcal{D}})$ (respectively $O'(M)$ and $O(\widetilde{M})$).

Problem 1. *Describe l.c. domains and compact sets which can be approximated from inside (from outside) by bounded l.c. domains with C^2-boundary.*

Let $z \in \partial\mathcal{D}$, $0 \in \mathcal{D}$ and let $F(z)$ denote the set of $u \in \mathbb{C}^n$ such that the plane $\{\zeta : \langle u, \zeta\rangle = 1\}$ passes through z and does not intersect $\mathcal{D}$.

Conjecture 1. *A bounded l.c. domain $\mathcal{D}$, $0 \in \mathcal{D}$ with the piecewise smooth boundary $\partial\mathcal{D}$ admits the approximation indicated in Problem 1 if and only if the sets $\Gamma(z)$ are connected for all $z \in \partial\mathcal{D}$.*

Let $F \in O'(\mathcal{D})$ (respectively, $F \in O'(M)$). The function $F_z\big[(1 - \langle z, w\rangle)^{-n}\big]$ is called the *Fantappié indicator;* here $z \in \mathcal{D}$, $w \in \widetilde{\mathcal{D}}$, $0 \in \mathcal{D}$ (respectively, $z \in M$, $w \in \widetilde{M}$, $0 \in M$). The function φ in (1) is the Fantappié indicator of the functional F. A l.c. domain $\mathcal{D}$ (a compact M) is called *strongly linearly convex,* if the mapping which establishes the correspondence between functionals and their Fantappié indicators is an isomorphism of spaces $O'(\mathcal{D})$ and $O(\widetilde{\mathcal{D}})$ (respectively, $O'(M)$ and $O(\widetilde{M})$). Similar definition has been introduced by A. Martineau (see references in [9]). Every convex domain or compact is strongly l.c. (see, for example, [9]). At last, the result from [2] discussed above means that the existence of approximation indicated in Problem 1 is sufficient for the strong linear convexity. Strongly l.c. sets have applications in certain problems of multidimensional complex analysis, such as decompositions of holomorphic functions into series of simplest fractions or into generalized Laurent series, the separation of singularities [1], [2], [5]. That is why the following problem is of interest.

Problem 2. *Give a geometrical description of strongly linearly convex domains and compact sets.*

Conjecture 2. *A domain (a compact set) is a strongly l.c. set if and only if there exists an approximation of this set indicated in Problem 1.*

It was shown in [5] that under some additional conditions, the intersection of any strongly l.c. compact with any analytic line contains only simply connected components. The next conjecture arose in Krasnoyarsk Town Seminar on the Theory of Functions of Several Complex Variables.

Conjecture 3. *A domain (a compact set) is a strongly l.c. set if and only if the intersection of this set with any analytic line is connected and simply connected.*

Let $\mathcal{D}$ be a bounded l.c. domain with the piecewise smooth boundary $\partial\mathcal{D}$. The set $\gamma = \{(\zeta, u) \in \mathbb{C}^{2n} : \zeta \in \partial\mathcal{D}, u \in \Gamma(\zeta)\}$ is called the *Leray boundary* of $\mathcal{D}$. Suppose that γ is a cycle. In this case it can be shown that for any function f holomorphic in $\mathcal{D}$ and continuous in $\operatorname{Clos}\mathcal{D}$, we have

$$f(z) = \int_\gamma f(\zeta)\omega(u, \zeta - z), \qquad z \in \partial\mathcal{D}. \tag{2}$$

This representation generalizes the integral formula indicated at the beginning of the note to the case of l.c. domains with non-smooth boundaries. If a l.c. domain $\mathcal{D}$ (a compact set M) can be approximated from inside (from the outside) by l.c. domains whose Leray boundaries are cycles then every continuous linear functional on $\mathcal{O}(\mathcal{D})$ ($\mathcal{O}(M)$) can be described by a formula analogous to (1) with $\gamma(\mathcal{D}_m)$ instead of $\partial\mathcal{D}_m$. Note that such a domain $\mathcal{D}$ (a compactum M) is strongly l.c. Therefore the following problem is closely connected with Problem 2.

PROBLEM 3. *Describe bounded l.c. domains whose Leray boundary is a cycle.*

This problem is important not only in connection with the description of linear continuous functionals on spaces of functions holomorphic in l.c. domains (on compacta). Formula (2) would have other interesting consequences (cf.[1], [2]).

CONJECTURE 4. *The classes of domains in Problems 1–3 coincide.*

REFERENCES

1. Aizenberg L. A., *The expansion of holomorphic functions of several complex variables in partial fractions*, Sib. Matem. Zhurnal **8** (1967), no. 5, 1124–1142. (Russian)
2. Aizenberg L. A., *Linear convexity in C^n and the distribution of the holomorphic functions*, Bull Acad. Polon. Sci., Ser. mat. **15** (1967), no. 7, 487–495. (Russian)
3. Aizenberg L. A., Trutnev V. M., *On a method for the Borel summation of n-fold power series*, Sib. Matem. Zhurnal **12** (1971), no. 6, 1398–1404 (Russian); English transl. in Siberian Math. J. **12** (1971), no. 6, 1011–1015.
4. Aizenberg L. A., Gubanova A. S., *The domains of holomorphy of functions with real or nonnegative Taylor coefficients*, Teor. Funkts., Funktsion. Analiz i Prilozh. **15** (1972), 50–55. (Russian)
5. Trutnev V. M., *Properties of functions that are holomorphic on strongly linearly convex set*, Properties of holomorphic functions of several complex variables, Krasnoyarsk, 1973, pp. 139–155. (Russian)
6. Aizenberg L. A., Yuzhakov A. P., Makarova L. Ya., *Linear convexity in $\mathbb{C}^n$*, Sib. Matem. Zhurnal **9** (1968), no. 4, 731–746. (Russian)
7. Leray J., *Le calcul différentiel et intégral sur une variété analytique complexe*, Ediz. Cremonese, Rome, 1965 (French); Russian transl. in 1961, Inostr. Lit., Moscow.
8. Aizenberg L. A., *Integral representations of functions which are holomorphic in convex region of $\mathbb{C}^n$ space*, Doklady Akad. Nauk SSSR **151** (1963), 1247–1249 (Russian); English transl. in Soviet Math. Dokl. **4** (1963), no. 4, 1149–1152.
9. Aizenberg L. A., *The general form of a linear continuous functional in spaces of functions that are holomorphic in convex domains of $\mathbb{C}^n$*, Doklady Akad. Nauk SSSR **166** (1966), 1015–1018 (Russian); English transl. in Soviet Math. Dokl. **7** (1966), no. 1, 198–202.

UL. KIRENSKOGO, D.3A, KV.77,
660074, KRASNOYARSK-74
RUSSIA

COMMENTARY BY THE AUTHOR

A solution of Problem 2 given in [10], [11] shows that Conjecture 3 is true. The definition of the strong linear convexity (s.l.c.) due to Martineau differs from the definition in the text only by the power (-1) (instead of $(-n)$) in the indicatrix formula. The two definitions turned out to be equivalent.

Yu. B. Zelinskii has shown in [12], [13] that the second conditions of Conjecture 3 and Conjecture 1 are equivalent. They mean the acyclicity of all sections of the domain by analytic planes of fixed dimension k, $1 \leqslant k < n$, and coincide with the s.l.c. [1]. These conditions form a precise "complex analogue" of the usual convexity [11], [12].

But the standard convex machinery cannot be generalized to this context. In particular, an ε-contraction of a s.l.c. domain is no longer s.l.c. in general.

Using the results of [10] one can show that the sections of s.l.c. domains are not too tortuous. This observation yields examples of unbounded s.l.c. domains non-approximable by bounded l.c. domains with smooth boundaries.

References

10. Znamenskii S. V., *A geometric criterion for strong linear convexity*, Funktsion. Anal. i Prilozhen. **13** (1979), no. 3, 83–84 (Russian); English transl. in Funct. Anal. and its Appl. **13** (1979), no. 3, 224–225.
11. Znamenskii S. V., *The equivalence of different definitions of strongly linear convexity*, International Conference on Complex Analysis and Applications. Varna, September 20–27, 1981, p. 30. (Russian)
12. Zelinskii Yu. B., *On the strongly linear convexity*, International Conference on Complex Analysis and Applications. Varna, September 20–27, 1981, p. 198.
13. Zelinskii Yu. B., *On geometrical criteria for strong linear convexity*, Doklady Akad. Nauk SSSR **261** (1981), no. 1, 11–13 (Russian); English transl. in Soviet Math. Dokl. **24** (1981), no. 3, 449–451.

1.19
v.old

ON THE UNIQUENESS OF THE SUPPORT OF AN ANALYTIC FUNCTIONAL

V. M. TRUTNEV

The symbol $H(E)$ will denote the space of all functions analytic on the (open or compact) set E, $E \subset \mathbb{C}^n$, endowed with the usual topology. Elements of the dual space $H'(\mathcal{D})$ (here and below $\mathcal{D}$ stands for an *open* set) are called analytic functionals (= a.f.). A. Martineau has introduced the notions of the carrier (porteur) and of the support of an a.f.

A compact set K, $K \subset \mathcal{D}$, is called *a carrier* of an a.f. T if T admits a continuous extension onto $H(K)$, or equivalently [2] if T is continuously extendable onto $H(\omega)$ for an arbitrary open ω, $\mathcal{D} \supset \omega \supset K$. Every a.f. has at least one carrier.

Let $\mathcal{A}$ be a family of compact subsets of $\mathcal{D}$ such that if $\{A_\alpha\}$ is a subfamily of $\mathcal{A}$ linearly ordered by inclusion then $\cap A_\alpha \in \mathcal{A}$. A compact set K, $K \in \mathcal{A}$ is called *an $\mathcal{A}$-support* of the analytic functional T if K is a carrier of T and K is minimal (with respect to the inclusion relation) among all carriers of T in $\mathcal{A}$. If $\mathcal{D}$ has a fundamental sequence of compact sets from $\mathcal{A}$ then any analytic functional has an $\mathcal{A}$-support but in general the $\mathcal{A}$-support is not unique. It is possible to consider various families of compact subsets of $\mathcal{D}$, e.g. the family of all compact subsets of $\mathcal{D}$, the family of $H(\mathcal{D})$-convex compact sets, the family of all convex compact sets (in this case an $\mathcal{A}$-support is called a convex support or a C-support).

Any analytic functional T on $\mathbb{C}^1$ has a unique C-support but T can have many polynomially convex (= *pc*) supports. (If for example

$$T(f) = \int_0^1 f(z)\,dz, \qquad f \in H(\mathbb{C}^1),$$

then any simple arc connecting 0 and 1 is a *pc*-support of T.)

PROBLEM. *Describe convex compact sets $K(\subset \mathbb{C}^n)$ such that K is the unique C-support of any analytic functional C-supported by K.*

C. O. Kiselman [3] has obtained for $n = 1$ necessary and sufficient conditions for a compact set to be a unique *pc*-support. For $n > 1$ a compact set with C^2-boundary is a unique *pc*-support [4]. Kiselman has proved in [4] that a convex compact set with smooth boundary is a unique C-support. A stronger result is due to Martineau [5]: a convex compact K is a unique C-support if any extreme point p of K ($p \in \operatorname{Extr} K$) belongs to a unique complex supporting hyperplane (with respect to the complex affine manifold $V(K)$ generated by K).

Our problem is stated for the above two families of compacts only, though it is interesting for other families as well. Using the ideas of Martineau one can prove the following

THEOREM. *A convex compact set $K(\subset \mathbb{C}^n)$ is a unique C-support if the set of all its supporting hyperplanes is the closure of the set of hyperplanes h with the following property:* $\operatorname{Extr} K \cap h$ *contains a point lying in a unique complex supporting hyperplane with respect to $V(K)$.*

It is probable that the sufficient condition of the theorem is also necessary.

REFERENCES

1. Martineau A., *Sur les fonctionneles analytiques et la transformation de Fourier–Borel*, J. Analyse Math. **9** (1963), 1–164.
2. Björk J.-E., *Every compact set in $\mathbb{C}^n$ is a good compact set*, Ann. Inst. Fourier **20** (1970), no. 1, 493–498.
3. Kiselman C. O., *Compact d'unicité pour les fonctionnelles analytiques en une variable*, C. R. Acad. Sci. Paris **266** (1969), no. 13, A661–A663.
4. Kiselman C. O., *On unique supports of analytic functionals*, Arkiv för Math. **16** (1965), no. 6, 307–318.
5. Martineau A., *Unicité du support d'une fonctionnelle analytique: un théorème de C. O. Kiselman*, Bull. Soc. Math. France **92** (1968), 131–141.

KRASNOYARSK STATE UNIVERSITY
UL. MAERCHAKA 6
KRASNOYARSK 660075
RUSSIA

Chapter 2

BANACH ALGEBRAS

Edited by

H. G. Dales
Department of Pure Math.
School of Mathematics
University of Leeds
Leeds LS2 9JT
England

A. Ya. Helemskii
Department of Mechanics
and Mathematics
Moscow State University
117234 Moscow
Russia

INTRODUCTION

Our chapter covers the vast topic of Banach algebras. In 1960, C.E. Rickart published the classic text, *General theory of Banach algebras*, but by now the subject encompasses so much material that it would require many volumes to cover in some detail all the major manifestations of the theory. Let us first mention an area that has been excluded from this chapter, the theory of C^*-algebras. Of course, every C^*-algebra is a Banach algebra, but it seems that the subject of C^*-algebras has a life of its own outside general Banach algebra theory, and so, with regret, there is almost no mention of this important area of mathematics in the present chapter.

The order of the sections which are included here is quite natural, and is rather close in spirit to that of the 1984 version of this book.

We begin with general questions which can be posed for all or for large classes of Banach algebras. An innovation is to first consider the homology and cohomology of Banach algebras as an aspect of their general structure. Two seminal papers in this theory are the memoir [4] of B. E. Johnson and the paper [6] of J. L. Taylor, which were both published in 1972. The topic has been studied from a somewhat different perspective by a school in Moscow, and an account is given in the monograph of A. Ya. Helemskii, published in 1986 in Russian and in 1989 in English ([3]). The topic now has important ramifications in many parts of our subject, including in C^*-algebra theory, where many very deep theorems have recently been proved. It seems necessary, in order to explain the terminology in the general theory, to give a somewhat longer account than is needed for most problems; see 2.1 for the general theory, and 2.2 for some questions which arise about the structure of C^*-algebras. Key notions in this theory are several different concepts of a "homologically trivial" Banach algebra; among them that of an *amenable* algebra appears to be most important. Amenability is a seemingly ubiquitous concept that cuts across many other classifications in Banach algebra theory; in 2.3, P. C. Curtis poses some questions about the existence of specific types of amenable Banach algebras.

The notion of a (bounded) approximate identity in a Banach algebra $\mathfrak{A}$ plays a surprisingly important role in the structure theory of Banach algebras. The key theorem is Cohen's factorization theorem, which we state in the form: If $\mathfrak{A}$ has a bounded left approximate identity and $a_n \to 0$ in $\mathfrak{A}$, then there exists $b \in \mathfrak{A}$ and $c_n \to 0$ in $\mathfrak{A}$ with $a_n = bc_n$ $(n \in \mathbb{N})$. These ideas lead to significant applications, and they have an intrinsic interest because there are several variations of the key notions whose inter-relationships should be explored: the concepts arise in 2.3, and are developed further in two closely related sections, 2.4 and 2.5.

The subject of *automatic continuity theory* is rather special in our subject because it explores the extent to which the existence of certain algebraic structures implies the existence of analytic or topological structures. The prototypic result is Johnson's uniqueness of norm theorem: a semisimple Banach algebra has a unique complete algebra norm. (See [1], for example; a delightful, simple proof of this result is given by T. J. Ransford in [5].) The topic is also characterized by its links with mathematical logic: indeed certain key questions are now known to be independent of the usual axioms of set theory, ZFC ([2]). Many results in automatic continuity theory are known, but some key questions about well-known algebras are still unresolved. In 2.6, homomorphisms from (non-commutative) C^*-algebras are considered, and in 2.7 the relation between the commutative C^*-algebras ℓ^∞ and c, and their quotients by prime ideals, is explored.

A class of commutative, radical Banach algebras that has attracted much attention in recent years, the class consisting of the radical convolution algebras $\mathfrak{A} = L^1(\mathbb{R}^+, \omega)$, is described in 2.8, and the fundamental question of whether every derivation from $\mathfrak{A}$ into a Banach $\mathfrak{A}$-module is automatically continuous is raised.

Let a be an element of a Banach algebra $\mathfrak{A}$. Then the definition of $\sigma(a)$, the *spectrum* of a with respect to $\mathfrak{A}$, goes back to the beginnings of our subject, and very soon it was known that there was a natural definition of $f(a)$ for a function f analytic on a neighbourhood of $\sigma(a)$ in $\mathbb{C}$. A much deeper generalization considers the *joint spectrum* $\sigma(a_1, \dots, a_n)$ of n (commuting) elements of $\mathfrak{A}$ and the definition of $f(a_1, \dots, a_n)$ in $\mathfrak{A}$ for $f = f(z_1, \dots, z_n)$ an analytic function on a neighbourhood of $\sigma(a_1, \dots, a_n)$ in $\mathbb{C}^n$. A seminal paper in this subject is [7]; see also the text [8]. Sections 2.9, 2.10 and 2.11 present problems in this area.

We now turn to sections which link the general theory to specific classes of Banach algebras. In the Appendix to his 1960 text, Rickart considered three such classes, and these are still the three classes which we consider as the most important today.

First Banach algebra theory is linked to complex function theory and topology through the theory of Banach function algebras and, in particular, uniform algebras. These are uniformly closed subalgebras of the algebra $C(\Omega)$ of all continuous functions on a compact space Ω. The main examples that have been considered for a long time are related to the disc algebra $A(\overline{\mathbb{D}})$ and the algebra $H^\infty(\mathbb{D})$ of all bounded, analytic functions on the open unit disc $\mathbb{D}$ in $\mathbb{C}$. Questions on function algebras are contained in sections 2.14 – 2.17.

Second, as we have said before, Banach algebra theory is linked to the immense topic of the theory of operators on Banach and Hilbert spaces: we include here just three very special sections in this area (2.18 – 2.20).

Finally we finish with three sections (2.21 – 2.23) about the two main types of Banach algebras traditionally associated with a locally compact group G—the *group algebra* $L^1(G)$ and the *measure algebra* $M(G)$. The latter example may be the most intractable specific Banach algebra that is studied, and many fundamental questions about it remain mysterious.

At the very last moment problems 2.24–2.32 were added.

References

1. Bonsall F. F., Duncan J., *Complete Normed Algebras*, Springer-Verlag, New York, 1973.
2. Dales H. G., Woodin W. H., *An Introduction to Independence for Analysts*, London Mathematical Society Lecture Note Series, vol. 115, Cambridge University Press, 1987.
3. Helemskii A. Ya., *The Homology of Banach and Topological Algebras* (1989), Kluwer Academic Publishers, Dordrecht.
4. Johnson B. E., *Cohomology in Banach Algebras*, Mem. Amer. Math. Soc., vol. 127, 1972.
5. Ransford T. J., *A short proof of Johnson's uniqueness-of-norm theorem*, Bulletin London Math. Soc. **21** (1989), 487–488.
6. Taylor J. L., *Homology and cohomology for topological algebras*, Advances in Mathematics **9** (1972), 137–182.
7. Taylor J. L., *A general setting for a multi-operator functional calculus*, Advances in Mathematics **9** (1972), 183–252.
8. Vasilescu F. H., *Analytic Functional Calculus and Spectral Decompositions*, Reidel, Dordrecht, 1982.

2.1

31 PROBLEMS OF THE HOMOLOGY OF THE ALGEBRAS OF ANALYSIS

A. YA. HELEMSKII

The area of mathematics which is indicated by the title (its another name is *topological homology*) lies in the boundary layer between the theories of topological (in particular, Banach) algebras and homological algebra. The area, which was represented by four or five papers in the 1960's (following the pioneering work of Kamowitz [62] in 1962), has since greatly expanded. Accordingly, the number of open problems, which always has a tendency to grow in a developing area, has become much more numerous. Therefore this paper of necessity noticeably exceeds the average size recommended for papers in this volume. I shall try to recount, with appropriate comments, some of the rather large number of accumulated problems which seem to me sufficiently interesting and important. Some related questions are deliberately left outside the scope of this paper. I mean some applications of homological theory, and also the automatic continuity of cocycles; there are people—for example, Dales [15], Johnson [54], Ringrose [82]—who know these things much better. Some specific problems about homological properties of operator algebras are also excluded: the present volume contains a paper by Z. A. Lykova which concerns just these problems. Finally, I do not consider (purely algebraic) homological invariants of topological algebras; recent papers of Wodzicki [95–98] have achieved a big advance in this area.

The fundamentals of topological homology are the subject of the monographs [40] and [50], and also of special chapters in the textbooks [2] and [43]. The subject is also the main topic of articles [36], [38], [58], [92], and of the surveys [41], [42], [45], [54], [81], and [82].

The reader will notice that not all problems in our discussion are presented in special paragraphs and with an individual number. This is the case for only some of the problems which seem to be sufficiently "ripe", and can be precisely and concretely formulated. Many of the other problems necessarily have a somewhat vague character, and they are actually rather suggestions about desirable investigations.

Let us first fix some notations. **A-mod**, **mod-A**, and **A-mod-A** are the categories of Banach left, right, and bi- (two-sided) modules, respectively, over a Banach algebra A. When there is no danger of misunderstanding, the same notations will be used for some categories related to more general polynormed algebras and modules. For a given A, A_+, A^{op}, and A^{env} are, respectively, the unitization, the opposite algebra, and the enveloping Banach algebra (that is $A_+\hat{\otimes}A_+^{\mathrm{op}}$) of A.

When speaking about algebras or, perhaps, modules of continuous functions, we shall use the standard notations $C(\Omega)$, $C_b(\Omega)$, and $C_0(\Omega)$ respectively for all, all bounded and all vanishing at infinity functions of this class on Ω. For a Banach space E, $\mathcal{B}(E)$, $\mathcal{K}(E)$, and $\mathcal{N}(E)$ are, respectively, the Banach algebras (or modules) of all continuous, all compact, and all nuclear operators on E; the latter is equipped with the nuclear norm. For a locally compact group G, $L^1(G)$ is the algebra (module) of Haar integrable functions on G with the convolution product. For a smooth manifold M, $C^\infty(M)$ is the Fréchet algebra of (infinitely) smooth functions on M, with the topology of uniform convergence of all partial derivatives. Finally, for a Stein analytic space S, $\mathcal{O}(S)$ is the Fréchet algebra of holomorphic functions on S, with the topology of uniform convergence on compact sets.

1. Some fundamentals, and problems of a general character

Topological homology, as well as "pure" homology, is founded on three principal concepts: projectivity, injectivity, and flatness. The most important of the derived notions is that of cohomology (and homology) groups. These groups, which are essential for applications, are the subject of at least half of all the papers in the whole area. It should be mentioned, however, that these groups can be defined independently of other notions (as really happened [49], [58], [62]).

Let A be a Banach algebra. A Banach left A-module X is called *projective* (respectively, *injective*), if, for every diagram $X \overset{\varphi}{\longleftarrow} Y$ (respectively, $X \overset{\varphi}{\longrightarrow} Y$) in **A-mod**, φ is a retraction (respectively, coretraction) in this category provided it is a retraction (coretraction) in the category of Banach spaces. Analogous definitions apply for other types of Banach modules.

A complex in each category of Banach modules is called *admissible* if it splits as a complex of Banach spaces. Every morphism which is part of an admissible complex is itself admissible.

A left (respectively, right, bi-) Banach A-module X is *flat* if, for every admissible complex $\mathcal{Y}$ in **mod-A** (respectively, **A-mod**, **A-mod-A**), the complex $\mathcal{Y}\hat{\otimes}_A X$ of Banach spaces is exact. An equivalent definition is the following: X is flat if its dual Banach module X^* is injective in the corresponding category of modules. Projective modules are certainly flat.

For various categories of polynormed modules over polynormed algebras the definitions of projectivity and injectivity are the same modulo obvious modifications. As to flatness, only the definition in terms of tensor products still makes sense in general.

Every admissible complex over (respectively, under) X is called a *resolution* (respectively, *coresolution*) of X. A resolution of X is called *projective* if each module in the resolution, save perhaps X itself, is projective. In a similar way, injective coresolutions (and also flat resolutions) are defined.

Every object in the categories of Banach (bi)modules has projective (and hence flat) resolutions and injective coresolutions. In the categories corresponding to more general classes of polynormed modules, every object still has a projective resolution. However, the possible absence in these categories of so-called cofree objects makes the question of existence of injective coresolutions rather delicate. (See [92] for a detailed discussion.) The following problem shows how little we actually know.

Problem 1. *Is it true that for every (non-normed) Fréchet algebra A an arbitrary Fréchet A-module has at least one injective coresolution?*

Probably, such a conjecture is too optimistic; here is its pessimistic counterpart.

Problem 2. *Does there exist a Fréchet algebra A for which there is no non-zero injective A-module?*

Possible candidates for such an algebra are $\mathcal{O}(U)$ for domains U in $\mathbb{C}^n$; for these no such module has been found. We only know that non-zero finite-dimensional $\mathcal{O}(U)$-modules are not injective.

Let A be a Banach algebra, $X, Y \in$ **A-mod**, $0 \leftarrow X \leftarrow \mathcal{P}$, and $0 \rightarrow Y \rightarrow \mathcal{T}$ be a projective resolution and an injective coresolution of the corresponding modules. We apply the respective morphism functors to these resolutions, and consider the resulting complexes ${}_Ah(\mathcal{P}, Y)$ and ${}_Ah(X, \mathcal{T})$ of Banach spaces. The fundamental theorem of homological algebra (in its "Banach" variant) asserts that the cohomologies of both complexes do not depend on a particular choice of (co)resolutions, and coincide with each other up to a topological isomorphism of semi-normed spaces. The (same) n-th cohomology of these complexes is denoted by $\operatorname{Ext}^n_A(X, Y)$. If $X \in$ **mod-A** and $Y \in$ **A-mod**, then the homologies of the complexes $X \underset{A}{\hat{\otimes}} \mathcal{Q}$ and $\mathcal{P} \underset{A}{\hat{\otimes}} Y$, where $0 \leftarrow X \leftarrow \mathcal{P}$ and $0 \leftarrow Y \leftarrow \mathcal{Q}$ are projective—or, more generally, flat—resolutions of the respective modules, also do not depend on the choice of resolutions and coincide with each other. The n-th homology of these complexes is denoted by $\operatorname{Tor}^A_n(X, Y)$.

For other types of Banach modules, the spaces Ext and Tor are defined similarly; the same is true for other standard categories of polynormed modules with the exception of the "injective" definition of Ext. If we are discussing bimodules, the notations $\operatorname{Ext}^n_{A-A}(\cdot,\cdot)$ and $\operatorname{Tor}^{A-A}_n(\cdot,\cdot)$ will be used.

The following direction of investigations seems to be interesting and important for applications: *to discover and study situations in which, in order to calculate* Ext *or* Tor, *we use complexes which are not necessarily projective or, in the appropriate case, injective or flat—(co)resolutions.* Several such cases are known. For example, if Y is a dual module over a Banach algebra A, then $\operatorname{Ext}^n_A(X, Y)$ is the n-th cohomology of ${}_Ah(\mathcal{F}, X)$ for any flat pseudo-resolution $0 \leftarrow X \leftarrow \mathcal{F}$ ("pseudo" means that the dual of our complex, but not the complex itself, is assumed to be admissible) [38].

Outside the scope of Banach structures, new possibilities arise if some of the spaces in question are nuclear in the sense of Grothendieck. For example, let A be a unital Fréchet algebra, and let $0 \leftarrow X \leftarrow \mathcal{P}$ be an exact (not necessarily admissible) complex in **A-mod-A**, which consists of projective, unital nuclear Fréchet bimodules (except for X). Then, for every Fréchet A-bimodule X, the space $\operatorname{Tor}^{A-A}_n(X, A)$ is the n-th homology of $X \underset{A}{\hat{\otimes}} \mathcal{P}$, whereas for every Fréchet A-bimodule X which is a DF-space, $\operatorname{Ext}^n_{A-A}(A, X)$ is the n-th cohomology of ${}_Ah_A(\mathcal{P}, X)$. See Taylor [92] about this and several similar results.

We also know too little about the situations where a *given exact complex of (bi)modules generates the exact sequence for* Ext *or* Tor—except the obvious case of the admissibility of our complex. Several sufficient conditions are discussed in the same article [92].

Among all the various Ext (respectively, Tor), the spaces $\operatorname{Ext}^n_{A-A}(A_+, X)$ (respectively, $\operatorname{Tor}^{A-A}_n(X, A_+)$) are apparently the most important; here A_+ is considered as an A-bimodule, and X is a "variable" bimodule. The first (second) space is called the *n-dimensional cohomology* (respectively, *homology*) *group* of a Banach—or, according to the case, polynormed—algebra A with coefficients in X, and it is denoted by $\mathcal{H}^n(A, X)$ (respectively, $\mathcal{H}_n(A, X)$).

Historically, both of these groups were defined in terms of so-called standard complexes. Namely, let us consider the complexes

$$0 \to X \xrightarrow{\delta^0} C^1(A,X) \xrightarrow{\delta^1} \cdots \to C^n(A,X) \xrightarrow{\delta^n} C^{n+1}(A,X) \to \dots \qquad (\tilde{\mathcal{C}}(A,X))$$

and

$$0 \leftarrow X \xleftarrow{d^0} C_1(A,X) \xleftarrow{d^1} \cdots \leftarrow C_n(A,X) \xleftarrow{d^n} C_{n+1}(A,X) \leftarrow \dots \qquad (\underset{\sim}{\mathcal{C}}(A,X)),$$

where $C^n(A,X)$ is the space of n-linear operators from $A \times \cdots \times A$ to X, $C_n(A,X) = A\hat{\otimes}\dots\hat{\otimes}A\hat{\otimes}X$, δ^n is defined by

$$\delta^n f(a_1,\dots,a_{n+1}) = a_1 \cdot f(a_2,\dots,a_{n+1}) + \sum_{k=1}^{n}(-1)^k f(a_1,\dots,a_k a_{k+1},\dots,a_{n+1})$$
$$+ (-1)^{n+1} f(a_1,\dots,a_n)\cdot a_{n+1}$$

(and $\delta^0 x(a) = a\cdot x - x \cdot a$), and d_n is defined by

$$d_n(a_1 \otimes \cdots \otimes a_{n+1} \otimes x) = a_2 \otimes \cdots \otimes a_{n+1} \otimes a_1 \cdot x$$
$$+ \sum_{k=1}^{n}(-1)^k a_1 \otimes \cdots \otimes a_k a_{k+1} \otimes \cdots \otimes a_{n+1} \otimes x$$
$$+ (-1)^{n+1} a_1 \otimes \cdots \otimes a_n \otimes x \cdot a_{n+1}$$

(with $d^0(a \otimes x) = a\cdot x - x\cdot a$). Note that $(\underset{\sim}{\mathcal{C}}(A,X))^*$ is just $\tilde{\mathcal{C}}(A,X)$. It turns out that the n-th cohomology of $\tilde{\mathcal{C}}(A,X)$ is just $\mathcal{H}^n(A,X)$, whereas the n-th homology of $\underset{\sim}{\mathcal{C}}(A,X)$ is $\mathcal{H}_n(A,X)$.

(In the non-normed case some rather harmless nuances arise. They are connected with the choice of the type of continuity of multilinear operators, and the possible replacement of "$\hat{\otimes}$" by some other type of topological tensor product. As to the details, see, e.g., [40], [92]).

Various applications and interpretations of "Banach" and "polynormed" cohomology and homology are expounded in the books and articles already cited, and also, e.g., in [7], [37], [52], [62], and [93].

We now proceed to study the most important numerical characteristics of modules and algebras, the so-called homological dimensions.

The *homological dimension* of a Banach, or polynormed, left module X over a Banach, or polynormed, algebra A is the least n for which there exists a projective resolution

$$0 \leftarrow X \xleftarrow{\epsilon} P_0 \xleftarrow{\varphi_0} P_1 \xleftarrow{\varphi_1} P_2 \xleftarrow{\varphi_2} \dots \qquad (0 \leftarrow X \leftarrow \mathcal{P})$$

such that $P_k = 0$ for $k > n$; it is denoted by $dh_A X$. The same number is also the least n for which $\mathrm{Ext}_A^{n+1}(X,Y) = 0$ for all $Y \in$ **A-mod**, or, equivalently, for which $\mathrm{Ext}_A^k(X,Y) = 0$ for all $Y \in$ **A-mod** and $k \geqslant n+1$. (If there is no such n, we put $dh_A X = \infty$.) The homological dimension of a right module and of a bimodule are defined similarly; the latter is denoted by $dh_{A-A} X$.

The *bidimension* of A (denoted by $db\,A$) is defined as $dh_{A-A}\,A_+$. It is also the least n for which $\mathcal{H}^{n+1}(A,X) = 0$ for all $X \in \mathbf{A\text{-}mod\text{-}A}$, or, equivalently, for which $\mathcal{H}^k(A,X) = 0$ for all $X \in \mathbf{A\text{-}mod\text{-}A}$ and $k \geqslant n+1$ (this explains its other name: cohomological dimension). The *left global dimension* of A (denoted by $dg\,A$) is defined as $\sup\{\, dh_A\,X : X \in \mathbf{A\text{-}mod}\,\}$. A similar supremum, but with the participation only of irreducible modules, is the so-called *left small dimension of* A, denoted by $ds\,A$. (Thus, in the case of a commutative Banach algebra, we take our supremum over all one-dimensional or, equivalently, finite-dimensional modules.) As to some other equivalent definitions for all of these dimensions, see [40] and [43].

For arbitrary A, we have $ds\,A \leqslant dg\,A \leqslant db\,A$ [43]. Also, we have $ds\,A < dg\,A$ for many algebras; for example, $ds\,C_0(\mathbb{N}) = 1$, but $dg\,C_0(\mathbb{N}) = 2$ (cf. 3.1–3.2 below). At the same time, for all Banach algebras for which both "big" dimensions have been calculated, they coincide. Naturally, we ask:

Problem 3. *Do there exist Banach algebras A with $dg\,A < db\,A$?*

The same problem is open for more general classes of polynormed algebras with jointly continuous multiplication, in particular, for Arens–Michael algebras (in other words, complete locally multiplicatively-convex algebras) and for Fréchet algebras. However, the strict inequality is possible for pure algebras or, equivalently, for polynormed algebras with the strongest locally convex topology. For example, the algebra $\mathbb{C}(t)$ of all rational functions has—as a field—zero global dimension, but, since it is an infinite-dimensional linear space, its bidimension is not zero. Because of the obvious equality $db\,A = db\,A^{\mathrm{op}}$, every Banach algebra with $dg\,A \neq dg\,A^{\mathrm{op}}$ $(= \sup\{\, dh_A\,X : X \in \mathbf{mod\text{-}A}\,\})$ would certainly provide the required example. Again, in pure algebra such things are possible ([63]).

Homological dimensions of Banach algebras do possess some specific features which have no analogue for pure algebras, and even for many polynormed algebras which are in a sense close to Banach ones, as we shall see below. This phenomenon is connected with some peculiar properties of Banach structures, and first of all with the existence of non-complemented closed subspaces of Banach spaces. The main consequence of this effect can be seen in the "global dimension theorem": if A is a commutative Banach algebra with an infinite set of maximal ideals, then $dg\,A$ (and hence also $db\,A$) is strictly greater than 1. In particular, this means that 1 is a "forbidden value" for both dimensions in the class of Banach function algebras. To what degree do other classes of Banach algebras share this property? Let us formulate such a question more exactly.

Problem 4. *Is $dg\,A$ and/or $db\,A$ always strictly greater than 1 in the class of infinite-dimensional semisimple Banach algebras? Of infinite-dimensional Banach algebras with involution? Of infinite-dimensional C^*-algebras? Finally, of commutative, radical Banach algebras?*

Up to now, apart from Banach function algebras, the inequalities in question have been proved for those infinite-dimensional Banach algebras which are biprojective and have the approximation property ([86]); biprojective algebras will be discussed in 2.1. Other known cases are infinite-dimensional liminal ("CCR-") C^*-algebras ([73]) and non-idempotent (not coinciding with their topological square) radical Banach algebras ([26], [34]).

Also in the class of all Banach algebras, there are no forbidden values for homological dimensions. Every Banach space can be made into a Banach algebra in such a way that all three dimensions are 1, and those of the n-th projective tensor power of its unitization are n ([85]). However, these algebras are not semisimple and they are non-commutative; also, they are not radical algebras. Examples of non-normed function algebras of every prescribed small homological dimension, including 1, will be mentioned in 3.3.

There is a suspicion that the homological dimensions of Banach algebras have one more peculiarity: they behave "too regularly" under the operation of projective tensor product of algebras.

PROBLEM 5. *Let A_k $(1 \leqslant k \leqslant n)$ be unital Banach algebras. Is the "additivity formula" $dg(A_1 \hat{\otimes} \dots \hat{\otimes} A_n) = dg\,A_1 + \dots + dg\,A_n$ always valid? Is the similar formula for db and/or ds valid?*

In pure algebra, there exist rather simple counterexamples. For example, all three dimensions are 1 for the algebra of all sequences that are eventually constant, and for the tensor powers of this algebra. At the same time the already mentioned peculiarities of Banach structures prevent the construction of similar examples. Moreover, they enable one to prove the additivity formula for a number of concrete situations. Such formulae for db and dg were obtained in [23] and [65], respectively, for the case where each algebra is the unitization of a biprojective Banach function algebra. The additivity formula for ds is known even for more general cases when the small dimension of each A_k (assumed to be a Banach function algebra) is at most 1.

Recently Selivanov (unpublished) has shown that the above results about "big" dimensions can be deduced from the following general theorem: if A is a biprojective, commutative Banach algebra with an infinite set of maximal ideals, and B is an arbitrary unital Banach algebra, then $dg(A_+ \hat{\otimes} B) = dg\,A_+ + dg\,B = 2 + dg\,B$ (cf. 3.1, below); a similar formula holds for db. It is not difficult to see that all potential additivity formulae in Problem 5 are intimately connected with the (hypothetical) formula

$$dh_{A_1 \hat{\otimes} A_2} X_1 \hat{\otimes} X_2 = dh_{A_1} X_1 + dh_{A_2} X_2.$$

Here the right-hand side of the equality concerns arbitrary Banach algebras and left modules, and $X_1 \hat{\otimes} X_2$ is considered as a left module over $A_1 \hat{\otimes} A_2$. Although this additivity on the "module level" seems to be more doubtful than on the "algebra level", counter-examples have still not been found.

Homological dimensions of modules and algebras, as discussed above, are often equipped with the adjective "projective" since they were defined by means of projective resolutions. But some other "injective" dimensions, defined by means of injective resolutions, and especially "flat" resolutions, are also of interest. These dimensions are often called "weak" dimensions. In the definitions, the role which is played by Ext (in particular, cohomology groups) for the projective dimensions, is now assumed by Tor (in particular, homology groups). We shall restrict ourselves to one—perhaps, the most important—of these.

The *weak bidimension* of a polynormed algebra A (denoted by $db_w\,A$) is the least n for which $\mathcal{H}_{n+1}(A, X) = 0$ and $\mathcal{H}_n(A, X)$ is Hausdorff for all $X \in$ **A-mod-A**. In the case of a Banach algebra this value coincides with $\min\{n : \mathcal{H}^{n+1}(A, X) = 0$ for all *dual* A-bimodules $X\}$. (In both definitions, the vanishing of the $(n+1)$-dimensional groups can be replaced by the vanishing of all k-dimensional groups for $k \geqslant n+1$). The principal problem concerning the weak bidimension is that of the validity of an assertion which could play the same role as the theorem on global dimension. We shall postpone its exact statement to the end of Section 3.

2. Problems about homologically best algebras and modules

2.1. Various manifestations of projectivity. A Banach (or polynormed) algebra A is called *contractible* if A_+ is a projective A-bimodule, that is, if $db\, A = 0$ or, in other words (see Section 1), $\mathcal{H}^n(A, X) = 0$ for all $X \in$ **A-mod-A** and $n \geqslant 1$ (hence the term "contractible", inherited from topology). As to other equivalent definitions (see, e.g., [43]) we shall present one which provides an effective method of checking the property: A is contractible if and only if it is unital, and the canonical morphism $\pi : a \otimes b \mapsto ab$, $A \hat{\otimes} A \to A$, is a retraction in **A-mod-A**. A (bi)module of every type over a contractible algebra is projective. Every complemented (that is having, as a subspace, a topological direct complement) closed left (respectively, right) ideal of a contractible algebra has a right (respectively, left) identity.

Contractibility is an extremely rigid condition. Finite-dimensional semisimple algebras are contractible. If A is a Banach algebra, the converse is true under some extra assumptions that some space related to A has a variant of the approximation property ([14], [86], [92], [99]). Nevertheless the following important problem is still open in its full generality.

Problem 6. *Consider the following hierarchy of properties of Banach algebras: A is semisimple and finite-dimensional $\Longrightarrow$ A is contractible $\Longrightarrow$ every left Banach A-module is projective $\Longrightarrow$ every irreducible Banach left A-module is projective. Is it true that all these implications, or at least some of them, are in fact equivalences?*

(For the second implication the posed question is obviously a part of Problem 3).

The equivalence of all four properties is known for commutative Banach algebras ([29]) and for C^*-algebras and L^1-algebras of locally compact groups ([86]). Note that a counterexample to the converse of the first implication would imply the existence of a simple, contractible Banach algebra of a rather exotic nature: every non-trivial Banach module over such an algebra would not possess the approximation property as a Banach space (Selivanov, unpublished).

The well-known algebraic description of classically semisimple algebras as algebras with projective irreducible modules (see, e.g., [68]) enables us to observe the intimate connection of this problem with the following problem about projective modules.

Problem 7. *Let A be a unital Banach algebra, and let I be a closed left ideal such that the (cyclic) A-module A/I is projective. Is it true that there always exists a decomposition $A = I \oplus J$, where J is another closed ideal in A? In other words, does I always have a right identity?*

A positive answer to this problem just in the case where I is a maximal ideal would already guarantee the equivalence of all four properties in Problem 6. As well as in the obvious case, when I has a Banach complement in A, the required decomposition is known in the case where I is maximal and A/I has the approximation property (Selivanov; cf. [40]).

The analogous problem to 6 posed for more general polynormed algebras, has a negative answer. Even for commutative Arens–Michael algebras each of the conditions $db\,A = 0$ and $dg\,A = 0$ is equivalent to the fact that A is the algebra of all functions on some set (of arbitrary cardinality), with the topology of pointwise convergence. This result was announced in [92] as a particular case of a theorem on general Arens–Michael algebras, but the argument had a gap. Nevertheless the mentioned result is true [40]. However its natural non-commutative analogue is still open, and can be formulated as follows.

PROBLEM 8. *Let an Arens–Michael algebra A have zero global dimension or even be contractible. Does this imply that A is a topological direct product of some family of full matrix algebras?*

A Banach (or polynormed) algebra A is called *biprojective* if A is a projective A-bimodule. This condition is weaker than that of contractibility; in particular, a biprojective algebra is not bound to be unital, although it does always coincide with its topological square. There is also an equivalent definition, which provides an effective tool to check the property: A is biprojective if the canonical morphism $\pi : A\hat{\otimes}A \to A$ is a retraction in **A-mod-A**.

For many classes of algebras of analysis the standard question of describing biprojective algebras is completely solved. The algebra $L^1(G)$ is biprojective if G is compact ([33], [40], [80]); a C^*-algebra is biprojective if and only if it is a c_0-sum of a certain family of full matrix algebras ([33], [87]); in particular, $C_0(\Omega)$ is biprojective if Ω is discrete. (In general, the maximum ideal space of each commutative, biprojective Banach algebra is discrete [40].) The algebra $\mathcal{N}(E)$ is biprojective if and only if E has the approximation property [88]. The latter example is very important: algebras $\mathcal{N}(E)$, after proper generalization, play a role in the description of the structure of all biprojective Banach algebras with the approximation property ([87]).

It was the following property of biprojective algebras which was actually the initial stimulus to their investigation: for such an A, necessarily $\mathcal{H}^n(A, X) = 0$ for $n \geqslant 3$. In other words, $db\,A \leqslant 2$, and hence $dg\,A \leqslant 2$ ([34], see also [36]). Thus, the homological dimension of a left module X over a biprojective Banach algebra A can assume only the values 0, 1, or 2. In the case where A is commutative, has a bounded approximate identity, and is infinite-dimensional, one can describe modules with $dh_A\,X = 2$ purely in terms of Banach geometry: they are exactly those X for which their so-called essential submodule $\overline{A \cdot X}$ has no Banach complement in X ([64]). As an example, $dh_{c_0}\,c_b = 2$.

A Banach, or polynormed, algebra is called *left hereditary* if every closed left ideal is projective; *right hereditary* algebras are defined similarly. A biprojective Banach algebra with a left (right) bounded approximate identity is always left (right) hereditary [40]. The algebra $C_0(\Omega)$ is hereditary if and only if Ω is hereditary paracompact [31]. Every separable C^*-algebra (e.g., $\mathcal{K}(H)$ for separable H) is left and right hereditary. On the other hand, every infinite-dimensional von Neumann algebra (for example, $\mathcal{B}(H)$) and, more generally, every AW^*-algebra, has a non-projective closed left ideal ([72]). *It would be interesting to give a full characterization of hereditary algebras among C^*-algebras.*

The following problem about operator algebras on more general spaces again brings us to Banach geometry.

PROBLEM 9. *Is the algebra $\mathcal{K}(E)$ left and/or right hereditary for every Banach space E?*

Is it at least projective as a left and/or right Banach $\mathcal{K}(E)$-module?

Lykova [71] has shown that $\mathcal{K}(E)$ is projective as a right module provided E has an elastic basis; if, further, E is reflexive, then $\mathcal{K}(E)$ is also projective as a left $\mathcal{K}(E)$-module.

At the same time, it is known that, for an infinite-dimensional E, the algebra $\mathcal{K}(E)$ is not biprojective in just the same situation for which the biprojectiveness of $\mathcal{N}(E)$ is proved: namely, when E has the approximation property [87]. *Most likely, $\mathcal{K}(E)$ is not biprojective for any infinite-dimensional E; however, as yet, there is no complete proof of this conjecture.*

One more question about projective ideals, which was for a time open, is now solved. Let A be a commutative Banach algebra, and let I be a projective maximal ideal. Then the linear space $I/\overline{I^2}$ is either zero- or one-dimensional, and the first alternative occurs provided that I belongs to the Shilov boundary of the maximal ideal space of A ([34]). Answering a problem in [40], Pugach [79] has proved that the converse is also true: if I does not belong to the Shilov boundary, then $I/\overline{I^2}$ is one-dimensional. As a development of results of Pugach in [78], it would be interesting to know more about the behavior of $\operatorname{codim}_I \overline{I^2}$, where I is a maximal ideal of a given finite *positive* homological dimension, and about the causes of this behaviour.

Our final problem about projective modules concerns the polynormed algebras $\mathcal{O}(U)$. These are in a certain sense closest to the "pure" polynomial algebras. It is known that every projective module over a polynomial algebra is free (the positive solution of the problem of Serre, by Suslin and Quillen). Is it true that every projective—this time in the sense of homological topology—$\mathcal{O}(U)$-module is free as a polynormed module? (Now "freedom" means that it is topologically isomorphic to a module of holomorphic functions on U with values in some polynormed space.) Here, unlike the polynomial algebras, the one-dimensional case already presents difficulties.

2.2. Various manifestations of injectivity and of flatness. A polynormed algebra A is called *amenable* if A_+ is a flat A-bimodule. This means exactly that $db_w A = 0$ or, equivalently, $\mathcal{H}_1(A, X) = 0$ and $\mathcal{H}_0(A, X)$ is Hausdorff (i.e., $\operatorname{Im} d_0$ is closed) for all $X \in$ **A-mod-A**. In the latter definition, one can replace the vanishing of $\mathcal{H}_1(A, X)$ by the vanishing of $\mathcal{H}_k(A, X)$ for all $k \geqslant 1$. I would like to know, to what extent the condition on the zero-dimensional homology groups to be Hausdorff is necessary.

PROBLEM 10. *Do there exist non-amenable polynormed algebras with $\mathcal{H}_k(A, X) = 0$ for all $X \in$ **A-mod-A** and $k \geqslant 1$? Do there exist Banach algebras with this property?*

The similar question on the "module level"—about the connection of flatness with Tor spaces—has a positive solution. As an example, the l_1-module $\mathbb{C} = (l_1)_+/l_1$ is not flat, but $\operatorname{Tor}_n^{l_1}(X, \mathbb{C}) = 0$ for all $X \in$ **mod**-l_1 and $n \geqslant 1$ [38].

Within the class of Banach algebras, there are a number of further equivalent definitions of amenability (for the main ones, see, e.g., [38]). So, for example, a Banach algebra A is amenable if and only if:

- A has a bounded approximate identity, and is a flat A-bimodule;
- $\mathcal{H}^1(A, X) = 0$ for every *dual* A-bimodule X (in other words, every continuous derivation into a dual Banach A-bimodule is inner)[1];

[1] This is just the original definition of amenability, due to Johnson [50]. The approach in terms of

- $\mathcal{H}^k(A, X) = 0$ for every dual Banach A-bimodule X and $k \geqslant 1$;
- A has a bounded approximate identity, and the canonical dual morphism $\pi^* : A^* \to (A \hat{\otimes} A)^*$ is a coretraction in **A-mod-A**.

(The last criterion provides one of the most effective tools to check amenability.)

If A is an amenable polynormed algebra, then all (bi)modules of every type over A are flat. It would be interesting to know, *whether there exist non-amenable, polynormed (Banach, C^*-) algebras, over which all left modules are flat.*

Every complemented, or even weakly complemented (this means that its orthogonal complement is complemented in the dual space), closed left ideal of an amenable Banach algebra has a bounded right approximate identity; the same is true if we replace "left" by "right", and vice versa.

Amenable Banach algebras have a complete description in terms of approximate identities. Let I^Δ be the kernel of the canonical morphism $\pi_+ : a \otimes b \mapsto ab$, $A_+ \hat{\otimes} A_+ \to A_+^{\mathrm{op}}$, so that I^Δ is a left ideal in A^{env}. Then A is amenable if and only if I^Δ has a bounded right approximate identity (see, e.g., [48]). The proof we know uses some special properties of Banach structures such as the Banach–Alaoglu theorem. However, all notions in the statement of the criterion make sense for general polynormed algebras. Naturally the following problem arises.

PROBLEM 11. *Is the given criterion for amenability in terms of approximate identities valid for all polynormed algebras? Is this true at least for Arens–Michael algebras?*

Note that the posed question is a particular case of the following one. Let A be a unital polynormed algebra, and let I be a complemented left ideal. *Is it true that the cyclic A-module A/I is flat if and only if I has a bounded right approximate identity?* In the Banach case both conditions are equivalent provided that I is weakly complemented in A ([38], [89]). On the other hand, without any geometrical conditions, the answer is negative. As an example, we recall that for the Wiener algebra and, more generally, for the L^1-algebra of an arbitrary commutative, non-compact group there exist ideals surely devoid of approximative identities ("Malliavin ideals"). At the same time, all modules, and among them cyclic ones, over these algebras are flat because our algebras are amenable.

We note in this connection that bounded approximate identities play essentially the same role for amenability and flatness as "ordinary" identities do for contractibility and projectivity. For the latter properties, however, examples similar to those just presented are still not known; a hypothetical algebra which has a left ideal without a right identity, surely cannot be a direct sum of matrix algebras (cf. Problem 6).

How do we characterize amenable algebras within concrete classes of "algebras of analysis"? It was proved in the seminal memoir of Johnson [50] that the algebra $L^1(G)$ is amenable if and only if the group G is amenable in the traditional meaning of harmonic analysis, that is, it has a left-invariant mean on $C_b(G)$, or, equivalently, on $L^\infty(G)$ (hence the very term "amenable algebra"). In [50] and in [59], it was established that all $C_0(\Omega)$ are amenable. (For another proof, see [38].) Sheinberg [91] has shown that, apart from $C_0(\Omega)$, there are no amenable uniform algebras.

---

the flatness originates in [32].

The description of amenable C^*-algebras took more time. In the beginning, Kadison and Ringrose [59] established the amenability of all uniformly hyperfinite algebras (their argument applies to all approximatively finite-dimensional C^*-algebras), and Johnson [50] did the same for all postliminal ("GCR-") algebras. Then Bunce [3] proved that, for a discrete group G, the algebra $C_r^*(G)$ is amenable if and only if G itself is amenable; this result provided the first examples of non-amenable C^*-algebras.

Up to this time, the class of nuclear C^*-algebras has acquired a considerable popularity (see Lance [69] as a pioneering work, and Kadison and Ringrose [60] for an up-to-date exposition). These algebras are characterized in many ways; in particular, as those whose von Neumann envelope is injective in the sense of Connes [5]. With the help of this fact, Connes [12] proved that an amenable C^*-algebra is nuclear (which, in particular, implies that $\mathcal{B}(H)$ is not amenable), and conjectured that the converse is also true. This conjecture turned out to be a strong nut; it was cracked by Haagerup [28], who thus established that a C^*-algebra is amenable if and only if it is nuclear. Later simpler proofs that a nuclear C^*-algebra is amenable were obtained ([18], [44]). All of them, however, use a strong tool, invented by Haagerup—his estimate of the norm of a bilinear functional on C^*-algebras (the "Grothendieck–Pisier–Haagerup inequality" [27]).

As for standard operator algebras on Banach spaces, the following problem is still unsolved.

PROBLEM 12. *For which Banach spaces E is the algebra $\mathcal{K}(E)$ amenable?*

For many spaces, e.g., $C_0(\Omega)$ and L^p $(1 \leqslant p \leqslant \infty)$, this is indeed the case. On the other hand, the necessary presence of a bounded approximate identity in an amenable algebra (see above) tells us that E must have some good geometry—at any rate, it must have the bounded compact approximation property [16]. But this condition is not sufficient: recently Groenbaek and Willis have shown that $\mathcal{K}(E)$ is not amenable for $E = c_0 \oplus l_1$ (to appear). Perhaps, the right answer will be in terms of the existence in $\mathcal{K}(E)$ of an approximate identity with some additional properties.

The following problem of Johnson [50], which was proposed even before the clarification of the situation for $\mathcal{B}(H)$ (see above), is still also open: *Do there exist infinite-dimensional spaces E for which $\mathcal{B}(E)$ is amenable?*

As to $\mathcal{N}(E)$, it is never amenable when E is infinite-dimensional ([25], [88]).

We shall conclude the discussion of amenability with the following problem, raised by Curtis (see the present book) and also posed in the postscript to [40].

PROBLEM 13. *Do there exist (non-zero) amenable, commutative radical Banach algebras?*

The known examples of radical algebras, including those with bounded approximate identities, are not amenable. For example, $L^1[0,1]$ with convolution product is not amenable because it has non-idempotent complemented ideals; one can easily show that it is also non-biprojective. *(Notice that the existence of biprojective, radical, commutative Banach algebras is also in question.)*

The case of $\mathcal{B}(H)$ apparently reflects the general phenomenon that for von Neumann algebras amenability is too rigid a condition. For these algebras, Connes [11] introduced a variant of the concept. His definition is given in terms of the following class of modules. Kadison and Ringrose [58] called a dual bimodule X over an operator C^*-algebra

A *normal*, if, for every $x \in X$, the operators $A \to X$, $a \mapsto a \cdot x$, $x \cdot a$, are continuous relative to the ultraweak topology in A and the weak*-topology in X. We shall call an operator C^*-algebra, in particular, a von Neumann algebra, *amenable-after-Connes*, if $\mathcal{H}^1(A, X) = 0$ for all *normal* dual A-bimodules X (see [44], [55] for equivalent definitions in terms of cohomology of arbitrary dimensions, and also of so-called normal cohomology). As to algebras amenable in the previous sense, we shall now call them *amenable-after-Johnson*, for the sake of precision.

As Connes himself showed ([12]), the class of von Neumann algebras which are amenable "in his sense" coincides with the class of injective—again "in the sense of Connes"—von Neumann algebras. One of many equivalent definitions of these algebras (see [17], for example) states that such an algebra A is the image of a projection in $\mathcal{B}(H)$ of norm 1 (which turns out to be automatically a morphism of A-bimodules [4]). The two varieties of amenability are intimately connected: a C^*-algebra is amenable-after-Johnson if and only if its von Neumann envelope is injective, and hence amenable-after-Connes.

Earlier it was mentioned that a contractible algebra can be characterized in terms of its projectivity as a bimodule, and an amenable-after-Johnson algebra in terms of its flatness as a bimodule, that is, in terms of the injectivity of its dual bimodule. It was asked in [40] whether a characterization of algebras amenable-after-Connes is possible in similar terms. The answer was given by a theorem in [44] which, in particular, shows that a von Neumann algebra is amenable-after-Connes if and only if its *predual* bimodule is injective. As to this latter property, one can check it as follows. Let $(A \times A)_*$ be a sub-A-bimodule in $(A \hat{\otimes} A)^*$ consisting of separately weak*-continuous bilinear functionals, and let $\pi_* : A_* \to (A \times A)_*$ be the restriction of π^* (see above) to $A_* \subseteq A^*$. Then A_* is injective if and only if π_* is a coretraction in **A-mod-A** (cf. the similar role of π and π^* in the verification of the projectivity of A and the injectivity of A^*). Both results have their natural analogues for arbitrary operator C^*-algebras.

The proof of the stated characterization of algebras which are amenable-after-Connes is considerably more complicated than that of parallel characterizations of contractibility and of amenable-after-Johnson. The main difficulty is that there are not many normal bimodules, and, in particular, the dual to $(A \times A)_*$ is, to all appearance, not normal. (This fact is announced in [19], but no proof is presented.) The argument could be much simplified if one could give a positive answer to a problem which is of independent interest. It concerns another bimodule, the dual of which is certainly normal this time.

Let A be a von Neumann algebra. Let us consider the so-called Haagerup norm

$$\|u\|_h = \inf\left\{ \left\|\sum x_j x_j^*\right\|^{1/2} \left\|\sum y_j^* y_j\right\|^{1/2} : u = \sum x_j \otimes y_j \right\}$$

on the algebraic tensor product $A \otimes A$. Now let us introduce the Banach A-bimodule $(A \times A)^h = (A \otimes A, \|\cdot\|_h)^*$, and consider its closed sub-bimodule $(A \times A)^h_*$ consisting of separately weak $*$-continuous bilinear functionals.

PROBLEM 14. *Is the A-bimodule $(A \times A)^h_*$ injective?*

(It is known that the A-bimodules $(A \hat{\otimes} A)^*$ and $(A \times A)_*$ are injective [44].)

A polynormed algebra A is called *biflat* if the A-bimodule A is flat. In the Banach case there is also the following equivalent definition: the dual canonical morphism $\pi^* : A^* \to$

$(A\hat{\otimes}A)^*$ is a coretraction in **A-mod-A**. One of the reasons for our interest in biflat Banach algebras is that such an algebra has the property that $\mathcal{H}^n(A, X) = 0$ for all *dual* A-bimodules X and all $n \geqslant 3$.

Every biprojective algebra is obviously biflat, and an amenable Banach algebra is just a biflat one which has a bounded approximate identity. An interesting question of a somewhat vague character is as follows: *how much wider is the class of biflat algebras than "combinations" of biprojective and amenable algebras?* We know no example of a biflat algebra which is not obtained from the two latter classes with the help of some standard operation like direct sum, tensor product, etc.

Let us touch upon the homological properties of a Banach space E which is considered as a natural left module over some algebra of operators acting on the space. Questions about such properties are rather important; for example, the known equality $\mathcal{H}^n(A, \mathcal{B}(E)) = \mathrm{Ext}^n_A(E, E)$ ([29], [61]) implies that, for $n \geqslant 1$, the indicated cohomology groups vanish whenever the A-module E is either projective or injective. (Notice that E is certainly projective if A contains all finite-dimensional operators.) Of all the problems of this kind the following is, perhaps, the most important.

PROBLEM 15. *Let A be an operator C^*-algebra on a Hilbert space H. Is the left A-module H always injective?*

(It is not always projective: $C[0,1]$ acting on $L^2[0,1]$ provides a counter-example.)

Notice that the injectiveness of the module H is equivalent to its flatness for every self-adjoint Banach operator algebra in H and, in particular, for every C^*-algebra.

The problem, similar to 15, but stated for the important class of nest algebras, and also for some of their generalizations, was solved by Golovin [21]. He shows that, for such algebras, the module H is always injective and flat. Also, he has given a criterion for the projectivity of this module in terms of an order structure of a set of invariant subspaces of the algebra in question. When considering nest algebras, questions about the homological properties of their radicals arise. Golovin [22] gave a criterion for the projectivity of such a radical in the case of a discrete nest. Practically nothing is known in this connection about radicals of algebras corresponding to continuous nests. It is doubtful whether they are projective, but some of them are, perhaps, flat.

3. PROBLEMS ABOUT HOMOLOGICAL DIMENSION

3.1. Dimensions of Banach function algebras. Some general questions about these dimensions were already discussed in Section 1. Now we shall concentrate on the behavior of homological dimension in some specific classes of Banach and polynormed algebras. The following problem, which is intimately connected with the classical theory of commutative Banach algebras, is one of the oldest and most important.

PROBLEM 16. *What is the set of values taken by the global dimension, and what is the set of values taken by the bidimension, for the class of Banach function algebras?*

Denote by M_g the first, and by M_b the second of the indicated sets. (These sets may be equal.) At the present time it is known that each set must necessarily have one of the following descriptions:

1) all even numbers and ∞;

2) all even numbers, all odd numbers save some initial segment $1, \dots, 2n+1$ ($n \geqslant 0$), and ∞.

In particular, if in fact $3 \in M_g, M_b$, then there is only one "forbidden" value for both dimensions, namely 1. Therefore the following "subproblem" of the stated problem deserves special attention.

PROBLEM 17. *Does there exist a Banach function algebra with both global dimension and bidimension—or at least with global dimension alone—equal to* 3*?*

Let us recount how the information related to these problems was accumulated. First of all, it was shown in [29] that in the class of commutative Banach algebras, the equality $dg\,A = 0$, as well as $db\,A = 0$, characterizes the algebras $\mathbb{C}^n$ with coordinatewise product; hence, in particular, $0 \in M_g, M_b$. Further, the theorem on global dimension [34] (see also [36]) already mentioned in Section 1 gave, in particular, the information that $1 \notin M_g, M_b$. The same theorem, together with the upper estimates of dimensions of biprojective algebras [33], [34] mentioned in 2.1 gives the following result: both dimensions are equal to 2 for biprojective, commutative Banach algebras with infinite spectrum (and hence for their unitizations). Thus $2 \in M_g, M_b$ for such algebras as $C_0(\mathbb{N})$, l_1, and the Fourier algebra on $\mathbb{Z}$.

The first example of A with $dg\,A = db\,A = 2n$, where $2n$ is a prescribed even number, beginning with 4, was suggested by Selivanov [86]; it is the n-th projective tensor power of $(l_1)_+$. Later it was proved that such a result still holds if l_1 be replaced by an arbitrary biprojective, infinite-dimensional Banach function algebra (Golovin and Helemskii [23] for db, Krichevets [65] for dg). Meanwhile Moran [74] considered algebras $C(\Omega_\alpha)$, where Ω_α is the segment of the transfinite line up to an ordinal α, with the order topology. He showed that $ds\,C(\Omega_\alpha) > n$ whenever $\alpha > \aleph_n$. This implies that for sufficiently large α we have $ds\,C(\Omega_\alpha) = \infty$, and thus $\infty \in M_g, M_b$. Finally, one can deduce from Selivanov's results on dimensions of tensor products (see Section 1) that both M_g and M_b have the following property: if n belongs to such a set, the same is true of $n+2$. Together with the previous facts, this gives the information that there are only the two possibilities for both sets mentioned above.

It was not by chance that the small dimension did not figure in Problems 16 and 17. It has been known for a long time that this dimension can assume arbitrary positive integer values and ∞. For finite values, this follows from the fact that the homological dimension of the one-dimensional modules over the "polydisc algebra" $A(\mathbb{D}^n)$ assumes all values from 1 to n [30]. (There is another algebra sharing the same property: it is the n-multiple Varopoulos algebra $C(\Omega)\hat{\otimes}\dots\hat{\otimes}C(\Omega)$, where Ω is a connected, metrizable compact set [39].) On the other hand, it has been known since 1973 that the homological dimension of each non-trivial, one-dimensional module over the Fourier algebra of a non-discrete abelian group is equal to ∞. (This follows from a combination of the Sheinberg's theorem on dimensions of group algebras, which we shall present later, with some earlier observations from [35].)

3.2. Dimensions of algebras of continuous functions on compact sets. The following analogue of Problem 16, posed for a special class of function algebras, and intimately connected with questions of general topology and set theory, is of an independent interest.

PROBLEM 18. *What is the set of values taken by the global dimension and by the bidimension in the class of algebras $C(\Omega)$, where Ω is a compact set. In particular, does there exist a compact set Ω such that both "big" dimensions of the algebra $C(\Omega)$—or at least the global dimension alone—are equal to 3?*

Probably for both sets of values there are only the two possibilities discussed above for general function algebras. Nevertheless, the only things which are known in this case are that both sets contain all even numbers and ∞, and that they do not contain the number 1.

Our discussion of Problem 16 already contained the information that both $dg\,C(\Omega)$ and $db\,C(\Omega)$ are equal to 0 if Ω is finite, to 2 if $\Omega = \mathbb{N}_+$, to ∞ if $\Omega = \Omega_\alpha$ for a sufficiently large ordinal α, and at the same time that both of them are never equal to 1. As to arbitrary even dimensions, the tensor products mentioned in the discussion do not solve the problem since they do not belong to the class $C(\Omega)$. The required examples were found by Krichevets [66]. Namely, take K to be the one-point compactification of a discrete topological space of cardinality exceeding all $\aleph_n$ $(n = 1, 2, \dots)$; if K^n is the n-th cartesian power of K, then $dg\,C(K^n) = db\,C(K^n) = 2n$. It was proved in the same paper [66] that $ds\,C(K^n) = n$. Therefore (taking account of what was known earlier about Ω_α) it was established that the small dimension takes all values in $\mathbb{N} \cup \{\infty\}$, not only in the class of all function algebras, but already in the class of algebras $C(\Omega)$.

The following problem—this time concerning an individual algebra—seems to be especially challenging because of its apparent simplicity. It is an old question, which was raised more than once in various papers and books (cf. [15], [40], [43], and [50]).

PROBLEM 19. *What are $dg\,A$ and $db\,A$ for $A = C[0,1]$?*

Where is the difficulty? Taking an algebra A and trying to compute, say, $db\,A$, we must—explicitly or implicitly—use resolutions of A_+ (or of A itself, provided it is unital) as an A-bimodule. The point is that the estimate $db\,A \leqslant n$ is equivalent to the condition that in every such resolution

$$0 \leftarrow A_+ \text{ (or } A) \xleftarrow{d_{-1}} P_0 \xleftarrow{d_0} \dots \xleftarrow{d_{n-2}} P_{n-1} \xleftarrow{d_{n-1}} P_n \leftarrow \dots$$

the A-bimodule $\ker d_{n-2}$ must be projective. In the completely investigated case $A = c$ (i.e., $C(\mathbb{N}_+)$), our algebra is the unitization of the biprojective algebra $c_0 = C_0(\mathbb{N})$ and has, in this capacity, a special and very convenient resolution of the length 2 ([33]). However, in the case when Ω is a connected compact set, the only known projective resolution for $C(\Omega)$ is the so-called bar resolution, and one has to work just with this one (modulo some close variations). The bimodule P_{n-1} which occurs in this resolution for $A = C[0,1]$ happens to be the Varopoulos algebra $C[0,1]\hat{\otimes}\dots\hat{\otimes}C[0,1]$, consisting of functions on the $(n+1)$-dimensional cube $[0,1]^{n+1}$. This bimodule has a rather complicated structure, since there is no transparent description of its functions (somewhere between the 1-smooth functions and all continuous ones). The sub-bimodule $\ker d_{n-2}$ of P_{n-1} is still less understandable. It is true that one can deduce from the theorem on global dimension in some roundabout way that $\ker d_{n-2}$ is not projective provided $n = 1$. However, the endeavour to solve the question of its projectivity for larger n—

at least without use of some radically new ideas[2]—apparently encounters considerable difficulties.

It is interesting to compare the discussed problem on the dimensions of $C(\Omega)$ with another one. The latter can be stated in a very similar way, but one can achieve much more advance in its investigation.

Recall that for $A = C(\Omega)$ the bilinear product operator $(a, b) \mapsto ab$, $A \times A \to A$ can be canonically identified not only with a certain operator from $A\hat{\otimes}A$ to A (as is the case for all Banach algebras), but also with an operator from $A\hat{\hat{\otimes}}A$ to A. In other words, $C(\Omega)$ is an $\hat{\hat{\otimes}}$-algebra in the sense of Taylor [92] or, as we shall say, a *strict* Banach algebra. When considering only those (bi)modules over A which are similarly "strict" (cf. "$\hat{\hat{\otimes}}$-modules" from [92]), one can develop a parallel homological theory and define, in particular, "strict" homological dimensions. In this "strict" theory the bar resolution of $C(\Omega)$ is much simpler than in the "ordinary" one: it consists of bimodules of the form $C(\Omega \times \cdots \times \Omega)$. Using this fact and some strong tools of general topology, Kurmakaeva [67] has obtained the following nice-sounding result: the strict bidimension of $C(\Omega)$, where Ω is an infinite compact set, is equal to 1 if and only if Ω is metrizable (and hence it is more than 1 otherwise). In particular, for $\Omega = [0, 1]$, this value, as well as the strict global dimension, is just 1. It is still an open question, *whether the strict global dimension of $C(\Omega)$ with a non-metrizable Ω is always more than* 1.

To conclude out discussion of the algebras $C(\Omega)$, let us return to their ("ordinary") small dimensions. As well as the already mentioned examples, which provide every possible value of this dimension, the following general result is known: $ds\, C(\Omega) \leqslant 1$ if and only if the complement of any one-point subset of Ω is paracompact [31], and hence, in particular, $ds\, C(\Omega) = 1$ (cf. Problem 19). Nevertheless, the mechanism of the dependence of $ds\, C(\Omega)$ on topological properties of Ω is not exactly clear; one can only guess that this number is related to some "degree of non-paracompactness" of the subsets indicated above. Perhaps much will be clarified if an answer to the following quite concrete question is obtained.

PROBLEM 20. *What are $ds\, A$, $dg\, A$ and $db\, A$ for $A = c_b$ (that is, for $A = C(\beta\mathbb{N})$ where $\beta\mathbb{N}$ is the Stone-Cech compactification of a discrete countable set)?*

One can deduce from the above-mentioned role of the paracompactness property that all these dimensions are more than 1.

3.3. Miscellany. Compared with the commutative case, we know much less about the values taken by both "big" homological dimensions in the class of all—generally speaking, non-commutative—semisimple Banach algebras. In fact, we know nothing about the forbidden values for either dimension (cf. Problem 4). The same is the state of affairs concerning the most important class of these algebras, C^*-algebras. As to some particular algebras, it has been possible in a number of cases to compute their dimensions using their biprojectivity (cf. above). As examples, we have $dg\, A = db\, A = 2$ and $ds\, A = 1$ for such A as $\mathcal{N}(H)$, $L^1(G)$, and $C^*(G)$, where G is in both cases a compact group; the same is true for any C^*-algebra with a discrete spectrum and

[2] According to some unconfirmed rumors A. Connes is now able to prove that $db\, C[0, 1] = \infty$—supposedly with the essential use of so-called *entire cyclic cohomology*, which was recently invented by himself.

finite-dimensional irreducible representations. At the same time what is striking is the absence of any information concerning the extremely important algebras of the following problem.

PROBLEM 21. *What are the "big" dimensions of $\mathcal{K}(H)$? What are all three dimensions for $\mathcal{B}(H)$?*

It is easy to show that $ds\,\mathcal{K}(H) = 1$. Apart from this, the known equality $dh_{\mathcal{K}(H)}\,\mathcal{B}(H) = 2$ implies that $db\,\mathcal{K}(H) \geqslant dg\,\mathcal{K}(H) \geqslant 2$. As to $\mathcal{B}(H)$, the presence of non-projective ideals in this algebra (see Section 2) gives rise to the conjecture that all these dimensions are more than 1. To prove this it would be sufficient to display a maximal left ideal in $\mathcal{B}(H)$ which is simultaneously non-projective and complemented.

Let us pay special attention to the L^1-algebras of non-compact, locally compact groups (the compact case was already discussed above). Here the following theorem due to Sheinberg provides the principal information: if G contains at least one closed, amenable, non-compact subgroup, then all three dimensions of $L^1(G)$ are equal to ∞, and the same is true for the homological dimension of the so-called *augmenting* $L^1(G)$-module, that is, $\mathbb{C}$ with the exterior multiplication $f \cdot z = \left(\int_G f(t)\,dt\right)z$ ($f \in L^1(G)$, $z \in \mathbb{C}$). (This theorem was proved in [90] under the assumption that G itself is amenable. However, the argument still applies in the indicated case.)

PROBLEM 22. *Is it true that $db\,L^1(G) = dg\,L^1(G) = ds\,L^1(G) = dh_{L^1(G)}\,\mathbb{C} = \infty$ for arbitrary non-compact, locally compact groups G?*

Notice that the great majority of known groups surely satisfy the condition of Sheinberg's theorem. However, this condition is not ubiquitous. One can show, for example, that the discrete groups discovered by Ol'shanskii [100] are not amenable, but that all their subgroups are finite. Similar exotic groups certainly require some special treatment.

Very little is known about the homological dimensions of radical Banach algebras. In particular, I have no example of such an algebra with a positive finite value for *ds*, *dg* or *db*. If an algebra is nilpotent, all these dimensions are equal to ∞. The same is true for the algebras $l_1(\omega)$ with a sufficiently rapidly decreasing weight $\omega = \omega(n)$, e.g., for $\omega(n) = \exp(-n^{1+\epsilon})$ with $\epsilon > 0$ ([26]). Apart from this, all three dimensions of a non-idempotent (not coinciding with its topological square) commutative radical Banach algebra are more than 1 [26], [40].

As to concrete radical algebras, it would be very interesting to compute the homological dimensions of the Volterra algebra $L^1[0,1]$ (with convolution product).

The existence of forbidden values for some homological dimensions, as discussed above, is a purely "Banach" phenomenon. It does not occur when we proceed to more general topological algebras, even to the class of Arens–Michael algebras. Taylor [93] as early as 1972 showed that the bidimension of $\mathcal{O}(U)$, where U is a domain of a holomorphy in $\mathbb{C}^n$, as well as of $C^\infty(V)$, where V is an open set in $\mathbb{R}^n$, is equal to n. One can easily deduce (see [40]) that the global and the small dimensions of these algebras are also n. This means, in particular, that every non-negative integer is a value of all three dimensions of some commutative, Arens–Michael algebra, moreover of a metrizable one. Later Ogneva [76] considerably strengthened the second of these results of Taylor. She showed that, if M is a smooth manifold of a topological dimension of n, then all three homological dimensions of the algebra $C^\infty(M)$ are also equal to n. But the question about the similar strengthening of the first result of Taylor is open.

PROBLEM 23. *Let S be an arbitrary Stein manifold of a complex dimension n. Is it true that all three homological dimensions of the algebra $\mathcal{O}(S)$ (or some of them at least) are equal to n?*

In all appearance, the "holomorphic" case requires new ideas compared with the "smooth" case. The point is that the theorem of Ogneva heavily relies on the following result (also due to herself): the polynormed $C^\infty(M)$-module $C^\infty(V)$, where V is an arbitrary map in M, is projective. There is no similar result for holomorphic functions. For example, the polynormed $\mathcal{O}(\mathbb{C})$-module $\mathcal{O}(\mathbb{D})$, where $\mathbb{D}$ is an open disc in $\mathbb{C}$, is not projective (M. Putinar, not published).

Hitherto, our discussion concerned projective homological dimensions. As to weak dimensions, they have been much less investigated. Therefore it seems somewhat premature to state similarly detailed problems about these dimensions. We shall only emphasize that one should begin with an endeavour to answer the following question, whose exact formulation was promised in Section 1.

PROBLEM 24. *Do there exist Banach function algebras A with $db_w\, A = 1$ (in other words, non-amenable Banach function algebras A with $\mathcal{H}^2(A, X) = 0$ for all dual Banach A-bimodules X)?*

A similar question was originally stated by Johnson [50] and then, many years later, by Effros and Kishimoto [20]. However in both papers the question concerned all Banach algebras. But the answer is certainly positive in such generality: every biprojective, non-unital algebra with a one-sided identity has the desired property. Recently Selivanov (not published) has shown that a biflat Banach algebra A has $db_w\, A = 1$ if it has a one-sided, but no two-sided, bounded approximate identity. As a corollary, we have $db_w\, A = 2$ for every biflat Banach algebra which has neither left nor right bounded approximate identity.

4. PROBLEMS ABOUT (CO)HOMOLOGY GROUPS WITH SPECIAL CLASSES OF COEFFICIENTS

4.1. Cohomology with symmetric coefficients; simplicial (co)homology; cyclic (co)homology. An A-bimodule X is called *symmetric* if $a \cdot x = x \cdot a$ for all $a \in A$, $x \in X$. Kamowitz in his pioneering paper [62] of 1962 already considered such bimodules, and established that $\mathcal{H}^1(C(\Omega), X) = \mathcal{H}^2(C(\Omega), X) = 0$ for every compact set Ω and every symmetric $C(\Omega)$-bimodule X. Later, Johnson [50] extended this result from $C(\Omega)$ to all commutative, amenable Banach algebras (for a further generalization in terms of the functor Ext, see [38]).

PROBLEM 25. *Is the equality $\mathcal{H}^n(A, X) = 0$, where X is a symmetric Banach bimodule over a commutative, amenable Banach algebra A, valid for all, or at least for some, $n \geqslant 3$? Is it true at least for $A = C(\Omega)$?*

Notice that the second part of this problem is apparently the oldest among all open questions in topological homology (cf. [15], [40], and [50]).

Bade, Curtis and Dales [1] showed that, in the case of commutative A, the condition $\mathcal{H}^1(A, X) = 0$ for all symmetric X is equivalent to the apparently weaker condition $\mathcal{H}^1(A, A^*) = 0$. Algebras (not necessarily commutative) satisfying the latter condition, are called *weakly amenable*. Groenbaek [24] gave a nice intrinsic characterization of weakly amenable, commutative Banach algebras parallel to the characterization of "usual" amenability in terms of bounded approximate identities (see 2.2 above). Namely, such an algebra is weakly amenable if and only if the diagonal ideal I^{Δ} of its enveloping Banach algebra is idempotent. (Another proof of this criterion was given by Runde [83].)

PROBLEM 26. *Is the criterion of Groenbaek valid for non-commutative Banach algebras?*

There are many more weakly amenable algebras than amenable ones. As examples, all C^*-algebras (Haagerup [27]) and all L^1-algebras of locally compact groups (Johnson, to appear) are weakly amenable; see [1] for other interesting examples.

The cohomology groups of a Banach algebra A with coefficients in its dual bimodule A^*—we have just discussed the one-dimensional examples—are called *simplicial* cohomology groups of A. As distinct from the case of usual amenability, the weak amenability of an algebra does not, generally speaking, imply the vanishing of its higher simplicial cohomology. For example, all l_p $(1 \leqslant p < \infty)$ with coordinatewise multiplication are weakly amenable, but we surely have $\mathcal{H}^2(l_p, l_p^*) \neq 0$ at least for $1 < p < 3$ (Aristov, not published). However, the following problem is open.

PROBLEM 27. *Is it true that the simplicial cohomology groups of arbitrary C^*-algebra vanish in all positive dimensions?*

This is certainly so provided our C^*-algebra is nuclear (cf. 2.2). Apart from this, Christensen and Sinclair [9] gave a positive answer for C^*-algebras without continuous traces (that is, for algebras A such that $\mathcal{H}^0(A, A^*) = 0$, or, equivalently, such that there is no $f \neq 0$ in A^* with $f(ab) = f(ba)$ for all $a, b \in A$). Notice that the condition of the vanishing of $\mathcal{H}^n(A, A^*) = 0$ for all $n \geqslant 0$ is equivalent to the condition of the vanishing of the so-called simplicial homology groups $\mathcal{H}_n(A, A)$, again for all $n \geqslant 0$. The latter property was established by Wodzicki [96] for stable C^*-algebras, i.e., for algebras isomorphic to their tensor product with $\mathcal{K}(H)$ for a separable H.

The problem of the vanishing of simplicial cohomology in dimensions 2 *and more of L^1-algebras of locally compact groups is also still open.*

The simplicial cohomology of a Banach algebra is intimately connected with its *cyclic cohomology*. The latter is an important new type of cohomology, discovered in the purely algebraic situation by Connes [13]. At the same time Tzygan [94] independently introduced a dual concept, *cyclic homology*. Namely, let us take the standard complex $\tilde{C}(A, A^*)$ (see Section 1), and consider the subspaces $CC^n(A, A^*)$ of $C^n(A, A^*)$ $(n = 0, 1, \dots)$ consisting of the so-called cyclic cochains, that is, of n-linear operators which, after their canonical identification with $(n+1)$-linear functionals on A, satisfy the identity

$$f(a_0, \dots, a_n) = (-1)^n f(a_1, \dots, a_n, a_0).$$

These subspaces form a subcomplex in $\tilde{C}(A, A^*)$. The n-th cohomology of this subcomplex, denoted by $\mathcal{HC}^n(A)$, is called the *n-dimensional cyclic cohomology group* of a given polynormed or Banach algebra A. In a similar way, *cyclic homology groups* of A are defined as the homology of a certain factor complex of $\underset{\sim}{C}(A, A)$; they are denoted by $\mathcal{HC}_n(A)$.

A complicating factor in the study of cyclic (co)homology is that there is no—so far at least—simple expression for them in terms of derived functions. In this respect it differs from simplicial (co)homology, which can be presented as $\mathrm{Ext}^n_{A-A}(A, A^*)$ (respectively, $\mathrm{Tor}^{A-A}_n(A, A)$) (see Section 1). Nevertheless recently cyclic, and also simplicial, (co)homology has been expressed as a somewhat more complicated Banach Ext (Tor); this was done with the help of the construction of a Banach algebra associated with a small category [47]. However, there is a much more effective tool for computing cyclic (co)homology. It is the *Connes–Tzygan exact sequence*, which connects the cyclic (co)homology of many algebras with their simplicial one. A criterion for the existence of such a sequence is given in [46]; it is an analogue of the purely algebraic condition of so-called H-unitality in the sense of Wodzicki [95]. As a corollary, for a number of Banach algebras, and among them for all biflat algebras, the following description of their cyclic (co)homology was obtained:

$$\mathcal{HC}^n(A) \simeq \{ f \in A^* : f(ab) = f(ba) \}$$

and $\mathcal{HC}_n(A) = A/\overline{[A, A]}$ provided that n is even, whereas $\mathcal{HC}^n(A) = \mathcal{HC}_n(A) = 0$ provided that n is odd [46]. As an example, for $A = \mathcal{N}(H)$, our cohomology and homology groups are $\mathbb{C}$ in even dimensions, and they all vanish in odd dimensions.

PROBLEM 28. *Does the above description of cyclic (co)homology also hold for all C^*-algebras?*

This problem is "almost equivalent" to the previous one, that is, if the answer to Problem 27 is positive, the same is true for the latter problem; on the other hand, the converse is also true under some natural additional assumptions on the form of the topological isomorphism between $\mathcal{HC}^0(A)$ and $\mathcal{HC}^n(A)$ for $n = 2, 4, \ldots$. In particular, this observation implies that cyclic cohomology, as well as homology, of C^*-algebras without bounded trace vanishes in all dimensions.

4.2. Cohomology with coefficients in the algebra itself, in larger algebras, and in its ideals. Among the most important cohomology groups one can distinguish $\mathcal{H}^n(A, A)$ for $n \geqslant 1$. They are very useful in studying automorphisms of Banach algebras (see, e.g., [43]), lifting problems of derivations ([81]), and stability properties of Banach algebras under small perturbations of their multiplication [97]. The vanishing of these groups in positive dimensions has been established in a number of cases. They include $C(\Omega)$ for a metrizable, compact space Ω ([52]), $\mathcal{B}(E)$ for any Banach space E ([61]), hyperfinite von Neumann algebras ([55]). As for $A = \mathcal{K}(H)$ and $A = \mathcal{N}(H)$, it is obvious that $\mathcal{H}^1(A, A) \neq 0$, whereas $\mathcal{H}^n(A, A) = 0$ for $n \geqslant 2$. Recall from 2.1 that $\mathcal{H}^n(\mathcal{N}(E), \mathcal{N}(E)) = 0$ for a Banach space E with the approximation property and for $n \geqslant 3$. Recently Selivanov [88] has shown that $\mathcal{H}^3(\mathcal{N}(E), \mathcal{N}(E)) \neq 0$ whenever E does not have the approximation property, and at the same time $\mathcal{H}^2(\mathcal{N}(E), \mathcal{N}(E)) = 0$ for all E.

Thanks to Kadison and Sakai [57], [84], it has been known since 1966 that $\mathcal{H}^1(A, A) = 0$ for any von Neumann algebra. More intriguing is the following problem, which was stimulated by papers of Kadison and Ringrose [57], [58] and was explicitly stated in [82].

PROBLEM 29. *Does the equality $\mathcal{H}^n(A, A) = 0$ hold for each $n \geqslant 1$ and for each von Neumann algebra A?*

Johnson [51] has displayed a non-hyperfinite factor of type II_1 with $\mathcal{H}^2(A, A) = 0$.

Now let A be an operator C^*-algebra on a Hilbert space H. The following problem, considered by Christensen in [6] for $n = 1$ and in [7] for all $n \geqslant 1$, also remains open in its full generality.

PROBLEM 30. *Is it true that $\mathcal{H}^n(A, \mathcal{B}(H)) = 0$ for all $n \geqslant 1$, or at least for $n = 1$?*

One of the motivations for this problem is that a positive answer for $n = 1$ is necessary for a positive solution of the known "similarity problem": whether every representation of a C^*-algebra on a Hilbert space is similar to one of its $*$-representations? In his papers, Christensen had established the vanishing of $\mathcal{H}^1(A, \mathcal{B}(H))$ in a number of cases, among them for all algebras possessing a cyclic vector. Further, $\mathcal{H}^n(A, \mathcal{B}(H)) = 0$ for all $n \geqslant 1$ when the ultraweak closure of A is hyperfinite, and hence when A is amenable-after-Connes (cf. 2.2). Finally, as was mentioned in 2.2, the same equality holds when H is either projective (say, $A = \mathcal{B}(H)$) or injective.

It is important to note that both Problems 29 and 30 can be solved (in full generality) if one replaces ordinary cohomology in their statement by *completely bounded cohomology*. This new invariant of C^*-algebras was introduced in a paper of Christensen, Effros and Sinclair [8], in which they gave a positive answer to the "completely bounded" version of Problem 30. As to Problem 29, an analogous positive result was obtained in the subsequent paper of Christensen and Sinclair [10]. In both cases the study of Problems 29 and 30 themselves was advanced. It was shown that $\mathcal{H}^n(A, A) = 0$ ($n \geqslant 1$) provided a central direct summand of type II_1 of the algebra A is isomorphic to its tensor product by a certain hyperfinite type II_1 factor [10]. At the same time $\mathcal{H}^n(A, \mathcal{B}(H)) = 0$ ($n \geqslant 1$) provided that the ultraweak closure of A is either a properly infinite von Neumann algebra, or is a tensor product with a hyperfinite type II_1 factor [8].

The natural problem which arises after replacing C^*-algebras by nest algebras in Problem 30 has also been solved positively. Indeed, the vanishing of $\mathcal{H}^n(A, X)$ ($n \geqslant 1$), where A is a nest algebra, is ensured for the following two classes of bimodules: 1) X is an ultraweakly closed A-bimodule situated between A and $\mathcal{B}(H)$ (Lance [69], Nielsen [75]); 2) $X = \mathcal{B}(Y, H)$, where Y is an arbitrary left Banach A-module (Golovin [21]). Both of these classes obviously contain $\mathcal{B}(H)$.

Apart from Problems 29 and 30, a problem concerning a third class of coefficients, which was stated by Johnson (cf. [56]), remained open for a long time. It was as follows: Is it true that $\mathcal{H}^1(A, \mathcal{K}(H)) = 0$ for any von Neumann algebra A on a Hilbert space H? After a conspicuous advance in [56], a positive answer was given by Popa [77]. However, it still remains unclear *whether this equality is valid for higher cohomology*, i.e., after replacing 1 by $n \geqslant 2$.

Our final problem, also belonging to Johnson [50], concerns group algebras of a locally compact group G. Let $M(G) = C_0(G)^*$ be the convolution algebra of Radon measures on G, considered as a Banach $L^1(G)$-bimodule.

PROBLEM 31. *Is it true that $\mathcal{H}^n(L^1(G), M(G)) = 0$ for all $n \geqslant 1$, or at least for $n = 1$?*

Since $M(G)$ is a dual bimodule, the equality holds for all amenable G (cf. 2.2). Recently, Johnson (to appear) has shown that an operator $f : L^1(G) \to M(G)$ belonging to $\ker \delta^1$ in the standard complex $\tilde{C}(L^1(G), M(G))$ (see Section 1) is of the form $\delta^0 \mu$ for $\mu \in M(G)$ whenever the image of f lies in $L^1(G) \subseteq M(G)$.

REFERENCES

1. Bade W. G., Curtis P. C., Jr., Dales H. G., *Amenability and weak amenability for Beurling and Lipschitz algebras*, Proc. London Math. Soc. (3) **55** (1987), 359–377.
2. Bonsall F. F., Duncan J., *Complete Normed Algebras*, Springer, New York, 1973.
3. Bunce J. W., *Finite operators and amenable C^*-algebras*, Proc. Amer. Math. Soc. **56** (1976), 145–151.
4. Bunce J. W., Paschke W. L., *Derivations on a C^*-algebra and its double dual*, J. Functional Analysis **37** (1980), 235–247.
5. Choi M. D., Effros E. G., *Nuclear C^*-algebras and injectivity: the general case*, Indiana Univ. Math. J. **26** (1977), 443–446.
6. Christensen E., *Extensions of derivations*, II, Math. Scand. **50** (1982), 111–122.
7. Christensen E., *Derivations and their relation to perturbations of operator algebras*, "Operator Algebras and Applications" (Kadison R. V., ed.), Proc. of Symp. In Pure Math., Amer. Math. Soc., Part 2, 1982, pp. 261–274.
8. Christensen E., Effros E. G., Sinclair A. M., *Completely bounded maps and C^*-algebraic cohomology*, Invent. Math. **90** (1987), 279–296.
9. Christensen E., Sinclair A. M., *On the vanishing of $H^n(A, A^*)$ for certain C^*-algebras*, Pacific J. Math. **137** (1989), 55–63.
10. Christensen E., Sinclair A. M., *On the Hochschild cohomology for von Neumann algebras*, Kobenhavns Univ. Mat. Inst., Preprint Ser. 32, 1988.
11. Connes A., *Classification of injective factors*, Ann. of Math. **104** (1976), 73–115.
12. Connes A., *On the cohomology of operator algebras*, J. Functional Analysis **28** (1978), 248–253.
13. Connes A., *Non-commutative differential geometry*, Parts I and II, I.H.E.S. **62** (1985), 157–360.
14. Curtis P. C., Jr., Loy R. J., *The structure of amenable Banach algebras*, J. London Math. Soc. (1989).
15. Dales H. G., *Automatic continuity: a survey*, Bull. London Math. Soc. **10** (1978), 129–183.
16. Dixon P. G., *Left approximate identities in algebras of compact operators on Banach spaces*, Proc. Royal Soc. Edinburgh, 104A (1986), 169–175.
17. Effros E. G., *Aspects of non-commutative order*, "C^*-Algebras and Applications to Physics", Lecture Notes in Math., Springer-Verlag, Berlin **650** (1978), 1–40.
18. Effros E. G., *Amenability and virtual diagonals for von Neumann algebras*, J. Functional Analysis **78** (1988), 137–153.
19. Effros E. G., *Advances in quantized functional analysis*, Proc. of ICM, Vol. II (1986), 906–916.
20. Effros E. G., Kishimoto A., *Module maps and Hochschild-Johnson cohomology*, Indiana Univ. Math. J. **36** (1987), 157–176.
21. Golovin Yu. O., *Homological properties of Hilbert modules over nest operator algebras and over their generalizations*, Mat. Zametki **41** (1987), 769–775. (Russian)
22. Golovin Yu. O., *Homological properties of some modules over reflexive operator algebras*, Proceedings of the 19th All-Union Algebraic Conference, L'vov, 9-11th September 1987, 69. (Russian)
23. Golovin Yu. O., Helemskii A. Ya., *The homological dimension of certain modules over a tensor product of Banach algebras*, Vest. Mosk. Univ. ser. mat. mekh. **1** (1977), 54–61 (Russian); English transl. in Moscow Univ. Math. Bull. **32** (1977), 46–52.
24. Groenbaek N., *A characterization of weakly amenable Banach algebras*, Studia Math. **94** (1989), 149–162.
25. Groenbaek N., *Amenability and weak amenability of tensor algebras and algebras of nuclear operators* (to appear).

26. Gumerov R. N., *Homological dimension of radical algebras of Beurling type with rapidly decreasing weight*, Vest. Mosk. Univ. ser. mat. mekh. **5** (1988), 18–22. (Russian)
27. Haagerup U., *The Grothendieck inequality for bilinear forms on C^*-algebras*, Adv. in Math. **56** (1985), 93–116.
28. Haagerup U., *All nuclear C^*-algebras are amenable*, Invent. Math. **74** (1983), 305–319.
29. Helemskii A. Ya., *The homological dimension of normed modules over Banach algebras*, Mat. Sbornik **81** (1970), 430–444 (Russian); English transl. in Math. USSR-Sb. **10** (1970), 399–411.
30. Helemskii A. Ya., *The homological dimension of Banach algebras of analytic functions*, Mat. Sbornik **83** (1970), 222–233 (Russian); English transl. in Math. USSR-Sb. **12** (1970), 221–233.
31. Helemskii A. Ya., *A description of relatively projective ideals in the algebras $C(\Omega)$*, Dokl. Akad. Nauk SSSR **195** (1970), 1286–1289 (Russian); English transl. in Soviet Math. Dokl **1** (1970), 1680–1683.
32. Helemskii A. Ya., *A certain class of flat Banach modules and its applications*, Vest. Mosk. Univ. ser. mat. mekh. **27** (1972), 29–36. (Russian)
33. Helemskii A. Ya., *A certain method of computing and estimating the global homological dimension of Banach algebras*, Mat. Sbornik **87** (1972), 122–135 (Russian); English transl. in Math. USSR-Sb. **16** (1972), 125–138.
34. Helemskii A. Ya., *The global dimension of a Banach function algebra is different from one*, Funktsional. Anal. i Prilozhen. **6** (1972), 95–96 (Russian); English transl. in Functional Anal. Appl. **6** (1972), 166–168.
35. Helemskii A. Ya., *A locally compact abelian group with trivial two-dimensional Banach cohomology is compact*, Dokl. Akad. Nauk SSSR **203** (1972), 1004–1007 (Russian); English transl. in Soviet Math. Dokl. **13** (1972), 500–504.
36. Helemskii A. Ya., *The lowest values taken by the global homological dimensions of functional Banach algebras*, Trudy sem. Petrovsk. **3** (1978), 223–242 (Russian); English transl. in Amer. Math. Soc. Trans. (1984).
37. Helemskii A. Ya., *Homological methods in the holomorphic calculus of several operators in Banach space after Taylor*, Uspekhi Matem. Nauk **36** (1981), 127–172 (Russian); English transl. in Russian Math. Surveys **36** (1981).
38. Helemskii A. Ya., *Flat Banach modules and amenable algebras*, Trudy MMO **47** (1984), 179–218. (Russian)
39. Helemskii A. Ya., *Homological characteristics of Banach algebras*, Doctoral dissertation, Moscow State Univ., Moscow, 1973. (Russian)
40. Helemskii A. Ya., *The Homology of Banach Algebras and Topological Algebras*, Moscow University Press, 1986 (Russian); English transl.: Kluwer Academic Publishers, Dordrecht, 1989.
41. Helemskii A. Ya., *Homology in Banach and polynormed algebras: some results and problems*, "Operator Theory: Advances and Applications", vol. 42, Birkhäuser Verlag, Basel, 1990.
42. Helemskii A. Ya., *Some remarks about ideas and results of topological homology*, Proc. Centre for Math. Anal., Australian National University **21** (1989), 203–238.
43. Helemskii A. Ya., *Banach and Polynormed Algebras: General Theory, Representations, Homology*, Nauka, Moscow, 1989 (Russian); English transl.: Oxford University Press, 1992 (to appear).
44. Helemskii A. Ya., *Homological algebra background of the "amenability after A. Connes": injectivity of the predual bimodule*, Mat. Sbornik (1989), 1680–1690. (Russian)
45. Helemskii A. Ya., *From the topological homology: algebras with various conditions of the homological triviality*, "Algebraic Problems of the Analysis and Topology", Voronezh University Press, 1990, pp. 77–96 (Russian); English transl. in Lecture Notes in Math., Springer Verlag (to appear).
46. Helemskii A. Ya., *Banach cyclic (co)homology and the Connes-Tzygan exact sequence*, J. London Math. Soc. (1991) (to appear).
47. Helemskii A. Ya., *Banach cyclic (co)homology as Banach derived functor*, St.-Petersburg Math. J. (to appear). (Russian)
48. Helemskii A. Ya., Sheinberg M. V., *Amenable Banach algebras*, Funktional. Anal. i Prilozhen. **13** (1979), 42–48 (Russian); English transl. in Functional Anal. Appl. **13** (1979), 32–37.
49. Hochschild G., *On the cohomology groups of an associative algebra*, Ann. of Math. **46** (1945), 58–76.
50. Johnson B. E., *Cohomology in Banach algebras*, Mem. Amer. Math. Soc. **127** (1972).

51. Johnson B. E., *A class of II_1-factors without property P but with zero second cohomology*, Arkiv mat. **12** (1974), 153–159.
52. Johnson B. E., *Perturbations of Banach algebras*, Proc. London Math. Soc. **34** (1977), 439–458.
53. Johnson B. E., *Introduction to cohomology in Banach algebras*, "Algebras in Analysis" (Williamson J. H., ed.), Acad. Press, London, 1975.
54. Johnson B. E., *Low dimensional cohomology of Banach algebras*, "Operator Algebras and Applications" (Kadison R. V., ed.), Proc. of Symp. in Pure Math., Providence, Amer. Math. Soc., Part 2, 1982, pp. 253–259.
55. Johnson B. E., Kadison R. V., Ringrose J. R., *Cohomology of operator algebras*, III, *Reduction to normal cohomology*, Bull. Soc. Math. France **100** (1972), 73–96.
56. Johnson B. E., Parrot S. K., *Operators commuting with a von Neumann algebra modulo the set of compact operators*, J. Functional Analysis **11** (1972), 39–61.
57. Kadison R. V., *Derivations of operator algebras*, Ann. of Math. **83** (1966), 280–293.
58. Kadison R. V., Ringrose J. R., *Cohomology of operator algebras* I, *Type* I *von Neumann algebras*, Acta Math. **126** (1971), 227–243.
59. Kadison R. V., Ringrose J. R., *Cohomology of operator algebras* II, *Extended cobounding and the hyperfinite case*, Ark. Mat. **9** (1971), 55–63.
60. Kadison R. V., Ringrose J. R., *Fundamentals of the Theory of Operator Algebras*, Vol. II, Acad. Press, London, 1986.
61. Kaliman S. I., Selivanov Yu. V., *On the cohomology of operator algebras*, Vest. Mosk. Univ. ser. mat. mekh. **29** (1974), no. 5, 24–27. (Russian)
62. Kamowitz H., *Cohomology groups of commutative Banach algebras*, Trans. Amer. Math. Soc. **102** (1962), 352–372.
63. Kaplansky I., *On the dimension of modules and algebras,* X, Nagoya Math. J. **13** (1958), 85–88.
64. Krichevets A. N., *On the connection of homological properties of some Banach modules and questions of the geometry of Banach spaces*, Vest. Mosk. Univ. ser. mat. mekh. **2** (1981), 55–58 (Russian); English transl. in Moscow Univ. Math. Bull. **36** (1981), 68–72.
65. Krichevets A. N., *Calculation of the global dimension of tensor products of Banach algebras and a generalization of a theorem of Phillips*, Matem. Zametki **31** (1982), 187–202 (Russian); English transl. in Math. Notes **31** (1982), 95–104.
66. Krichevets A. N., *On the homological dimension of $C(\Omega)$*, VINITI Preprint, 9012-V86, 1986. (Russian)
67. Kurmakaeva E. Sh., *On the strict homological bidimension of the algebras*, submitted to Mat. Zametki. (Russian)
68. Lambek I., *Rings and Modules*, Blaisdell, Waltham, Mass., 1966.
69. Lance E. C., *On nuclear C^*-algebras*, J. Functional Analysis **12** (1973), 157–176.
70. Lance E. C., *Cohomology and perturbations of nest algebras*, Proc. London Math. Soc. no. 3, **43** (1981), 334–356.
71. Lykova Z. A., *On conditions for projectivity of Banach algebras of completely continuous operators*, Vest. Mosk. Univ. ser. mat. mekh. **4** (1979), 8–13 (Russian); English transl. in Moscow Univ. Math. Bull. **334** (1979), 6–11.
72. Lykova Z. A., *On homological characteristics of operator algebras*, Vest. Mosk. Univ. ser. mat. mekh. **1** (1986), 8–13. (Russian)
73. Lykova Z. A., *A lower estimate of the global homological dimension of infinite-dimensional CCR-algebras*, Uspekhi Mat. Nauk **41** (1986), no. 1, 197–198 (Russian); English transl. in Russian Math. Surveys **41** (1986), no. 1, 233-234.
74. Moran W., *The global dimension of $C(X)$*, J. London Math. Soc. (2) **17** (1978), 321–329.
75. Nielsen J. P., *Cohomology of some non-selfadjoint operator algebras*, Math. Scand. **47** (1980), 150–156.
76. Ogneva O. S., *Coincidence of homological dimensions of Fréchet algebra of smooth functions on a manifold with the dimension of the manifold*, Funktsional. Anal. i Prilozhen. **20** (1986), 92–93. (Russian)
77. Popa S., *The commutant modulo the set of compact operators of a von Neumann algebra*, J. Functional Analysis **71** (1987), 393–408.

78. Pugach L. I., *Homological properties of functional algebras and analytic polydiscs in their maximal ideal spaces*, Rev. Roumaine Math. Pure and Appl. **31** (1986), 347–356. (Russian)
79. Pugach L. I., *Projective ideals of functional algebras and the Shilov boundary*, "Problems of the Theory of Groups and Homological Algebra", Yaroslavl, 1987. (Russian)
80. Racher G., *On amenable and compact groups*, Monatshefte für Math. **92** (1981), 305–311.
81. Ringrose J. R., *Cohomology of operator algebras*, "Lectures in Operator Algebras", Lecture Notes in Math. **247** (1972), 355–434, Springer-Verlag, Berlin.
82. Ringrose J. R., *Cohomology theory for operator algebras*, "Operator Algebras and Applications" (Kadison R. V., ed.), Proc. of Symp. in Pure Math., Providence, Amer. Math. Soc., Part 2, 1982, pp. 229–252.
83. Runde V., *A functorial approach to weak amenability for commutative Banach algebras*, Glasgow Math. J. **34** (1992).
84. Sakai S., *Derivations of W^*-algebras*, Ann. of Math. **83** (1966), 287–293.
85. Selivanov Yu. V., *The values assumed by the global dimension in certain classes of Banach algebras*, Vest. Mosk. Univ. ser. mat. mekh. **30** (1975), no. 1, 37–42 (Russian); English transl. in Moscow Univ. Math. Bull. **30** (1975), 30–34.
86. Selivanov Yu. V., *Banach algebras of small global dimension zero*, Uspekhi Matem. Nauk **31** (1976), 227–228. (Russian)
87. Selivanov Yu. V., *Biprojective Banach algebras*, Izv. Akad. Nauk SSSR ser. matem. **43** (1979), 1159–1174 (Russian); English transl. in Math. USSR-Izv. **13** (1979).
88. Selivanov Yu. V., *Homological characterization of the approximative property for Banach spaces*, J. Glasgow Math. Soc. (to appear).
89. Sheinberg M. V., *Homological properties of closed ideals that have a bounded approximate unit*, Vest. Mosk. Univ. ser. mat. mekh. **27** (1972), no. 4, 39–45 (Russian); English transl. in Moscow Univ. Math. Bull. **27** (1972), 103–108.
90. Sheinberg M. V., *On the relative homological dimension of group algebras of locally compact groups*, Izv. Akad. Nauk SSSR ser. mat. **37** (1973), 308–318 (Russian); English transl. in Math. USSR-Izv. **7** (1973), 307–317.
91. Sheinberg M. V., *A characterization of the algebra $C(\Omega)$ in terms of cohomology groups*, Uspekhi Matem. Nauk **32** (1977), 203–204. (Russian)
92. Taylor J. L., *Homology and cohomology for topological algebras*, Adv. Math. **9** (1972), 137–182.
93. Taylor J. L., *A general setting for a multi-operator functional calculus*, Advances in Math. **9** (1972), 183–252.
94. Tzygan B. L., *Homology of matrix Lie algebras over rings and Hochschild homology*, Uspekhi Mat. Nauk **38** (1983), 217–218. (Russian)
95. Wodzicki M., *The long exact sequence in cyclic homology associated with an extension of algebras*, C. R. Acad. Sci. Paris, Vol. 306 (1988), 399–403.
96. Wodzicki M., *Vanishing of cyclic homology of stable C-algebras*, C. R. Acad. Sci. Paris, Vol. 307 (1988), 329–334.
97. Wodzicki M., *Homological properties of rings of functional-analytic type*, Proc. National Academy Science, USA (to appear).
98. Wodzicki M., *Resolution of the cohomology comparison problem for amenable Banach algebras*, submitted to Invent. Math.
99. Liddel M. J., *Separable topological algebras* I, Trans. Amer. Math. Soc. **195** (1974), 31–59.
100. Ol'shanskii A. Yu., *On the question of the existence of an invariant mean on a group*, Uspekhi Matem. Nauk **35** (1980), 199–200. (Russian)

Department of Mechanics and Mathematics
Moscow State University
117234 Moscow
Russia

2.2

THE HOMOLOGY OF C^*-ALGEBRAS

Z. A. Lykova

General questions on the homology of algebras of analysis are given in the preceding paper of Helemskii. Here we concentrate on the case of C^*-algebras. We recall that some of the main concepts of homological theory for Banach algebras are given in [4], [5].

1. Let A be a Banach algebra, and let X be a left Banach A-module. We shall denote the global homological dimension of A by $dg\, A$, the homological bidimension of A by $db\, A$, and the homological dimension of X by $dh\, X$.

Consider the general question: *Let A be an infinite-dimensional C^*-algebra. Are $dg\, A$ and/or $db\, A$ always greater than one?*

It is known that, if A is commutative, or, if A is a CCR-algebra, then $db\, A \geqslant dg\, A > 1$ ([3], [6], [12]).

In particular we concentrate on the following question.

QUESTION 1. *Let A be an infinite-dimensional GCR-algebra. Are $dg\, A$ and/or $db\, A$ always greater than 1?*

One can see that this question can be reduced to the following

QUESTION 2. *Let A be a GCR-algebra, and let T be an infinite-dimensional, irreducible representation of A. Is $dg(A/\operatorname{Ker} T)$ always greater than 1?*

If this is true, then we can use the estimate $dg\, B > 1$ for infinite-dimensional CCR-algebras B and a lemma from [6], which allows us to estimate the global homological dimension of a given C^*-algebra by the same characteristic of its quotient algebras. These enable us to prove that $dg\, A > 1$ for an infinite-dimensional GCR-algebra A.

As to Question 2, we know that, for an infinite-dimensional irreducible representation T of a GCR-algebra A on a Hilbert space H the C^*-algebra $A/\operatorname{Ker} T$ contains the C^*-algebra of compact operators $K(H)$ and that $dg\, K(H) > 1$.

So, we come to the question:

QUESTION 3. *Let A be a C^*-algebra and let I be a closed two-sided ideal of A such that $dg\, I > 1$. Is $dg\, A$ greater than 1?*

For another important class of C^*-algebras, the von Neumann algebras, we ask also:

QUESTION 4. *Let A be an infinite-dimensional von Neumann algebra. Are $dg\, A$ and/or $db\, A$ always greater than 1?*

For commutative algebras of this class, the proof that $dg\, A > 1$ uses the existence of a complemented, non-projective, closed ideal I of A such that $dg\, A \geqslant dh\, A/I > 1$. In the non-commutative case, it is shown in [7, Theorem 2] that there exists a non-projective, closed left ideal I of A. It seems to be difficult, however, to show that this ideal I is complemented. So, it is not clear whether the estimate $dh\, A/I > 1$ is true.

On the other hand, let us recall [15, Theorem 1] that, for a commutative C^*-algebra A with an identity and a closed ideal I, the estimate $dh\, A/I \leqslant 1$ implies that I is projective. This method of proof has some complications connected with the construction of a skeleton of a non-commutative C^*-algebra. The attempt to introduce the family of functions which resemble the skeleton was done in [12, §3]. One naturally asks the following question.

QUESTION 5. *Let A be a C^*-algebra with an identity, and let I be a closed left ideal of A. Is it true that the estimate $dh\, A/I \leqslant 1$ always implies that I is projective?*

2. Following [4, IV.2.9], a Banach algebra A is said to be hereditary (on the left) if every closed left ideal of A is projective. In the case where A is a commutative C^*-algebra, A is hereditary if and only if the spectrum of A is hereditarily paracompact [4, IV.3.8]. It is known also ([9, Theorem 3]) that infinite-dimensional AW^*-algebras and, in particular, von Neumann algebras are not hereditary. Separable C^*-algebras are hereditary [7, Theorem 1]. It is easy to see that there exist non-separable, hereditary C^*-algebras.

QUESTION 6. *Let A be a C^*-algebra. Is it true that A is separable if and only if the C^*-tensor product $A \otimes_{\min} A \otimes_{\min} A$ is hereditary?*

We recall that this result is true for a commutative C^*-algebra A [15, Theorem 4] or [9, Theorem 5].

QUESTION 7. *Let A be a C^*-subalgebra of the C^*-algebra $B(H)$ of all continuous operators on a separable Hilbert space H. Is it true that A is separable if and only if the C^*-tensor product $A \otimes_{\min} A$ is hereditary?*

It is known ([9, Theorem 4]) that this result is true when A is commutative. As to the non-commutative case, it is shown in [9, Theorem 5] that, for a separable C^*-algebra A, the C^*-tensor products $A \otimes_{\min} A \otimes_{\min} A$ and $A \otimes_{\min} A$ are hereditary. On the other hand, one can see that, if either $A \otimes_{\min} A \otimes_{\min} A$ or $A \otimes_{\min} A$ (where A is from Question 7) is hereditary, then every commutative C^*-subalgebra of A is separable. However, this does not imply directly that A is separable, since there are examples of non-separable C^*-algebras with only separable, commutative C^*-subalgebras [1].

QUESTION 8. *Do there exist non-hereditary C^*-algebras all of whose commutative C^*-subalgebras are hereditary?*

We know that every commutative C^*-subalgebra of a hereditary C^*-algebra is also hereditary [9, Theorem 2].

3. Let us now consider questions on the relative homological theory over a commutative Banach algebra B with an identity (see [8], [13]). Let A be a Banach B-algebra. We shall denote the global B-homological dimension of A by $dg_B\, A$, the B-homological bidimension of A by $db_B\, A$, and the centre of A_+ by $Z(A_+)$, where A_+ is the unitization of A. In the case where $B = Z(A_+)$, we add an adjective "central" to the corresponding definitions.

QUESTION 9. *What is the structure of a central biprojective C^*-algebra?*

In [8] we obtained a description of the structure of a central biprojective C^*-algebra with an identity: it was proved that such a C^*-algebra is a finite direct sum of homogeneous C^*-algebras of finite rank with an identity. In the special case of separable C^*-algebras, this statement was proved in [14] in equivalent terms with the help of a different approach. As to C^*-algebras A, not necessarily possessing an identity, it was proved in [2] that a central biprojective C^*-algebra such that $\overline{Z(A)A} = A$ is a c_0-sum of some family of homogeneous C^*-algebras of finite rank.

QUESTION 10. *Let A be a central biprojective C^*-algebra. Are $dg_{Z(A_+)} A$ and/or $db_{Z(A_+)} A$ equal to one?*

We recall that an infinite-dimensional biprojective C^*-algebra is central biprojective, and its central homological dimensions $dg_{Z(A_+)} A$ and $db_{Z(A_+)} A$ are equal to one [10, Theorem 2.1]. This situation is contrary to the state of affairs in the "traditional" theory ($B = \mathbb{C}$) of the Banach cohomology. The bidimension and global homological dimension of any infinite-dimensional biprojective C^*-algebra are equal to two [4, Ch. 5]. We know also that $dg_{Z(A_+)} A \leqslant db_{Z(A_+)} A \leqslant 2$ for a central biprojective C^*-algebra A ([8, Theorem 2.3]).

QUESTION 11. *Does there exist a C^*-algebra A with $dg_{Z(A_+)} A = n$ and/or $db_{Z(A_+)} A = n$ for $n \geqslant 2$ (with $dg_B A = n$ and/or $db_B A = n$ for some closed subalgebra B of $Z(A_+)$ and for $n \geqslant 2$)?*

It is shown in [10, Theorem 3.1] that for any CCR-algebra A having at least one infinite-dimensional irreducible representation, we have $db_{Z(A_+)} A \geqslant db_{Z(A_+)} A \geqslant 2$. Also, it is proved in [11, Theorem 3.1] that there exists a unital, semisimple Banach algebra A and a closed subalgebra B of $Z(A)$ such that $dg_B A = db_B A = n$ for $n \geqslant 1$.

4. Let A_1 and A_2 be C^*-algebras, and let $A_1 \hat{\otimes} A_2$ be a projective tensor product of A_1 and A_2 ([15, 3.4.13]).

QUESTION 12. *Let A_1 be a nuclear C^*-algebra, and let A_2 be a C^*-algebra. Is $A_1 \hat{\otimes} A_2$ always semisimple?*

We recall [16] that, if A_1 and A_2 are commutative, semisimple Banach algebras such that at least one of them has, as a Banach space, the approximation property, then $A_1 \hat{\otimes} A_2$ is semisimple.

REFERENCES

1. Akeman Ch. D., Doner J. E., *A non-separable C^*-algebra with only separable abelian C^*-subalgebras*, Bull. London Math. Soc. **11** (1979), 279–284.
2. Bugaev A. I., *On the structure of central biprojective C^*-algebras*, submitted to Vest. Mosk. Univ. ser. mat. mekh.. (Russian)
3. Helemskii A. Ya., *The homological dimension of Banach algebras of analytic functions*, Mat. Sb. **83** (1970), 222–233; English translation in Math. USSR-Sb. **12** (1970), 221–233.
4. Helemskii A. Ya., *The Homology of Banach Algebras and Topological Algebras*, Moscow University Press, 1986 (Russian); English translation: Kluwer Academic Publishers, Dordrecht, 1980.
5. Helemskii A. Ya., *Banach and Polynormed Algebras: General Theory, Representations, Homology*, Nauka, Moscow, 1989 (Russian); English translation: Oxford University Press, 1992 (to appear).

6. Lykova Z. A., *A lower estimate of the global homological dimension of infinite-dimensional CCR-algebras*, Uspekhi Mat. Nauk **41** (1986), no. 1, 197–198 (Russian); English transl. in Russian Math. Surveys **41** (1986), no. 1, 233-234.
7. Lykova Z. A., *On homological characteristics of operator algebras*, Vest. Mosk. Univ. ser. mat. mekh. **1** (1986), 8–13. (Russian)
8. Lykova Z. A., *Structure of Banach algebras with trivial central cohomology*, J. Operator Theory (to appear).
9. Lykova Z. A., *Connection between homological characteristics of a C^*-algebra and its commutative sub-C^*-algebras*, VINITI preprint, 1985, 19 p. (Russian)
10. Lykova Z. A., *On non-zero values of central homological dimension of C^*-algebras*, Glasgow Math. J. (to appear).
11. Lykova Z. A., *Examples of Banach algebras over a commutative Banach algebra B with given B-homological dimension*, submitted to The Proceedings of the II International Conference on Algebra, Barnaul, USSR (1991).
12. Lykova Z. A., *A lower estimate of the global homological dimension of infinite-dimensional CCR-algebras*, The Proceedings of the OATE 2 Conference, Craiova, Romania (1989) (to appear).
13. Phillips J., Raeburn I., *Central cohomology of C^*-algebras*, J. London Math. Soc. (2) **28** (1983), 363–372.
14. Phillips J., Raeburn I., *Voiculescu's double commutant theorem and the cohomology of C^*-algebras*, preprint, University of Victoria, Canada, 1989.
15. Selivanov Yu. V., *The homological dimension of cyclic Banach modules and a homological characterization of metrizable compacta*, Matem. zametki **17** (1975), 301–305 (Russian); English transl. in Math. Notes **17** (1975), 174–176.
16. Tomiyama J., *Tensor products of commutative Banach algebras*, Tohôku Math. J. **12** (1960), 143–154.

Chair of Algebra and Analysis
Moscow Institute of Electronic Machine Industry
B.Vuzovskii 3/12
Moscow 109028
Russia

2.3

AMENABLE AND WEAKLY AMENABLE COMMUTATIVE BANACH ALGEBRAS

P. C. CURTIS, JR.

Following B. E. Johnson [6], a Banach algebra $\mathfrak{A}$ is said to be *amenable* if all bounded derivations from $\mathfrak{A}$ to each Banach dual $\mathfrak{A}$-bimodule $\mathfrak{X}$ are inner. In the case where $\mathfrak{A}$ is $L^1(G)$, where G is a locally compact group, then $\mathfrak{A}$ is an amenable Banach algebra if and only if G is an amenable group. Amenable Banach algebras must have bounded aproximate identities; all quotient algebras of amenable algebras are amenable; and a closed twosided ideal $\mathfrak{J}$ of an amenable algebra is itself amenable if and only if $\mathfrak{J}$ possesses a bounded approximate identity. Amenable algebras need not be semisimple, and, in the commutative case, the quotient algebras of $L^1(G)$ provide a large class of examples of non-semisimple amenable algebras. Let G be a locally compact abelian group, and identify $L^1(G)$ with the Fourier algebra $A(\Gamma)$ on the dual group Γ. Let E be a compact subset of Γ. Then E *fails to satisfy spectral synthesis* if the ideal

$$I(E) = \{ f \in A(\Gamma) : f(E) = 0 \}$$

properly contains $\overline{J(E)}$, where $J(E) = \{ f \in A(\Gamma) : f = 0$ in the neighbourhood of $E \}$ and the closure is taken in the norm topology of $A(\Gamma)$. Set $A(E) = A(\Gamma)/I(E)$, $A_E = A(\Gamma)/\overline{J(E)}$ and $R_E = I(E)/\overline{J(E)}$, so that R_E is exactly the radical of $A(E)$. The following is the fundamental question.

QUESTION. *Are there any non-zero, radical, amenable, commutative Banach algebras?*

In particular we ask:

QUESTION. *Is an algebra R_E of the above form ever amenable?*

A closed ideal in $\mathfrak{J}$ in an amenable algebra is amenable (i.e. has a bounded approximate identity) if and only if $\mathfrak{J}^\perp$ has a closed, complementary subspace in the dual space $\mathfrak{A}^*$ of the algebra $\mathfrak{A}$ ([5]). Thus a related question is whether the radical of a commutative, amenable algebra $\mathfrak{A}$ can have a closed complementary subspace in $\mathfrak{A}$.

A Banach algebra $\mathfrak{A}$ is *weakly amenable* if all bounded derivations from the algebra to its dual module $\mathfrak{A}^*$ are inner. In the commutative case, an equivalent condition is that all derivations from $\mathfrak{A}$ to all commutative Banach $\mathfrak{A}$-modules are zero ([2]). For commutative $\mathfrak{A}$, weak amenability implies that $\overline{\mathfrak{A}^2} = \mathfrak{A}$; all quotient algebras of weakly amenable algebras are weakly amenable; and closed ideals $\mathfrak{J}$ are weakly amenable if and only if $\overline{\mathfrak{J}^2} = \mathfrak{J}$. If $E \subset \Gamma$ fails to satisfy spectral synthesis and if, further, $A(E) = C(E)$ (so that E is a *Helson* set), then it has recently been shown that $\overline{R_E^2} = R_E$, provided that E is metrizable ([3]).

This latter result uses techniques introduced by Bachelis and Saeki, who proved in [1] that quotient algebras $A(E)$ have the property that, whenever two continuous homomorphisms φ and ψ of $A(E)$ into a commutative Banach algebra $\mathfrak{B}$ are congruent modulo the radical of $\mathfrak{B}$ (i.e., whenever $(\varphi - \psi)(\mathfrak{A}) \subset \operatorname{rad}\mathfrak{B}$), then $\varphi = \psi$.

QUESTION. *Is the above homomorphism property satisfied by all non-radical, commutative, amenable Banach algebras? Which weakly amenable (perhaps all?) commutative Banach algebras which are not entirely radical, have this property?*

If Γ is compact, this property holds for $A(\Gamma)$. It was shown in [4] that $L^1(\mathbb{R})$ has this homomorphism property, and it has been recently shown ([7]) using analytic semigroup techniques that $L^1(\mathbb{R}^n)$ and $L^1(\mathbb{R}^n, \omega)$ for suitable weight function ω also have this homomorphism property.

REFERENCES

1. Bachelis G. F., Saeki S., *Banach algebras with uncomplemented radical*, Proc. Amer. Math. Soc. **100** (1987), 271–274.
2. Bade W. G., Curtis P. C., Jr., Dales H. G., *Amenability and weak amenability in Beurling and Lipschitz algebras*, Proc. London Math. Soc. **55** (1987), 359–377.
3. Bade W. G., Curtis P. C., Jr., *Radical algebras associated with sets of non spectral synthesis*, Preprint.
4. Curtis P. C., Jr., *Complementation problems concerning the radical of a commutative amenable Banach algebra*, Proc. Centre for Math. Anal., Australian National University **21** (1989), 56–60.
5. Hemelskii A. Ya., *Flat Banach modules and amenable algebras*, Trans. Moscow Math. Soc.; Amer. Math. Soc. Translations (1984/5), 199-224.
6. Johnson B. E., *Cohomology in Banach algebras*, Memoir Amer. Math. Soc. **127** (1972).
7. White M. C., *Strong Wedderburn decomposition of Banach algebras containing analytic semigroups*, Preprint.

DEPARTMENT OF MATHEMATICS
UNIVERSITY OF CALIFORNIA
LOS ANGELES
CA 90024
USA

2.4

IDEALS IN BANACH ALGEBRAS

MARIUSZ WODZICKI

Let I be a closed left ideal in a unital Banach algebra B. Let us consider the following properties of I:

(a) I has a bounded right approximate unit;

(b) I satisfies the *right triple factorization property* TF_{right} of [4, §3] (cf., also [7] and [9]): for any finite collection $a_1, \ldots, a_m \in I$, there exist such $b_1, \ldots, b_m, c, d \in I$ that

$$a_i = b_i cd \quad (1 \leqslant i \leqslant m)$$

and the left annihilators in I of the elements c and d are equal (the left annihilator $\ell(a)$ of an element $a \in I$ is defined as $\ell(a) = \{x \in I \mid ax = 0\}$);

(c) I has the *factorization property*, i.e., every $x \in I$ factorizes as $x = yz$ for some $y, z \in I$;

(d) I is *left universally flat*, i.e., for every unital ring containing I as a left ideal, I is a flat left R-module [7];

(e) I is an *H-unital $\mathbb{C}$-algebra*, i.e., the Bar complex $(B_*(I), b')$,

$$B_q(I) = I \otimes_{\mathbb{C}} \cdots \otimes_{\mathbb{C}} I \quad (q \text{ times}; q \geqslant 1)$$

$$b'(a_1 \otimes \cdots \otimes a_q) = \sum_{i=1}^{q-1} (-1)^{i-1} a_1 \otimes \cdots \otimes a_i a_{i+1} \otimes \cdots \otimes a_q,$$

is acyclic [5], [6, p. 600];

(f) $I = I^2 = \left\{ \sum_{j=1}^{m} x_j y_j : x_j, y_j \in I, \;\; m \in \mathbb{Z}_+ \right\}$;

(g) I is a *Banach H-unital algebra* (called *strongly H-unital* in [5, Remark (3)], i.e., the completed Bar complex $(B_*^{\text{cont}}(I), b')$, where $B_q^{\text{cont}}(I) = I \widehat{\otimes}_\pi \cdots \widehat{\otimes}_\pi I$ (q times; $q \geqslant 1$), is acyclic;

(h) I is a *pure* Banach subspace of B, i.e., the conjugate space $I^\perp = (B/I)^*$ is complemented in B^*;

(i) I is a *flat left B-module* (cf., [1, V.8, p.163]).

One has similar properties for right ideals of B.

The above nine properties are related to each other via the following diagram of implications

$$\begin{array}{ccccccccc}
(g) & \Leftarrow & (a) & \Rightarrow & (b) & \Rightarrow & (c) & \Rightarrow & (f) \\
 & & \Downarrow & & \Downarrow & & & & \\
 & & (h) & & (d) & \Rightarrow & (e) & \Rightarrow & (f) \\
 & & & & \Downarrow & & & & \\
 & & & & (i) & & & &
\end{array}$$

PROBLEM. *For a given unital Banach algebra B classify its closed one-sided ideals according to any one of the properties* (a)–(i) *above.*

Comments.

(1) Every closed ideal in the disc algebra $\mathcal{A}(\mathbb{D})$ is flat (property (i)) while properties (a)–(g) are equivalent for closed ideals of $\mathcal{A}(\mathbb{D})$.
(2) Every closed left ideal in an arbitrary C^*-algebra possesses all nine properties.
(3) Every two-sided, though not necessarily closed, ideal in an arbitrary von Neumann algebra is flat both as a left and as a right module. In addition, properties (b)–(f) (and their counterparts for right ideals) are equivalent for two-sided (and not necessarily closed) ideals in any von Neumann algebra ([8]).
(4) Properties (a) and (h) are equivalent for closed left ideals in any amenable Banach algebra ([2], [3, Theorem 3]).
(5) Little seems to be known in the very interesting case of ideals in the group algebra $L^1(G)$, where G is a locally compact group (in this case $B = \mathbb{C} \times L^1(G)$ is the Banach algebra obtained by adjoining the unit to $L^1(G)$).

REFERENCES

1. MacLane S., *Homology*, Grundlehren der math. Wiss., vol. 114, 3rd corr. printing, Springer-Verlag, 1975.
2. Helemskii A. Ya., Scheinberg M. V., *Amenable Banach algebras*, Functional Ann. Appl. **13** (1979), 32–37.
3. Scheinberg M. V., *Homological properties of closed ideals with a bounded approximate unit*, Vestnik MGU **27** (1972), 39–45 (Russian).
4. Suslin A. A., Wodzicki M., *Excision in algebraic K-theory*, Ann. Math. **135** (to appear).
5. Wodzicki M., *The long exact sequence in cyclic homology associated with an extension of algebras*, C.R. Acad. Sci., Paris 306 (1988), 399–403.
6. Wodzicki M., *Excision in cyclic homology and in rational algebraic K-theory*, Ann. Math. **129** (1989), 591–639.
7. Wodzicki M., *Homological properties of rings of functional-analytic type*, Proc. Nat. Acad. Sci., USA **87** (1991), 4910–4911.
8. Wodzicki M., unpublished notes, Berkeley 1990.
9. Wodzicki M., *Resolution of the cohomology comparison problem for amenable Banach algebras*, Invent. Math. **106** (1991).

DEPARTMENT OF MATHEMATICS
UNIVERSITY OF CALIFORNIA
BERKELEY
CA 94720
USA

FACTORIZATION IN BANACH ALGEBRAS

G. A. WILLIS

Throughout, A will denote a Banach algebra, X a left Banach A-module, and $A \cdot X = \{a \cdot x : a \in A, x \in X\}$. In the case where $X = A$, $A \cdot A$ will be denoted by A^2. Another A-module which will arise is

$$c_0(A) \equiv \{ (a_n)_{n=1}^{\infty} : a_n \in A \text{ and } \lim_{n\to\infty} \|a_n\| = 0 \}.$$

A has a *left bounded approximate identity* (abbreviated, l.b.a.i.) if there is a bounded net $(e_\lambda)_{\lambda\in\Lambda}$ contained in A such that, for every a in A, $\lim_{\lambda\to\infty} \|a - e_\lambda a\| = 0$. It is clear that, if A has a l.b.a.i., then $\overline{A^2} = A$ and $\overline{A \cdot c_0(A)} = c_0(A)$, but the factorization theorem of Cohen, as extended by Hewitt (see [3] and [9]), tells us much more.

THEOREM (Cohen, Hewitt). *Let A be a Banach algebra with a l.b.a.i. Then, for any left Banach A-module X, $A \cdot X$ is equal to its closed linear span.*

This theorem has found many applications in Banach algebra theory, where it is frequently necessary to know that $A = A^2$ or that $c_0(A) = A \cdot c_0(A)$. Here is a list of factorization conditions, many of which have been required in particular instances.

(1) A has a l.b.a.i.
(2) $A = \overline{\text{lin}\, A^2}$ and $\overline{\text{lin}\, A \cdot X} = A \cdot X$ for every left Banach A-module X.
(3) $c_0(A) = A \cdot c_0(A)$.
(4) There is an $M \geqslant 1$ such that, for every positive integer n, whenever $a_1, a_2, \ldots, a_n$ are in A there are $b, c_1, c_2, \ldots, c_n$ in A with $a_i = bc_i$ and $\|b\|\,\|c_i\| \leqslant M\,\|a_i\|$ for each i.
(5) The same statement as (4), but for $n = 1, 2$ only, i.e., pairs in A have a common factor.
(6) The same statement as (4), but for $n = 1$ only, i.e., every element of A is a product in A.
($5 + n$; $n \geqslant 2$) There is an $M \geqslant 1$ such that for every a in A there are $b_1, \ldots, b_n$ and $c_1, \ldots, c_n$ in A with

$$a = \sum_{i=1}^{n} b_i c_i \quad \text{and} \quad \sum_{i=1}^{n} \|b_i\|\,\|c_i\| \leqslant M\,\|a\|.$$

Each of these conditions implies the succeeding ones. Examples show that many of these implications cannot be reversed. The examples generally are either not separable or not commutative, and results in [2], [4] and [11] show that separability and commutativity can be very important for factorization questions. However, it is shown in [14] that some of the converse implications fail even for commutative, separable algebras. That paper answers some of the questions posed in the problem section of [1], but several interesting questions remain.

QUESTION 1. *Does* (2) *imply* (1)*?*

This question has not attracted much attention hitherto because Cohen's factorization theorem is usually applied to A or $c_0(A)$. Most work has been on whether factorization in either of these modules implies that A has a l.b.a.i. It is now known that (3) does not imply (1) (or (2)) even in the most favourable circumstances when A is commutative and separable. (See [10], [13] and [14].) Hence, that (2) implies (1) under reasonably general conditions is probably the best possible converse to Cohen's factorization theorem we can still hope for.

QUESTION 2. *Does* (6) *imply* (3)*?*

The most used factorization condition is (3) and so it could be very useful to know whether it is implied by one of the apparently weaker conditions. The only knowledge we have, (see [12]), is that there is a commutative, non-separable algebra which satisfies (6) but not (5).

A special case is whether the algebra $\ell^1(\mathbb{Q}^+)$ with convolution product satisfies (6). This algebra is discussed in [8]. It is a commutative, separable algebra and it may be seen, by appealing to Titchmarsh's convolution theorem, that the functions

$$\delta_1 \quad \text{and} \quad \sum_{n=1}^{\infty} \left(\frac{1}{2}\right)^n \delta_{\frac{1}{n}}$$

do not have a common factor. Intuition suggests that any function f in $\ell^1(\mathbb{Q}^+)$ whose support is a sequence converging to zero should be difficult to factor because the supports of products are generally 'thicker' than this. However it can be shown that there is a function which is a product in $\ell^1(\mathbb{Q}^+)$ and whose support is a sequence converging to zero. This suggests that the question as to whether $\ell^1(\mathbb{Q}^+)$ satisfies (6) may be very delicate.

Note that (5), that is, factorization of pairs, implies that n-tuples have a common factor, but does not seem to imply the existence of a constant M controlling the size of the factors. Also, there seem to be no Banach algebras known which satisfy (4) or (5) but not (3). These considerations offer some hope that (4) or (5) implies (3) but I think that is unlikely. Some examples, even non-separable or non-commutative ones, would be welcome.

QUESTION 3. *Does* $(5+n+1)$ *imply* $(5+n)$*?*

A commutative, separable Banach algebra A is constructed in [14] which satisfies (7) but not (6), that is, every element of A is the sum of two products, but A contains an element which is not a product. The obstruction to being able to factor elements is topological and relies on the non-vanishing of $H^1(\Phi_A, \mathbb{Z})$, where Φ_A denotes the maximal ideal space of A. It is possible that this is not just a feature of the example, but is an indication of a deeper connection between factorization and topology. One way to understand this better would be to decide whether $(5+n+1)$ implies $(5+n)$ for $n \geqslant 2$. Perhaps the higher cohomology groups have a role to play in this problem.

There are some other indications of a connection between factorization and topology: an example of Peter Dixon in [5] uses topological techniques. Furthermore, it is shown in [14] that (4) implies (1) in the class of commutative, completely regular Banach algebras A such that Φ_A is discrete. It would be very interesting to know whether [5] or even, and especially, [7] imply [1] under these conditions.

References

1. Bachar J. M. et al. (eds.), *Radical Banach Algebras and Automatic Continuity*, Lecture Notes in Mathematics, vol. 975, Springer-Verlag, Berlin, Heidelberg, New York, 1983.
2. Christensen J. P. R., *Codimension of some subspaces in a Fréchet algebra*, Proc. Amer. Math. Soc. **57** (1976), 276–278.
3. Cohen P. J., *Factorization in group algebras*, Duke Math. J. **26** (1959), 199–205.
4. Dixon P. G., *Non-separable Banach algebras whose squares are pathological*, J. Functional Analysis **26** (1977), 190–200.
5. Dixon P. G., *An example for factorization theory in Banach algebras*, Proc. Amer. Math. Soc. **86** (1982), 65–66.
6. Dixon P. G., *Factorization and unbounded approximate identities in Banach algebras*, Math. Proc. Camb. Philos. Soc. **107** (1990), 557–571.
7. Doran R. S. and Wichmann J., *Approximate Identities and Factorization in Banach Modules*, Lecture Notes in Mathematics, vol. 768, Springer-Verlag, Berlin, Heidelberg, New York, 1979.
8. Gronbaek N., *Weighted discrete convolution algebras*, "Radical Banach Algebras and Automatic Continuity", [1].
9. Hewitt E., *The ranges of certain convolution operators*, Math. Scand. **15** (1964), 147–155.
10. Leinert M., *A commutative Banach algebra which factorizes but has no approximate units*, Proc. Amer. Math. Soc. **55** (1976), 345–346.
11. Loy R. J., *Multilinear mappings and Banach algebras*, J. London Math. Soc. **14** (1976), no. 2, 423–429.
12. Ouzomgi S. I., *A commutative Banach algebra with factorization of elements but not of pairs*, Proc. Amer. Math. Soc. (to appear).
13. Paschke W. L., *A factorable Banach algebra without bounded approximate unit*, Pacific J. Math. **46** (1973), 249–252.
14. Willis G. A., *Examples of factorization without bounded approximate units*, Proc. Lond. Math. Soc. (to appear).

School of Mathematics
University of Leeds
Leeds LS2 9JT
England

and

Australian National University
GPO box 4
Canberra, ACT 2601
Australia

2.6
old

HOMOMORPHISMS FROM C^*-ALGEBRAS

H. G. DALES

Let A and B be Banach algebras. A basic automatic continuity problem is to give algebraic conditions on A and B which ensure that each homomorphism from A into B is necessarily continuous.

An important tool in investigations of this problem is the separating space: if $\theta: A \to B$ is a homomorphism, then the separating space of θ is

$$\mathfrak{S}(\theta) = \{ b \in B : \text{there is a sequence } (a_n) \subset A \text{ with } a_n \to 0 \text{ and } \theta a_n \to b \}.$$

Of course, θ is continuous if and only if $\mathfrak{S}(\theta) = \{0\}$. The basic properties of $\mathfrak{S}(\theta)$ are described in [6].

Consider the general question: *If $b \in \mathfrak{S}(\theta)$, what can one say about $\sigma(b)$, the spectrum of b in the Banach algebra B?*

First let us note that, if $b \in \mathfrak{S}(\theta)$, then $b \in \operatorname{Ker}\varphi$ for each character φ on B. For such a character φ is necessarily continuous, the character $\varphi \circ \theta$ is continuous on A, and so $\varphi(b) = \lim \varphi(\theta a_n) = \lim(\varphi \circ \theta)(a_n) = 0$. Thus, if B is commutative, it follows that $\sigma(b) = \{0\}$.

Is the same result true in the non-commutative case? An element b of a Banach algebra is a quasi-nilpotent if $\sigma(b) = \{0\}$, and so our question is the following.

QUESTION 1. *Let $\theta: A \to B$ be a homomorphism, and let $b \in \mathfrak{S}(\theta)$. Is b necessarily a quasi-nilpotent element of B?*

It can be shown that $\sigma(b)$ is always a connected subset of $\mathbb{C}$ containing the origin (see [6, 6.16]), but nothing further seems to be known in general.

The question was raised as Question 5′ in [3], and it is shown there that the question is equivalent to the following. Let $\theta: A \to B$ be a homomorphism, and suppose that $\overline{\theta(A)}$ is semisimple. *Is θ necessarily continuous?*

It is shown by Aupetit in [2] that, if A and B are unital Banach algebras, if $\theta: A \to B$ is a homomorphism, and if $b \in \mathfrak{S}(\theta)$ then $\nu(\theta a) \leqslant \nu(b + \theta a)$ for all $a \in A$. (Here, ν denotes the spectral radius.) Thus, if $b \in \mathfrak{S}(\theta) \cap \theta(A)$, then $\nu(b) = 0$ and b is a quasi-nilpotent. However, it is not in general true that the set of quasi-nilpotents in a Banach algebra is closed (see [1]), and so we cannot immediately conclude from this result that each $b \in \mathfrak{S}(\theta)$ is quasi-nilpotent.

Quite probably, there is a counter-example to Question 1. However, let us concentrate on the case in which both A and B are C^*-algebras.

QUESTION 2. *If, in Question 1, both A and B are C^*-algebras, can we then conclude that b is necessarily quasi-nilpotent?*

The continuity ideal of a homomorphism $\theta\colon A \to B$ is the set

$$\mathfrak{I}(\theta) = \{\, a \in A : \theta(a)\mathfrak{S}(\theta) = \mathfrak{S}(\theta)\theta(a) = \{0\}\,\}.$$

It was proved by Johnson ([4], see [6, 12.1]) that, if A is a C^*-algebra, then $\mathfrak{I}(\theta)$ is a two-sided ideal in A and that its closure $\overline{\mathfrak{I}(\theta)}$ has finite codimension in A. Next, Sinclair ([5, Theorem 4.1]) showed that if A and B are both C^*-algebras, and if $\theta\colon A \to B$ is a homomorphism with $\overline{\theta(A)} = B$, then $\theta|\overline{\mathfrak{I}(\theta)}$ can be decomposed as $\mu + \lambda$, where μ is a continuous homomorphism and $\lambda\colon \overline{\mathfrak{I}(\theta)} \to \mathfrak{S}(\theta)$ is a discontinuous homomorphism (or $\lambda = 0$). Now $\overline{\mathfrak{I}(\theta)}$ and $\mathfrak{S}(\theta)$ are closed ideals in A and B, respectively, and so both are C^*-algebras. Moreover, the range $\lambda(\overline{\mathfrak{I}(\theta)})$ is a dense subalgebra of $\mathfrak{S}(\theta)$ and so, by our above remarks, consists of quasi-nilpotent elements. Thus $\mathfrak{S}(\theta)$ is a C^*-algebra with a dense subalgebra consisting of quasi-nilpotents. No such C^*-algebra is known, and I would like it to be true that no such C^*-algebra exists. So we come to the sharpest form of our original question.

QUESTION 3. *Is there a C^*-algebra, other than $\{0\}$, which has a dense subalgebra consisting of quasi-nilpotent elements?*

If no such C^*-algebra exists, then the homomorphism λ in Sinclair's theorem must be zero, and so the element b in Question 2 must indeed be quasi-nilpotent.

REFERENCES

1. Aupetit B., *Propriétés spectrales des algèbres de Banach*, Lecture Notes in Math., vol. 735, 1979.
2. Aupetit B., *The uniqueness of the complete norm topology in Banach algebras and Banach–Jordan algebras*, J. Functional Analysis **47** (1982), 1–6.
3. Dales H. G., *Automatic continuity: a survey*, Bull. London Math. Soc. **10** (1978), 129–183.
4. Johnson B. E., *Continuity of homomorphisms of algebras of operators II*, J. London Math. Soc. **1** (1969), no. 2, 81–84.
5. Sinclair A. M., *Homomorphisms from C^*-algebras*, Proc. London Math. Soc. **29** (1974), no. 3, 435–452; Corrigendum **32** (1976), 322.
6. Sinclair A. M., *Automatic continuity of linear operators*, London Math. Soc. Lecture Note Series, vol. 21, Cambridge University Press, Cambridge, 1976.

SCHOOL OF MATHEMATICS
UNIVERSITY OF LEEDS
LEEDS LS2 9JT
ENGLAND

2.7

DISCONTINUOUS HOMOMORPHISMS FROM ALGEBRAS OF CONTINUOUS FUNCTIONS

H. G. DALES

1. Let Ω be a (non-empty, Hausdorff) compact space, and let $C(\Omega)$ be the algebra of all continuous, complex-valued functions on Ω. Then $C(\Omega)$ is a Banach algebra with respect to the uniform norm $|\cdot|_\Omega$, where

$$|f|_\Omega = \sup\{\,|f(x)| : x \in \Omega\,\}.$$

The first question about the continuity of homomorphisms from $C(\Omega)$ (throughout, we are considering homomorphisms into some other Banach algebra) was asked by Kaplansky in 1949 ([8]). Let $\|\cdot\|$ be any algebra norm on $C(\Omega)$. Then Kaplansky proved that $\|f\| \geqslant |f|_\Omega$ $(f \in C(\Omega))$, and he asked if, necessarily, $\|\cdot\|$ is equivalent to$|\cdot|_\Omega$. The question is equivalent to asking whether all homomorphisms from $C(\Omega)$ are automatically continuous. Kaplansky's question has been resolved, but a number of intriguing questions in this area remain open, and I would like to indicate some of these questions here.

Let me recapitulate some steps in the solution of Kaplansky's problem. First, we require some notation. Take $x \in \Omega$. Then

$$M_x = \{\, f \in C(\Omega) : f(x) = 0\,\},$$
$$J_x = \{\, f \in C(\Omega) : f|U = 0 \quad \text{for some neighbourhood } U \text{ of } x\,\}.$$

A *radical homomorphism* is a non-zero homomorphism θ from some M_x into a Banach algebra such that $\theta|J_x = 0$.

Bade and Curtis made the first successful attack on Kaplansky's question in 1960 ([1]). They proved that an arbitrary homomorphism from $C(\Omega)$ is necessarily continuous on a dense subalgebra of $C(\Omega)$, and that, if there is a discontinuous homomorphism, then there is a radical homomorphism from some maximal ideal M_x of $C(\Omega)$.

Independently, Esterle ([5]) and Sinclair (see [9]) proved in 1975 that it further follows that there is a radical homomorphism from M_x whose kernel is a prime ideal. Hence, there is a non-maximal, prime ideal P in $C(\Omega)$ such that $C(\Omega)/P$ is a normed algebra. It was proved independently in 1976 by myself ([2]) and by Esterle ([6]), that, if the Continuum Hypothesis (CH) be assumed, then, for each infinite compact space Ω, there is a non-maximal, prime ideal P in $C(\Omega)$ such that $C(\Omega)/P$ is normable, and hence that there is a discontinuous homomorphism from $C(\Omega)$. We shall come to the question for which non-maximal, prime ideals P, $C(\Omega)/P$ is normable shortly.

Let me first comment on the role of CH in the above result. It is not true that it is a theorem of ZFC that there is a discontinuous homomorphism from each $C(\Omega)$: Solovay and Woodin have proved that there is a model of ZFC (in which CH is false) in which every homomorphism from each space $C(\Omega)$ is continuous, and so Kaplansky's question is independent of ZFC. This result is expounded in the book [3]. (As an advertisement, may I say that this book was written for analysts?)

However the solution of Kaplansky's problem is not equivalent to CH: recently, Woodin ([10]) has proved that there is a model of ZFC in which CH is false (in fact, the cardinality of the continuum in this model is equal to $\aleph_2$), but in which there is a discontinuous homomorphism from $C(\Omega)$ for every infinite compact space Ω.

2. We now come to the first open question: *are all (infinite) compact spaces Ω equivalent for Kaplansky's problem?* (For this question to make sense, we should be working in the theory ZFC itself.) There are two fairly easy known results, both given in [3]. We write ℓ^∞ and c for the spaces of all bounded and all convergent sequences, respectively, so that $\ell^\infty \cong C(\beta\mathbb{N})$ and $c \cong C(\mathbb{N} \cup \{\infty\})$.

THEOREM. (i) *Assume that there is a discontinuous homomorphism from ℓ^∞. Then there is a discontinuous homomorphism from $C(\Omega)$ for each infinite, compact space Ω.*

(ii) *Assume that there is a discontinuous homomorphism from $C(\Omega)$ for some compact space Ω. Then there is a discontinuous homomorphism from c.*

Thus to complete the circle, it would be nice to prove that the existence of a discontinuous homomorphism from c entails the existence of one from ℓ^∞. We formulate the question in terms of prime ideals.

QUESTION 1. *Assume that there is a non-maximal, prime ideal P in c such that there is an algebra norm on c/P. Is there a prime ideal Q in ℓ^∞ such that there is an algebra norm on ℓ^∞/Q ?*

The above question can be reformulated in terms of ultrafilters on $\mathbb{N}$ (see [3, p.47]), and it is then clear that there are two or three variants of the question, given in [3, p.47 and p.53]. I can resolve none of these questions: the algebraic structure of ℓ^∞ is surprisingly complicated.

3. Let θ be a discontinuous homomorphism from $C(\Omega)$. It follows from the work of Bade and Curtis that there is an ideal L in $C(\Omega)$ which is maximal with respect to the property that $\theta|L$ is continuous (in fact, L is the so-called *continuity ideal* of θ), so that

$$L = \{ f \in C(\Omega) : g \mapsto \theta(fg) \text{ is continuous} \},$$

and that L is contained in only finitely many maximal ideals. It also follows from the work of Esterle and of Sinclair that L is an intersection of prime ideals. On the other hand, if I is a *finite* intersection of prime ideals, then there is a discontinuous homomorphism from $C(\Omega)$ whose continuity ideal is exactly I (in $ZFC + CH$). A positive answer to the following question would again give a pleasing result that completes a circle.

QUESTION 2. *Is the continuity ideal of a discontinuous homomorphism from $C(\Omega)$ always a* **finite** *intersection of prime ideals?*

This question has a positive answer in certain cases, for example when $C(\Omega) = \ell^\infty$ or c ([5, §5]), but is open when $\Omega = [0,1]$.

4. Let P be a non-maximal, prime ideal in $C(\Omega)$, and let $A_P = C(\Omega)/P$. We should like to know just when A_P is normable (in the theory $ZFC+GCH$). The results referred to above show that A_P is normable in the case where the cardinality of $A_P, |A_P|$, is $2^{\aleph_0} = \aleph_1$. In the opposite direction, we have a result of Esterle ([5, 7.1]). For $a, b \in A_P \setminus \{0\}$, we say that $a \sim b$ if there exists $n \in N$ such that $|a| \leqslant n\,|b|$ and $|b| \leqslant n\,|a|$. Then $\sim$ is an equivalence relation on A_P. The equivalence classes are called the *archimedean classes.* Set $\Gamma_P = (A_P \setminus \{0\})/\sim$ (so that Γ_P is the *value semigroup* of A_P). Esterle showed that, if A_P is normable, then $|\Gamma_P| \leqslant \aleph_1$, and this implies that $|A_P| \leqslant \aleph_2$ ([4]). Thus the only remaining case is the following.

QUESTION 3. *Let P be a non-maximal, prime ideal in some $C(\Omega)$ such that $|A_P| = \aleph_2$ and $|\Gamma_P| = \aleph_1$. Is A_P normable?*

It is far from clear, in fact, that an ideal P satisfying the conditions of the question exists, but one is exhibited in [4]; in this latter work, the above question is placed in a more general setting involving totally ordered fields.

REFERENCES

1. Bade W. G., Curtis P. C., Jr., *Homomorphisms of commutative Banach algebras*, Amer. J. Math. **82** (1960), 589–608.
2. Dales H. G., *A discontinuous homomorphism from $C(X)$*, Amer. J. Math. **101** (1979), 647–734.
3. Dales H. G., Woodin W. H., *An Introduction to Independence for Analysts*, London Mathematical Society Lecture Note Series, vol. 115, Cambridge University Press, 1987.
4. Dales H. G., Woodin W. H., *Totally ordered fields with additional structure*, (in preparation).
5. Esterle J., *Semi-normes sur $C(K)$*, Proc. London Math. Soc. **36** (1978), no. 3, 27–45.
6. Esterle J., *Injection de semi-groupes divisibles dans des algèbres de convolution et construction d'homomorphismes discontinus de $C(K)$*, Proc. London Math. Soc. **36** (1978), no. 3, 59–85.
7. Esterle J., *Homomorphismes discontinus des algèbres de Banach commutatives séparables*, Studia Math. **56** (1979), 119–141.
8. Kaplansky I., *Normed algebras*, Duke Math. J. **16** (1949), 399–418.
9. Sinclair A. M., *Automatic Continuity of Linear Operators*, London Mathematical Society Lecture Note Series, vol. 21, Cambridge University Press, 1976.
10. Woodin W. H., *A discontinuous homomorphism from $C(X)$ without CH*, submitted.

SCHOOL OF MATHEMATICS
UNIVERSITY OF LEEDS
LEEDS LS2 9JT
ENGLAND

CONTINUITY OF DERIVATIONS OF RADICAL CONVOLUTION ALGEBRAS

W. G. BADE

Let $\mathbb{R}^+ = [0,\infty)$, and let ω be a continuous positive function on $\mathbb{R}^+$ satisfying

$$\omega(0)=1, \quad \omega(s+t) \leqslant \omega(s)\omega(t) \quad (s,t \in \mathbb{R}^+)$$

and

$$\lim_{t\to\infty} \omega(t)^{1/t} = 0\,.$$

We call ω a *radical weight function.* For example, ω might be the function $\omega\colon t \mapsto e^{-t^2}$. We denote by $L^1(\mathbb{R}^+,\omega)$ the Banach space of equivalence classes of complex-valued, measurable functions on $\mathbb{R}$ for which

$$\|f\| = \int_0^\infty |f(x)|\omega(t)\,dt < \infty\,.$$

Then $A = L^1(\mathbb{R}^+,\omega)$ is a radical Banach algebra under convolution multiplication

$$(f * g)(t) = \int_0^t f(t-s)g(s)\,ds \quad (t \in \mathbb{R}^+).$$

For a function φ set $\check{\varphi}(t) = \varphi(-t)$.

We represent the dual A' of A as the space $L^\infty(\mathbb{R}^-,\check{\omega}^{-1})$ of classes of measurable functions φ on $\mathbb{R}^- = (-\infty,0]$ for which

$$\|\varphi\| = \operatorname{ess\,sup}\left\{\frac{|\check{\varphi}(t)|}{\omega(t)} : t \in \mathbb{R}^+\right\} < \infty,$$

with the pairing $\langle f,g\rangle = \int_{\mathbb{R}^+} f(s)\check{\varphi}(s)\,ds \;\; (f \in A,\ \varphi \in A')$.

Then A' is the dual module of A for the module operation

$$f\cdot\varphi = (f*\varphi)|\mathbb{R}^- \quad (f \in A,\ \varphi \in A').$$

PROBLEM 1. *Let ω be a radical weight function. Is every derivation from $A = L^1(\mathbb{R}^+,\omega)$ into the Banach A-module A' necessarily continuous?*

The answer to this question is not known for any radical weight ω. However, module derivations from these convolution algebras do have certain striking continuity properties. Partial results for this problem are given in [1], where the corresponding problem is also discussed for the algebras $\ell^1(\mathbb{S}^+,\omega)$, where $\mathbb{S}$ is a dense subgroup of $\mathbb{R}$. A positive solution to Problem 1 could reveal new principals and techniques in automatic continuity theory.

We briefly discuss a standard way of constructing discontinuous derivations. In a general setting, let $\mathfrak{A}$ be a commutative Banach algebra and let $\mathfrak{M}$ be a Banach $\mathfrak{A}$-module. Suppose that $a \in \mathfrak{A}$. Thus $\mathfrak{M}$ is *a-divisible* if, for each $\varphi \in \mathfrak{M}$, there exists $\psi \in \mathfrak{M}$ such that $a \cdot \psi = \varphi$. The module $\mathfrak{M}$ is *$\mathfrak{A}$-divisible* if it is a-divisible for each $a \in \mathfrak{A} \setminus \{0\}$; $\mathfrak{M}$ is *torsion free* if $a \cdot \psi \neq 0$ whenever $a \neq 0$ and $\psi \neq 0$. If $\mathfrak{A}$ is an integral domain and $\mathfrak{M}$ contains a non-zero $\mathfrak{A}$-divisible submodule $\mathfrak{N}$, then there is a discontinuous derivation from $\mathfrak{A}$ to $\mathfrak{M}$. For if D is a derivation from a subalgebra $\mathfrak{B}$ of $\mathfrak{A}$ into $\mathfrak{N}$ and $a \in \mathfrak{A} \setminus \mathfrak{B}$, then D can be extended to a derivation $\overline{D}\colon \mathfrak{B}[a] \to \mathfrak{N}$, where $\mathfrak{B}[a]$ is the subalgebra generated by $\mathfrak{B} \cup \{a\}$. If a is algebraic over $\mathfrak{B}$, the extension $\overline{D}$ is unique. If a is transcendental over $\mathfrak{B}$, then $\overline{D}(a)$ can be chosen to be any element of $\mathfrak{N}$. By using this arbitrary choice for transcendental elements and Zorn's Lemma, a discontinuous derivation from $\mathfrak{A}$ to $\mathfrak{N}$ may be constructed. In the paper [2], this method was used to construct discontinuous derivations from Banach algebras of formal power series such as the algebras $\ell^1(\mathbb{Z}^+,\omega)$. Unfortunately, this method cannot be applied to the algebras $L^1(\mathbb{R}^+,\omega)$; although they are integral domains, it is proved in [1] that there is no non-zero $L^1(\mathbb{R}^+,\omega)$-divisible submodule in any Banach $L^1(\mathbb{R}^+,\omega)$-module.

A reduction of the problem is possible under certain conditions on the weight ω. Let ω be unicellular and regular, and let $g \in A \setminus \{0\}$. Then a derivation $S\colon A \to A'$ is continuous if and only if the derivation $g \cdot D$, defined by $(g \cdot D)(f) = g \cdot D(f)$ $(f \in A)$ is continuous. The weight ω is *unicellular* if every closed ideal in A is one of the standard ideals $I_t = \{f \in A : \alpha(f) \geqslant t\}$, $0 \leqslant t < \infty$. Here $\alpha(f) = \inf \operatorname{supp} f$. We say that ω is *regulated* if $\lim_{s\to\infty} \omega(s+t)/\omega(s) = 0$ for each $t > 0$. For example, the weight $\omega(t) = e^{-t^2}$ $(t \in \mathbb{R}^+)$ is both unicellular [3] and regulated. If one takes for g the characteristic function u of $\mathbb{R}^+$, then $D\colon A \to A'$ is continuous if and only if $u \cdot D$ is continuous, for any radical weight ω.

Finally we describe some concrete continuity results for derivations. All functions defined on $\mathbb{R}^+$ or $\mathbb{R}^-$ will be considered to be zero on the complementary interval. Given a function F on $\mathbb{R}$, we define the function xF by $(xF)(t) = tF(t)$ $(t \in \mathbb{R})$. A basic fact is the description of all continuous derivations from A to A', due to Groenbaek [4]. If $D\colon A \to A'$ is continuous, there exists a measurable function ψ on $\mathbb{R}^-$ such that $Df : (xf * \psi)|\mathbb{R}^-$ $(f \in A)$. We say that ψ *represents* D. Let W be the space of all such functions ψ. It is proved in [1] that a measurable function ψ on $\mathbb{R}^-$ belongs to W if and only if the function $x\psi$ belongs to A'. We define $V = A' \cap W$.

Special continuity properties hold for derivations D for which $D(A) \subseteq V$. The premultiplied derivations $g \cdot D$ can have this property. For if g satisfies

$$\sup_{a>0} \frac{a}{\omega(a)} \int_{\mathbb{R}^+} |g(t)|\omega(t+a)\, dt < \infty ,$$

then one can show that $g \cdot A' \subseteq V$ for every D.

Suppose that $D\colon A \to A'$ is a derivation and $D(A) \subseteq V$. Then one can find an element $k \in A$ such that $k \cdot D$ is continuous on a dense subalgebra of A. Let h be an element of A such that $xh \in A$ and such that the subalgebra $P(h)$ of all convolution polynomials on h is dense in A. Let $k = xh$ and consider the derivation $k \cdot D$. Since $D(h) \in V$, the map $E \colon f \to xf \cdot D(h)$ is a continuous derivation and $(k \cdot D)(h) = E(h)$. Thus $k \cdot D$ agrees with E and is continuous on the dense subalgebra $P(h)$. As an example, one can take for h the function u. Since $u^{*n}(t) = t^{n-1}/(n-1)!$ $(n \in \mathbb{N}, t \in \mathbb{R}^+)$, $P(u)$ is the dense subalgebra of all ordinary polynomials in the variable t. The condition $U \cdot A' \subseteq V$ is that

$$(\dagger) \qquad \sup_{a>0} \frac{a}{\omega(a)} \int_{\mathbb{R}+} \omega(t+a)\, dt < \infty ;$$

which is satisfied, for example, by $\omega(t) = e^{-t^2}$. Let ω be a weight satisfying (†) and let $g(t) = t^2$. Then for any derivation $D \colon A \to A'$, the derivation $g \cdot D$ is continuous on the dense subalgebra of polynomials in t.

References

1. Bade W. G., Dales H. G., *Continuity of derivations from radical convolution algebras*, Studia Math. **95** (1989), 59–91.
2. Bade W. G., Dales H. G., *Discontinuous derivations from algebras of power series*, Proc. London Math. Soc. **59** (1989), no. 3, 133–152.
3. Domar Y., *Extensions of the Titchmarsh convolution theorem with applications in the theory of invariant subspaces*, Proc. London Math. Soc **46** (1983), no. 3, 288–300.
4. Groenbaek N., *Commutative Banach algebras, module derivations, and semigroups*, J. London Math. Soc. **40** (1989), no. 2, 137–157.

Department of Mathematics
University of California
Berkeley
CA 94720
USA

2.9
old

THE SPECTRUM OF AN ENDOMORPHISM IN A COMMUTATIVE ALGEBRA

E. A. GORIN, A. K. KITOVER

The theorem of H. Kamowitz and S. Scheinberg [1] establishes that the spectrum of a non-periodic automorphism $T\colon A \to A$ of a semisimple commutative Banach algebra A (over $\mathbb{C}$) contains the unit circle $\mathbb{T}$. Several rather simple proofs of the theorem have been obtained besides the original one (e.g., [2], [3]) and various generalizations have been found (see e.g., [4], [5], [6]). It is easy to show that under the conditions of the theorem the spectrum is connected. At the same time all "positive" information is exhausted, apparently, by these two properties of the spectrum. There are examples (see [7], [8], [9]) demonstrating the absence of any kind of symmetry structure in the spectrum even if we suppose that the given algebra is regular in the sense of Shilov.

Let for instance K be a compact set in $\mathbb{C}$ lying in the annulus $\{z \in \mathbb{C} : 1/4 \leqslant |z| \leqslant 4\}$ and containing $\{z \in \mathbb{C} : 1/2 \leqslant |z| \leqslant 2\}$ and equal to the closure of its interior $\operatorname{int} K$. Denote by A the family of all functions continuous on K and holomorphic on $\operatorname{int} K$. Clearly, A equipped with the usual uniform norm on K, and with the pointwise operations is a Banach algebra. The spectrum of multiplication by the "independent variable" on A evidently coincides with K. On the other hand, the conditions imposed on K imply that A is a Banach algebra (without unit) with respect to the convolution

$$(f * g)(z) = \frac{1}{2\pi i} \int_{|\zeta|=1} f(\zeta)\, g(\overline{\zeta} z)\, \frac{d\zeta}{\zeta}$$

corresponding to multiplication of the Laurent coefficients. This algebra being semisimple, its maximal ideal space can be identified with the set of all integers. Adjoining a unit to A thus turns it into a regular algebra. Obviously the above mentioned operator on A is an automorphism.

1. *Are there other necessary conditions on the spectrum of a non-periodic automorphism of a semisimple, commutative Banach algebra besides the two mentioned above?*

In particular, is it obligatory for the spectrum to have interior points when it differs from $\mathbb{T}$? It is known in such cases (see [7], [8]) that the set of interior points may not be dense in the spectrum and may not be connected either.

Let M_A be the maximal ideal space of a commutative, semisimple Banach algebra A. An automorphism T of A induces an automorphism of the algebra $C = C(M_A)$. The essential meaning of the Kamowitz-Scheinberg theorem is that $\sigma_C(T) \subset \sigma_A(T)$. It is natural from this point of view to study the inclusion $\sigma_C(L) \subset \sigma_A(L)$ for a more general class of operators L. The case of weighted automorphisms $L\colon La \stackrel{\text{def}}{=} u \cdot Ta$, with u an invertible element of A, had, for example, been studied in [10]. It turns out that the inclusion does not hold for this class of operators.

2. *Does the spectrum of $L = uT$, constructed for a non-periodic automorphism T, contain any circle centered at the origin?*

If it does, then we obtain an instant generalization of the theorem of Kamowitz and Scheinberg.

The spectrum of operators looking like L acting on the algebra of all continuous functions on a compact set, has a complete description [11]. If A is also a uniform algebra, then $\sigma_C(L) = \sigma_A(L)$. Thus $\sigma_A(L) = \sigma_B(L)$ provided that L is a weighted automorphism of two uniform algebras A and B having the same maximal ideal spaces.

3. *Let A be a closed subalgebra of a semisimple, commutative Banach algebra B, and let $M_A = M_B$. Let L be a weighted automorphism of A and B simultaneously. Is it then true that $\sigma_A(L) = \sigma_B(L)$?*

We conjecture that this question has a negative answer.

The spectrum of an endomorphism apparently does not have any particular properties even if we suppose that A is a uniform algebra. Given two compacta S_1 and S_2 it is easy to obtain an endomorphism with the spectrum either $S_1 \cup S_2$ or $S_1 \cdot S_2$. The only obvious property of spectra is that λ^n belongs to the spectrum when λ ranges over its boundary and n over the set of non-negative integers.

4. *Let K be a compact subset of $\mathbb{C}$ satisfying $\lambda^n \in K$ for $n = 1, 2, \ldots$ and for all points λ in the boundary of K. Is there an endomorphism of a uniform algebra whose spectrum is equal to K?*

Spectra of endomorphisms of uniform algebras (and even those for weighted endomorphisms) can be described pretty well under the additional assumption that the induced mapping of the maximal ideal space keeps the Shilov boundary invariant (see [12], where one can find references to preceding papers of Kamowitz). Roughly speaking, things in this case are going as well as in the case of Banach algebras of functions continuous on a compact set. The situation changes dramatically when the boundary, or only a part of it, penetrates the interior. In such circumstances, it is common to begin with the consideration of classical examples. Let $\mathbb{D}$ be the unit disc in $\mathbb{C}$, let $A(\mathbb{D})$ be the algebra (disc algebra) of all functions continuous on the closure of $\mathbb{D}$ and holomorphic on $\mathbb{D}$, and let $H^\infty(\mathbb{D})$ be the algebra of all functions bounded and holomorphic on $\mathbb{D}$. Both of the algebras $A(\mathbb{D})$ and $H^\infty(\mathbb{D})$ are equipped with the uniform norm.

5. *Every endomorphism of $A(\mathbb{D})$ induces a natural endomorphism of H^∞. Do the spectra of these endomorphisms coincide?*

In this connection it is worthwhile to note that the answer to an analogous question concerning the algebra of all continuous functions on a compact set and the algebra of all bounded functions on the same set is in the affirmative [13]. The proof of this result uses, however, a full (though comparatively simple) analysis of the possible spectral pictures depending on the dynamics generated by the endomorphism.

The interesting papers [14], [15], [16] of Kamowitz (see also [6], [9]) deal with the spectra of the endomorphisms of $A(\mathbb{D})$ whose induced mappings do not preserve the boundary of $\mathbb{D}$. In the non-degenerate case the spectrum has a tendency to fill out the disc. Discrete and continuous spirals as well as compacta bounded by such spirals may nevertheless appear as the spectra of an endomorphism. (But only the spirals can appear in the case of Möbius transformations.)

6. *Is the spectrum of an endomorphism of the disc algebra a semigroup (with respect to multiplication in $\mathbb{C}$)? What kind of semigroups can arise as spectra?*

7. *Is it possible to say something concerning the spectra of endomorphisms of natural multidimensional generalizations of the disc algebra?*

Note that in the one-dimensional case, the theory of Denjoy-Wolff and the interpolation theorem of Carleson-Newman are often involved in the question.

The problem of describing spectra for weighted automorphisms is closely related with an analogous one for the so-called *shift-type* operators which have been studied by A. Lebedev [17] and A. Antonevich [18].

Let A be a uniform algebra of operators on a Banach space X. An invertible operator U on X is called a *shift-type* operator if $UAU^{-1} = A$. Usually X is a Banach space of functions and A is a subalgebra of the algebra of multipliers for X. The transformation $a \to U \cdot aU^{-1}$ determines an automorphism T of A which induces the mapping $\varphi\colon M_A \to M_A$. It is assumed that:

(i) the set of φ-periodic points is of first category in the Shilov boundary ∂A;
(ii) the spectrum of $U\colon X \to X$ is contained in the unit circle;
(iii) each invertible operator $a\colon X \to X$ for $a \in A$ is invertible as an element of A.
(iv) the topological spaces M_A and ∂A have the same stock of clopen (closed and open) φ-invariant subsets.

Then $\sigma_X(aU) = \sigma_A(aT)$ for all a in A [19].

We conjecture that Condition (iv) is superfluous. If this were true it would be possible (in view of [11]) to obtain a complete description of $\sigma_X(aU)$. It is reasonable to ask the same question for other algebras A besides the uniform ones.

References

1. Kamowitz H., Scheinberg S., *The spectrum of automorphisms of Banach algebras*, J. Funct. Anal. **4** (1969), 268–276.
2. Johnson B. E., *Automorphisms of commutative Banach algebras*, Proc. Amer. Math. Soc. **40** (1973), 497–499.
3. Levi R. N., *New proof of the theorem on automorphisms of Banach algebras*, Vestnik MGU, ser. math. mech. **4** (1972), 71–72. (Russian)
4. Levi R. N., *On automorphisms of Banach algebras*, Funct. Anal. and Appl. **6** (1972), no. 1, 16–18. (Russian)
5. Levi R. N., *On the joint spectrum of some commuting operators*, Kandidat Thesis, Moscow, 1973. (Russian)
6. Gorin E. A., *How the spectrum of an endomorphism of the disc algebra looks like*, Zapiski Nauchn. Semin. Leningrad Branch of Inst. Math. **126** (1983), 55–68. (Russian)
7. Scheinberg S., *The spectrum of an automorphism*, Bull. Amer. Math. Soc. **78** (1972), 621–623.
8. Scheinberg S., *Automorphisms of commutative Banach algebras*, Problems in analysis, Princeton University Press, Princeton, 1970, pp. 319–323.
9. Gorin E. A., *On the spectrum of endomorphisms of uniform algebras*, Proceedings of Theoretical and Applied Problems of Mathematics, Tartu (1980), 108–110. (Russian)
10. Kitover A. K., *On the spectrum of automorphisms with a weight and the theorem of Kamowitz and Scheinberg*, Funct. Anal. and appl. **13** (1979), no. 1, 70–71. (Russian)
11. Kitover A. K., *Spectral properties of automorphisms with a weight in uniform algebras*, Zapiski Nauchn. Semin. Leningrad Branch of Inst. Math. **92** (1979), 288–293. (Russian)
12. Kitover A. K., *Spectral properties of homomorphisms with weight in algebras of continuous functions and applications*, Proceedings of Scient. Sem. Leningrad Branch of Inst. Math. **107** (1982), 89–103. (Russian)

13. Kitover A. K., *On operators in C^1 which are indicated by smooth maps*, Funct. Anal. and Appl. **16** (1982), no. 3, 61–62. (Russian)
14. Kamowitz H., *The spectra of endomorphisms of the disc algebra*, Pacific J. Math. **46** (1973), 433–440.
15. Kamowitz H., *How the spectrum of an endomorphism of the disc algebra looks like*, Pacific J. Math. **66** (1976), 433–442.
16. Kamowitz H., *Compact operators of the form uC_φ*, Pacific J. Math. **80** (1979), 205–211.
17. Lebedev A. V., *On operators of the type of the weighted translation*, Kandidat Thesis, Minsk, 1980. (Russian)
18. Antonevich A. B., *Operators with a translation which are induced by an action of a compact Lie group*, Siberian Journal of Math. **20** (1979), no. 3, 467–478. (Russian)
19. Kitover A. K., *Operators of substitution with a weight in Banach modules over uniform algebras*, Dokl. Akad. Nauk SSSR **271** (1983), 528–531 (Russian); English transl. in Soviet Math. Dokl. **28** (1983), 110–113.

UL. OSTROVITYANOVA 20, 135
MOSCOW 117321
RUSSIA

2395, 43D AVENUE, APT 1,
SAN FRANCISCO,
USA

COMMENTARY BY H. KAMOWITZ

There has been essentially no progress on the questions raised in this problem. A related paper to [19] is [20].

In another direction, there have been some results on compact and weighted composition endomorphisms of Banach algebras and their spectra. See [21–26].

The problem of determining the spectra of compact and weighted composition operators on function algebras seems to remain open. Although the answer is known for $C(X)$, the disc algebra, and H^∞ of the disc the answer does not appear to be known for H^∞ of complicated domains, or for $R(K)$.

REFERENCES

20. Kamowitz H., *The spectra of a class of operators on the disc algebra*, Indiana Univ. Math. J. **27** (1978), 581–610.
21. Ohno S., Wada J., *Compact homomorphisms on function algebras*, Tokyo J. Math. **4** (1981), 105–112.
22. Takagi H., *Compact weighted composition operators on function algebras*, Tokyo J. Math. **11** (1988), 119–129.
23. Uhlig H., *The eigenfunctions of compact weighted endomorphisms of $C(X)$*, Proc. Amer. Math. Soc. **98** (1986), 89–93.
24. Kamowitz H., *Compact endomorphisms of Banach algebras*, Pacific J. Math. **89** (1980), 313–325.
25. Kamowitz H., Scheinberg S., Wortman D., *Compact endomorphisms of Banach algebras* II, Proc. Amer. Math. Soc. **107** (1989), 417–421.
26. Kamowitz H., *Compact weighted endomorphisms of $C(X)$*, Proc. Amer. Math. Soc. **83** (1981), 517–521.

11 JACOBS TERRACE
NEWTON CENTER
MASSACHUSETTS 02159
USA

2.10

ONE-SIDED SPECTRAL CALCULUS

JAROSLAV ZEMÁNEK

Let A be a complex Banach algebra with identity. Let a be a left invertible element in A, so that $ba = 1$ for some b in A. Let $r(a)$ denote the left invertibility radius of a, that is, the supremum of all $\alpha > 0$ such that the elements $a - \lambda$ are left invertible for $|\lambda| < \alpha$. An elementary estimate [6, Theorem 1.4.6] shows that $r(a) \geqslant \|b\|^{-1}$. Since $r(a^n) = r(a)^n$ by [7], the spectral radius formula yields the inequality $r(a) \geqslant \rho(b)^{-1}$ valid for each left inverse b of a, where $\rho(b)$ denotes the spectral radius of b.

CONJECTURE 1. $r(a) = \sup\{\, \rho(b)^{-1} : \ ba = 1\,\}$.

Of course, this is true if a is invertible. If the element a is not invertible, then the set of its left inverses is infinite: it is precisely the set $b + A(1 - ab)$, where b is any left inverse of a (observe that $ca = 1$ implies that $c = b + c(1 - ab)$). Define

$$s(a) = \sup\{\, \rho(b)^{-1} : \ ba = 1\,\}.$$

Since $ba = 1$ implies that $b^n a^n = 1$, we have

$$s(a) \leqslant s(a^n)^{1/n} \leqslant s(a^{2n})^{1/2n} \leqslant s(a^{4n})^{1/4n} \leqslant \ldots \leqslant r(a)$$

for each $n = 1, 2, \ldots$. Moreover, the surjectivity radius formula [3, Theorem 1] applied to the operator of right multiplication by a on A gives the following result.

THEOREM 1. $r(a) = \lim_{n\to\infty} s(a^n)^{1/n}$.

In contrast to the geometric quantities appearing in various asymptotic spectral formulae [7], the quantity $s(a)$ is of spectral origin, so that the next property could be expected.

CONJECTURE 2. $s(a^n) = s(a)^n$ for $n = 1, 2, \ldots$.

This together with Theorem 1 would prove Conjecture 1, and conversely, Conjecture 1 implies Conjecture 2.

It turns out that these problems are related to the question of the existence of global left pseudo-resolvents which was studied in [4]. To simplify the notation, we assume that $r(a) = 1$, and write $\mathbb{D}$ for the open unit disk in the complex plane. A theorem of Allan [1] ensures the existence of an analytic function $g(\cdot)$ on $\mathbb{D}$, with values in A, such that

$$g(\lambda)(a - \lambda) = 1 \tag{1}$$

for all λ in $\mathbb{D}$. We can write

$$g(\lambda) = b_0 + b_1\lambda + b_2\lambda^2 + b_3\lambda^3 + \ldots, \tag{2}$$

where the coefficients b_i are in A. It is not obvious, however, whether or not (1) can be obtained with the additional requirement that

$$g(\lambda) - g(\mu) = (\lambda - \mu) g(\lambda) g(\mu) \tag{3}$$

for all λ and μ in $\mathbb{D}$. The importance of this condition, the classic first resolvent equation [2], is apparent from the following characterization and subsequent remarks.

THEOREM 2. *Let g be an analytic function on $\mathbb{D}$, with values in A, as specified in (2). Then the following are equivalent:*

(i) *$b_n = b_0^{n+1}$ for $n = 1, 2, \ldots$;*
(ii) *g satisfies condition (3) for all λ and μ in $\mathbb{D}$;*
(iii) *g satisfies condition (3) for $\mu = 0$ and all λ in $\mathbb{D}$.*

Proof. We note that (i) implies (ii) by routine calculation, and (ii) implies (iii) trivially. To show that (iii) implies (i) it suffices to verify, by induction, that the derivatives of g satisfy the relations

$$g^{(n)}(0) = n g^{(n-1)}(0) g(0) \tag{4}$$

for $n = 1, 2, \ldots$. This is clear for $n = 1$. Differentiating equation (3) n times yields

$$g^{(n)}(\lambda) = \left[n g^{(n-1)}(\lambda) + \lambda g^{(n)}(\lambda)\right] g(0). \tag{5}$$

Subtracting (4) from (5), we get

$$g^{(n)}(\lambda) - g^{(n)}(0) = n\left[g^{(n-1)}(\lambda) - g^{(n-1)}(0)\right] g(0) + \lambda g^{(n)}(\lambda) g(0),$$

hence

$$g^{(n+1)}(0) = n g^{(n)}(0) g(0) + g^{(n)}(0) g(0) = (n+1) g^{(n)}(0) g(0).$$

This shows the induction step, and the proof is complete.

The equivalence of (i) and (ii) was noticed in [2, Theorem 5.9.1]; condition (iii) seems to be easier to handle as it is expressed in terms of one variable only.

CONJECTURE 3. *Let K be a compact subset of $\mathbb{D}$. Then there exists an analytic function g on a neighbourhood of K, with values in A, satisfying (1) and (3) on K.*

Local germs of the function required are easy to construct, but the problem is that some of them do not extend to larger domains [5]. Nevertheless, Theorem 2 shows that, if two such germs coincide at a point, then they coincide on a neighbourhood of that point, because the corresponding Taylor coefficients are, in both cases, powers of the same element.

THEOREM 3. *Conjectures* 1, 2, *and* 3 *are equivalent.*

Proof. The compact set K is contained in a disk $(1+\varepsilon)^{-1}\mathbb{D}$ for some $\varepsilon > 0$. Granting Conjecture 1 to be true, we can choose a left inverse b_0 of a with $\rho(b_0) < 1 + \varepsilon$, and define $g(\lambda)$, for λ in $(1+\varepsilon)^{-1}\mathbb{D}$, by (2) with $b_n = b_0^{n+1}$ for $n = 0, 1, 2, \ldots$. Then both (1) and (3) are checked on $(1+\varepsilon)^{-1}\mathbb{D}$ by direct calculation.

Conversely, let g be as in Conjecture 3 with $K = (1+\varepsilon)^{-1}\mathbb{D}$ for some $\varepsilon > 0$. Then the Cauchy integral formula applied to (2) yields the estimate

$$\|b_n\| \leqslant M(1+\varepsilon)^n ,$$

where $M = \sup\{\, \|g(\lambda)\| : \lambda \in K \,\}$ is finite, and $b_n = b_0^{n+1}$ by Theorem 2. Hence $\rho(b_0) \leqslant 1 + \varepsilon$ by the spectral radius formula, and $b_0 a = 1$ by (1). This implies Conjecture 1.

We conclude by noting that a particular case of Conjecture 3 was verified in [5].

References

1. Allan G. R., *Holomorphic vector-valued functions on a domain of holomorphy*, J. London Math. Soc. **42** (1967), 509–513.
2. Hille E., Phillips R. S., *Functional Analysis and Semi-groups*, Colloq. Publ., vol. 31, Amer. Math. Soc., 1957.
3. Makai E., Jr., Zemánek J., *The surjectivity radius, packing numbers and boundedness below of linear operators*, Integral Equations Operator Theory **6** (1983), 372–384.
4. Miller J. B., *One-sided spectral theory and functional calculus*, Analysis Paper **48**, Monash University, 1985.
5. Miller J. B., *One-sided spectral theory of the unilateral shift*, Analysis Paper **57**, Monash University, 1987.
6. Rickart C. E., *General Theory of Banach Algebras*, van Nostrand, 1960.
7. Zemánek J., *Open semigroups in Banach algebras*, This collection, Problem 2.30.

Institute of Mathematics
Polish Academy of Sciences
00-950 Warszawa, P.O. Box 137
Poland

2.11

FOUR PROBLEMS CONCERNING JOINT SPECTRA

W. ŻELAZKO

Denote by $B(X)$ the algebra of all bounded endomorphisms of a complex Banach space X. There are several objects called joint spectra, such as the Harte spectrum σ_H, the Taylor spectrum σ_T, left and right spectra σ_l and σ_r, commutant and bicommutant spectra σ' and σ'', the approximate point spectrum σ_π and the defect spectrum σ_δ (see [2], [4], [5], and [7]). In order to treat all these objects in a uniform way the author proposed in [7] the following axioms. Suppose that to each n-tuple $(T_1, \ldots, T_n)$ of pairwise commuting operators in $B(X)$ there corresponds a non-void compact subset $\sigma_s(T_1, \ldots, T_n)$ of $\mathbb{C}^n$ such that

(i) $$\sigma_s(T_1, \ldots, T_n) \subset \prod_{i=1}^{n} \sigma(T_i),$$

where $\sigma(T)$ is the usual spectrum of an operator T. From this axiom it follows that $\sigma_s(T) \subset \sigma(T)$ for all T in $B(X)$. The second axiom is that

(ii) $$\sigma_s(T) = \sigma(T)$$

for all T in $B(X)$.

Let $P_{n,k}$ be a polynomial map from $\mathbb{C}^n$ to $\mathbb{C}^k$, i.e., a map of the form

$$P_{n,k}(z_1, \ldots, z_n) = \big(p_1(z_1, \ldots, z_n), \ldots, p_k(z_1, \ldots, z_n)\big),$$

where the p_i are polynomials in n complex variables. The most essential axiom (the spectral mapping property) is the following one:

(iii) $$\sigma_s\big(P_{n,k}(T_1, \ldots, T_n)\big) = P_{n,k}\big(\sigma_s(T_1, \ldots, T_n)\big)$$

for all n-tuples $(T_1, \ldots, T_n)$ of pairwise commuting operators in $B(X)$ and all polynomial maps $P_{n,k}$ $(n, k = 1, 2, \ldots)$.

A particular case of the spectral mapping property is the following axiom (the translation property)

(iv) $$\sigma_s(T_1 + \alpha_1 I, \ldots, T_n + \alpha_n I) = \sigma_s(T_1, \ldots, T_n) + (\alpha_1, \ldots, \alpha_n)$$

for all commuting n-tuples and all vectors $(\alpha_1, \ldots, \alpha_n)$ in $\mathbb{C}^n$.

Call σ_s a *spectrum* if it satisfies axioms (i), (ii) and (iii). Examples: σ_H and σ_T. Call it a *subspectrum* if it satisfies (i) and (iii). Examples: σ_l, σ_r, σ_π, σ_δ. Call it a *spectroid* if it satisfies only (i) and (iv). Examples: σ' and σ''.

Our first problem is connected with the following fact, proved in [7]: for each X there is a maximal spectrum on $B(X)$, i.e., a spectrum σ_m such that $\sigma_s(T_1, \ldots, T_n) \subset \sigma_m(T_1, \ldots, T_n)$ for all commuting n-tuples and all subspectra σ_s.

PROBLEM 1. *Describe σ_m on $B(H)$, where H is an infinite-dimensional Hilbert space.*

CONJECTURE. *In this case we have $\sigma_m = \sigma_T$.*

In fact we do not know what σ_m is for any concrete infinite-dimensional Banach space X.

PROBLEM 2. *Which spectra are suitable for the holomorphic functional calculus?*

It is known [6] that σ_T is such a spectrum, while σ_H fails to have a holomorphic functional calculus (see [1], application 5.30).

Let H be an infinite-dimensional Hilbert space. Słodkowski [3] constructed on $B(H)$ a countable family of joint spectra.

PROBLEM 3. *Does there exist an uncountable family of joint spectra on $B(H)$?*

PROBLEM 4. *Is it possible to have infinitely many joint spectra on $B(X)$ for an infinite dimensional Banach space X?*

Problem 4 is connected with the following (perhaps folklore) question. Does there exist an infinite dimensional Banach space X such that every operator in $B(X)$ is of the form $K + \lambda I$, where K is a compact operator and λ is a complex scalar? If such a space exists then there is only one joint spectrum on $B(X)$, i.e., all joint spectra coincide.

REFERENCES

1. Curto R. E., *Applications of several complex variables to multiparameter spectral theory*, in Surveys of some recent results in operator theory, Pitman Research Notes in Math. Ser., vol. 192, Boston, 1988, pp. 25–90.
2. Harte R., *Invertibility and Singularity for Bounded Linear Operators*, Dekker, 1988.
3. Słodkowski Z., *An infinite family of joint spectra*, Studia Math. **61** (1977), 239–255.
4. Słodkowski Z., Żelazko, W., *On joint spectra of commuting families of operators*, Studia Math. **50** (1974), 127–148.
5. Taylor J. L., *A joint spectrum for several commuting operators*, J. Functional Anal. **6** (1970), 172–191.
6. Taylor J. L., *The analytic functional calculus for several commuting operators*, Acta Math. **125** (1970), 1–38.
7. Żelazko W., *An axiomatic approach to joint spectra I*, Studia Math. **64** (1979), 249–261.

INSTITUTE OF MATHEMATICS
POLISH ACADEMY OF SCIENCES
00–950 WARSZAWA, P.O. BOX 137
POLAND

2.12
v.old

ALGEBRAIC EQUATIONS WITH COEFFICIENTS IN COMMUTATIVE BANACH ALGEBRAS AND SOME RELATED PROBLEMS

E. A. GORIN

The proposed questions have arisen in the seminar of V. Ya. Lin and the author on Banach Algebras and Complex Analysis at the Moscow State University.

In what follows A is a commutative Banach algebra (over $\mathbb{C}$) with unity and connected maximal ideal space M_A. For $a \in A$, $\widehat{a}$ denotes the Gelfand transform of a.

A polynomial $p(\lambda) = \lambda^n + a_1\lambda^{n+1} + \ldots + a_n$, where $a_i \in A$, is said to be *separable* if its discriminant d is invertible (i.e., for every ξ in M_A the roots of $\lambda^n + \widehat{a}_1(\xi)\lambda^{n-1} + \ldots + \widehat{a}_n(\xi)$ are simple); p is said to be *completely reducible* if it can be expanded into a product of polynomials of degree one. The algebra is called *weakly algebraically closed* if all separable polynomials of degree greater than one are reducible over it.

In many cases there exist simple (necessary and sufficient) criteria for all separable polynomials of a fixed degree n to be completely reducible. A criterion for $A = C(X)$, with a finite cell complex X, consists in the triviality of all homomorphisms $\pi_1 : X \to B(n)$, $B(n)$ being the Artin braid group with n threads [1]. If (and only if) $n \leqslant 4$ this is equivalent to $H^1(X, \mathbb{Z}) = 0$ (which is formally weaker). The criterion fits as a sufficient one for arbitrary arcwise connected, locally arcwise connected spaces X.

It can be deduced from the implicit function theorem for commutative Banach algebras that, if the polynomial with coefficients $\widehat{a}_i$ is reducible over $C(M_A)$, then the same holds for the original polynomial p over A. On the other hand (cf. [2], [3]) for arbitrary integers k and n with $4 < k \leqslant n$, there exists a pair of uniform algebras $A \subset B$, with the same maximal ideal space, such that $\dim B/A = 1$, all separable polynomials of degree $\leqslant n$ are reducible over A, but there exists an irreducible (over B) separable polynomial of degree k.

Indeed let G_k be the collection of all separable polynomials $p(\lambda) = \lambda^k + z_1\lambda^{k-1} + \ldots + z_k$ with complex coefficients $z_1, \ldots, z_k$, endowed with the complex structure induced by the natural embedding into $\mathbb{C}^k$, $p \mapsto (z_1, \ldots, z_k)$. Define X as the intersection of G_k, the submanifold $\{ z_1 = 0, d(z) = 1 \}$ and a ball $\{ z : \|z - z^0\| \leqslant C(n) < \infty \}$. Note that X is a finite complex. The algebra B is the uniform closure on X of polynomials in $z_1, \ldots, z_k$ and A consists of all functions in B with an appropriate directional derivative at an appropriate point equal to zero. With the parameters properly chosen, (A, B) is the pair which we are looking for (the proof uses the fact that the set of holomorphic functions on an algebraic manifold which do not take values 0 and 1 is finite, as well as some elementary facts of Morse theory and Montel theory of normal families that enable us to control the Galois group).

Do there exist examples of the same nature with A weakly algebraically closed? We do not even know any example in which A is weakly algebraically closed and $C(M_A)$ is not. A refinement of the construction in [4] and [5] may turn out to be sufficient.

If X is an arbitrary compact space such that division by 6 is possible in $H^1(X,\mathbb{Z})$, then all separable polynomials of degree 3 are reducible over $C(X)$. The situation is more complicated for polynomials of degree 4: there exists a metrizable compact space X of dimension two such that $H^1(X,\mathbb{Z})=0$ but some separable polynomial of degree 4 is irreducible over $C(X)$ ([6]). On the other hand, the condition that all elements of $H^1(X,\mathbb{Z})$ are divisible by $n!$ is necessary and sufficient for all separable polynomials of degree $\leqslant n$ to be completely reducible, provided that X is a homogeneous space of a connected compact group (and in some other cases). These types of results are of interest, e.g., for the investigation of polynomials with almost periodic coefficients.

Is it possible to describe "all" spaces X (not necessarily compact) for which the problem of complete reducibility over $C(X)$ of the separable polynomials can be solved in terms of one-dimensional cohomologies? In particular, is the condition $H^1(X,\mathbb{Z})=0$ sufficient in the case of a (compact) homogeneous space of a connected Lie group? (Note that the answer is affirmative for the homogeneous spaces of complex Lie groups and for the polynomials with holomorphic coefficients [9]).

Though the question of the complete reducibility of separable polynomials in its full generality seems to be transcendental, there is an encouraging classical model, i.e., the polynomials with holomorphic coefficients on Stein (in particular algebraic) manifolds. Note that the known sufficient conditions [9] for holomorphic polynomials are essentially weaker than in general case.

The peculiarity of holomorphic function algebras is revealed in a very simple situation. Consider the union of m copies of the annulus $\{z : R^{-1} < |z| < R\}$ identified at the point $z=1$. It can be shown that a separable polynomial of prime degree n with coefficients holomorphic on these spaces, and with discriminant $d=1$ is reducible if $R \geqslant R_0(m,n)$, the fact that n is prime being essential in the case where $m \geqslant 2$ [10]. If $m=1$, n can be arbitrary [2], and we denote by $R_0(n)$ the corresponding least possible constant. Now, if n is even, then $R_0(n)=1$, and so the holomorphicity assumption is superfluous. However $R_0(k\ell) \geqslant C(k)^{k\ell}$ if k and ℓ are odd, with $C(k)>1$ for $k \geqslant 3$. At the same time $R_0(n) \leqslant C^n$ for all n. These results, as well as the fact that $R_0(p)^{1/p} \to 1$ as p tends to infinity along the set of prime numbers, have been proved in [10]. *Nevertheless the exact asymptotic of $R_0(p)$ remains unknown. It is unknown even whether $R_0(p) \to \infty$ as $p \to \infty$.*

If X is a finite cell complex with $H^1(X,\mathbb{Z})=0$, then each completely reducible separable polynomial over $C(X)$ is homotopic in the class of all such polynomials to one with constant coefficients (the reason for this is that $\pi_q(G)=0$ for $q>1$). Let $X=M_A$ and consider a polynomial completely reducible over A.

Is it possible to realize the homotopy within the class of polynomials over A?

Such a possibility is equivalent, as a matter of fact, to the holomorphic contractibility of the universal covering space $\widetilde{G}_n$ for G_n. It is known [11] that $\widetilde{G}_n=\mathbb{C}^2\times V^{n-2}$, V^{n-2} being a bounded domain of holomorphy in $\mathbb{C}^n$ homeomorphic to a cell [12]. In $\mathbb{C}^n$ there are contractible but non-holomorphically contractible domains [12], though examples of bounded domains of such a sort seem to be unknown (that might be an additional reason to study the above question). Evidently $\widetilde{G}_3=\mathbb{C}^2\times\mathbb{D}$ is holomorphically contractible.

Is the same true for $\widetilde{G}_n$ with $n \geqslant 4$?

There are some reasons to consider also transcendental equations $f(w)=0$, where $f\colon A \to A$ is a Lorch holomorphic mapping (i.e., f is Fréchet differentiable and its derivative is the operator of multiplication by an element of A). In [13] the cases when equations of this form reduce to algebraic ones have been treated (in this sense the standard implicit function theorem is nothing but a reduction to a linear equation).

REFERENCES

1. Gorin E. A., Lin V. Ya., *Algebraic equations with continuous coefficients and certain questions in the algebraic theory of braids*, Matem. Sbornik **78** (1969), no. 4, 579–610. (Russian)
2. Gorin E. A., Lin V. Ya., *On separable polynomials over commutative Banach algebras*, Dokl. Akad. Nauk SSSR **218** (1974), no. 3, 505–508 (Russian); English transl. in Soviet Math. Dokl. **15** (1974), 1357–1361.
3. Gorin E. A., *Holomorphic functions over algebraic manifolds and the reducibility of separable polynomials over certain commutative Banach algebras*, 7th All-Union Conference on Topology (Minsk), 1977, p. 55. (Russian)
4. Gorin E. A., Karakhanyan M. I., *Certain remarks about algebras of continuous functions on locally connected compacta*, 7th All-Union Conference on Topology (Minsk), 1977, p. 56. (Russian)
5. Karakhanyan M. I., *On algebras of continuous functions over locally connected compacta*, Funct. Analysis and its Applications **12** (1978), no. 2, 93–94. (Russian)
6. Lin V. Ya., *On fourth-order polynomials over algebras of continuous functions*, Functional Analysis and its Applications **8** (1974), no. 4, 89–90. (Russian)
7. Zyuzin Yu. V., *Algebraic equations with continuous coefficients over a homogeneous space*, Vestnik MGU ser. mat. mekh. (1972), no. 1, 51–53. (Russian)
8. Zyuzin Yu. V., Lin V. Ya., *Non-branching algebraic extensions of commutative Banach algebras*, Matem. Sbornik **91** (1973), no. 3, 402–420. (Russian)
9. Lin V. Ya., *Algebroid functions and holomorphic elements of the homotopy groups of complex manifolds*, Dokl. Akad. Nauk SSSR **201** (1971), no. 1, 28–31 (Russian); English transl. in Soviet Math. Dokl. **12** (1971), 1608–1612.
10. Zyuzin Yu. V., *Irreducible separable polynomials with holomorphic coefficients on a certain class of complex spaces*, Matem. Sbornik **102** (1977), no. 4, 159–591. (Russian)
11. Kaliman I., *Holomorphic universal covering spaces of polynomials without multiple roots*, Funktsional. Anal. i Prilozhen. **9** (1975), no. 1, 71. (Russian)
12. Hirchowitz A., *A propos de principe d'Oka*, C.R. Acad. Sci. Paris **272** (1971), A792–A794.
13. Gorin E. A., Fernández C. Sánchez, *On transcendental equations in commutative Banach algebras*, Funktsional. Anal. i Prilozhen. **11** (1977), no. 1, 63–64 (Russian); English transl. in Funct. Anal. Appl. **11** (1977), 53–55.

FACULTY OF MATHEMATICS
MOSCOW LENIN PEDAGOGICAL INSTITUTE
MOSCOW
RUSSIA

COMMENTARY BY THE AUTHOR

Bounded contractible but non-holomorphically contractible domain of holomorphy in $\mathbb{C}^2$ have been constructed in [14]. All other questions, including that of contractibility of the Teichmüller space $\widetilde{G}_n$, seem to remain open.

A detailed exposition of a part of [13] can be found in [15].

REFERENCES

14. Zaidenberg M. G., Lin V. Ya., *On holomorphically non-contractible bounded domains of holomorphy*, Dokl. Akad. Nauk SSSR **249** (1979), no. 2, 281–285 (Russian); English transl. in Soviet Math. Dokl. **28** (1983), 200–204.

15. Fernández C. Sánchez, Gorin E. A., *Variante del teorema de la función implícita en álgebras de Banach commutativas*, Revista Ciencias Matemáticas (Univ. de La Habana, Cuba) **3** (1983), no. 1, 77–89.

Further Commentary by the Author

It was shown (in the thesis of V. V. Petunin) that

$$\log R_0(n) < 10^7 n.$$

This is a constructive version of a Zyuzin's theorem. The proof is parallel to the original Zyuzin's proof, but the theory of normal families is replaced by some specific estimates of the Kobayashi metric. Moreover, V. V. Petunin proved that for any $\varepsilon > 0$

$$\log R_0(2n+1) > (2n+1)^{1-\varepsilon}, \qquad n > n(\varepsilon).$$

In particular, the growth of $R_0(p)$ may be observed along prime p's. Petunin's thesis contains some progress concerning the problem of holomorphic contractibility. It was developed further in works of M. G. Zaidenberg, Sh. I. Kaliman and V. Ya. Lin.

Unfortunately, for some transcendental reasons Petunin's results appeared in inaccessible publications.

References

16. Petunin V. V., *Investigation of reducibility of polynomials over algebras of analytic functions*, Candidate Thesis, MIEM, Moscow, 1987. (Russian)
17. Petunin V. V., *On the reducibility of polynomials of prime degree over the algebra of holomorphic functions on a circular annulus*, Theory of functions and its appl. (collection of works of students and postgraduates), MGU, Moscow, 1986, pp. 73–74. (Russian)
18. Petunin V. V., *On the reducibility of polynomials of prime degree over the algebra of holomorphic functions on a twice-connected domain*, Information Problems Inst. AN SSSR, Moscow, 19 pp. (1987), Dep. VINITI 10.06.87, 4193–B87. (Russian)

2.13
old

GENERALIZED DERIVATIONS AND SEMIDIAGONALITY

V. S. Shul'man

Let A_1 and A_2 be bounded linear operators on a Hilbert space H, and let $\Delta = \Delta(A_1, A_2)$ be an operator on the space $B(H)$ of all bounded operators on H defined by $\Delta(X) = A_1X - XA_2$ ($X \in B(H)$). If $A_1 = A_2$, $\Delta(A_1, A_2)$ is a derivation of $B(H)$. That is why the operators $\Delta(A_1, A_2)$ are sometimes called *generalized derivations*. Put $\widetilde{\Delta} = \Delta(A_1^*, A_2^*)$. A question of whether

$$\operatorname{Ker} \widetilde{\Delta} \cap \operatorname{Im} \Delta = \{0\} \tag{1}$$

has been raised by various people (see [1], [2], [3]). Equality (1) is true for normal operators A_1, A_2, whence the Fuglede-Putnam theorem follows (see [4]). This equality means that $\operatorname{Ker} \widetilde{\Delta}\Delta = \operatorname{Ker} \Delta$ and so $\operatorname{Ker} \widetilde{\Delta} = \operatorname{Ker} \Delta\widetilde{\Delta} = \operatorname{Ker} \widetilde{\Delta}\Delta = \operatorname{Ker} \Delta$. In [3] it is proved that (1) holds when $A_1 = A_2$ is a cyclic subnormal operator or a weighted shift with non-vanishing weights.

Let $p \in [1, \infty]$. An operator A in $B(H)$ is called *p-semidiagonal* if its modulus of p-quasidiagonality

$$qd_p(A) = \liminf \{ \|PA - AP\| : P \in \mathcal{P} \}$$

($\mathcal{P}$ being the set of all finite rank projections) is finite. Denote by $\mathcal{M}_p$ the class of $\mathfrak{S}_p$-perturbations of direct sums of p-semidiagonal operators. In [5] it is proved that (1) is true if one of A_j belongs to $\mathcal{M}_1$. Although that result covers a rather extensive class of generalized derivations ($\mathcal{M}_1$ contains all normal operators with one-dimensional spectra and their nuclear perturbations, weighted shifts of an arbitrary multiplicity and polynomials of such shifts and Bishop's operators), it is not applicable to many generalized derivations with normal coefficients. Namely, a normal operator belongs in general to $\mathcal{M}_2$, but not to $\mathcal{M}_1$. It seems reasonable to try to replace the hypothesis of 1-semidiagonality of one of the A_j's by 2-semidiagonality of both.

Question 1. *Does (1) hold if A_1, $A_2 \in \mathcal{M}_2$?*

Question 2. *Does there exist an operator not in $\mathcal{M}_2$?*

An affirmative answer to the following question would solve Question 2 (see [5]).

Question 3. *Do there exist $A \in B(H)$ and $X \in \mathfrak{S}_2$ such that $AX - XA$ is a non-zero projection?*

References

1. Johnson B. E., Williams J., *The range of normal derivations*, Pacific. J. Math. **58** (1975), 105–122.
2. Williams J., *Derivation ranges: open problems*, in Top Modern Oper. Theory, 5 Int. Conf. Oper. Theory, Timisoara, Birkhäuser, 1981, pp. 319–328.
3. Yang Ho, *Commutants and derivation ranges*, Tohoku Math. J. **27** (1975), 509–514.
4. Putnam C., *Commutation Properties of Hilbert Space Operators and Related Topics*, vol. 36, Springer-Verlag, Ergebnisse, 1967.
5. Shul'man V. S., *On multiplication operators and traces of commutators*, Zapiski Nauchn. semin. LOMI **135** (1984), 182–194 (Russian); English transl. in J. Soviet Math. **31** (1985), 2749–2757.

Department of Mathematics
Pedagogical Institute
Mayakovskogo st. 6
Vologda 160600
Russia

Commentary by D. Voiculescu

In my paper [6] I show that there are operators which are not in $\mathcal{M}_p$ for any $p < \infty$. This answers to Question 2.

Reference

6. Voiculescu D., *A note on quasidiagonal operators*, Topics in operator theory, Operator Theory: Adv. and Appl., vol. 32, Birkhäuser, Basel – Boston, 1988, pp. 265–274.

2.14
v.old

PROBLEMS PERTAINING TO THE ALGEBRA OF BOUNDED ANALYTIC FUNCTIONS

T. W. GAMELIN

Here is a list of problems concerning my favourite algebra, the algebra $H^\infty(V)$ of bounded analytic functions on a bounded open subset V of the complex plane. Some of the problems are old and well-known, while some have arisen recently. We will restrict our discussion of each problem to the barest essentials. For references and more details, the reader is referred to the expository account [1], where a number of these same problems are discussed. The maximal ideal space of $H^\infty(V)$ will be denoted by $\mathcal{M}(V)$, and V will be regarded as an open subset of $\mathcal{M}(V)$. The grandfather of problems concerning $H^\infty(V)$ is the following.

PROBLEM 1 (CORONA PROBLEM). *Is V dense in $\mathcal{M}(V)$?*

The Corona Theorem of L. Carleson gives an affirmative answer when V is the open unit disc $\mathbb{D}$.

In the cases in which $\mathcal{M}(V)$ has been described reasonably completely, there are always analytic discs in $\mathcal{M}(V) \setminus V$, but never a higher dimensional analytic structure.

PROBLEM 2. *Is there always an analytic disc in $\mathcal{M}(V) \setminus V$? Is there ever an analytic bidisc in $\mathcal{M}(V)$?*

The Shilov boundary of $H^\infty(V)$ will be denoted by Γ_V. A function $f \in H^\infty(V)$ is *inner* if $|f| = 1$ on Γ_V. There is a plethora of inner functions in $H^\infty(V)$, but the following question remains unanswered.

PROBLEM 3. *Do the inner functions separate the points of $\mathcal{M}(V)$?*

An affirmative answer in the case of the unit disc was obtained by K. Hoffman, R. G. Douglas, and W. Rudin [2, p. 316].

The Shilov boundary Γ_V is extremely disconnected. Its Dixmier decomposition takes the form $\Gamma_V = T \cup Q$, where T and Q are closed disjoint sets, $C(T) \simeq L^\infty(\nu)$ for a normal measure on T, and Q carries no nonzero normal measures. The next problem is to identify the normal measure ν. There is a natural candidate at hand. Let λ_z be the "harmonic measures" on $\mathcal{M}(V)$. These are certain naturally-defined probability measures on $\mathcal{M}(V) \setminus V$ that satisfy $f(z) = \int f d\lambda_z$ for $f \in H^\infty(V)$.

PROBLEM 4. *Can the normal measure ν on T be taken to be the restriction of harmonic measure to T?*

There are a number of problems related to the linear structure of $H^\infty(V)$. It is not known, for instance, whether $H^\infty(V)$ has the approximation property, even when V is the unit disc. As a weak-star closed subalgebra of $L^\infty(dxdy|V)$, $H^\infty(V)$ is a dual space. The following problem ought to be accessible by the same methods used to study Γ_V.

PROBLEM 5. *Does $H^\infty(V)$ have a unique predual?*

T. Ando [3] and P. Wojtaszczyk [4] have shown that any Banach space B with dual isometric to $H^\infty(\mathbb{D})$ is unique (up to isometry). However, Wojtaszczyk shows that various non-isomorphic B's have duals isomorphic to $H^\infty(\mathbb{D})$. An extension of the uniqueness result is obtained by J. Chaumat [5].

The weak-star continuous homomorphisms in $\mathcal{M}(V)$ are called *distinguished* homomorphisms. The evaluations at points of V are distinguished homomorphisms, and there may be other distinguished homomorphisms. Related to Problem 2 is the following

PROBLEM 6. *Does each distinguished homomorphism lie on an analytic disc in $\mathcal{M}(V)$?*

The coordinate function z extends to a map $Z\colon \mathcal{M}(V) \to \overline{V}$. If $\zeta \in \partial V$, then the fibre $\mathcal{M}_\zeta = Z^{-1}(\{\zeta\})$ contains at most one distinguished homomorphism.

PROBLEM 7. *Suppose that there is a distinguished homomorphism $\varphi \in \mathcal{M}_\zeta$ and suppose that Γ is an arc in V terminating at ζ. If $f \in H^\infty(V)$ has a limit along Γ, does that limit coincide with $f(\varphi)$?*

J. Garnett [6] has obtained an affirmative answer when Γ is appropriately smooth.

The next problem is related to Iversen's Theorem on cluster values, and to the work in [7]. Define $V_\zeta = \Gamma_V \cap \mathcal{M}_\zeta$. Denote by $R(f,\zeta)$ the range of $f \in H^\infty(V)$ at $\zeta \in \partial V$, consisting of those values assumed by f on a sequence in V tending to ζ. An abstract version of Iversen's Theorem asserts that $f(V_\zeta)$ includes the topological boundary of $f(\mathcal{M}_\zeta)$, so that $f(\mathcal{M}_\zeta)\setminus f(V_\zeta)$ is open in $\mathbb{C}$. The problem involves estimating the defect of $R(f,\zeta)$ in $f(\mathcal{M}_z)\setminus f(V_\zeta)$.

PROBLEM 8. *If every point of ∂V is an essential singularity for some function in $H^\infty(V)$, does $f\big(\mathcal{M}_\zeta \setminus [f(V_\zeta)\cup R(f,\zeta)]\big)$ have zero logarithmic capacity for each $f \in H^\infty(V)$?*

The remaining problems pertain to the algebra $H^\infty(R)$, where R is a Riemann surface. We assume that $H^\infty(R)$ separates the points of R. Then there is a natural embedding of R into $\mathcal{M}(R)$.

PROBLEM 9. *Is the natural embedding $R \to \mathcal{M}(R)$ a homeomorphism of R and an open subset of $\mathcal{M}(R)$?*

PROBLEM 10. *If Γ is a simple closed curve in R that separates R, does Γ separate $\mathcal{M}(R)$?*

The preceding problem arises in the work of M. Hayashi [8], who has treated Widom surfaces in some detail. For this special class of surfaces, Hayashi obtains an affirmative answer to the following problem.

PROBLEM 11. *Is Γ_R extremely disconnected?*

REFERENCES

1. Gamelin T. W., *The algebra of bounded analytic functions*, Bull. Amer. Math. Soc. **79** (1973), 1095–1108.
2. Douglas R. G., Rudin W., *Approximation by inner functions*, Pacific J. Math. **31** (1969), 313–320.
3. Ando T., *On the predual of H^∞*, Special Issue dedicated to Wladislaw Orlicz on the occasion of his 75th birthday, Comment. Math., Special Issue **1** (1978), 33–40.

4. Wojtaszczyk P., *On projections in spaces of bounded analytic functions with applications*, Studia Math. **65** (1979), 147–173.
5. Chaumat J., *Unicité du prédual*, C.R. Acad. Sci. Paris, Sér A-B 288 (1979), A411–A414.
6. Garnett J., *An estimate for line integrals and an application to distinguished homomorphisms*, Illinois J. Math. **19** (1975), 537–541.
7. Gamelin T. W., *Cluster values of bounded analytic functions*, Trans. Amer. Math. Soc. **225** (1977), 295–306.
8. Hayashi M., *Linear extremal problems on Riemann surfaces*, preprint.

DEPARTMENT OF MATHEMATICS
UNIVERSITY OF CALIFORNIA
LOS ANGELES
CA 90024
USA

2.15

FINITELY GENERATED BANACH ALGEBRAS

J. WERMER

A Banach algebra $\mathcal{A}$ is called *finitely generated* if there exists a finite set of elements $x_1, \dots, x_n$ in $\mathcal{A}$ such that the smallest closed subalgebra of $\mathcal{A}$ which contains $x_1, \dots, x_n$ is $\mathcal{A}$ itself.

QUESTION 1. *Let E be a compact set in $\mathbb{C}$, and let A_E be the algebra of all continuous functions on the Riemann sphere which are analytic on $\mathbb{C} \setminus E$ and at ∞. For $f \in A_E$ put*

$$\|f\| = \max_E |f|.$$

Then A_E is a Banach algebra in this norm. Is A_E finitely generated?

QUESTION 2. *Let α be an irrational positive number. Let A_α denote the algebra of all continuous functions f on the torus $\mathbb{T}^2$ with Fourier series on $\mathbb{T}^2$:*

$$f \sim \sum_{n+m\alpha \geqslant 0} c_{nm} e^{in\theta} e^{im\varphi}.$$

Let $\|f\| = \sup\{\,|f(\theta)| : \theta \in \mathbb{T}^2\,\}$ *for* $f \in A_\alpha$. (These algebras have been studied by Arens–Singer, Helson–Lowdenslager, and others.) *Is A_α finitely generated?*

Background material for these two questions is contained in the references. For example, the algebra A_E is described in [2, Chapter II, §1], and the algebra A_α is described in [3, §18].

REFERENCES

1. Browder A., *Introduction to Function Algebras*, W. A. Benjamin, New York, 1969.
2. Gamelin T. W., *Uniform Algebras*, Prentice-Hall, Englewood Cliffs, New Jersey, 1969.
3. Stout E. L., *The Theory of Uniform Algebras*, Bogden and Quigley, Tarrytown-on-Hudson, New York, 1971.

DEPARTMENT OF MATHEMATICS
BROWN UNIVERSITY
PROVIDENCE
RHODE ISLAND 02912
USA

2.16
v.old

SETS OF ANTISYMMETRY AND SUPPORT SETS FOR $H^\infty + C$

D. SARASON

Let X be a compact Hausdorff space and A a closed subalgebra of $C(X)$ which contains the constants and separates the points of X. A subset S of X is called a *set of antisymmetry* for A if any function in A which is real-valued on S is constant on S. This notion was introduced by E. Bishop [1] (see also [2]), who established the following fundamental results:

(i) X can be written as the disjoint union of the maximal sets of antisymmetry for A;

(ii) if S is a maximal set of antisymmetry for A, then the restriction algebra $A|S$ is closed;

(iii) if f is in $C(X)$ and $f|S$ is in $A|S$ for every maximal set of antisymmetry S for A, then f is in A.

A closed subset of X is called a *support set for* A if it is the support of a representing measure for A (i.e., a Borel probability measure on X which is multiplicative on A). It is trivial to verify that every support set for A is a set of antisymmetry for A. However, there is in general no closer connection between these two classes of sets. This is illustrated by B. Cole's counter-example to the peak point conjecture [3, Appendix], which is an algebra $A \neq C(X)$ such that X is the maximal ideal space of A and such that every point of X is a peak point of A. For such an algebra, the only support sets are singletons, but not every set of antisymmetry is a singleton (by (iii)).

The present problem concerns a naturally arising algebra for which there does seem to be a close connection between maximal sets of antisymmetry and support sets. However, the evidence at this point is circumstantial and the precise connection remains to be elucidated. Let L^∞ denote the L^∞-space of Lebesgue measure on $\mathbb{T}$. Let H^∞ be the space of boundary functions on $\mathbb{T}$ for bounded holomorphic functions in $\mathbb{D}$, and let C denote $C(\mathbb{T})$. It is well known that $H^\infty + C$ is a closed subalgebra of L^∞ ([4]), so we may identify it, under the Gelfand transformation, with a closed subalgebra of $C(M(L^\infty))$, where $M(L^\infty)$ denotes the maximal ideal space of L^∞ (with its Gelfand topology). In what follows, by a set of antisymmetry or a support set, we shall mean these notions for the case $X = M(L^\infty)$ and A the Gelfand transform of $H^\infty + C$. Also, we shall identify the functions in L^∞ with their Gelfand transforms.

The first piece of evidence for the connection alluded to above is the following result from [5]: If f is in L^∞ and $f|S$ is in $(H^\infty + C)|S$ for each support set S, then f is in $H^\infty + C$. This is an ostensible improvement of part (iii) of Bishop's theorem in the present special situation. It is natural to ask whether it is an actual improvement, or whether it might not be a corollary to Bishop's theorem via some hidden connection between maximal sets of antisymmetry and support sets. The proof of the result is basically classical analysis and so offers no clues about the latter question. The question is motivated, in part, by a desire to understand the result from the viewpoint of abstract function algebras.

A second piece of evidence comes from [6], where a sufficient condition is obtained for the semi-commutator of two Toeplitz operators to be compact. The condition can be formulated in terms of support sets, and it is ostensibly weaker than an earlier sufficient condition of Axler [7] involving maximal sets of antisymmetry. Again, it is natural to ask whether the newer result is really an improvement of the older one, or whether the two are actually equivalent by virtue of a hidden connection between maximal sets of antisymmetry and support sets. As before, the proof offers no clues.

As a final piece of evidence, one can add the following unpublished results of K. Hoffman:

(1) If two support sets for $H^\infty + C$ intersect, then one of them is contained in the other;

(2) There exist maximal support sets for $H^\infty + C$.

All of the above makes one suspect that each maximal set of antisymmetry for $H^\infty + C$ can be built up in a "nice" way from support sets. It would not be surprising to learn that each maximal set of antisymmetry is a support set. At any rate, there is certainly a connection worth investigating.

Comments of 1984. The structure of the maximal sets of antisymmetry for $H^\infty + C$ remains mysterious, although a little progress has occurred. P. M. Gorkin in her dissertation [8] has the very nice result that $M(L^\infty)$ contains singletons which are maximal sets of antisymmetry for $H^\infty + C$. Such singletons are of course also maximal support sets. The author's paper [9] contains a result which is probably relevant to the problem.

References

1. Bishop E., *A generalization of the Stone-Weierstrass theorem*, Pacific J. Math. **11** (1961), 777–783.
2. Glicksberg I., *Measures orthogonal to algebras and sets of antisymmetry*, Trans. Amer. Math. Soc. **105** (1962), 415–435.
3. Browder A., *Introduction to Function Algebras*, W. A. Benjamin Inc., 1969.
4. Sarason D., *Algebras of functions on the unit circle*, Bull. Amer. Math. Soc. **79** (1973), 286–299.
5. Sarason D., *Functions of vanishing mean oscillation*, Trans. Amer. Math. Soc. **207** (1975), 391–405.
6. Axler S., Chang S.-Y., Sarason D., *Products of Toeplitz operators*, Int. Equat. Oper. Theory **1** (1978), 285–309.
7. Axler S., Doctoral Dissertation, University of California, Berkeley, 1975.
8. Gorkin P. M., *Decompositions of the maximal ideal space of L^∞*, Doctoral Dissertation, Michigan State University, East Lansing, 1982.
9. Sarason D., *The Shilov and Bishop decompositions of $H^\infty + C$*, Conference on Harmonic Analysis in Honour of Antoni Zygmund,II, Wadsworth, Belmont, 1983, pp. 464–474.

Department of Mathematics
University of California
Berkeley
California 94720
USA

Chapter Editors' Note

For discussion of this problem, see the following problem of P. M. Gorkin.

2.17

RESULTS ON ANTISYMMETRIC SETS AND A QUESTION ABOUT GLEASON PARTS

PAMELA GORKIN

1. Gleason parts in $M(H^\infty)$. Let H^∞ denote the algebra of bounded analytic functions on the open unit disc D and let $M(H^\infty)$ denote the maximal ideal space of H^∞. Recent results related to Sarason's problem 2.16 have caused more attention to be focused on certain other subsets of $M(H^\infty)$. For $\phi, \tau \in M(H^\infty)$, the pseudohyperbolic distance between them, denoted $\rho(\phi, \tau)$, is defined by

$$\rho(\phi, \tau) = \sup\{\, |\phi(f)| : f \in H^\infty, |f| < 1 \text{ on } D, \tau(f) = 0\,\}.$$

The subsets we will look at are the Gleason parts. For each $\phi \in M(H^\infty)$ we define the Gleason part of ϕ, denoted $P(\phi)$, by

$$P(\phi) = \{\, \tau \in M(H^\infty) : \rho(\phi, \tau) < 1\,\}.$$

It is easy to see that any two parts are either equal or disjoint.

In Sarason's article (2.16), he considered the algebra L^∞ with respect to Lebesgue measure on T. Letting H^∞ also denote the space of boundary functions on T for bounded holomorphic functions on D and C the space of continuous functions on T, Sarason looks at the Bishop decomposition of $M(L^\infty)$ into maximal antisymmetric sets for $H^\infty + C$ and the relationship of these sets to support sets of representing measures for $H^\infty + C$ (see Sarason's article for definitions). Motivated by a result in [5], operator theoretic results for Toeplitz operators on the Hardy space and by an unpublished result of K. Hoffman, which says that if two support sets intersect then one is contained in the other, Sarason conjectures that maximal antisymmetric sets are somehow made up of support sets. When we look at algebras of functions on the disc or Toeplitz operators on the Bergman space, Gleason parts seem to play the role that support sets and antisymmetric sets play for algebras of functions on the unit circle T.

Let dA denote normalized area measure on D. The Bergman space L^2_a is the space of analytic functions on D which are in $L^2(D, dA)$. In [17] D. Zheng gave necessary and sufficient conditions for the semi-commutator of two Toeplitz operators on the Bergman space with bounded harmonic symbols to be compact. This result can be formulated in terms of Gleason parts (just as the results in [6] and [16] can be stated in terms of support sets). In ([10], [11], [12], [13]) one looks at a function's behavior on each Gleason part to determine whether or not it belongs to a certain algebra, in much the same way as the Bishop decomposition theorem can be used to determine whether or not a function is in a certain algebra. To determine the reason for the similarities in the results we are getting, it would help if we had an analogue of K. Hoffman's unpublished result mentioned above. An answer to the following would be helpful:

PROBLEM. *If the closures of two Gleason parts (other than D) intersect, must there exist a Gleason part $P(\phi)$ with $\phi \notin D$ such that the union of the two parts is contained in $\overline{P(\phi)}$?*

2. An update on problem 2.16. There has been some progress made on Problem 2.16 in recent years, but this problem remains open. Most of the progress has occurred by considering the Shilov decomposition of $M(L^\infty)$.

The largest C^*-subalgebra of $H^\infty + C$ will be denoted (as usual) by QC. The sets in the Shilov decomposition are the maximal level sets of QC. In [9], D. Sarason showed that, in fact, the Bishop decomposition of $M(L^\infty)$ is strictly finer than the Shilov decomposition. More examples of this sort appear in [8] where it was also shown, by considering limits of certain sequences of points in $M(L^\infty)$, that there exist one-point maximal antisymmetric sets. Since these are also support sets, the examples of one point maximal antisymmetric sets provide more evidence that there is a relationship between maximal antisymmetric sets and support sets.

Assuming the continuum hypothesis, K. Izuchi [14] showed that for any interpolating Blaschke product b there is a point x in the zero set of b such that the support set of the representing measure for x is a QC level set. Such a support set must be a maximal antisymmetric set. It would be nice to have a proof of this fact that did not depend on the continuum hypothesis.

If we restrict our attention to interpolating Blaschke products with the condition that the zero sequence $\{z_n\}$ satisfy

$$\lim_{n\to\infty} \prod_{m:m\neq n} \frac{|z_m - z_n|}{|1 - \overline{z_n} z_m|} = 1,$$

then we get more information. These interpolating Blaschke products have been studied extensively and are referred to as *thin* or *sparse* Blaschke products. Izuchi showed that if x is a point in the zero set of a thin Blaschke product b which is a cluster point of a countable subset of the zeros of b, then the support set of the representing measure for x is not a QC level set. The most recent result on this problem that I know of is also due to K. Izuchi. In [15, Theorem 5.3] Izuchi studied sequences of QC level sets which are called *strongly discrete*. Choosing one point in each such QC level set, he showed that any cluster point is a maximal antisymmetric set and is not a QC level set.

REFERENCES

(References 1–9 are given in Sarason's article 2.16)

10. Axler S., Gorkin P., *Algebras on the disk and doubly commuting multiplication operators*, Trans. Amer. Math. Soc. **309** (1988), 711–722.
11. Gorkin P., *Gleason parts and COP*, J. Funct. Anal. **83** (1989), 44–49.
12. Gorkin P., *Algebras of bounded functions on the disc*, Proceedings of the Conference on Function Spaces, Marcel Dekker (to appear).
13. Gorkin P., Izuchi K., *Some counterexamples in subalgebras of $L^\infty(D)$*, preprint.

14. Izuchi K., *QC-level sets and quotients of Douglas algebras*, J. Funct. Anal. **65** (1986), 293–308.
15. Izuchi K., *Countably generated Douglas algebras*, Trans. Amer. Math. Soc. **299** (1987), 171–192.
16. Volberg A. L., *Two remarks concerning the theorems of S. Axler, S.-Y. A. Chang and D. Sarason*, J. Operator Theory **7** (1982), 209–218.
17. Zheng D. C., *Hankel and Toeplitz operators on the Bergman space*, J. Funct. Anal. **83** (1989), 98–120.

Department of Mathematics
Bucknell University
Lewisburg PA 17837
USA

2.18

FILTRATIONS OF C^*-ALGEBRAS

D. VOICULESCU

Let A be a unital C^*-algebra endowed with a *filtration* $(F_n)_{n\in\mathbb{N}}$. By this we mean that the F_n's are vector subspaces of A such that:

(a) $F_0 \subset F_1 \subset F_2 \subset \ldots$;
(b) $F_0 = \mathbb{C}1$, $F_mF_n \subset F_{m+n}$, and $F_m = F_m^*$ for all $m, n \in \mathbb{N}$;
(c) $\bigcup\{F_n : n \in \mathbb{N}\}$ is dense in A.

A filtration of A is called *subexponential* if the F_n's have finite dimension and if

$$\lim_{n\to\infty} (\dim F_n)^{1/n} = 1.$$

The problem we propose is inspired by the case of groups. It is known that, if G is a group of subexponential growth, then G is amenable. Here the subexponential growth means that $G = \bigcup\{S_k : k \in \mathbb{N}\}$, where $\{e\} = S_0 \subset S_1 \subset \ldots$, $S_k = S_k^{-1}$, $S_mS_n \subset S_{m+n}$ and

$$\lim_{n\to\infty} |S_n|^{1/n} = 1.$$

Clearly, if in the C^*-algebra $C^*(G)$ of the discrete group G, we define F_n to be the linear span of the unitaries corresponding to the elements of S_n, then we get a filtration of $C^*(G)$, and this filtration is clearly subexponential. On the other hand, the amenability of G is known to be equivalent to the nuclearity of $C^*(G)$. This suggests the following problem.

PROBLEM. *If A is a C^*-algebra with a subexponential filtration, does it follow that A is nuclear?*

For other facts on filtrations of C^*-algebras, see [1, §5].

REFERENCE

1. Voiculescu D., *On the existence of quasicentral approximate units relative to normed ideals*, Part I, J. Funct. Anal. **90** (1990), 1–36.

DEPARTMENT OF MATHEMATICS
UNIVERSITY OF CALIFORNIA
BERKELEY
CA 94720
USA

2.19
old

A QUESTION INVOLVING ANALYTIC FAMILIES OF OPERATORS

RICHARD ROCHBERG

Let A be a uniform algebra with Shilov boundary X. (The case when A is the disk algebra and $X = \mathbb{T}$ is an interesting example for this purpose.) Suppose that we are given a linear operator S which maps A into $C(X)$ and has small norm. Suppose further that the image $(I + S)(A)$ is a subalgebra (here I is the inclusion of A into $C(X)$).

QUESTION. *Is there an analytic family of linear operators $S(z)$ defined for z in $\overline{\mathbb{D}}$ so that $(I + S(z))(A)$ is a subalgebra of $C(X)$ for each z in $\overline{\mathbb{D}}$ and so that $S(1) = S$?*

The hypotheses are related to questions of deformation of the structure of A; see [2] for details and examples. In cases where $S(z)$ can be obtained, the differential analysis of $S(z)$ connects the deformation theory of A with the cohomology of A. (See [1].) For instance, $S'(0)$ would be a continuous derivation of the algebra A into the A-module $C(X)/A$. Such considerations lead rapidly to questions about operators on spaces such as $VMO(= C(\mathbb{T})/\text{disk algebra})$. (See [3] for an example.)

REFERENCES

1. Johnson B. E., *Low dimensional cohomology of Banach algebras*, Proc. Symp. Pure Math. **38** (1982), 253–259.
2. Rochberg R., *Deformation of uniform algebras.*, Proc. Lond. Math. Soc. **39** (1979), no. 3, 93–118.
3. Rochberg R., *A Hankel type operator arising in deformation theory*, Proc. Symp. Pure Math. **35** (1979), 457–458.

DEPARTMENT OF MATHEMATICS
WASHINGTON UNIVERSITY
BOX 1146
ST. LOUIS, MO 63130
USA

2.20
old

OPERATOR ALGEBRAS IN WHICH ALL FREDHOLM OPERATORS ARE INVERTIBLE

N. Ya. Krupnik, A. S. Markus, I. A. Fel'dman

Let $\mathcal{L}(X)$ be the algebra of all bounded linear operators on the Banach space X. An operator $A \in \mathcal{L}(X)$ is called a *Fredholm operator* if

$$\dim \operatorname{Ker} A < \infty \text{ and } \dim X / \operatorname{Im} A < \infty.$$

It is well known that the operator of multiplication by a function (in L^p or C) is invertible if it is a Fredholm operator. The same is true for multidimensional Wiener-Hopf operators in $L^p(\mathbb{R}^n_+)$, for their discrete analogues in $\ell^p(\mathbb{Z}^n_+)$, and for operators in $L^p(\mathbb{T})$ of the form

$$(A\varphi)(t) = a(t)\varphi(t) + b(t)\varphi(d(t)) \qquad (a, b \in L^\infty(\mathbb{T})),$$

where d is a homeomorphism of the circle $\mathbb{T}$ onto itself. This property is valid for the elements of uniformly closed algebras (it is supposed that all algebras under consideration contain the identity operator) generated by the above operators as well (see [1] and the literature cited therein). The usual scheme of the proof consists of two stages. First we prove the invertibility of Fredholm operators in the non-closed algebra $\mathcal{A}$ generated by the initial operators (using the linear expansion [2] it is reduced to the same operators but with matrix coefficients or kernels [3]). Then we have to extend this statement in some way to the uniform closure $\operatorname{clos} \mathcal{A}$ of the algebra $\mathcal{A}$.

Question 1. *Let every Fredholm operator from the algebra $\mathcal{A}$ ($\subset \mathcal{L}(X)$) be invertible. Is every Fredholm operator in the algebra $\operatorname{clos} \mathcal{A}$ invertible?*

In the examples, the passage from $\mathcal{A}$ to $\operatorname{clos} \mathcal{A}$ becomes easier if $A^{-1} \in \operatorname{clos} \mathcal{A}$ for each invertible operator $A \in \mathcal{A}$.

Question 2. *Let every Fredholm operator $A \in \mathcal{A}$ be invertible, and let $A^{-1} \in \operatorname{clos} \mathcal{A}$. Is every Fredholm operator in the algebra $\operatorname{clos} \mathcal{A}$ invertible?*

We point out two cases, when the answer to Question 1 is positive [1].

1°. The algebra $\mathcal{A}$ is commutative (or $\mathcal{A} \subset \mathcal{L}(X^n)$ and $\mathcal{A}$ consists of operator matrices with elements in some commutative algebra $\mathcal{A}_0 \subset \mathcal{L}(X)$).

2°. The space X is a Hilbert space and $\mathcal{A}$ is a symmetric algebra.

The answer to Question 2 is positive if one of the following conditions is satisfied (see [1]).

3°. The algebra $\operatorname{clos} \mathcal{A}$ is semisimple.

4°. The system of minimal invariant subspaces of the algebra $\mathcal{A}$ is complete in X.

5°. The algebra $\operatorname{clos} \mathcal{A}$ does not contain nil ideals consisting of finite-dimensional operators.

We call a non-zero invariant subspace minimal if it does not contain any other such subspaces. A two-sided ideal is called a nil ideal if all its elements are nilpotent.

Either of the conditions 3°, 4° implies condition 5°. For 3° this is obvious, and for 4° this follows from [4] (compare with [5]).

References

1. Krupnik N. Ya., Fel'dman I. A., *On the invertibility of certain Fredholm operators*, Izv. Akad Nauk MSSR ser. fiz. tekh. i mat. nauk **2** (1982), 8–14. (Russian)
2. Gohberg I. Ts., Krupnik N. Ya., *Singular integral operators with piecewise continuous coefficients and their symbols*, Izv. Akad. Nauk SSSR ser. matem. **35** (1971), no. 4, 940–964 (Russian); English transl. in Math. USSR-Izv. **5** (1971), 955–979.
3. Krupnik N. Ya., Fel'dman I. A., *On the impossibility of introducing a matrix symbol for certain algebras of operators*, Linear Operators and Integral Equations, Shtiintsa, Kishinev, 1981, pp. 75–85. (Russian)
4. Lomonosov V. I., *On invariant subspaces of a family of operators which commute with a completely continuous operator*, Funktsional. Anal. i Prilozhen. **7** (1973), no. 3, 55–56. (Russian)
5. Markus A. S., Fel'dman I. A., *On algebras generated by operators with a one-sided inverse*, Research on Differential Equations, Shtiintsa, Kishinev, 1983, pp. 42–46. (Russian)

Department of Mathematics
Bar Ilan University
Ramat Gan
Israel

Department of Mathematics and Computer Sciences
Ben Gurion University of the Negev
P.O. Box 653 Beer-Sheva 84105
Israel

Department of Mathematics and Computer Sciences
Ben Gurion University of the Negev
P.O. Box 653 Beer-Sheva 84105
Israel

2.21
v.old

ON THE COHEN-RUDIN CHARACTERIZATION OF HOMOMORPHISMS OF MEASURE ALGEBRAS

SATORI IGARI

Let $L(\mathbb{T})$ be the Lebesgue space and $M(\mathbb{T})$ the set of all bounded regular Borel measures on the unit circle $\mathbb{T}$. Then $M(\mathbb{T})$ is a commutative Banach algebra with the convolution product and the norm of total variation, and $L(\mathbb{T})$ is embedded in $M(\mathbb{T})$ as a closed ideal. A subalgebra N of $M(\mathbb{T})$ is said to be an *L-subalgebra* if it is a closed subalgebra of $M(\mathbb{T})$ and $\mu \in N$ and $\nu \ll \mu$ (that is, ν is absolutely continuous with respect to μ) implies $\nu \in N$.

Let $\Delta'(N)$ be the set of all homomorphisms of N to the complex numbers (which might be trivial). Then, by Yu. Schreider [1], for every $\psi \in \Delta'(N)$, there corresponds a unique generalized character $\{\, \psi_\mu : \mu \in N \,\}$ or zero system such that

$$\psi(\mu) = \int_{\mathbb{T}} \psi_\mu(t)\, d\mu(t), \qquad \mu \in N.$$

In the following, we shall use the same notation ψ for $\{\psi_\mu\}$. A generalized character $\psi = \{\, \psi_\mu : \mu \in N \,\}$ satisfies, by definition,

(i) $\psi_\mu \in L^\infty(|\mu|)$ and $\mu - \operatorname{ess\,sup} |\psi_\mu| > 0$,
(ii) $\psi_\mu = \psi_\nu$ $\nu -$ a.e. if $\nu \ll \mu$,
(iii) $\psi_{\mu * \nu}(s + t) = \psi_\mu(s)\psi_\nu(t)$ for $\mu \times \nu -$ a.e. (s,t).

Let Ψ be a homomorphism of N to $M(\mathbb{T})$. Then the mapping $\nu \to (\Psi\nu)\,\widehat{}\,(n)$, $\nu \in N$, defines a homomorphism for every integer n, where '$\widehat{\ }$' denotes the Fourier-Stieltjes transform

$$\widehat{\mu}(n) = \int_{\mathbb{T}} e^{-int}\, d\mu(t).$$

Thus there exists a generalized character $\psi(n) = \{\, \psi_\nu(n,t) : \nu \in N \,\}$ or zero system such that

$$(\Psi\nu)\,\widehat{}\,(n) = \int_{\mathbb{T}} \psi_\nu(n,t)\, d\nu(t), \qquad n \in \mathbb{Z}. \tag{1}$$

Let $\{a_n\}_{n \geqslant 0}$ be a sequence of integers such that $a_n \geqslant 2$ and $a_n > 2$ for infinitely many n. Put

$$d_n = 2\pi \prod_{m=1}^{n} a_m^{-1}.$$

Let

$$\mu = \mathop{*}_{n=1}^{\infty} \frac{1}{2}(\delta(0) + \delta(d_n))$$

be a Bernoulli convolution product, where $\delta(a)$ is a Dirac measure concentrated on a point a. We fix such a μ and denote by $N(\mu)$ the smallest L-subalgebra containing μ.

THEOREM ([2], [3]). *Let M be an L-subalgebra $L(\mathbb{T})$ or $N(\mu)$ and Ψ be a homomorphism of M to $M(\mathbb{T})$. Suppose* (A) $|\psi(n)|^2 = |\psi(n)|$ *(i.e.,* $|\psi_\nu(n,t)|^2 = |\psi_\nu(n,t)|$ $\nu-$a.e. *for all n in $\mathbb{Z}$ and ν in M). Then we have:*

(a) *a positive integer m and a finite subset* $R = \{n_{m+1}, n_{m+2}, ..., n_\ell\}$ *of* $\mathbb{Z}$;
(b) $\varphi_j \in \Delta'(M)$ $(j = 1, 2, ..., \ell)$;
(c) $\pi_j \in \Delta'(M)$ *with* $|\pi_j|^2 = |\pi_j|$ $(j = 1, 2, ..., m)$ *such that*

$$\psi(n) = \begin{cases} \varphi_j, & n = n_j \in R \\ \sum_{j=1}^{m} \pi_j^n \varphi_j \; C_{m\mathbb{Z}+j}(n), & n \notin R \end{cases} \tag{2}$$

where C_E denotes the characteristic function of the set E.

Conversely, if $\{\psi(n)\}$ is a sequence in $\Delta'(M)$ satisfying (A), (a), (b), *and* (c), *then the mapping Ψ given by* (1) *is a homomorphism of M to $M(\mathbb{T})$.*

When $M = L(\mathbb{T})$, then $\Delta'(M) = \{ e^{int} : n \in \mathbb{Z} \} \cup \{0\}$ and the condition (A) is obviously satisfied. For this case the theorem is due to W. Rudin. The other case, when $M = N(\mu)$, has been proved by S. Igari, Y. Kanjin. Since $L(\mathbb{T})$ is an ideal, our theorem holds good for

$$M = L(\mathbb{T}) \oplus N(\mu).$$

We remark that we cannot expect the conditions (a), (b) and (c) without the hypothesis (A) (cf. [2]).

PROBLEM 1. *For what kind of L-subalgebra M does the above theorem hold good?*

PROBLEM 2. *Let $M = M(\mathbb{T})$ and Ψ be a homomorphism of $M(\mathbb{T})$ to $M(\mathbb{T})$. Let $\{\psi(n)\}$ be a sequence of $\Delta'(M(\mathbb{T}))$ given by* (1) *and assume that $\{\psi_\nu(n)\}$ satisfies the condition* (A) *for a measure ν. Then characterize ν such that the conditions* (a), (b) *and* (c) *hold for $\{\psi_\nu(n)\}$.*

REFERENCES

1. Schreider Yu. A., *Structures of maximal ideals in rings of measures with convolution*, Matem. Sbornik **27** (1950), 297–318. (Russian)
2. Rudin W., *The automorphisms and the endomorphisms of the group algebra of the unit circle*, Acta Math. **95** (1976), 39–56.
3. Igari S., Kanjin Y., *The homomorphisms of the measure algebras on the unit circle*, J. Math. Soc. Japan **31** (1979), 503–512.

MATHEMATICAL INSTITUTE
TOHÔKU UNIVERSITY
SENDAI 980
JAPAN

2.22
v.old

ANALYTICITY IN THE GELFAND SPACE OF THE ALGEBRA OF $L^1(\mathbb{R})$ MULTIPLIERS

G. BROWN, W. MORAN

We shall be concerned with spectral properties of the Banach algebra of those bounded linear operators on $L^1(\mathbb{R})$ which commute with translations. However, it is convenient to represent the action of each operator by convolution so that the object of study becomes the algebra $M(\mathbb{R})$ of bounded regular Borel measures on $\mathbb{R}$. The general problem to be considered is the classification of the analytic structure of the Gelfand space Δ of $M(\mathbb{R})$, despite the fact that Δ is sometimes regarded as the canonical example of a "horrible" maximal ideal space from the point of view of complex analysis (cf., [1, p. 9]). Some encouraging progress has been made in recent years and it will be possible to pose some specific questions which should be tractable.

We refer to Taylor's monograph [2] for a survey of work up to 1973. (Miller's conjectured characterization of the Gleason parts of Δ has since been verified in [3]), and for further details concerning general theory of convolution measure algebras. In particular, we follow Taylor in representing Δ as the semigroup of continuous characters on a compact semigroup S (the so-called *structure semigroup* of $M(\mathbb{R})$) and in transferring measures in $M(\mathbb{R})$ to measures on S. In this formulation an element f of Δ acts as a homomorphism according to the rule

$$f(\mu) = \int_S f(s)\, d\mu(s).$$

Every member f of Δ then has a canonical polar decomposition $f = |f|h$, where $|f|, h \in \Delta$ and h has idempotent modulus. If f itself does not have idempotent modulus (a possibility which corresponds to the Wiener-Pitt phenomenon and was first noted by Schreider [4]), then the map $z \mapsto |f|^z h$, for $\mathrm{Re}(z) > 0$, demonstrates analyticity in Δ. From that observation Taylor showed that the Shilov boundary ∂ of $M(\mathbb{R})$ is contained in $\operatorname{clos}\theta$, where

$$\theta = \{ f \in \Delta : |f| = |f|^2 \}.$$

He posed the converse question, which is still unresolved. Subsequent work tends to suggest a negative answer, so that we propose:

CONJECTURE 1. $\theta \setminus \partial \neq \emptyset$.

It should be noted that the result $\partial \subset \operatorname{clos}\theta$ remains valid for abstract convolution measure algebras, and that it is very easy to find convolution measure algebras for which $\theta \setminus \partial \neq \emptyset$. It is also possible to find natural L-subalgebras of $M(\mathbb{R})$ itself for which the corresponding conjecture is true. (An *L-subalgebra* is a closed subalgebra A which contains all measures absolutely continuous with respect to any measure in A.) Thus, a disproof would depend on not only a new phenomenon peculiar to $M(\mathbb{R})$, but one which is specific to the full algebra. In addition we established a weak form of the conjecture in [5] by showing that a certain idempotent $\mathbb{I}_d$ (see below) fails to be a strong boundary point for $M(\mathbb{R})$ (although it is a strong boundary point for the L-subalgebra of discrete measures). It should also be noted that Johnson [6], proved that $\Delta \setminus \partial \neq \emptyset$, but the techniques used to prove this result and its subsequent refinements depend essentially on the use of elements lying outside θ. A natural strategy is to embed $M(\mathbb{R})$ in a suitable super-algebra and prove the impossibility of extension of an appropriate homomorphism.

It appears to be almost as difficult to exhibit, in the opposite direction, large numbers of elements of θ which **do** belong to ∂. Before we describe some progress in this direction, let us introduce the notation $\mathbb{I}$ for the unit function in Δ and the notation $\mathbb{I}_d$ for the homomorphism given by

$$\mathbb{I}_d(\mu) = \int_{\mathbb{R}} d\mu_d\,,$$

where μ_d is the discrete part of μ. ($\mathbb{I}_d$ plays the role of the unit function for the subalgebra of discrete measures, which can be regarded as $M(\mathbb{R}_d)$, where $\mathbb{R}_d$ is the discrete real line.)

Let us define a partial order on Δ by saying that $f \leqslant g$ if

$$|f(s)|^2 \leqslant g(s)\overline{f(s)} \qquad (s \in S)\,.$$

We have shown in [7] that maximal elements are members of the Shilov boundary.

THEOREM 1.

(i) *If f is maximal in Δ, then f is a strong boundary point.*

(ii) *If $|f|$ is maximal in $\Delta \setminus \{\mathbb{I}\}$, then f belongs to ∂. If, moreover, the L-subalgebra*

$$\{\,\mu \in M(\mathbb{R}) : |f|\mu = \mu\,\}$$

is countably generated, then f is a strong boundary point.

It is obvious that maximal elements belong to θ, but it is not entirely trivial that there are many examples other than those homomorphisms induced by continuous characters of $\mathbb{R}$ (viz., extensions of non-zero homomorphisms of $L^1(\mathbb{R})$). To see that this is the case, consider in connection with (i), homomorphisms which are induced on discrete measures by discontinuous characters, and in connection with (ii), homomorphisms which annihilate some fixed member of $L^1(\mathbb{R})$.

The additional hypothesis in (ii) does not correspond to a specific obstruction and merely reflects the constructive nature of our proof. We have avoided a similar difficulty in (i) by an appeal to Rossi's local peak set theorem and it seems plausible that a similar device should be available here. A proof which reduced the uncountably generated case to the countably generated case by pure measure algebra techniques would be particularly interesting since this species of difficulty often arises. In any event we propose:

CONJECTURE 2. *If f is maximal in $\Delta \setminus \{\mathbb{I}\}$, then f is a strong boundary point.*

It would be useful to determine for specific subclasses of θ whether or not the elements are strong boundary points. The result that $\mathbb{I}_d$ is the center of an analytic disc was extended in [8] to cover the case of the idempotent corresponding to any single generator Raikov system. On the other hand, we show in [7] that $\mathbb{I}_d$ is accessible in the sense that it is the infimum of those maximal elements of $\Delta \setminus \{\mathbb{I}\}$ below which it lies. It is natural to expect that both results extend, although we feel that present techniques would require substantial development to prove the final conjecture:

CONJECTURE 3. *The idempotents corresponding to proper Raikov systems are accessible, but fail to be strong boundary points.*

We have chosen to present these problems from the standpoint of the development of the general theory. From a practical position the most useful results are those which exhibit classes of homomorphisms which belong to the Shilov boundary of L-subalgebras of $M(\mathbb{R})$, because such results give information on spectral extension. In fact, Theorem 1 is of this type because it remains valid for arbitrary convolution measure algebras (provided the technical hypothesis that $\mathbb{I}$ is a critical point is added to part (ii)). Variants of that theorem with the weaker conclusion that f belongs to the Shilov boundary but valid for a larger class of f would be of considerable interest.

REFERENCES

1. Gamelin T. W., *Uniform Algebras*, Prentice-Hall, New Jersey, 1969.
2. Taylor J. L., *Measure Algebras*, CBMS Regional Conference Ser. Math., vol. 16, American Math. Soc., Providence, 1973.
3. Brown G., Moran W., *Gleason parts for measure algebras*, Math. Proc. Cambridge Phil. Soc. **79** (1976), 321–327.
4. Schreider Yu. A., *On a particular example of a generalized character*, Matem. Sbornik **29** (1951), no. 2, 419–426. (Russian)
5. Brown G., Moran W., *Point derivations on $M(G)$*, Bull. London Math. Soc. **8** (1976), 57–64.
6. Johnson B. E., *The Šilov boundary of $M(G)$*, Trans. American Math. Soc. **134** (1968), 289–296.
7. Brown G., Moran W., *Maximal elements of the maximal ideal space of a measure algebra*, Math. Ann. **246** (1979), no. 2, 131–140.
8. Brown G., Moran W., *Analytic discs in the maximal ideal space of $M(G)$*, Pacific J. Math. **75** (1978), 45–57.

UNIV. OF NEW SOUTH WALES
SCHOOL OF MATHEMATICS
P. O. BOX 1, KENSINGTON (2033)
AUSTRALIA

DEPARTMENT OF MATHEMATICS
UNIVERSITY OF ADELAIDE
ADELAIDE
AUSTRALIA

2.23
old

TWO PROBLEMS CONCERNING SEPARATION OF IDEALS IN GROUP ALGEBRAS

W. ŻELAZKO

All algebras in this paper are commutative, complex, regular Banach algebras. In such algebras all ideals consist of joint topological divisors of zero, i.e., if I is a (not necessarily closed) ideal in a regular Banach algebra, then there is a net (z_α) of elements of the algebra in question, which does not tend to zero, but $\lim_\alpha z_\alpha x = 0$ for all x in I (cf. [1]). In this case we say that the net (z_α) *annihilates* the ideal I. We say that an ideal $I \subset A$ has the *separation property* if, for each x in $A \setminus I$, there is a net $(z_\alpha) \subset A$ annihilating I and such that the net $z_\alpha x$ does not tend to zero. It can be shown that in this case there exists one net (z_α) which works for all elements x in $A \setminus I$, and, in fact, $I = \{x \in A : z_\alpha x \to 0\}$ (cf. [2]). In the case where there exists such a bounded net we say that the ideal I has the *bounded separation property*. An ideal with the bounded separation property is necessarily closed. If A is a regular Banach algebra and F is a closed non-void subset of its maximal ideal space, then both the maximal and the minimal (non-closed) ideal with hull F have the separation property. However the bounded separation property may fail for the minimal closed ideal with the given hull, even if it possesses the separation property. It is also possible to construct a closed ideal in a regular Banach algebra which has the separation property and which is different from the intersection of all the maximal ideals containing it. Thus the nets provide a tool for separation and description of ideals. It is particularly interesting whether this tool works for the group algebras. In this context we pose the following problems.

QUESTION 1. *Let I be a closed ideal in $L_1(G)$ for an LCA group G. Does I possess the separation property?*

QUESTION 2. *Does there exist an LCA group G and a closed ideal I in $L_1(G)$ which has the bounded separation property and is not of the form* $(*)$ $I = \bigcap\{M \in \mathcal{M}_A : I \subset M\}$?

In fact we do not know any example of a closed ideal in a group algebra which has the separation property but which is not of the form $(*)$.

REFERENCES

1. Żelazko W., *On a certain class of non-removable ideals in Banach algebras*, Studia Math. **44** (1972), 87–92.
2. Żelazko W., *On domination and separation of ideals in commutative Banach algebras*, Studia Math. **71** (1981), 179–189.

INSTITUTE OF MATHEMATICS
POLISH ACADEMY OF SCIENCES
00-950 WARSZAWA, P.O. BOX 137
POLAND

2.24

ANALYTIC ALGEBRAS AND COMPACTIFICATIONS OF THE DISK

O. V. IVANOV

1. Denote by $\mathbb{D} = \{ z : |z| < 1 \}$ the open unit disk. A compact set $b\mathbb{D}$ is called a compactification of $\mathbb{D}$ if $\mathbb{D} \subset b\mathbb{D}$ and $\mathbb{D}$ is a dense subset of $b\mathbb{D}$. Let $b_0\mathbb{D}$ be the one-point compactification and $\beta\mathbb{D}$ be Cech compactification of $\mathbb{D}$. The description of $\beta\mathbb{D}$ as the maximal ideal space $M(C(\mathbb{D}))$ of the algebra $C(\mathbb{D})$ of all bounded continuous functions is well known. The set of all compactifications $\mathcal{B} = [b_0\mathbb{D}, \beta\mathbb{D}]$ is partially ordered. We write $b_1\mathbb{D} < b_2\mathbb{D}$ iff there exists a continuous map $\pi\colon b_2\mathbb{D} \to b_1\mathbb{D}$ with $\pi(z) = z$ for all $z \in \mathbb{D}$.

By the Gelfand–Raikov–Shilov theorem [1] any compactification $b\mathbb{D} \in \mathcal{B}$ can be identified with the maximal ideal space of some C^*-subalgebra of $C(\mathbb{D})$. The more surprising was the Corona theorem of Carleson [2], stating that $\mathbb{D}$ is dense in the maximal ideal space $M(H^\infty) = b_{H^\infty}\mathbb{D}$ of a *non* C^*-algebra $H^\infty = H^\infty(\mathbb{D})$. It is well known that the Corona theorem is true for some *analytic* algebras $\mathcal{A}$, $A \subset \mathcal{A} \subset H^\infty$ (A stands for the disk algebra) generated by Sarason algebras [3–6]. On the other hand there exist some examples of *analytic* algebras such that $M(\mathcal{A})$ is not a compactification of $\mathbb{D}$ [7,8]. The generalization of the Corona theorem to any *analytic* algebra is a very complicated problem.

Let $b_e\mathbb{D} = \overline{\mathbb{D}}$ be the Euclidean closure of the open disk $\mathbb{D}$.

PROBLEM 1. *Which compactifications $b\mathbb{D} \in [b_e\mathbb{D}, b_{H^\infty}\mathbb{D}]$ can be identified with the maximal ideal space $M(\mathcal{A})$ of an analytic algebra $\mathcal{A}$, $A \subset \mathcal{A} \subset H^\infty$, such that $M(\mathcal{A})$ is a compactification of $\mathbb{D}$?*

This problem is equivalent to the following one.

PROBLEM 1′. *Which uniform closed C^*-algebras W, $W \subset C(\mathbb{D})$, enjoy the following property: there exists an analytic algebra $\mathcal{A}_W$, $A \subset \mathcal{A}_W \subset H^\infty$, such that $M(\mathcal{A}_W)$ is a compactification of $\mathbb{D}$ and $W = \mathrm{alg}(\mathcal{A}_W, \overline{\mathcal{A}_W})$?*

PROBLEM 2. *Give a characterization of analytic algebras $\mathcal{A}$, $A \subset \mathcal{A} \subset H^\infty$, such that $M(\mathcal{A})$ is a compactification of $\mathbb{D}$.*

PROBLEM 3. *Let $\mathcal{A}$ be an analytic algebra, $A \subset \mathcal{A} \subset H^\infty$, and $M(\mathcal{A}) = M(H^\infty)$. Give a necessary and sufficient condition for $\mathcal{A}$ to coincide with H^∞.*

DAWSON CONJECTURE [7]. *Let $\mathcal{E}$ be a uniformly closed C^*-algebra on $\partial\mathbb{D}$, $C(\partial\mathbb{D}) \subset \mathcal{E} \subset L^\infty(\partial\mathbb{D})$. Then $M(H^\infty \cap \mathcal{E})$ is a compactification of $\mathbb{D}$.*

2. Denote by $\mathcal{H}_{H^\infty}$ the algebra of bounded continuous functions f, which have an angular limit $f^*(\zeta)$ for almost all $\zeta \in \partial\mathbb{D}$, and $f^* \in H^\infty$. A "two-sheeted" analog of the Corona theorem was proved in [9]: the maximal ideal space $M(\mathcal{H}_{H^\infty})$ contains two open disks $\mathbb{D} \cup \mathbb{D}$ and $\mathbb{D} \cup \mathbb{D}$ is dense in $M(\mathcal{H}_{H^\infty})$.

PROBLEM 4. *Does there exist a uniform algebra W, $W \subset C(\mathbb{D})$, such that $M(W)$ contains three or more copies of $\mathbb{D}$ whose union is dense in $M(\mathcal{H}_{H^\infty})$?*

REFERENCES

1. Gelfand I. M., Raikov D. A., Shilov G. E., *Commutative Normed Rings*, Chelsea, New York, 1964.
2. Garnett J.B., *Bounded Analytic Functions*, Academic Press, New York, 1981.
3. Tolokonnikov V. A., *Corona theorem in algebras of bounded analytic functions*, Sov. Inst. Tech. Inform (VINITI) **251** (1984), 1–60. (Russian)
4. Tolokonnikov V. A., *Generalized Douglas algebras*, Algebra i Analiz **3** (1991), no. 2, 231–252 (Russian); English transl. in Leningrad Math. J. **3** (1992), no. 2.
5. Mortini R., *Corona theorems for subalgebras of H^∞*, Mich. Math. J. **36** (1989), no. 2, 193–201.
6. Sundberg C., Wolff T. H., *Interpolating for QA_B*, Trans. Amer. Math. Soc. **276** (1983), 551–582.
7. Dawson D. W., *Subalgebras of H^∞*, Thesis, Indiana Univ., Bloomington, Indiana, 1975.
8. Scheinberg S., *Cluster sets and Corona theorems*, Lecture Notes in Math., vol. 604, 1977, pp. 103–106.
9. Ivanov O. V., *Generalized Douglas algebras and the Corona theorem*, Sibirsk. Mat. Zh. **32** (1991), no. 1, 37–42. (Russian)

SHAKESPEARE STR. 21, 19
DONETSK 50, 340050
UKRAINE

2.25

BOUNDARY BEHAVIOUR OF H^∞ ON AN INFINITELY CONNECTED DOMAIN

M. SAMOKHIN

Let D be an arbitrary domain of the extended complex plane, $\pi\colon \Delta \to D$ the universal covering mapping of the unit disk Δ onto D, $\mathcal{G}$ the group of covering transformation. Let $L^p(d\theta, \mathcal{G})$ be the subspace of the space $L^p(d\theta)$ on the unit circle, consisting of functions automorphic with respect to $\mathcal{G}$ and let $H^\infty(\Delta, \mathcal{G})$ be the analogous subspace of $H^\infty(\Delta)$.

PROBLEM 1. *Characterize the class of all domains D, for which the following statement is true: for any function $u \in L^\infty(d\theta, \mathcal{G})$, $u \geqslant 0$, $\ln u \in L^1(d\theta, \mathcal{G})$, there exists a function $H \in H^\infty(\Delta, \mathcal{G})$ such that the modulus of its radial boundary values coincides with u almost everywhere on $\partial\Delta$.*

Well known classical results imply that any simply connected domain with nondegenerate boundary or any finitely connected domain with Jordan boundary belongs to that class.

A restricted form of Problem 1 is of particular interest in case of infinitely connected domains. In this restricted problem the function u is supposed to be bounded away from zero ($u \geqslant \rho > 0$). Various characteristics and properties of the corresponding class of plane domains (so called "harmonic domains") are known. For example, harmonic domains $\mathcal{D}$ are characterized by the fact that Shilov boundary of $H^\infty(\mathcal{D})$ coincides, up to a natural homeomorphism, with Choquet boundary of the space of bounded harmonic functions. It should be noted that additional condition $u \geqslant \rho > 0$ cannot be eliminated: it is possible to construct a harmonic domain D and a function $u \in L^\infty(d\theta, \mathcal{G})$, $u \geqslant 0$, $\ln u \in L^1(d\theta, \mathcal{G})$ in such a way that there is no function in $H^\infty(\Delta, \mathcal{G})$ with modulus of its boundary values equal to u almost everywhere on $\partial\Delta$. So the problem now is to fill the gap between the class of harmonic domains and the class of domains considered in Problem 1. The latter one contains a wide class of Parreau–Widom type domains. Therefore it would be interesting to consider the following problem.

PROBLEM 2. *What are the relations between the class of domains of Parreau–Widom type and the class of domains considered in Problem 1?*

Problem 1 naturally leads to two other problems, though it looks like their solutions are far from being simple.

PROBLEM 3. *Solve Problem 1 not only for the class of plane domains, but also for the class of open Riemann surfaces.* (It is known that any Riemann surface of Parreau–Widom type belongs to the class of surfaces considered in this modification of Problem 1.)

Now, let a domain (resp., a Riemann surface) D belongs to the class considered in Problem 1 (resp., 3) and suppose the group $\mathcal{G}$ of covering transformation is known.

PROBLEM 4. *Find the way to construct the function H.*

For references, details and related topics see [1–3].

REFERENCES

1. Samokhin M. V., *On automorphic analytic functions with values of given modulus*, Mat. Sb. **101 (143)** (1976), 189–203 (Russian); English transl. in Math. USSR-Sb. **30** (1976).
2. Samokhin M. V., *On some questions connected with the problem of existence of automorphic analytic functions with given modulus of boundary values*, Mat. Sb. **111(153)** (1980), 557–578 (Russian); English transl. in Math. USSR Sb. **39** (1981).
3. Samokhin M. V., *On some boundary properties of bounded analytic functions and on the maximum of the modulus principle for multi-connected domains*, Mat. Sb. **135(177)** (1988), 497–513. (Russian)

DEPARTMENT OF MATHEMATICS
MOSCOW CIVIL ENGINEERING INSTITUTE
YAROSLAVSKOYE SHOSSE 26
129337 MOSCOW
RUSSIA

2.26

GLEASON PARTS AND PRIME IDEALS IN H^∞

RAYMOND MORTINI

1. Let H^∞ be the Banach algebra of all bounded analytic functions in the open unit disk $\mathbb{D} = \{ z \in \mathbb{C} : |z| < 1 \}$ and let $M(H^\infty)$ denote its maximal ideal space. For $m, x \in M(H^\infty)$, let $\rho(m, x) = \sup\{ |f(x)| : f \in H^\infty, \|f\| = 1, f(m) = 0 \}$, denote the pseudohyperbolic distance of the points m and x in $M(H^\infty)$. By Schwarz–Pick's lemma $\rho(z) = \left|\frac{z-w}{1-\bar{z}w}\right|$ if $z, w \in \mathbb{D}$. Let

$$P(m) = \{ x \in M(H^\infty) : \rho(m, x) < 1 \}$$

be the Gleason part of $m \in M(H^\infty)$. The Gleason parts of two points are either disjoint or equal.

A Gleason part P is called an *analytic disk* if there exists a continuous, bijective map L of $\mathbb{D}$ onto P such that $\hat{f} \circ L$ is analytic in $\mathbb{D}$ for every $f \in H^\infty$, where $\hat{f}$ denotes the Gelfand transform of $f \in H^\infty$. In his famous paper [6] K. Hoffman showed that any Gleason part $P(m)$ in $M(H^\infty)$ is either a single point or an analytic disk. Moreover, the latter occurs if and only if $m \in \mathbb{D}$ or lies in the (weak-$*$-) closure of an interpolating sequence in $\mathbb{D}$; that is, in the closure of a sequence $\{z_n\}$ satisfying

$$\inf_{k \in N} \prod_{\substack{n \in N \\ n \neq k}} \rho(z_n, z_k) \geqslant \delta > 0.$$

The corresponding maps L_m are called Hoffman maps and have the form

$$L_m(z) = \lim_\alpha \frac{z + z_\alpha}{1 + \bar{z}_\alpha z},$$

where (z_α) is a net in $\mathbb{D}$ converging to m in the weak-$*$-topology of $M(H^\infty)$.

Let G denote the set of all points in $M(H^\infty)$ whose Gleason parts are analytic disks. The elements of G will be called *nontrivial points*. The points whose Gleason parts reduce to a singleton, are called *trivial points*.

Budde [2] noticed that any L_m has a continuous extension L_m^* to the whole spectrum $M(H^\infty)$ of H^∞ defined by $L_m^*(x)(f) = x(\hat{f} \circ L_m)$. By the Corona Theorem this extension is of course unique. The following problem, however, remains open.

PROBLEM 1. *Let L_m be a homeomorphism of $\mathbb{D}$ onto $P(m)$. Does L_m admit a homeomorphic extension to all of $M(H^\infty)$?*

2. In [5] it is shown that under the previous assumption L_m has a homeomorphic extension to the set G of all nontrivial points in $M(H^\infty)$. In case the point m lies in the closure of a thin sequence, i.e. a sequence satisfying $\prod_{n \neq k} \rho(z_n, z_k) \to 1$ as $k \to \infty$, Problem 1 has a positive answer [6]. Moreover, in this case the algebra $H^\infty \circ L_m$ is known to coincide with H^∞. This raises the next question.

PROBLEM 2. *Let $P(m)$ be a part homeomorphic to $\mathbb{D}$. Are the algebras $A = H^\infty \circ L_m$ and $B = H^\infty | \overline{P(m)}$ Banach algebras with respect to the uniform norm? Do the closures $\overline{A}$ and $\overline{B}$ coincide with H^∞?*

3. In [4] Gorkin showed that for an arbitrary analytic disk $P(m)$ the algebras $\overline{A}$ and $\overline{B}$ are isometrically isomorphic, that the Corona Theorem holds in $\overline{A}$ and that the spectrum of $\overline{B}$ is $\overline{P(m)}$. For what points $m \in M(H^\infty)$ the spectrum $M(\overline{A})$ coincides with $M(H^\infty)$?

Let $S = \{x_n\}_{n \in N}$ be a sequence in $M(H^\infty)$. S is said to be an *interpolating sequence* (for H^∞) if for any bounded sequence $(w_n)_{n \in N}$ of complex numbers there exists a function $f \in H^\infty$ such that $\hat{f}(x_n) = w_n$ for all n. In [5] it is shown that a sequence $S = \{x_n\}_{n \in N}$ of points lying in a part $P(m)$ which is homeomorphic to $\mathbb{D}$ is interpolating if and only if it satisfies Carleson's condition

$$\prod_{n: n \neq k} \rho(x_n, x_k) \geqslant \delta > 0.$$

It is known that this condition, necessary for S to be an interpolating sequence in an arbitrary analytic disk, is not sufficient whenever S is a sequence in a nonhomeomorphic disk.

PROBLEM 3. *Give a complete characterization of the interpolating sequences for every analytic disk. Is $S \subseteq P(m)$ an interpolating sequence if and only if S satisfies Carleson's condition and has no cluster points in $P(m)$?*

Partial results have been obtained by Izuchi [8].

4. It is known that for any analytic disk $P(m)$ the set $I_m = \{ f \in H^\infty : \hat{f}|_{P(m)} \equiv 0 \}$ is a nonmaximal closed prime ideal in H^∞ [1,9]. The following conjecture of N. Alling is still open:

PROBLEM 4 [1]. *Every nonmaximal closed prime ideal in H^∞ has the form I_m for some m whose Gleason part is nontrivial.*

Several results which may support this conjecture can be found in [10]. For example, it is known that any nonmaximal closed prime ideal P is contained in the ideal I_m for some m not belonging to the Shilov boundary of H^∞.

5. In contrast to many other algebras of analytic functions on the unit disk, as e.g. the disk algebra, the algebra QA of bounded analytic functions of vanishing mean oscillation on $\partial\mathbb{D}$, there do exist nonmaximal prime ideals in H^∞ which are countably generated (see [9]).

We give the following conjecture.

PROBLEM 5. *Let P be a countably generated prime ideal in H^∞. Then either P has the form*

$$P = M(z_0) = \{ f \in H^\infty : f(z_0) = 0 \} \quad (z_0 \in \mathbb{D})$$

or

$$P = (S_\alpha, S_\alpha^{1/2}, S_\alpha^{1/3}, \dots),$$

where

$$S_\alpha(z) = \exp\left(-\frac{\alpha + z}{\alpha - z}\right), \quad |\alpha| = 1.$$

REFERENCES

1. Alling N., *Aufgabe 2.3*, Jahresbericht DMV **73** (1971/72), no. 2, 2.
2. Budde P., *Support sets and Gleason parts of $M(H^\infty)$*, Thesis, Univ. of California, Berkeley, 1982.
3. Gorkin P., *Prime ideals in closed subalgebras of L^∞*, Michigan Math. J. **33** (1986), 315–323.
4. Gorkin P., *Gleason parts and COP*, J. Funct. Anal. **83** (1989), 44–49.
5. Gorkin P., Lingenberg H.-M., Mortini R., *Homeomorphic disks in the spectrum of H^∞*, Indiana Univ. Math. J. **39** (1990), 961–983.
6. Hoffman K., *Bounded analytic functions and Gleason parts*, Ann. Math. **86** (1967), 74–111.
7. Izuchi K., *Interpolating sequences in a homeomorphic part of H^∞*, Proc. Amer. Math. Soc. **111** (1991), 1057–1065.
8. Izuchi K., *Interpolating sequences in the maximal ideal space of H^∞ (II)*, preprint.
9. Mortini R., *Zur Idealstruktur der Disk-Algebra $A(\mathbb{D})$ und der Algebra H^∞*, Dissertation, Univ. Karlsruhe, Germany, 1984.
10. Mortini R., *Closed and prime ideals in the algebra of bounded analytic functions*, Bull. Austral. Math. Soc. **35** (1987), 213–229.
11. Mortini R., *Countably generated prime ideals in algebras of analytic functions*, Complex Variables **14** (1990), 215–222.

MATHEMATISCHES INSTITUT I
UNIVERSITÄT KARLSRUHE
POSTFACH 6980
D-7500 KARLSRUHE 1
GERMANY

BANACH ALGEBRAS OF ANALYTIC FUNCTIONS

VADIM TOLOKONNIKOV

Some problems grouped around the corona problem for subalgebras of H^∞ and Douglas algebras are presented here. Some of these problems were formulated by other mathematicians, but it seems useful to mention them here once more.

1. The corona theorem and Gleason parts. Let us say that the *corona theorem* is true for a subalgebra X of the algebra H^∞ if

$$f_i \in X,\ i = 1, 2, \ldots, n,\ \sum_{i=1}^{n} |f_i(z)| \geqslant \delta > 0,\ z \in \mathbb{D} \implies \exists\, g_i \in X : \sum_{i=1}^{n} f_i g_i = 1.$$

Let us identify a point of $\mathbb{D}$ and the corresponding evaluation functional. Then the following equivalent reformulation of the corona theorem can be given: the unit disk $\mathbb{D}$ is dense in $\mathcal{M}(X)$, the algebra's X maximal ideal space. Another equivalent reformulation asserts that the canonical map π_X of the maximal ideal spaces is surjective, where

$$\pi_X : \mathcal{M}(H^\infty) \to \mathcal{M}(X), \quad f(\pi_X(m)) \stackrel{\text{def}}{=} f(m), \quad f \in X.$$

1.1. Let $\mathcal{M}P$ be the set of all nontrivial Gleason parts of $\mathcal{M}(H^\infty)$; if $P \in \mathcal{M}P$, then put $\partial P = \partial(H^\infty|_P)$, the Shilov boundary for the Banach algebra of restrictions of the functions from H^∞ to the part P. *Is it true that for every maximal ideal $m \in \mathcal{M}(H^\infty)$ with trivial Gleason part there exists a (unique?) part $P \in \mathcal{M}P$, such that $m \in \partial P$?* As the corona theorem is true for the algebra $H^\infty|_P$ [4], we have $\partial P \subset \operatorname{clos}(P)$, and a positive solution of Problem 1.1 will give a positive answer to the following question [4]:

1.1.1. *Is it true that for every maximal ideal $m \in \mathcal{M}(H^\infty) \setminus (\partial(H^\infty) \cup \mathbb{D})$ there exists a part $P \in \mathcal{M}P$, $P \neq \mathbb{D}$, such that m belongs to the closure of P?*

1.2. Let X be a closed subalgebra of H^∞ containing the disk algebra C_A, for which the corona theorem is not true. *Does there exist a nontrivial part P in $\mathcal{M}(X)$ so that $P \setminus \pi_X(\mathcal{M}(H^\infty)) \neq \emptyset$?* This is true for all counterexamples to the corona problem known to the author, and in all such examples the set $P \setminus \pi_X(\mathcal{M}(H^\infty))$ has an analytic structure.

1.3. *How to estimate a solution of the corona problem in H^∞?*

Let us introduce for $1 \leqslant p \leqslant \infty$ the spaces $H^\infty(\ell^p)$ of vector-functions $f = (f_1, f_2, \ldots, f_i, \ldots)$ such that $f_i \in H^\infty$ for $i = 1, 2, \ldots$,

$$\|f\|_{H^\infty(\ell^p)} \stackrel{\text{def}}{=} \sup \left\{ \left(\sum_{i=1}^{\infty} |f_i(z)|^p \right)^{1/p} : z \in \mathbb{D} \right\} < \infty,$$

and for $f \in H^\infty(\ell^p)$, $g \in H^\infty(\ell^q)$, $1/p + 1/q = 1$, put $\langle f, g\rangle \stackrel{\text{def}}{=} \sum_{i=1}^{\infty} f_i \cdot g_i$.

A solution of the corona problem for H^∞ can be estimated in terms of the following functions $C_{p,n}(\delta)$, $0 < \delta \leqslant 1$, $1 \leqslant p \leqslant \infty$, $n = 1,2,3,\ldots,\infty$:

$$C_{p,n}(\delta) \stackrel{\text{def}}{=} \sup\{\inf\{\|g\|_{H^\infty(\ell^q)} : \langle f,g\rangle = 1\} : \|f\|_{H^\infty(\ell^p)} \leqslant 1,\ f_{n+1} = f_{n+2} = \cdots = 0,\ \forall z \in \mathbb{D},\ \|f(z)\|_{\ell^p} \geqslant \delta\},$$

$$C_p(\delta) \stackrel{\text{def}}{=} C_{p,\infty}(\delta).$$

The following is known:

1. For $n < \infty$, $C_{p,n}$ is finite. This is exactly Carleson's famous corona theorem [3].
2. C_2 is finite. This result was proved in [12] and independently by the author in [14,15]. This result easily implies that C_{2k} for $k = 1,2,\ldots$ is finite too.
3. For $\delta \to 0$ the following inequalities are true:

$$C_2(\delta) \leqslant \text{const} \cdot \delta^{-2} \cdot \log(1/\delta) \text{ [22]},$$
$$C_2(\delta) \geqslant \text{const} \cdot \delta^{-2} \text{ [15]}.$$

Analogous upper estimates were proved in [1,11,13]; here "const" means some positive constants. The function $C_2(\delta)$ near the point $\delta = 1$ was estimated in [15]:

$$C_2(\delta) - 1 \asymp (1-\delta), \quad \delta \to 1.$$

4. C_∞ is finite and

$$C_\infty(\delta) \leqslant \text{const} \cdot \delta^{-\text{const}};$$
$$C_\infty(\delta) - 1 \asymp \log^{-1}((1-\delta)^{-1}), \quad \delta \to 1 \text{ [22]}.$$

This is the Uchiyama theorem, partly published in [23]; an analogous estimate for $C_{\infty,2}(\delta) - 1$, $\delta \to 1$ was given earlier in [7].

More information about these functions is needed; for example, it is interesting to know the answers to the following questions:

1.3.1. *Is it true that C_p is finite for all p?*

1.3.2. *What is the behavior of $C_2(\delta)$ for $\delta \to 0$?*

CONJECTURE. *The upper estimate from [22] gives the true order:*

$$C_2(\delta) \asymp C_{2,2}(\delta) \asymp \delta^{-2} \cdot \log(1/\delta).$$

2. The spaces ℓ^p and their multipliers. Let $\ell_{p,s}$, $1 \leqslant p \leqslant \infty$, $s \in \mathbb{R}$, be the space of distributions on the unit circle T satisfying $\sum_{-\infty}^{\infty} |\hat{f}(n)|^p (1+|n|)^s < \infty$, and let $M_{p,s}$ be the algebra of its multipliers. We write simply ℓ_p and M_p for $s = 0$.

2.1. *Is the corona theorem true for the algebra M_pA with $p \neq 2$?* More exactly, is the next statement true:

$$f_i \in M_pA,\ i = 1,2,\ldots n, \quad \sum_{i=1}^{n} |f_i(z)| \geqslant \delta > 0,\ z \in \mathbb{D} \implies \exists g_i \in M_pA : \sum_{i=1}^{n} f_i g_i = 1?$$

We suppose the answer is NO. Even the case $n = 1$ is unknown here. S. Igari [6] proved that M_p is not a filled subalgebra of L^∞:

$$\exists f \in M_p : f^{-1} \in L^\infty, \quad f^{-1} \notin M_p.$$

One can try Wolff's method to prove the corona theorem for M_pA. It consists of constructing some non-analytic solution, namely $F_i = \bar{f}_i / \sum_{i=1}^{n} |f_i|^2$, and seeking to modify it afterwards. But from the example of Igari we see that this non-analytic solution may fail to belong to M_p. So we ask the following question:

2.2. *Is the implication from 2.1 true under the additional assumption* $\sum_{i=1}^{n} |f_i(m)| \geqslant \geqslant \delta > 0$, $m \in \mathcal{M}(M_p)$?

An equivalent reformulation: Is $\mathbb{D} \cup j^*(\mathcal{M}(M_p))$ dense in $\mathcal{M}(M_pA)$? Let us note that it is unknown whether the restriction of j^* to $\mathcal{M}(M_p)$ is injective and whether the equality $\mathcal{M}(M_p) = \partial(\mathcal{M}(M_p))$ holds, where $\partial(\mathcal{M}(M_p))$ is the Shilov boundary of the algebra M_pA. The answer to all these questions seems to be YES.

2.3. Let I be an inner function from M_pA, $p \neq 2$. *Is it true that I is a Blaschke product* (a question of S. A. Vinogradov)?

2.3.1. *Is it true (under the same assumptions) that I is a Carleson–Blaschke product, that is a finite product of interpolating Blaschke products* (a question of I. E. Verbitskii)?

Probably, the theory of Douglas algebras would be useful for this problems.

2.4. *Is the space* $X_p \stackrel{\text{def}}{=} \operatorname{clos}_{L^\infty}(H^\infty + M_p)$ *an algebra?* (Then X_p is a Douglas algebra.) Let us note that the space $H^\infty + M_p$ is not always an algebra, consider $p = 1$ for example.

Let us define for $f \in L^\infty$ the action of the Toeplitz operator $T_f : T_f g = P_+(fg)$, g being an analytic polynomial, where P_+ is the Riesz orthogonal projection of L^2 onto H^2. If for a Banach space of analytic functions X, T_f can be continuously extended to a linear operator from the closure of polynomials in X to X, then we say that f belongs to the space $ST(X)$ (i.e. the space of symbols of Toeplitz operators).

2.5. *Is the equality* $ST(\ell_{p,s}A) = M_pA + M_{p,s}$ *true for all* $s \leqslant 0$? The space on the right is always smaller than that on the left. The equality is known to hold for $s = 0$ or $s < 1 - 1/p$ or for all $s \leqslant 0$ if $p = 2$, see [18,20] and the list of references therein.

The following two problems were indicated in the book [2]. Let us define the Hankel operator H_f for an analytic function $f : H_f(g) = (I - P_+)(fg)$, g being an analytic polynomial. Let us denote PC_0 the space of piecewise constant functions on the circle $\mathbb{T}$, and PC its closure with respect to the uniform norm (the norm of L^∞).

2.6. *Is it true that H_f acts from $\ell^p A$ to ℓ^p iff $f \in P_+(M_p)$?* Of course, this condition is sufficient. Some necessary conditions are known: $f \in \mathrm{BMOA}$ [18], $\|f_r'\|_{M_p} \leqslant \mathrm{const} \cdot (1-r)^{-1}$, i.e. f belongs to a space analogous to the Bloch space.

2.7. *Is it true that for all functions $f \in PC \cap M_p$ there exists a sequence of functions f_n from PC_0, bounded in M_p and tending to f in the uniform norm?*

It is easy to see that the usual approximation by a convolution cannot be applied in this case.

3. Douglas algebras and their generalization. Every closed subalgebra D of L^∞ containing H^∞ is called a Douglas algebra. For every such algebra D let us introduce

ID, the set of all inner functions invertible in D, and two Sarason algebras $QD \stackrel{\text{def}}{=} D \cap \overline{D}$ and CD, the closed symmetric subalgebra generated by ID. If Y is a space of distributions on T, then YA consists of functions with zero negative Fourier coefficients.

In [19] the Douglas algebra $D_b = H^\infty + \text{Mult}(L_b^\infty)$ was introduced, where L_b^∞ is the closure of $H^\infty + \overline{H^\infty}$ in L^∞; it was shown that the analytic Sarason algebra QDA is a Dirichlet algebra iff D is a subalgebra of D_b.

3.1. *Does every inner function invertible in D_b satisfy the Carleson condition (i.e. is representable as a finite product of interpolating Blaschke products)?*

In [5] the inner functions in the algebra $\text{Mult}(H^\infty + \overline{H^\infty})$, a subalgebra of $\text{Mult}(L_b^\infty)$, are described. They are Blaschke products with zeros satisfying "the uniform Frostman condition" $\sup_{z\in T}\left(\sum_n (1-|z_n|)/|1-z_n\bar{z}|\right) < \infty$, and main difficulties in [5] occur exactly in the proof that all inner functions in $\text{Mult}(H^\infty + \overline{H^\infty})$ satisfy the Carleson condition. In [19] more examples of Blaschke products from $\text{Mult}(L_b^\infty)$ are given, e.g. with $\lim\limits_{n\to\infty} \prod\limits_{n\neq m} |z_n - z_m|/|1 - z_n\bar{z}_m| = 1$.

3.2. *Is the stable rank of QDA equal to 1 for any Douglas algebra D?* In other words, is the following statement true:

$$\begin{aligned}&\forall f, g \in QDA : |f(z)| + |g(z)| \geqslant \delta > 0, \quad z \in \mathbb{D} \Longrightarrow \\ &\exists F, G \in QDA : Ff + Gg = 1, \quad F^{-1} \in QDA?\end{aligned}$$

The same question for CDA is open too. In the case $D = L^\infty$ (then $QDA = CDA = H^\infty$), a positive answer was given by Treil in [21]. A positive answer to Question 3.2 would give a positive answer to Question 3.1.

3.3. *Describe Douglas algebras such that all invertible inner functions are Blaschke products.* In [19] a description of Douglas algebras was given whose invertible inner functions satisfy the Carleson condition: this is the case if the algebra is generated by Blaschke products B such that for all $a \in \mathbb{D}$ the Frostman shifts $(B-a)/(1-B\bar{a})$ are Carleson Blaschke products. But a similar necessary condition is not sufficient for the case of Blaschke products. More precisely: it was proved in [9, proposition 4.1] that if we take as B a non-zero Frostman shift of singular inner function $\exp\big(-(1+z)/(1-z)\big)$ divided by any its Blaschke subfactor, then all Frostman shifts of B are Blaschke products, but in the Douglas algebra generated by B the function $\exp\big(-(1+z)/(1-z)\big)$ is invertible.

A subalgebra X of L^∞ containing H^∞ is called a generalized Douglas algebra (a GD-algebra) iff

$$f \in X,\ \|f\|_\infty < 1 \Longrightarrow \frac{1}{1-f} \in X.$$

Examples and properties of these algebras can be found in [18]. For example, the corona theorem is always true for $H^\infty \cap \bar{X}$ for any GD-algebra X.

3.4. *Is the algebra $H^\infty + \theta \cdot L^\infty$ a GD-algebra for every bounded outer function θ?* This question is connected with questions about multipliers of de Branges' spaces [8]. Analogous questions can be asked about the spaces $L^\infty \cap (H^p + \theta \cdot L^p)$, $1 \leqslant p < \infty$; whether this space is always an algebra is unknown here.

3.5. *Is the space $ST(H^\infty)$ a GD-algebra?* An equivalent question: Does the spectral radius in $ST(H^\infty)$ always coincide with the L^∞-norm? This question is open for the algebra of multipliers of Cauchy type integrals $ST(H^\infty) \cap H^\infty$ too (a question of S.A. Vinogradov). One more question: Does the spectral radius of any Toeplitz operator acting on H^∞ always coincide with the L^∞-norm? For H^p, $1 < p < \infty$, $p \neq 2$, this is not true, as can easily be derived from the following theorem: The essential spectrum of a Toeplitz operator with a piecewise continuous symbol can be obtained from the graph of the symbol by adding to every discontinuity point circle arcs such that the angle of view from the circle to discontinuities is $2\pi(1 - 1/p)$ [2].

3.6. *Is the space $ST(X)$ always an algebra?* To be more precise, let X be a Banach space of analytic functions in $\mathbb{D}$ with the following properties:

1. X is invariant under rotations of $\mathbb{T}$;
2. Polynomials are dense in X;
3. Toeplitz operators T_z and $T_{\bar{z}}$ act on X and their spectral radii equal one: $r(T_z) = r(T_{\bar{z}}) = 1$.

Then $ST(X) \subset L^\infty$ and $ST(X)$ contain all functions analytic in a neighborhood of T. The space $ST(X)$ is a subalgebra of L^∞ iff there exists a space Y of functions harmonic in $\mathbb{D}$ with Properties 1 and 2 mentioned above and $Y \supset X$, $P_+Y = X$, $ST(X) = \mathrm{Mult}(Y)$ [17]. In [21] some way to compute the symbol for closed operators was given (under the hypothesis that $ST(X)$ is an algebra and some additional assumptions).

In all examples known to the author the space $ST(X)$ is an algebra. So the exact form of Question 3.6 is as follows: *Is the space $ST(X)$ a subalgebra of L^∞ for any space X satisfying conditions 1–3?*

3.7. This question is about the comparison of two classes of analytic algebras for which the corona theorem is known.

THEOREM A [10]. *Let X be a closed subalgebra of H^∞ and suppose that*

(1) *$CDA \subset X \subset QDA$ for some Douglas algebras D;*
(2) *X has the Newman property for the Shilov boundary ∂X:*

$$\partial X = \{m \in \mathcal{M}(X) : |m(I)| = 1 \text{ for every inner function } I \text{ in } X\};$$

(3) *$f, I \in X$, I an inner function, $f\bar{I} \in H^\infty \implies f\bar{I} \in X$;*
(4) *X is a filled subalgebra of H^∞: $f \in X$, $f^{-1} \in H^\infty \implies f^{-1} \in X$.*

Then the corona theorem is true for X.

THEOREM B [16,10,18]. *Let X be a closed subalgebra of H^∞. If for some C^*-algebra V and a Douglas algebra D we have $CD \subset V \subset QD$, $X = VA$, then the corona theorem is true for X.*

From the formal point of view this result is weaker than Theorem A, but the answer to the following question is unknown to the author.

QUESTION. *Does there exist an algebra X satisfying the hypotheses of Theorem A and not satisfying the hypotheses of Theorem B?*

References

1. Berndtson B., Ransford T. J., *Analytic multifunctions, the $\bar{\partial}$-equation and a proof of the corona theorem*, Pacif. J. Math. **124** (1986), no. 1, 57–72.
2. Böttcher A., Silbermann B., *Analysis of Toeplitz Operators*, Akad.-Verlag, Berlin, 1989.
3. Carleson L., *Interpolation by bounded analytic functions and the corona problem*, Ann. Math. **76** (1962), 547–559.
4. Gorkin P., *Gleason parts and COP*, J. Funct. Anal. **83** (1989), 44–49.
5. Hruščëv S. V., Vinogradov S. A., *Inner functions and multipliers of Cauchy type integrals*, Arkiv för Mat. **19** (1981), no. 1, 23–42.
6. Igari S., *Function of L^p-multipliers*, Tohoku Math. J. **21** (1969), 304–320.
7. Jones P. W., *Estimates for the corona problem*, J. Funct. Anal. **39** (1981), 162–181.
8. Lotto B. A., Sarason D., *Multiplicative structure of De Branges spaces*, preprint, Los Angeles, 1990.
9. Morse H.S., *Destructible and indestructible Blaschke products*, Trans. Amer. Math. Soc. **272** (1980), no. 1, 247–253.
10. Mortini R., *Corona theorems for subalgebras of H^∞*, Michigan Math. J. **36** (1989), no. 2, 193–201.
11. Nikolskii N. K., *Treatise of the Shift Operator*, Springer-Verlag, 1986.
12. Rosenblum M., *A corona theorem for countably many functions*, Integr. Equat. Operat. Theory **3** (1980), no. 1, 125–137.
13. Slodkowski Z., *On bounded analytic functions in finitely connected domains*, Trans. Amer. Math. Soc. **300** (1987), no. 2, 721–736.
14. Tolokonnikov V., *Estimates in the Carleson Corona theorem and finitely generated ideals of the algebra H^∞*, Funktsional. Anal. i Prilozhen. **14** (1980), no. 4, 85–86. (Russian)
15. Tolokonnikov V., *Estimates in the Carleson Corona theorem, ideals of the algebra H^∞, a problem of Sz.-Nagy*, Zapiski nauchn. semin. LOMI **113** (1981), 178–190 (Russian); English transl. in J. of Soviet. Math. **22** (1983), 1814–1828.
16. Tolokonnikov V., *Corona theorems in algebras of bounded analytic functions*, Sov. Inst. Scien. Tech. Inform. (VINITI) **251** (1984), 1–60. (Russian)
17. Tolokonnikov V., *Hankel and Toeplitz operators in Hardy spaces*, Zapiski nauchn. semin. LOMI **141** (1985), 165–175 (Russian); English transl. in J. of Soviet. Math. **37** (1987), 1359–1364.
18. Tolokonnikov V., *Generalized Douglas algebras*, J. Algebra and Analysis **3** (1991), no. 2, 231–252. (Russian)
19. Tolokonnikov V., *Blaschke products satisfying the Carleson-Newman conditions and Douglas algebras*, J. Algebra i Analiz **3** (1991), no. 4, 185–196 (Russian); English transl. in Leningrad Math. J. **3** (1992), no. 4.
20. Tolokonnikov V., *Algebras of Toeplitz operators in spaces of smooth functions*, Matemat. Zametki **49** (1991), no. 3, 124–130. (Russian)
21. Treil S, *The stable rank of the algebra H^∞ equals* 1, preprint, 1990.
22. Uchiyama A., *Corona theorems for countably many functions and estimates for their solutions*, preprint, Univ. of Calif. at Los Angeles, 1981.
23. Uchiyama A., *The construction of certain BMO functions and the corona problem*, Pacif. J. Math. **99** (1982), no. 1, 183–204.

Torzhkovskaja str.2. korp.1 kv.76.
St.-Petersburg, 197342
Russia

2.28
old

EXTREMUM PROBLEMS

VLASTIMIL PTÁK

1. THE GENERAL MAXIMUM PROBLEM. *Given a unital Banach algebra A, a compact set F in the plane and a function f holomorphic in a neighbourhood of F, to find*

$$\sup\{|f(a)| : |a| \leqslant 1, \sigma(a) \subset F\}.$$

The problem was formulated and solved first [1, 2] in the case of

2. THE SPECIAL MAXIMUM PROBLEM. *Let n be a natural number, r a positive number less than one, to find, among all contractions T on n-dimensional Hilbert space whose spectral radius does not exceed r those (it turns out that there is essentially only one) for which the norm $|T^n|$ assumes its maximum.*

This is a particular case of the general problem for

$$A = B(H_n), \qquad F = \{z : |z| \leqslant r\}, \qquad f(z) = z^n.$$

The solution of Problem 2 was divided into two stages. The first step consists in replacing the awkward constraint that the spectral radius be $\leqslant r$ by a more restrictive one which makes the problem considerably easier: the operator is to be annihilated by a given polynomial. This is

3. THE FIRST MAXIMUM PROBLEM. *Let n be a natural number, p a polynomial of degree n with all roots inside the unit disc. Let $A(p)$ be the set of all contractions $T \in B(H_n)$ such that $p(T) = 0$. To find the maximum of $|T^n|$, more generally of $|f(T)|$ as T ranges in $A(p)$.*

We call this maximum $C(p, f)$.
Having solved the first maximum problem we have to solve

4. THE PROBLEM OF THE WORST POLYNOMIAL. *To find, among all polynomials with roots $\leqslant r$ in modulus that one for which $C(p, f)$ is maximal.*

For the case $f(x) = x^n$ the result of [2] shows that the worst polynomial is $p(z) = (z - r)^n$. The method used in [2] is based on some fairly complicated algebraic considerations and does not extend to functions other than x^n. It is not known whether the worst polynomial for f other than x^n has a root of multiplicity n whether the roots have to be concentrated on the boundary of the disc $|z| \leqslant r$. Thus it seems useful to study $C(p, f)$ as a function of the roots of p; a recent contribution to Problem 4 is [4]. A list of references up to 1979 is contained in the survey paper [3].

References

1. Pták V., *Norms and the spectral radius of matrices*, Czechosl. Math. J. **87** (1962), 553–557.
2. Pták V., *Spectral radius, norms of iterates and the critical exponent*, Lin. Alg. Appl. **1** (1968), 245–260.
3. Pták V., Young N. J., *Functions of operators and the spectral radius*, Lin. Alg. Appl. **29** (1980), 357–392.
4. Young N. J., *A maximum principle for interpolation in* H^∞, Acta Sci. Mat. Szeged, no. 1–2, **43** (1981), 147–152.

Institute of Mathematics
Czech Academy of Sciences
Žitná 25
11567 Praha 1
Czech Republic

2.29
old

MAXIMUM PRINCIPLES FOR QUOTIENT NORMS IN H^∞

N. J. YOUNG

A surprising variety of starting points can lead one to study norms in quotient algebras of H^∞ by a closed ideal. Well known examples are classical complex interpolation [4], canonical models of operators [2] and some problems of optimal circuit design [1]. It also arises from a maximum problem for matrices to which V. Pták was led by considerations relating to numerical analysis. The problem is to estimate the maximum value of $\|\psi(A)\|$, where $\psi \in H^\infty$, over all contractions A of spectral radius at most $r < 1$ on n-dimensional Hilbert space (Pták was mainly concerned with the case $\psi(A) = A^m$, some $m \in \mathbb{N}$). An account of this problem is given in [3].

The only known way of handling the spectral constraint is to replace it by the condition $p(A) = 0$, for some polynomial p, and then to vary p among the polynomials of degree n having all their zeros in $r(\operatorname{clos}\mathbb{D})$. After some calculations one is led to study the functional

$$F(p) = \|\psi + pH^\infty\|_{H^\infty/pH^\infty}$$

as a function of $p \in H^\infty$ for fixed ψ. In particular, as p varies over the class of polynomials described above, *at what p does F attain the maximum?* The result we should like to prove is that F attains its maximum at a polynomial of the form $p(z) = (z - \varepsilon r)^n$ for some $\varepsilon \in \mathbb{C}$ with $|\varepsilon| = 1$. Pták proved this was so in the case $\psi(z) = z^n$, and I proved that if ψ is a Blashke product of degree n then all zeros of an extremal polynomial p have modulus r: in fact, F is then the composition of a strictly increasing function and a plurisubharmonic function of the zeros of p [5]. Nothing, is known, however, about the most interesting case,

$$\psi(z) = z^m \qquad \text{with } m > n.$$

To formulate the problem concisely, let us say that the *maximum principle holds for* $f\colon \Omega \subset \mathbb{C} \to \mathbb{R}$ if for any compact set $K \subset \Omega$, the supremum of f on K is attained at some point of the boundary of K relative to Ω. And if M is a complex manifold, we shall say that the maximum principle holds for $F\colon M \to \mathbb{R}$ if, for any open set $\Omega \subset \mathbb{C}$ and any analytic function $G\colon \Omega \to M$, the maximum principle holds for $F \circ G$.

PROBLEM. *Let $\psi \in H^\infty$ and let $F\colon \mathbb{D}^n \to \mathbb{R}$ be defined by*

$$F(\alpha_1, \ldots, \alpha_n) = \|\psi + \varphi H^\infty\|_{H^\infty/\varphi H^\infty}$$

where

$$\varphi(z) = \prod_{i=1}^{n} (z - \alpha_i).$$

Does the maximum principle hold for F?

REFERENCES

1. Helton J. W., *Non-Euclidean functional analysis and electronics*, Bull. Amer. Math. Soc. **7** (1982), 1–64.
2. Nikolskii N. K., *Lectures on the shift operator*, Nauka, Moscow, 1980 (Russian); English translation: *Treatise of the shift operator*, Springer-Verlag, 1986.
3. Pták V., Young N. J., *Functions of operators and the spectral radius*, Linear Algebra and its Appl. **29** (1980), 357–392.
4. Sarason D., *Generalized interpolation in H^∞*, Trans. Amer. Math. Soc. **127** (1967), 179–203.
5. Young N. J., *A maximum principle for interpolation in H^∞*, Acta Sci. Math. **43** (1981), no. 1–2, 147–152.

DEPARTMENT OF MATHEMATICS
FYLDE COLLEGE
LANCASTER UNIVERSITY
LANCASTER
LA1 4YF
ENGLAND

2.30
old

OPEN SEMIGROUPS IN BANACH ALGEBRAS

Jaroslav Zemánek

Let A be a complex Banach algebra with identity, not necessary commutative. Let S be some open multiplicative semigroup in A. For an element a in A let $d(a)$ denote the distance from the point a to the closed set $A \setminus S$ (in other words, it is the radius of the largest open ball centered at a and contained in S), and let $r(a)$ be the supremum of all $\varepsilon \geqslant 0$ such that the elements $a - \lambda \cdot 1$ belong to S for $|\lambda| < \varepsilon$ (that is the radius of the largest open disk centered at a and contained in the intersection of S with the subspace spanned by a and 1). So we clearly have $r(a) \geqslant d(a)$. *For a variety of particular semigroups S we know that the formula*

$$r(a) = \lim_{n\to\infty} d(a^n)^{1/n}$$

is valid for every a in A. We list below the most important cases.

First of all, if $S = G(A)$, the group of invertible elements, the result follows from the spectral radius formula. Second, the formula is true when S is the semigroup of left (or right) invertible elements of A, cf. [6]. Third, it also holds when S is the complement of the set of left (right) topological divisors of zero in A, cf. [1].

Next the formula is true for various semigroups of the algebra $A = B(X)$ of bounded linear operators on a Banach space. In the case when S is the semigroup of surjective (or bounded from below) operators on X it was obtained in [1] by an analytic argument (in fact, this is equivalent to the third case mentioned before). Using an additional geometric device these results were applied in [6] to prove the above formula for the semigroup S of (upper or lower) semi–Fredholm operators on X, and hence it follows for the semigroup of Fredholm operators as well. In these cases the distance $d(T)$ admits other natural interpretations, namely, it coincides with (or is related to) certain geometric characteristics of the operator T (like the surjection modulus, the injection modulus, the essential minimum modulus [5], etc).

In each of the cases listed above an individual approach was needed to find a proof. The difficult steps are of an analytic character, bases on the theorems of G. R. Allan (1967) and J. Leiterer (1978) on analytic vector–valued solutions of linear equations depending analytically on a parameter. The main idea is well demonstrated in [1] though in that case a combinatorial argument is also available [2]. So there seems to be some motivation *for investigating the problem in general to seek a theorem which would contain all these particular results.*

Let us give some warnings. Let $S = G_1(A)$ be the principal component of the set of invertible elements in A. There are (noncommutative) Banach algebras A for which the group $G(A)/G_1(A)$ is finite, but not trivial, cf [4], [3]. in such a situation for every invertible element x not in $G_1(A)$ one can find a positive integer k such that x^k is in $G_1(A)$. Then we have $r(x) = 0$ but $r(x^k) > 0$ so that the formula cannot be true for this S and all a in A. Moreover, it is easy to see that $\lim_{n\to\infty} d(x^n)^{1/n}$ cannot exist for these x.

Let A be a commutative Banach algebra and let S be the open semigroup of all elements whose spectra are contained in the open unit disk. In this case we have $r(x) = d(x)$ for every x in S but $\lim_{n\to\infty} d(x^n)^{1/n} = 1 \geqslant r(x)$, and the last inequality may be strict. We arrive at the same conclusion if we replace the unit disk in the preceding definition by a q–multiple of it, with $0 < q < 1$. This suggests some analogy between our problem and the classical formula for the radius of convergence of a power series (the formula does not give the radius of the disk where we are considering the function but the radius of the disk where the given function can naturally be defined).

Thus some additional conditions should be imposed on S in general. For instance, the property "if $ab = ba$ belongs to S then both a and b are in S" shared by most of the semigroups for which the problem is solved in the affirmative. This condition ensures, by the way, that $r(x^n) = r(x)^n$ for all x in A and $n = 1, 2, \ldots$ (we note that $r(x^n) \geqslant r(x)^n$ is always true). *Is it important that the identity element be in S, or that S be connected, or "maximal"? Does $\lim_{n\to\infty} d(a^n)^{1/n}$ exist for a in an arbitrary open semigroup S? What is the meaning of it in general?*

I should like to thank Tom Ransford for a valuable discussion on this topic.

References

1. Makai E., Jr., Zemánek J., *The surjectivity radius, packing numbers and boundedness below of linear operators*, Integr. Eq. Oper. Theory **6** (1983).
2. Müller V., *The inverse spectral radius formula and removability of spectrum*, Časopis Pěst. Mat. **108** (1983), no. 4, 412–415.
3. Paulsen V., *The group of invertible elements in a Banach algebra*, Colloq. Math. **47** (1982).
4. Yuen Y., *Groups of invertible elements of Banach algebras*, Bull. Amer. Math. Soc. **79** (1973).
5. Zemánek J., *Geometric interpretation of the essential minimum modulus*, Operator Theory: Advances and Applications, vol. 6, Birkhäuser Verlag, Basel, 1982, pp. 225–227.
6. Zemánek J., *The semi–Fredholm radius of a linear operator*, Bull. Polish Akad. Sci. Math. **32** (1984), no. 1–2, 67–76.

Institute of Mathematics
Polish Academy of Sciences
00–950 Warszawa, P. O. Box 137
Poland

2.31
v.old

POLYNOMIAL APPROXIMATION

L. DE BRANGES

Let A be a uniformly closed algebra of continuous functions on a compact Hausdorff space S. Assume that A contains constants, that every continuous linear multiplicative functional on A is of the form $f \to f(s)$ for a unique element s of S, and that every element of A whose reciprocal belongs to A is of the form $\exp f$ for an element f of A.

Let σ be a positive measure on a Borel subsets of S, whose support contains more than one point, such that the closure of A in $L^\infty(\sigma)$, considered in its weak topology induced by $L^1(\sigma)$, contains no nonconstant real element. Assume that the functions of the form $f + \bar{g}$ with f and g in A are dense in $L^\infty(\sigma)$ in the same topology. It is CONJECTURED *that the closure of A in $L^\infty(\sigma)$ is isomorphic to the algebra of functions which are bounded and analytic in the unit disk.*

Positive measures on the Borel subsets of S are considered in the weak topology induced by the continuous functions on S. Two positive measures μ and ν are said to be *equivalent* (with respect to A) if the identity

$$\int f\, d\mu = \int f\, d\nu$$

holds for every element f of A. The closure of the set of measures which are absolutely continuous with respect to μ and equivalent to μ is a compact convex set, which is the closed convex span of its extreme points. Extremal measures are characterized by the density of the function of the form $f + \bar{g}$, with f and g in A, in $L^1(\mu)$. Let μ be an extremal measure and let B be the weak closure in $L^\infty(\mu)$ of the functions of the form $f + \bar{g}$ with f and g in A. It is CONJECTURED *that the quotient Banach space $L^\infty(\mu)/B$ is reflexive.*

For equivalent positive measures μ and ν, define μ to be *less than or equal to* ν if the inequality

$$\int \log |f|\, d\mu \leqslant \int \log |f|\, d\nu$$

holds for every element f of A. If μ is an extremal measure, it is CONJECTURED *that a greatest element σ exists in (the closure of) the set of measures which are absolutely continuous with respect to μ and equivalent to μ.* It is CONJECTURED *that the functions of the form $f + \bar{g}$ with f and g in A are weakly dense in $L^\infty(\sigma)$.*

REFERENCES

1. de Branges L., Trutt D., *Quantum Cesàro operators*, Topics in functional analysis (essays dedicated to M. G. Krein on the occasion of his 70 th birthday), Advances in Math., Suppl. Studies, vol. 3, Academic Press, New York, 1978, pp. 1–24.
2. de Branges L., *The Riemann mapping theorem*, J. Math. Anal. Appl. **66** (1978), no. 1, 60–81.

PURDUE UNIVERSITY
DEPARTMENT OF MATHEMATICS
LAFAYETTE, INDIANA 47907, U.S.A.

2.32
v.old

SUBALGEBRAS OF THE DISK ALGEBRA

J. WERMER

let A denote the disk algebra, i.e. the algebra of all functions continuous on $\operatorname{clos}\mathbb{D}$ and analytic on $\mathbb{D}$. Fix functions f and g in A. We denote by $[f, g]$ the closed subalgebra of A generated by f and g, i.e. the closure in A of the set of all functions

$$\sum_{n,m=0}^{N} c_{n,m} f^n g^m, \quad c_{n,m} \text{ constants.}$$

We ask: *when does* $[f, g] = A$?

Necessary conditions are

1) f, g together separate points of $\operatorname{clos}\mathbb{D}$.
2) For each a in $\mathbb{D}$, either $f'(a) \neq 0$ or $g'(a) \neq 0$.

1) and 2) together are not sufficient for $[f, g] = A$. Some regularity condition must be imposed on the boundary. We assume

3) f, g are smooth on $\mathbb{T}$, i.e. the derivatives f' and g' extend continuously to $\mathbb{T}$.

1), 2), 3), are not yet sufficient conditions. We add

4) For each a on $\mathbb{T}$, either $f'(a) \neq 0$ or $g'(a) \neq 0$.

In [1] R. Blumenthal showed

THEOREM 1. 1), 2), 3) *and* 4) *together are sufficient for* $[f, g] = A$.

Related results are due to J.-E. Björk, [2], and to Sibony and the author, [3].

Condition 4) is, however, not necessary, since for instance $[(z-1)^2, (z-1)^3] = A$ and conditions 1), 2), 3) hold here while 4) is not satisfied.

The problem arises to give a condition that replaces 4) which is both necessary and sufficient for $[f, g] = A$. In the special case $f = (z-1)^3$ this problem has been solved by J. Jones in [4] and his result is the following: let W^+ and W^- be the two subregions of $\operatorname{clos}\mathbb{D}$ which are identified by the map $(z-1)^3$. Put

$$z^* = 1 + e^{\frac{2\pi i}{3}}(z-1).$$

Then for z in W^+, z^* lies in W^- and $(z-1)^3$ identifies z and z^*. Let χ be an inner function on W^+ whose only singularity is at $z = 1$. Then for some t, $t > 0$,

$$\chi(z) = \exp\left\{ t\left(1 + \frac{1}{(z-1)e^{\pi i/3}}\right)^3 \right\}. \tag{1}$$

THEOREM 2 ([5]). *Let g be a function in A such that $f=(z-1)^3$ and g together satisfy 1), 2), 3). Then $[f,g]\neq A$ if and only if for some χ of the form* (1),

$$|g(z)-g(z^*)|\leqslant K|\chi(z)| \tag{2}$$

for all z in W^+, where K is some constant.

We propose two problems.

PROBLEM 1. *Prove an analogue of Theorem 2 for the case when f is an arbitrary function analytic in an open set which contains* $\operatorname{clos}\mathbb{D}$ *by finding a condition to replace* (2) *which together with* 1), 2), 3) *is necessary and sufficient for* $[f,g]\neq A$.

Furthermore, condition (2) implies that the Gleason distance from z to z^*, computed relative to the algebra $[f,g]$, approaches 0 rapidly as $z\to 1$, and so is inequivalent to the Gleason distance computed relative to the algebra A. Let B denote a closed subalgebra of A which separates the points of $\operatorname{clos}\mathbb{D}$ and contains the constants. Let ρ_B denote the Gleason distance induced on $\operatorname{clos}\mathbb{D}$ by B, i.e.

$$\rho_B(z_1,z_2)=\sup_{\substack{\Phi\in B\\ \|\Phi\|=1}}|\Phi(z_1)-\Phi(z_2)|;\qquad |z_1|\leqslant 1,\quad |z_2|\leqslant 1.$$

Let ρ denote the Gleason distance on $\operatorname{clos}\mathbb{D}$ induced by A.

PROBLEM 2. *Assume that*

(a) *The maximal ideal space of B is the disk* $\operatorname{clos}\mathbb{D}$.
(b) *There exists a constant K, $K>0$, such that*

$$\rho_B(z_1,z_2)\geqslant K\rho(z_1,z_2);\qquad |z_1|\leqslant 1,\quad |z_2|\leqslant 1.$$

Show that then $B=A$.

REFERENCES

1. Blumenthal R., *Holomorphically closed algebras of analytic functions*, Math. Scand. **34** (1974), 84–90.
2. Björk J.-E., *Holomorphic convexity and analytic structures in Banach algebras*, Arkiv för Mat. **9** (1971), 39–54.
3. Sibony N., Wermer J., *Generators for $A(\Omega)$*, Trans. Amer. Math. Soc. **194** (1974), 103–114.
4. Jones J., *Generators of the disc algebra (Dissertation)*, Brown University, June, 1977.
5. Wermer J., *Subalgebras of the disk algebra*, Colloque d'Analyse Harmonique et Complexe, Univ. Marseille I, Marseille, 1977.

BROWN UNIVERSITY
DEPARTMENT OF MATH.
PROVIDENCE, R. I., 02912
USA

S.2.33
old

TO WHAT EXTENT CAN THE SPECTRUM OF AN OPERATOR BE DIMINISHED UNDER AN EXTENSION?

B. BOLLOBÁS

Let X be a Banach space, T be a bounded linear operator on X (i.e. $T \in \mathcal{B}(X)$).

QUESTION 1. *Are there a Banach space Y containing X and an operator $S \in \mathcal{B}(Y)$ such that $T = S|X$ and the essential spectrum of T is exactly the spectrum* of S?*

A stronger of Question 1 is

QUESTION 2. *Given $X, T \in \mathcal{B}(X)$, can one find a Banach space Y containing X and a isometrical algebra homomorphism $\varphi : \mathcal{B}(X) \to \mathcal{B}(Y)$ such that $\varphi(S)|X = S\ \forall S \in \mathcal{B}(X)$ and the spectrum of $\varphi(T)$ is exactly the essential spectrum of T?*

DEPARTMENT OF PURE MATHEMATICS AND MATHEMATICAL STATISTICS
UNIVERSITY OF CAMBRIDGE
16 MILL LANE, CAMBRIDGE
CB2 1SB, ENGLAND

CHAPTER EDITORS' COMMENTARY

The solution to both the questions of this paper are contained in paper [1] by V. Müller of Prague.

Let $T \in \mathcal{B}(X)$, where X is a Banach space. Then Question 1 asked if there is a Banach space $Y \supset X$ and $S \in \mathcal{B}(Y)$ such that $S|X = T$ and $\sigma(S)$ is the essential spectrum of T. The answer to this question is "yes". Question 2 asked if there is a Banach space $Y \supset X$ and an isometric homomorphism $\varphi\colon \mathcal{B}(X) \to \mathcal{B}(Y)$ such that $\varphi(S)|X = S \quad (S \in \mathcal{B}(X))$ and $\sigma(\varphi(T))$ is the essential spectrum of T. Müller proves that, for any T whose essential spectrum is a proper subset of $\sigma(T)$, no such pair (Y, φ) exists. This latter result was also obtained by M. I. Ostrovskii (Kharkov), [2].

REFERENCES

1. Müller V., *Adjoining inverses to non-commutative Banach algebras and extensions of operators*, Studia Math. **91** (1988), 73–77.
2. Ostrovskii M. I., *Diminishing of the spectrum of an operator under extension*, Teor. Funktsii, Funktsion. Anal. i Prilozhen. **45** (1986), 96–97 (Russian); English transl. in J. Soviet Math. **48** (1990), no. 4, 450–451.

*Apparently, here the essential spectrum of S is the set of $\lambda \in \mathbb{C}$ such that $\|(S - \lambda I)x_n\| \to 0$ for a sequence $\{x_n\}$ in X with $\|x_n\| = 1$. – Ed.

Chapter 3

PROBABILISTIC PROBLEMS

Edited by

Jean-Pierre Kahane
Université Paris-Sud
Mathématiques - Bât. 425
91405 Orsay Cedex
France

INTRODUCTION

Since 1984 a number of papers and books were published on probability and different aspects of analysis. Let us mention only one important book: Probability and Banach spaces, by M. Ledoux and M. Talagrand (1991).

The present chapter consists firstly of "old" problems, already mentioned in L. N. 1043, sometimes with partial solutions or commentaries. We removed most commentaries contained in L. N. 1043, and we added a few new questions in section 3.4. We kept the same sections 3.1 to 3.6 as in L. N. 1043, so that the designation of a problem (say 3.1.1) is the same here and in the previous edition. We removed the section 3.7 (question of N. A. Sapogov, on the existence of a set $E \subset \mathbb{R}^n$ of positive and finite Lebesgue measure, such that the Fourier transform of the indicator of E vanishes on an open non-empty set, answered positively by P. P. Kargayev, as already mentioned in L. N. 1043).

Secondly, it contains a few new questions: by V. F. Gaposhkin in section 3.4 (problems 6 and 7), inspired by the strong law of large numbers; by J. M. Anderson on random power series (new section 3.7); by Y. Guivarc'h, starting with products of random matrices (section 3.8); by J.-P. Kahane on a kind of covering problem (section 3.9). These new questions are kind of a random sample of problems and puzzles arising at the border between probability and analysis.

3.1
v.old

SOME QUESTIONS ABOUT HARDY FUNCTIONS

H. P. McKEAN

The theory of Gaussian-distributed noise leads to a variety of substantial mathematical questions about Hardy functions. I will put the questions in a purely mathematical way; the reader is referred to [1] for the statistical interpretation and/or additional information.

1. Let Δ, $\Delta \geqslant 0$, be summable on the line and let

$$\int_{\mathbb{R}} \frac{\log \Delta}{1+x^2} dx = -\infty.$$

Then the exponentials e^{ixt}, $t \leqslant 0$, span $L^2(\mathbb{R}, \Delta dx)$, *but how is* e^{ixT} *for fixed* $T > 0$ *efficiently approximated by these functions?* See [1], § 4.2.

2. Let h, $h \in H^2_+$, be outer, let T, $T > 0$, be fixed, and let

$$K(x) = \frac{e^{-ixT}}{h(x)} \int_T^\infty e^{ixt} \overset{\vee}{h}(t)\, dt,$$

$\overset{\vee}{h}(t)$ being the inverse transform $\frac{1}{2\pi} \int_{\mathbb{R}} e^{-ixt} h(x) dx$.

What can be said about $\overset{\vee}{K}$? $\|\overset{\vee}{K}\|_2$ *cannot be* $< \infty$ *for all small* T; also, $\overset{\vee}{K}$ can be enormously singular ; see [1], § 4.4.

3. Let h, $h \in H^2_+$, be outer. The QUESTION is *to explain what makes the phase function* h^*/h *the ratio of two inner functions or the reciprocal of an inner function;* see [1], § 4.6. $e^{ixT} h^*/h$ is itself an inner function if and only if h is integral of exponential type $\leqslant T$.

4. Let h, $h \in H^2_+$, be outer. *When does* h^*/h *belong to the span of* $e^{ixt} H^\infty$, $t \leqslant 0$, *in* L^∞? See [1], § 4.12.

5. The following conditions are equivalent for outer h, $h \in H^2_+$: a) $e^{2ixT} h^*/h$ is the ratio of a function of class H^2_+ and a function of class H^2_-; b) $\int_{\mathbb{R}} | f/h |^2 \, dx < \infty$ for some integral function f of exponential type $\leqslant T$; c) $\left\| e^{2ixt} \frac{h^*}{h} \frac{x+i}{x-i} - k \right\|_\infty \leqslant 1$ *for some* k, $k \not\equiv 0$, $k \in H^\infty_+$; see [1], § 4.12. *What can be said about such functions* h? note that b) is a problem of "multiplying down" the function $1/h$ in the style of [2]. *What outer function satisfy* a), b), c) *for every* T, $T > 0$? *for no* T, $T > 0$? Note that h cannot satisfy c) for T, $T = 0$.

6. The phase function h^*/h is ubiquitous. *What can be said about it for the general outer function* h, $h \in H^2_+$?

References

1. Dym H., McKean H. P., *Gaussian Processes, Function Theory, and the Inverse Spectral Problem*, Academic Press, New York, 1976.
2. Beurling A., Malliavin P., *On Fourier transforms of measures with compact support*, Acta Math. **107** (1962), 291–302.

New York University
Courant Institute of Mathematical Sciences
251 Mercer Street
New York, N.Y. 10012
USA

Editors' note

The question discussed in section 5 is related to the paper [3].

Reference

3. Koosis P., *The logarithmic integral*, volume I, Cambridge Univ. Press, 1988; volume II, 1992.

Commentary by the Chapter's Editor

A partial solution of the Problem 1 is obtained by V. Bart and V. Havin (to be published in the Collection of works dedicated to the memory of M. G. Krein which is being prepared now in Kharkov). This solution applies to *bounded* weights Δ satisfying the conditions of the problem. It is based on a modification of the Carleman formula recovering an analytic function with the given trace on a subset of the boundary.

3.2
v.old

SOME ANALYTICAL PROBLEMS IN THE THEORY OF STATIONARY STOCHASTIC PROCESSES

I. A. IBRAGIMOV, V. N. SOLEV

1. Let $\xi(t)$ be a stationary Gaussian process with discrete or continuous time (see [1] for the definitions of basic notions of stochastic processes used here). Denote by $f(\lambda)$ the spectral density of ξ (in case of discrete time f is a non-negative integrable function on the unit circle $\mathbb{T}$; in case of continuous time $\mathbb{T}$ is replaced by the real line $\mathbb{R}$).

Let $L(f)$ be a Hilbert space of functions on $\mathbb{T}$ (or $\mathbb{R}$) with the inner product

$$(\varphi, \psi)_f = \int \varphi(\lambda)\overline{\psi(\lambda)} f(\lambda)\, d\lambda.$$

For $\tau \geqslant 0$, let $L_\tau^+(f)$ (resp. $L_\tau^-(f)$) be the subspaces of $L(f)$ generated by exponentials $e^{it\lambda}$ with $t \geqslant \tau$ (resp. $t \leqslant -\tau$). Let $\mathcal{P}_\tau^+$ and $\mathcal{P}_\tau^-$ be the orthogonal projections onto $L_\tau^+(f)$ and $L_\tau^-(f)$. Consider the operators

$$B_\tau = \mathcal{P}_\tau^- \mathcal{P}_0^+ \mathcal{P}_\tau^- .$$

These positive selfadjoint operators were introduced into the theory of stochastic processes in [2]. Many characteristics of processes can be expressed in their terms. In particular, important classes of Gaussian processes correspond to the following conditions on B_τ:

a) B_τ is compact for all (sufficiently large in case of continuous time) τ;
b) B_τ is nuclear for all (sufficiently large in case of continuous time) τ.

Since the finite-dimensional distributions of a Gaussian process are completely determined by its spectral density f, it would be desirable to describe properties of B_τ in terms of f.

2. Processes with discrete time.

THEOREM 1 ([3]). *The operators B_τ are compact if and only if the spectral density f can be represented in the form*

$$f(\lambda) = |P(e^{i\lambda})|^2 \exp(u(\lambda) + \tilde{v}(\lambda)).$$

Here P is a polynomial with roots on the unit circle and the functions u and v are continuous.

THEOREM 2 (I. A. Ibragimov, V. N. Solev, cf. [1]). *The operators B_τ belong to the trace class if and only if*

$$f(\lambda) = |P(e^{i\lambda})|^2 e^{a(\lambda)}.$$

Here P stands for a polynomial with roots on $\mathbb{T}$ and

$$a(\lambda) \sim \sum a_j e^{ij\lambda}, \qquad \sum |a_j|^2 \, |j| < \infty.$$

PROBLEM 1. *Under what conditions on spectral density f do the operators B_τ belong to the class $\mathfrak{S}_p$, $1 \leqslant p \leqslant \infty$, i.e. for what f*

$$\sum \lambda_{j\tau}^p < \infty,$$

where $\lambda_{j\tau}$ are eigenvalues of B_τ?

Problem 1 was solved by V. V. Peller, as already mentioned in L. N. 1043 (see commentary pp. 90–91, not reproduced here).

Theorems 1 and 2 deal with the extreme cases $p = \infty,\ 1$.

THEOREM 3 (I. A. Ibragimov, see [1]). *The estimate*

$$|B_\tau| = 0(\tau^{-r-d}) \text{ for } \tau \to \infty$$

(r is an integer, $0 < \alpha < 1$) holds if and only if

$$f(\lambda) = |P(e^{i\lambda})|^2 e^{g(\lambda)},$$

where P is a polynomial, g is an r-times differentiable function and $g^{(r)}$ satisfies the Lipschitz condition of order d.

The value $|B_\tau|$ is exponentially small for $\tau \to \infty$ if and only if the spectral density f is an analytic function.

3. Processes with continuous time. Nothing similar to Theorem 3 is known in that case.

PROBLEM 2. *Under what conditions does the value $|B_\tau|$ decrease with power or exponential rate as $\tau \to +\infty$?*

THEOREM 4 (I. A. Ibragimov, [1]). *Let $f(\lambda) = |\Gamma(\lambda)|^{-2}$, where Γ is an entire function of exponential type with roots $z_1, z_2, \ldots$. Then $\lim\limits_{\tau \to +\infty} |B_\tau| = 0$ if and only if*

$$\frac{\log|\Gamma(\lambda)|}{1+\lambda^2} \in L^1(\mathbb{R}) \tag{1}$$

$$\sup_{-\infty<\lambda<\infty} \sum_j |\Im\mathrm{m} \frac{1}{\lambda - z_j}| < \infty. \tag{2}$$

PROBLEM 3. *Investigate the case when $f(\lambda) = \frac{|\Gamma_1(\lambda)|^2}{|\Gamma_2(\lambda)|^2}$ and Γ_1, Γ_2 are entire functions of exponential type.*

This problem is essential for the analysis of the operator B_τ in the multivariate case [4].

Note in conclusion that Problem 1 can be easily reformulated for continuous time.

References

1. Ibragimov I. A., Rosanov Yu. A., *Gaussian stochastic processes*, "Nauka", Moscow, 1970. (Russian)
2. Gelfand I. M., Yaglom A. M., *Computation of the amount of information about a stochastic function contained in another such function*, Uspekhi Mat. Nauk **12** (1957), no. 1, 3–52. (Russian)
3. Sarason D., *An addendum to Past and Future*, Math. Scand. **30** (1972), 62–64.
4. Ibragimov I. A., *Complete regularity of multidimensional stationary processes with discrete time*, Dokl. Akad. Nauk SSSR **162** (1965), no. 5, 983–985 (Russian); English transl. in Soviet Math. Dokl. **6** (1965), no. 3, 783–785.

Steklov Mathematical Institute
St. Petersburg branch
Fontanka 27
St. Petersburg, 191011
Russia

Steklov Mathematical Institute
St. Petersburg branch
Fontanka 27
St. Petersburg, 191011
Russia

3.3
old

MODULI OF HANKEL OPERATORS, PAST AND FUTURE

S. V. Hruščëv, V. V. Peller

A discrete, zero-mean, stationary Gaussian process is a sequence $\{X_n\}_{n\in\mathbb{Z}}$ in the real L^2 space of a probability measure P, such that $\mathbb{E}X_n = \int X_n\,dP = 0$, $n \in \mathbb{Z}$; $\mathbb{E}(X_nX_m) = Q(n-m)$ i.e. depends only on $n-m$; every function in the linear span of the functions X_n has a Gaussian distribution. In prediction theory the Past is associated with the closed linear span (over $\mathbb{C}$) $\mathbb{G}_p$ of X_k, $k < 0$, and the Future with the span $\mathbb{G}_f$ of X_k, $k \geqslant 0$. These closed subspaces are usually considered as subspaces of the complex Hilbert space $\mathbb{G}$ spanned by the whole sequence $\{X_k\}_{k\in\mathbb{Z}}$.

Our problem concerns the description of all possible positions in $\mathbb{G}$ of the Future $\mathbb{G}_f$ with respect to the Past $\mathbb{G}_p$.

The sequence $\{Q(n)\}_{n\in\mathbb{Z}}$ being positive definite, there exists a finite positive Borel measure μ on $\mathbb{T}$ satisfying $Q(n) = \hat{\mu}(n)$, $n \in \mathbb{Z}$. The measure μ is called the spectral measure of $\{X_n\}_{n\in\mathbb{Z}}$. Clearly the mapping Φ defined by $\Phi X_n = z^n$, $n \in \mathbb{Z}$, can be extended to a unitary operator from $\mathbb{G}$ to $L^2(\mu)$. To avoid technical difficulties we consider henceforth all stationary sequences $\{X_n\}_{n\in\mathbb{Z}}$ in $\mathbb{G}$, not necessarily real. A stationary sequence is unitarily equivalent to a Gaussian process iff the spectral measure μ is invariant under the transform $z \to \bar{z}$ of $\mathbb{T}$.

Consider the set of all triples $(A, B, \mathcal{H})$ where A and B are closed subspaces of the complex separable infinite–dimensional Hilbert space $\mathcal{H}$ such that $\operatorname{clos}(A+B) = \mathcal{H}$. The triples $(A_1, B_1, \mathcal{H}_1)$ and $(A_2, B_2, \mathcal{H}_2)$ are said to be equivalent if there exists an isometry V of $\mathcal{H}_1$ onto $\mathcal{H}_2$ satisfying $VA_1 = A_2$, $VB_1 = B_2$. Let $\mathcal{F}$ be the set of all equivalence classes with respect to the introduced equivalence relation.

Problem 1. *Which classes in $\mathcal{F}$ contain at least one element $(\mathbb{G}_p, \mathbb{G}_f, \mathbb{G})$ corresponding to a stationary sequence $\{X_n\}_{n\in\mathbb{Z}}$?*

The class $\mathcal{F}$ admits a more explicit description. Let $\mathcal{P}_A$ denote the orthogonal projection onto the subspace A. Each triple $t = (A, B, \mathcal{H})$ defines the selfadjoint operator $\mathcal{P}_A\mathcal{P}_B\mathcal{P}_A$ and the numbers

$$n_+(t) = \dim(A^\perp \cap B), \qquad n_-(t) = \dim(A \cap B^\perp).$$

Lemma. *A triple $t_1 = (A_1, B_1, \mathcal{H})$ is equivalent to $t_2 = (A_2, B_2, \mathcal{H}_2)$ iff $\mathcal{P}_{B_1}\mathcal{P}_{A_1}\mathcal{P}_{B_1}$ and $\mathcal{P}_{B_2}\mathcal{P}_{A_2}\mathcal{P}_{B_2}$ are unitarily equivalent and $n_\pm(t_1) = n_\pm(t_2)$.*

Sketch of the proof. We may assume without loss of generality that $\mathcal{H}_1 = \mathcal{H}_2 = \mathcal{H}$, $B_1 = B_2 = B$, $\mathcal{P}_B\mathcal{P}_{A_1}\mathcal{P}_B = \mathcal{P}_B\mathcal{P}_{A_2}\mathcal{P}_B$ (note that $\dim B_1 = \dim B_2$ under the assumption of the Lemma). Given a subspace C in $\mathcal{H}$ let $C^\perp = \mathcal{H} \ominus C$. Consider the partial isometries V_1, V_2 determined by the polar decompositions

$$\mathcal{P}_{A_i}\mathcal{P}_B = V_i(\mathcal{P}_B\mathcal{P}_{A_i}\mathcal{P}_B)^{1/2}, \quad i = 1, 2. \tag{1}$$

Let $\mathcal{U}$ be an operator on $\mathcal{H}$ defined by

$$\mathcal{U}x = V_2V_1^*x, \quad x \in A_1 \ominus B^\perp; \qquad \mathcal{U}x = x, \quad x \in B;$$

$\mathcal{U} \mid A_1 \cap B^\perp$ is an arbitrary unitary operator from $A_1 \cap B^\perp$ onto $A_2 \cap B^\perp$. It is easy to see that $\mathcal{U}$ is defined correctly (if $x \in A_1 \cap B$ then $V_2V_1^*x = x$). Also clearly $\mathcal{U}$ maps isometrically A_1 onto A_2. It remains to verify that $\mathcal{U}$ is a unitary operator on $\mathcal{H}$. Clearly it suffices to show that $(\mathcal{U}x, y) = (x, y)$ for $x \in A_1$, $y \in B$. If $x \in A_1 \cap B^\perp$, this is evident. Now we can consider only vectors x of the form $x = \mathcal{P}_{A_1}z$, $z \in B$. It follows from (1) that

$$(\mathcal{U}x, y) = (V_2V_1^*\mathcal{P}_{A_1}\mathcal{P}_B z, y) = (V_2(\mathcal{P}_B\mathcal{P}_{A_2}\mathcal{P}_B)^{1/2}z, y) = (\mathcal{P}_B\mathcal{P}_{A_1}\mathcal{P}_B z, y) = (x, y). \quad \bullet$$

For every $k, m \in \mathbb{Z}_+ \cup \{\infty\}$ and for every selfadjoint operator T on a Hilbert space $\mathcal{H}_1$ such that $\mathbb{O} \leqslant T \leqslant I$, $\operatorname{Ker} T = \{\mathbb{O}\}$ there exists $t = (A, B, \mathcal{H}) \in \mathcal{F}$ satisfying $n_+(t) = k$, $n_-(t) = m$ and such that $\mathcal{P}_B\mathcal{P}_A\mathcal{P}_B \mid B \ominus A^\perp$ is unitarily equivalent to T.

Indeed, without loss of generality $n_\pm(t) = 0$. By the well-known Naimark theorem, in $\mathcal{H}_1 \oplus \mathcal{H}_1$ there exists a projection $\mathcal{P}_A$ defined by

$$\mathcal{P}_A = \begin{pmatrix} T & T^{1/2}(I-T)^{1/2} \\ T^{1/2}(I-T)^{1/2} & I - T \end{pmatrix}.$$

Put $B = \mathcal{H}_1 \oplus \{\mathbb{O}\}$ and $\mathcal{H} = \operatorname{clos}(A + B)$. Then clearly $\mathcal{P}_B\mathcal{P}_A\mathcal{P}_B \mid B = T$.

By Szegö's alternative either $\mathbb{G}_p = \mathbb{G}_f = \mathbb{G}$ or $\mathbb{G}_p \neq \mathbb{G}$ and then $d\mu = \mid h \mid^2 dm + d\mu_s$, where h is an outer function in H^2 and μ_s is a singular measure on $\mathbb{T}$. We have $\Phi^{-1}(L^2(\mu_s)) \subset \mathbb{G}_p \cap \mathbb{G}_f$ and therefore only the case $\mu_s \equiv 0$ is interesting.

Recall that for a bounded function φ on $\mathbb{T}$ the Hankel operator H_φ on H^2 is defined by $H_\varphi f = (I - \mathbb{P}_+)\varphi f$, where $\mathbb{P}_+$ is the orthogonal projection from L^2 onto H^2.

Consider the Hankel operator $H_{\bar{h}/h}$ with the unimodular symbol $\varphi = \bar{h}/h$. It is easy to see that $\mathcal{P}_{\mathbb{G}_f}\mathcal{P}_{\mathbb{G}_p}\mathcal{P}_{\mathbb{G}_f} \mid \mathbb{G}_f$ is unitarily equivalent to $H^*_{\bar{h}/h}H_{\bar{h}/h}$ (see [1], Lemma 2.6).

The modulus of an operator T on Hilbert space is the selfadjoint non-negative operator $(T^*T)^{1/2}$. Problem 1 is therefore intimately connected with the problem of description of the moduli of Hankel operators up to the unitary equivalence.

PROBLEM 2. *Which operator can be the modulus of a Hankel operator?*

There are two necessary conditions for an operator to be the modulus of a Hankel operator which imply evident restrictions on triples equivalent to $(\mathbb{G}_p, \mathbb{G}_f, \mathbb{G})$.

For any $\varphi \in L^\infty$ the operator $(H_\varphi^* H_\varphi)^{1/2}$ *is not invertible* because $\lim_{n\to\infty} \|H_\varphi z^n\| = 0$. It follows that $\mathbb{G}_f$ does contain an orthogonal basis $\{e_n\}_{n\geqslant 0}$ with $\lim_n \|\mathbb{P}_{\mathbb{G}_0} e_n\| = 0$ (i.e. an orthogonal sequence "almost independent" with respect to the Past) provided $\mathbb{G}_p \neq \mathbb{G}_f$.

The kernel $\operatorname{Ker}(H_\varphi^* H_\varphi)^{1/2}$ is *either trivial or infinite dimensional.* Indeed, being invariant under multiplication by z by Beurling's theorem, it is either trivial or equal to θH^2 for an inner function θ.

Note that for $t = (\mathbb{G}_p, \mathbb{G}_f, \mathbb{G})$ we always have $n_+(t) = n_-(t)$. Indeed passing to the spectral representation Φ we see that $J \colon f \to \bar{z}\bar{f}$ is an isometry of $\mathbb{G}$ (over $\mathbb{R}$) with $J\mathbb{G}_p = \mathbb{G}_f$, $J\mathbb{G}_f = \mathbb{G}_p$. So we have either $n_+(t) = n_-(t) = 0$ or $n_+(t) = n_-(t) = \infty$.

Under the a priori assumption that the angle between $\mathbb{G}_p$ and $\mathbb{G}_f$ is positive Problem 1 can be in fact reduced to Problem 2.

Indeed, if the angle between $\mathbb{G}_p$ and $\mathbb{G}$ is positive then by the Helson–Szegö theorem the spectral measure μ is absolutely continuous and $\mathcal{P}_{\mathbb{G}_f}\mathcal{P}_{\mathbb{G}_p}\mathcal{P}_{\mathbb{G}_f}$ is unitarily equivalent to $H^*_{\bar{h}/h}H_{\bar{h}/h}$, $d\mu = |h|^2\,dm$. On the other hand if $R = H^*_\varphi H_\varphi$ and $\|R\| < 1$, then there exists an outer function h in H^2 with $H_\varphi = H_{\bar{h}/h}$ (see [2]). Put $\mathcal{H} = L^2(|\,h\,|^2)$, $B = \operatorname{span}_{L^2(|h|^2)}\{z^n : n \geqslant 0\}$, $A = \operatorname{span}_{L^2(|h|^2)}\{z^n : n \leqslant 0\}$. Clearly $\{z^n\}_{n\in\mathbb{Z}}$ is a stationary sequence with the Future B and the Past A.

It follows from the above considerations that in the case of non-zero angle between A and B the problem of the existence of a stationary sequence with the Future B and the Past A can be reduced to the existence of a Hankel operator whose modulus is unitarily equivalent to $\mathcal{P}_B\mathcal{P}_A\mathcal{P}_B$.

In connection with Problems 1 and 2 we can propose two conjectures.

CONJECTURE 1. *Let $\{s_n\}_{n\geqslant 0}$ be a non-increasing sequence of positive numbers and let $\lim_n s_n = 0$. Then there exists a Hankel operator whose singular numbers(*) $s_n(H_\varphi)$ satisfy*

$$s_n(H_\varphi) = s_n, \qquad n \geqslant 0.$$

CONJECTURE 2. *Let T be a compact selfadjoint operator such that $\operatorname{Ker} T$ is either trivial or infinite dimensional. Then there exists a Hankel operator H_φ satisfying $T = (H^*_\varphi H_\varphi)^{1/2}$.*

It can be shown that the last conjecture is equivalent to the following one.

CONJECTURE 2′. *Given a triple $t = (A, B, \mathcal{H}) \in \mathcal{F}$ such that $\mathcal{P}_A\mathcal{P}_B\mathcal{P}_A$ is compact and $n_+(t) = n_-(t)$ is either 0 or ∞ there exists a stationary sequence $\{X_n\}_{n\in\mathbb{Z}}$ in $\mathcal{H}$ whose Future is B and Past is A.*

We can also propose the following qualitative version of Conjecture 1.

THEOREM. *Let $\epsilon > 0$ and $\{s_n\}_{n\geqslant 0}$ be a non-increasing sequence of positive numbers. Then there exists a Hankel operator H_φ satisfying*

$$\frac{1}{1+\epsilon}s_n \leqslant s_n(H_\varphi) \leqslant (1+\epsilon)s_n, \quad n \geqslant 0.$$

Proof. Let b be an interpolating Blaschke product having zeros $\{\zeta_n\}_{n\geqslant 0}$ with the Carleson constant δ (see e.g. [4]). Consider the Hankel operators of the form $H_{f\bar{b}}$, $f \in H^\infty$. Then we have (see [4], Ch. VIII)

$$bH_{f\bar{b}}P_b = f(S_b), \tag{2}$$

where S_b is the compression of the shift operator S to $K_b \overset{\text{def}}{=} H^2 \ominus bH^2$, $S_b g = P_b zg$, $g \in K_b$, P_b is the orthogonal projection onto K_b, b is multiplication by b, $f(S_b) \overset{\text{def}}{=} P_b fg$, $g \in K_b$. Since b is an interpolating Blaschke product, there exists a function f in H^∞ satisfying $f(\zeta_n) = s_n$, $n \geqslant 0$.

(*)See definition of singular numbers in [3].

It follows from (2) that $s_n(H_{f\bar{b}}) = s_n(f(S_b))$, $n \geqslant 0$. Consider the vectors $e_n = \frac{b}{z-\zeta_n}|\zeta_n|(1-|\zeta_n|^2)^{1/2}$. We have $f(S_b)e_n = f(\zeta_n)e_n$ (see [4], Ch. VI) and there exists an invertible operator V on K_b such that the sequence $\{Ve_n\}_{n\geqslant 0}$ is an orthogonal basis of K_b, moreover if $1-\delta$ is small enough then we can choose V so that $\|V\|\cdot\|V^{-1}\| \leqslant 1+\epsilon$ (see [4], Ch. VII). The result follows from the obvious estimates

$$\frac{1}{\|V\|\cdot\|V^{-1}\|}s_n \leqslant s_n(f(S_b)) \leqslant \|V\|\cdot\|V^{-1}\|s_n. \quad \bullet$$

Conjecture 1 can be interpreted in terms of rational approximation. It follows from the theorems of Nehari and Adamian–Arov–Krein (see [1]) that for a function f in $\mathrm{BMO}_A \stackrel{\mathrm{def}}{=} \mathbb{P}_+ L^\infty$ we have

$$s_n(\Gamma_\varphi) = r_n(\varphi) \stackrel{\mathrm{def}}{=} \mathrm{dist}_{\mathrm{BMO}_A}(\varphi, \mathcal{R}_n),$$

where Γ_φ is the operator on ℓ^2 with the matrix $\{\hat{\varphi}(n+k)\}_{n,k\geqslant 0}$, $\mathcal{R}_n$ is the set of rational functions with at most n poles outside $\operatorname{clos}\mathbb{D}$ (including possible poles at ∞) counting multiplicities. Conjecture 1 is equivalent to the following one.

CONJECTURE 1$'$. *Let $\{r_n\}_{n\geqslant 0}$ be a non-increasing sequence, $\lim_{n\to\infty} r_n = 0$. Then there exists f in BMO_A such that $r_n(f) = r_n$, $n \geqslant 0$.*

If the conjecture is true then it would give an analogue of the well-known Bernstein theorem [5] for polynomial approximation. Note in this connection that Jackson-Bernstein type theorems for rational approximation in the norm BMO_A were obtained in [5], [7], [8].

We are grateful to T. Wolff for valuable discussions.

EULER INTERNATIONAL MATHEMATICAL INSTITUTE
PESOCHNAYA NAB. 10
ST. PETERSBURG, 197022
RUSSIA

STEKLOV MATHEMATICAL INSTITUTE
ST. PETERSBURG BRANCH
FONTANKA 27
ST. PETERSBURG, 191011
RUSSIA

AND

DEPARTMENT OF MATHEMATICS
KANSAS STATE UNIVERSITY
MANHATTAN
KANSAS, 66506
USA

COMMENTARY BY THE CHAPTER'S EDITOR

Both Conjectures are proved (see [9], [10]).

REFERENCES

1. Peller V. V., Hruščëv S. V., *Hankel operators, best approximations and stationary Gaussian processes*, Uspekhi Mat. Nauk **37** (1982), no. 1, 53–124. (Russian)
2. Adamian V. M., Arov D. Z., Krein M. G., *Infinite Hankel matrices and generalized Carathéodory-Fejer and Riesz problems*, Funkts. Anal. i Prilozh. **2** (1968), no. 4, 1–17 (Russian); English transl. in Funct. Anal. and its Appl. **2** (1968), no. 1, 1–18.

3. Gohberg I. C., Krein M. G., *Introduction to the theory of linear non-selfadjoint operators in Hilbert space*, "Nauka", Moscow, 1965. (Russian)
4. Nikolskii N. K., *Treatise on the shift operators*, "Nauka", Moscow, 1980 (Russian); English transl. Springer-Verlag, Berlin, New York, 1986.
5. Bernstein S. N., *On the inverse problem of the theory of best approximation of continuous functions*, Collected works, vol. 2, Izdat. Akad. Nauk SSSR, Moscow, 1954, pp. 292–294. (Russian)
6. Peller V. V., *Hankel operators of class $\mathfrak{S}_p$ and their applications (rational approximation, Gaussian processes, the problem of majorizing operators)*, Mat. Sbornik **113** (1980), no. 4, 538–581 (Russian); English transl. in Math. USSR Sbornik **41** (1982), no. 4, 443–479.
7. Peller V. V., *Hankel operators of the Schatten – von Neumann class $\mathfrak{S}_p$, $0 < p < 1$*, LOMI Preprints, E-6-82, Leningrad, 1982.
8. Semmes S., *Trace ideal criteria for Hankel operators, $0 < p < 1$*, Preprint, 1982.
9. Vasyunin V. I., Treil S. R., *Inverse spectral problem for the modulus of a Hankel operator*, Algebra i Analiz **1** (1989), no. 4, 56–67. (Russian)
10. Treil S. R., *Inverse problem for the modulus of a Hankel operator and balanced realizations*, Algebra i Analiz **2** (1990), no. 2, 158–182. (Russian)

3.4
old

SOME PROBLEMS RELATED TO THE STRONG LAW OF LARGE NUMBERS FOR STATIONARY PROCESSES

V. F. GAPOSHKIN

Let $(\xi_k : k \in \mathbb{Z})$ be a process in $L^2(\Omega, \mathcal{A}, P)$ stationary in the wide sense with $\mathbb{E}\xi_k = 0$ and $\mathbb{E} \mid \xi_k \mid^2 = \sigma^2 > 0$.

Denote the correlation function of the process by

$$R(n) = \mathbb{E}\bar{\xi}_k \xi_{k+n} \qquad \forall (k, n)$$

and let

$$\xi_k = \int_{-\pi}^{\pi} e^{ik\lambda} Z(d\lambda)$$

be its spectral representation. Here $Z(d\lambda)$ stands for the stochastic spectral measure of the process (ξ_k); $Z(d\lambda)$ is a process with orthogonal increments.

It is well-known that the strong law of large numbers (hereafter abbreviated as SLLN) holds for all processes stationary in the strict sense, that is, the limit of the means $\sigma_n = n^{-1} \sum_{k=1}^{n} \xi_k$ exists a.e. But there exist processes stationary in the wide sense such that the means σ_n converge in $L^2(\Omega)$ and diverge a.e. (see [1], [2]). SLLN criteria are given in [2].

THEOREM (Gaposhkin [2]). *In the above notation*

$$\lim_{n \to \infty} \left[\sigma_n - \int_{|\lambda| \leqslant 2^{-[\log_2 n]}} Z(d\lambda) \right] = 0. \tag{1}$$

Thus SLLN holds iff the limit

$$\lim_{m \to \infty} \int_{|\lambda| \leqslant 2^{-m}} Z(d\lambda) \tag{2}$$

exists a.e.

The theorem implies the following: if

$$R(n) = O\big((\log \log n)^{-2-\epsilon}\big), \; n \to \infty, \tag{3}$$

then SLLN holds provided $\epsilon > 0$, while for $\epsilon = 0$ it does not hold in general.

In all known counterexamples

$$\sup_k \mathbb{E}|\xi_k|^p = \infty \qquad (p > 2).$$

PROBLEM 1. *Is the condition*

$$\exists p > 2 : \sup_k \mathbb{E}|\xi_k|^p < \infty$$

(may be with the supplementary condition $\lim R(n) = 0$*) sufficient for the SLLN?*

PROBLEM 2. *Is the condition*

$$\sup_k \|\xi_k\|_\infty < \infty$$

(may be with the above supplementary condition) sufficient for the SLLN?

PROBLEM 3. *If the answers to problems 1 and 2 are negative, we may ask: are there stationary processes* (ξ_k) *satisfying*

$$\sup \|\xi_k\|_\infty < \infty, \quad R(n) = O\big((\log \log n)^{-2}\big)$$

while SLLN does not hold? Or condition (3) *can be relaxed for* L^∞*-bounded processes?*

All processes stationary in the strict sense obey SLLN, and so the Theorem implies the existence of limit (2) as well.

PROBLEM 4. *Why does limit* (2) *exist for stationary (in the strict sense) processes?*

Analogous problems are of interest not only for unitary operators determining stationary processes but also for normal operators in $L^2(X)$ (see an ergodic theorem of this kind in [3]). Here is one of possible problems in this direction.

PROBLEM 5. *Let* T *be a normal operator in* $L^2(X, \mu)$*,* μ *being a* σ*-finite measure. Suppose* $\|T\| = 1$*,* $f \in L^2(X)$*,* $\sup_k |T^k f(x)| \leqslant C$ a. e. *Does*

$$\lim_{n\to\infty} n^{-1} \sum_{k=1}^{n} T^k f(x)$$

exist a. e.*?*

COMMENTARY BY THE AUTHOR

The Problem 4 is partly solved by J. Campbell and K. Petersen (see [4]). These authors proved that the stochastical spectral measure $Z(d\lambda)$ of stationary (in the strong sense) process (ξ_k), $\xi_k \in L^2$, have the following property: *if* (ϵ_n) *is a fixed sequence of positive constants,* $\epsilon_n \to 0$*, then*

$$\lim_{n\to\infty} \int_{|\lambda|\leqslant\epsilon_n} Z(d\lambda) = 0 \text{ a.e.}$$

Addendum by the Author

Let us also consider the following problem.

Let $\xi(t)$ be a homogeneous (in the wide sense) random field, and $\mathbb{E}\xi(t) = 0$. $\mathbb{E}|\xi(t)|^2 = \sigma^2$; $t \in \mathbb{R}^n$. We denote by $R(\tau)$ the correlation function of the field, and

$$T = (T_1, \cdots, T_k) \qquad |T| = \min_k T_k, \qquad D_T = \{t : 0 \leqslant t_k \leqslant T_k; 1 \leqslant k \leqslant n\},$$

$$|D_T| = T_1 T_2 \cdots T_n, \qquad S_T = |D_T|^{-1} \int_{D_T} \xi(t)\, dt.$$

It is said that the n-parameter SLLN is *true* if

$$\lim_{|T| \to \infty} = 0 \text{ a. e.} \tag{4}$$

It is known that the condition

$$R(\tau) = O(\log \log |\tau|)^{-2n-\epsilon} \quad \text{as } |\tau| \to \infty; \quad \epsilon > 0,$$

is sufficient for the n-parameter SLLN (see [5,6]).

Problem 6. *Prove a criterium of n-parameter SLLN in terms of the stochastical spectral measure $Z(d\lambda)$ of the field.*

The n-parameter Dunford–Schwartz ergodic theorem implies that the property (4) is fulfilled for all homogeneous in the strong sense fields.

Problem 7. *Prove an analogue of the result of J. Campbell and K. Petersen for the stochastic spectral measure of homogeneous in the strong sense fields.*

References

1. Blanc-Lapierre A., Tortrat A., *Sur la loi forte des grands nombres*, C. R. Acad. Sc. Paris **267 A** (1968), 740–743.
2. Gaposhkin V. F., *Criteria for the strong law of large numbers for some classes of weakly stationary processes and homogeneous random fields*, Teor. Verojatn. i Primenen. **22** (1977), no. 2, 295–319. (Russian)
3. Gaposhkin V. F., *Individual ergodic theorem for normal operators in L_2*, Funkts. Anal. i Prilozhen. **15** (1981), no. 1, 18–22 (Russian); English transl. in Funct. Anal. and its Appl. **15** (1981), no. 1, 14–18.
4. Campbell J., Petersen K. The spectral measure and Hilbert transform of a measure-preserving transformation, Trans. Amer. Math. Soc. **313** (1989), 121–129.
5. Gaposhkin V. F., *On the n-parameter SLLN for homogeneous random fields*, Uspekhi mat. nauk **36** (1981), no. 4, 197–198. (Russian)
6. Klesov O. I., *The SLLN for homogeneous random fields*, Teor. Veroyatn. i Mat. Statist. **25** (1981), 29–40. (Russian)

Dept. of Applied Mathematics
MIIT,
Obraztsova str. 15,
Moscow 103055
Russia

3.5
old

THE THEORY OF MARKOV PROCESSES FROM THE STANDPOINT OF THE THEORY OF CONTRACTIONS

A. M. VERSHIK

Let (X, μ) be a Lebesgue space. A contraction P on $L^2(X, \mu)$ is called a Markov operator if P is order-positive and preserves the constants. In other words P is Markov if $P\mathbf{1} = \mathbf{1}$, $P^*\mathbf{1} = \mathbf{1}$ and P is positive. The integral representation of such an operator is given by a bistochastic measure ν on $X \times X : \nu \geqslant 0,\ \nu(X \times X) = 1$

$$(Pf)(x) = \int_X f(y)\nu(x, dy), \quad \iint_{AX} \nu(x, dy)\, dx = \mu(A), \quad \iint_{XB} \nu(dx, y)\, dy = \mu(B).$$

Markov operators form a convex semigroup with a zero (=projection onto the constants) and with a unit. This is a functional equivalent of the semigroup of multivalued maps, admitting an invariant measure, of (X, μ) onto itself. A detailed account of an analogous view–point see in [1].

A Markov operator gives rise in a natural way to a stationary Markov process with the state space X, the initial measure μ and the two-dimensional distribution ν (see above). In the space $\mathcal{X} = \prod_{-\infty}^{\infty} X_i$ $(X_i \equiv X)$ of realizations of the process a Markov measure $M_\nu = M$ appears. The left shift T in $(\mathcal{X}, M)$ generates a unitary operator U_T on $L^2(\mathcal{X}, M)$, a unitary dilation of P (non–minimal in general).

The main problem of the theory of Markov processes is the investigation of metric properties of the shift T in terms of the Markov operator P. The classical theory virtually used spectral properties of P only. This is insufficient for metric problems, P being nonselfadjoint.

Modern tools of the theory of contractions seem not have been used for this aim and we want to draw attention to this point (see also [1]). The connection between the contractions theory, their dilations, the scattering theory on the one hand and Markov processes theory on the other can be usefully applied in both directions.

1. Problems about past. It is easy to check that a Markov process is forward (back) mixing in the sense of Kolmogorov iff P belongs to the class $C_{0\cdot}$ (resp. $C_{\cdot 0}$), see notation in [2]. The opposite class C_1 includes two subcases. The first one is of no interest and corresponds to an isometric P and to deterministic processes. The second one, namely, the case of a completely non-isometric contraction, is very interesting. Its very existence is far from being obvious (for Markov operators), an example was given by R. Rosenblatt [3]. An important theorem (see [2]) asserts that the corresponding process, being non-deterministic, is quasisimilar to a deterministic one. Our PROBLEM is as follows: *use the technique of the theory of contractions to study mixing criteria of various kinds, deterministic and quasideterministic, the exactness* [1], *the bernoullity etc.* A powerful tool for these topics is the characteristic function of a Markov contraction. No adequate metric analogue of this notion seems to be found (e.g. how can one connect this function with the bistochastic measure ν)?

2. Non-linear dilations. The theory of Markov processes implicitly includes some constructions unfamiliar in the theory of contractions. We have already mentioned that a unitary operator U_T acting in $L^2(\mathcal{X}, M)$ is not the minimal dilation of the Markov operator P. The minimal dilation can be easily described in these terms. It coincides with the restriction of U_T to $\mathcal{L} = \text{span}\{L^2(\mathcal{X}_i) : i \in \mathbb{Z}\}$, $L^2(\mathcal{X}_i)$ being the subspace of $L^2(\mathcal{X}, M)$ consisting of functions depending on the i–th coordinate of $\{x_n\} \in X$ only. The subspace $\mathcal{L}$ is the subspace of all linear functionals of realizations ("one-particle" subspace). Thus the theory of minimal dilations corresponds to the linear theory of Markov processes whereas the dilation U_T has to be interpreted as a "non-linear" one (clearly U_T is a linear operator acting on non-linear functionals of realizations).

The investigation of the pair (P, U_T) ("a Markov operator plus a non-linear dilation") is of interest for the theory of contractions connecting it with methods and notions of the metric theory of processes (mixing, bernoullity etc.) E. g. the problem of the isomorphism of two Markov processes is analogous with the problem of existence of the wave operator in scattering theory. The enthropy yields an invariant of the dilation etc. It would be interesting to define the non-linear dilation for an arbitrary (non-positive) contraction.

3. C^*-algebra generated by Markov operators. Let us mention a more special problem: *to describe the C^*–envelope of the set of all Markov operators.* This algebra does not coincide with the algebra of all operators. (G. Lozanovsky gave a nice (unpublished) example: the distance between the Fourier transform as an operator in $L^2(\mathbb{R})$ and the set of all regular operators (=differences of positive operators) is one). It seems likely that a direct description of elements of this algebra can be given in terms of the order. This C^*–algebra plays an important rôle in the theory of grouppoids.

References

1. Vershik A. M., *Multivalued mappings with invariant measure (polymorphisms) and Markov operators*, Zap. Nauchn. Sem. LOMI **72** (1977), 26–61 (Russian); English transl. in J. Soviet Math. **23** (1983), no. 3, 2243–2266.
2. Sz.-Nagy B., Foiaş C., *Analyse harmonique des opérateurs de l'espace de Hilbert*, Acad. Kiado, Budapest, 1967.
3. Rosenblatt M., *Stationary Markov Processes*, Berlin, 1971.

Steklov Mathematical Institute
St. Petersburg branch
Fontanka 27
St. Petersburg, 191011
Russia

Commentary

See also

4. Vershik A. M., *Measurable realizations of groups of automorphisms, and integral representations of positive operators*, Sibirsk. Mat. Zh. **28** (1987), no. 1, 52–60; English transl. in Siberian Math. J. **28** (1987), no. 1, 36–43.

3.6
v.old

EXISTENCE OF MEASURES WITH GIVEN PROJECTIONS

V. N. SUDAKOV

Let F denote a measurable subset of the unit cube $Q \subset \mathbb{R}^d$, π_k being the canonical operator of projection onto the K–th axis, $k = 1, \ldots, d$ and X_k the side of Q situated on the k–th axis. Consider the linear operator π transforming every finite measure m on F into the system $(m_1, m_2, \ldots, m_d)$ of its marginals, $m_k A \stackrel{\text{def}}{=} m(\pi_k^{-1} A)$, $A \subset X_k$. It is often of importance to know *whether a given system* $(m_1, \ldots, m_d)$ *belongs to the image under* π *of a natural class of probability measures on* F. Some partial results are known, mostly for $d = 2$. E. g. for a class $\mathcal{F}$ of subsets F of Q defined in terms of measure spaces (X_k, m_k), $k = 1, 2$, an existence criterion of a probability measure on F with marginals (m_1, m_2) is as follows: the required measure exists iff no subset of $X_1 \times X_2$ of the form $F \cap (A \times B)$, where $A \subset X_1$, $B \subset X_2$ and $m_1 A + m_2 B > 1$, is a union of subsets N_1, N_2 with $m_1(\pi_1 N_1) = m_2(\pi_2 N_2) = 0$ (see [1]; a smaller class of closed set was considered in [2]). The class $\mathcal{F}$ is in particular characterized by the following property: for every $F \in \mathcal{F}$ the set of all measures on F with given marginals is compact in the topology of convergence on sets of the form $A \times B$. A similar condition (where the non-decomposability of sets $F \cap (A \times B)$ is replaced by $\text{mes}_2(F \cap (A \times B)) > 0$) is a criterion of existence of a probability measure on F with marginals (m_1, m_2) subordinated to the Lebesgue type (i. e. absolutely continuous with respect to the Lebesgue measure mes_2), [1].

Analogous conditions fail to be sufficient for $d > 2$, and the corresponding criterion is unknown. For $d = 2$ it is not known whether the Lebesgue type can be replaced by any other type in the last sentence of the previous paragraph. For $d > 2$ there is no existence criterion for a positive measure on the cube Q with given marginals whose density function with respect to Lebesgue measure is majorized by the density function of a given probability measure on Q (see the discussion in [1]).

REFERENCES

1. Sudakov V. N., *Geometric problems in the theory of infinite-dimensional probability distributions*, Proc. Steklov Inst. Math. **1979** (1976), no. 2. (Russian)
2. Strassen V., *Probability measures with given marginals*, Ann. Math. Stat. **36** (1965), no. 2, 423–439.

STEKLOV MATHEMATICAL INSTITUTE
ST. PETERSBURG BRANCH
FONTANKA 27
ST. PETERSBURG, 191011
RUSSIA

COMMENTARY BY THE AUTHOR

Consider a finite or countable family of probability distributions $\{m_k,\ k \in K\}$. The answer to the question as to whether there exists such a family $\{\xi_k,\ k \in K\}$ of random variables, each ξ_k being distributed according to m_k, that for every pair (k_1, k_2) the equality

$$\varkappa(m_{k_1}, m_{k_2}) \stackrel{\text{def}}{=} \min\Big\{ \iint |x-y|\,dm \;:\; m\pi_x^{-1} = m_{k_1}, \quad m\pi_y^{-1} = m_{k_2} \Big\} = \|\xi_{k_1} - \xi_{k_2}\|_{L^1}$$

holds, depends on existence of a probability measure m with marginals m_k on the set

$$F = \big\{ x \in \mathbb{R}^k \;:\; U_{k_1 k_2}(x_{k_1}) - U_{k_1 k_2}(x_{k_2}) = |x_{k_1} - x_{k_2}| \quad \forall\, k_1, k_2 \in K \big\}.$$

Here $\varkappa$ is the Kantorovich distance and $U_{k_1 k_2}$ stands for related potential function (see e.g. [1]):

$$U_{k_1 k_2}(\lambda) = \int \operatorname{sign}\big(m_{k_1}(-\infty, t] - m_{k_2}(-\infty, t]\big)\, dt.$$

One can show that for such a special type of subsets there always exists a measure with given marginals $\{m_k,\ k \in K\}$, so that the family $\{\xi_k\}$ under discussion does exist.

3.7

RANDOM POWER SERIES

J. M. Anderson

For a given sequence $\{a_n\}$ with $\sum_0^\infty |a_n|^2 < \infty$, let

$$f(z,\alpha) = \sum_{n=0}^{\infty} a_n e^{i\alpha_n} z^n,$$

where α_n, $n = 0, 1, 2, \ldots$ are the Steinhaus functions (independent r.v.s. with uniform distribution on $[0, 2\pi]$). For a given Banach space X, let $r(X)$ denote the set of those sequences $\{a_n\}$ such that almost surely $f(z,\alpha) \in X$. A long time ago, Salem and Zygmund showed [4], p. 291 that if $\sum_1^\infty |a_n|^2 \log n (\log\log n)^{1+\delta} < \infty$ for some $\delta > 0$ then a.s. $f(z,\alpha) \in C(T)$. This result has been strikingly improved by Marcus and Pisier and now $r(C(T))$ is known. For the details, and a nice exposition of this see [3], p. 231, Theorem 7.

Problem. *Determine $r(B)$ and $r(BMOA)$, where B denotes the Bloch space.*

It was shown in [1] that if $\sup_n \left(\sum_{\nu\in I_n} |a_\nu|^2 \log\nu\right) < \infty$, where I_n is the dyadic block $2^n \leqslant \nu \leqslant 2^{n+1}$ then $\{a_n\} \in r(B)$. A weaker result was stated in [1], but this is what was actually proved. Also Duren showed in [2] that if $\sum |a_\nu|^2 \log \nu < \infty$ then a.s. $\int_b^1 (1-r)\left[M_\infty(r, f'(z,\alpha))\right]^2 dr < \infty$ so that $\{a_n\} \in r(BMOA)$. These results are both deduced from the above work of Salem and Zygmund and neither seems to come to the heart of the matter. Perhapsthe techniques of Marcus and Pisier can be correspondingly adapted.

References

1. Anderson J. M., Clunie J., Pommerenke Ch., *On Bloch functions and normal functions*, J. Reine Angew. Math. **270** (1974), 12–37.
2. Duren P. L., *Random series and bounded mean oscillation*, Michigan Math. J. **32** (1985), 81–86.
3. Kahane J.-P., *Some random series of functions*, 2nd ed. Cambridge studies in advanced mathematics n° 5, Cambridge, 1985.
4. Salem R., Zygmund A., *Some properties of trigonometric series whose terms have random signs*, Acta Math. **91** (1954), 245–301.

Mathematics Department
University College
London WCIE 6BT, U.K.

3.8

PRODUCTS OF RANDOM MATRICES AND SPECTRAL PROPERTIES OF GENERALIZED TRANSFER OPERATORS

Y. GUIVARC'H

We consider the unit disk $D = \{z \in \mathbb{C};\ |z| < 1\}$, its boundary $\mathbb{T} = \{t \in \mathbb{C};\ |t| = 1\}$, the Poisson kernel $P(z,t) = \frac{1-|z|^2}{|t-z|^2}$ and the group G of conformal automorphisms of D. We denote by m the Lebesgue measure on $\mathbb{T}$ and we consider for $\lambda \in \mathbb{C}$, $g \in G$, the "cocycle" $\sigma^\lambda(g,t) = P^\lambda(g^{-1} \cdot o, t) = \left[\frac{dg^{-1}m}{dm}(t)\right]^\lambda$.

We denote by the single symbol ρ_λ, the two representations of G in the space $C(\mathbb{T})$ [resp. $M(\mathbb{T})$] of continuous functions [resp. measures] which are defined by the formulas (dual to each other):

$$\rho_\lambda(g)[\varphi](t) = \sigma^\lambda(g^{-1}, t)\varphi(g^{-1} \cdot t) \qquad \varphi \in C(\mathbb{T}), \quad g \in G$$
$$\rho_\lambda(g)[\nu] = \int \delta_{g \cdot t}\sigma^\lambda(g,t)\, d\nu(t) \qquad \nu \in M(\mathbb{T}).$$

We denote by p a probability measure on G and we consider the operators P_λ defined on $C(\mathbb{T})$ and $M(\mathbb{T})$ by the formulas

$$P_\lambda\varphi(t) = \int \rho_\lambda(g)\,[\varphi]dp(g)$$
$$P_\lambda\nu = \int \rho_\lambda(g)\,[\nu]\, dp(g).$$

If p has a density with respect to the Haar measure on G and has compact support, the operator P_λ is defined by a kernel, is compact in $C(\mathbb{T})$, and its spectrum is discrete. In general we can formulate the following problem.

PROBLEM 1. *For which pairs (λ, p) is the spectral radius of P_λ, on a suitable non trivial subspace of $C(\mathbb{T})$, given to by an isolated point of the spectrum.*

The operators P_λ are close to the transfer operators of the one-dimensional statistical mechanics [7] and the parameter λ plays the role of temperature so that some form of compactness can be expected. These operators occur naturally in the study of limit theorems for the product of random matrices chosen according to p. In this context the characteristic functions of relevant random variables are expressed by the iteration of P_λ $(i\lambda \in \mathbb{R})$ [4].

On the other hand, these operators occur also in the study of the convolution equation:

$$\int f(gh)\, dp(h) = kf(g) \qquad (k \in \mathbb{C}). \tag{1}$$

The connection is as follows: suppose $\lambda \in \mathbb{R}$, $k(\lambda) > 0$ and ν_λ is a probability measure on $\mathbb{T}$ such that: $P_\lambda \nu_\lambda = k(\lambda)\nu_\lambda$ and define, for $\varphi \in C(\mathbb{T})$, $\bar{\varphi}(g) = \rho_\lambda(g)\nu_\lambda[\varphi]$. Then $\bar{\varphi}$ is a solution of (1). This is a generalized Poisson formula which has been thoroughly studied by H. Furstenberg [2] in the general context of semi-simple Lie groups: for $k > 0$, there exist $\lambda \in \mathbb{R}$ such that the positive solutions of (1) are given by the above Poisson formula in the case where p has a density with compact support. Consequently an interplay between the methods of potential theory and the thermodynamic formalism can be expected.

A partial answer to Problem 1 is given by the

THEOREM [4]. *Suppose p is irreducible with compact support. Then there exist $\epsilon > 0$ such that for $\Re e\,\lambda \geqslant -\epsilon$, the spectral radius of P_λ is given by an eigenvalue which is a pole of the resolvent of P_λ on a space of Hölder functions.*

Such a result is also true without the restriction $\Re e\,\lambda \geqslant -\epsilon$, in the case where a non trivial compact interval of $\mathbb{T}$ is preserved under the action of the support of p. Another case of interest is the one where the support of p is contained in a discrete subgroup of G.

Let us consider now the situation where the support of p is contained in a discrete subgroup of G without fixed points in $\mathbb{T}$. We suppose that this subgroup Γ is generated by the support of p. Denote by L_Γ the set of limit points of Γ; this is a closed Γ-invariant subset of $\mathbb{T}$ with positive Hausdorff dimension d, and there is on L_Γ a canonical measure π called Patterson measure [8]. This measure has dimension d and for example coincide with Lebesgue measure if $L_\Gamma = \mathbb{T}$. Here we restrict to $\lambda = 0$ and consider the unique p-invariant measure ν on L_Γ: $\sum_{\gamma \in \Gamma} p(\gamma)\gamma \cdot \nu = \nu$.

PROBLEM 2. *Describe the class of p such that ν is equivalent to π.*

This class is non trivial as shown by the following result which extend a previous result of H. Furstenberg in case $\pi = m$ [3]. For a construction of such a p in the context of potential theory see [1]

THEOREM [6]. *If Γ is finitely generated, there exist p such that ν is equivalent to π.*

In a somewhat different context, the discussion of regularity of p-invariant measures occur in the study of Bernoulli convolutions [5]. In general very few is known on these measures ν which are used for example to express law of large numbers for product of random matrices. These problems admit natural extensions in the context of semi-simple Lie groups especially the hyperbolic group $SO(n,1)$ and the linear group $S\ell(n,\mathbb{R})$.

REFERENCES

1. Ancona A., *Théorie du potentiel sur les graphes et les variétés.*, Lecture Notes in Math., vol. 1427.
2. Furstenberg H., *Translation invariant cones of functions*, Bull. Amer. Math. Soc. **71** (1965), 271–326.
3. Furstenberg H., *Random walks and discrete subgroups of Lie groups*, Adv. in Prob. and related topics I (1970), Dekker, New York, 1–63.
4. Guivarc'h Y., Raugi A., *Products of random matrices, convergence theorems*, Contemporary Math. AMS **50** (1986), 31–54.
5. Kahane J.-P., *Sur la distribution de certaines séries aléatoires*, Colloque Théorie des nombres (1969 Bordeaux), Bull. SMF Mémoire **25** (1971), 119–122.

6. Lalley S. P., *Renewal theorems in symbolic dynamics with application to geodesic flows*, Acta Math. **163** (1989), 1–55.
7. Ruelle D., *An extension of the theory of Fredholm determinants*, Publ. Math. IHES **72** (1990).
8. Sullivan D., *The density at infinity of a discrete group of hyperbolic notions*, Publ. Math. IHES **50** (1979), 171–202.

IRMAR
Université Rennes I
Campus de Beaulieu
35042 Rennes Cedex
France

3.9

A KIND OF COVERING PROBLEM

J.-P. KAHANE

Given a series of positive and uniformly bounded functions $f_n(t)$ defined on $\mathbb{T} = \mathbb{R}/\mathbb{Z}$, one considers the random series

$$\text{(S)} \qquad \sum_{1}^{\infty} f_n(t - \omega_n)$$

where ω_n is a sequence of independent random variables whose common distribution is the Lebesgue (= Haar) measure on $\mathbb{T}$. Clearly

$$\sum_{1}^{\infty} \int f_n(t)\, dt = \infty$$

is a necessary and sufficient condition for (S) to diverge almost surely almost everywhere.

PROBLEM. *Give necessary conditions and sufficient conditions for* (S) *to diverge almost surely everywhere.*

When $f_n = 1_{A_n}$, indicator function of a subset of $\mathbb{T}$, this is the covering problem by random translates of $\mathbb{T}$. Historical comments can be found in [1], examples in [2].

REFERENCES

1. Kahane J.-P., *Recouvrements aléatoires*, Gazette des mathématiciens.
2. Fan Ai-Hua, Kahane J.-P., *Rareté des intervalles recouvrant un point dans un recouvrement aléatoires*, *Ann. Inst. H. Poincaré*, série Prob. & Stat., à paraître.

UNIVERSITÉ PARIS-SUD
MATHÉMATIQUES - BÂT. 425
91405 ORSAY CEDEX
FRANCE

Chapter 4

HOLOMORPHIC OPERATOR FUNCTIONS

Edited by

I. Gohberg
Raymond and Beverly Sackler
Faculty of Exact Sciences
School of Math. Sciences
Ramat-Aviv 69978
Israel

M. A. Kaashoek
Faculteit Wiskunde en Informatica
Vrije Universiteit
De Boelelaan 1081a
1081 HV Amsterdam
The Netherlands

4.1

THE SPECTRAL NEVANLINNA – PICK PROBLEM AND SCALAR COMPLEX INTERPOLATION

N. J. YOUNG

Suppose given points $z_1, \ldots z_n \in \mathbb{D}$ and complex $k \times k$ matrices $W_1, \ldots, W_n$ (where $n, k \in \mathbb{N}$). The *spectral Nevanlinna–Pick problem* is the following:

Ascertain whether there exists an analytic $k \times k$ matrix-valued function F on $\mathbb{D}$ such that

$$F(z_j) = W_j, \qquad 1 \leqslant j \leqslant n,$$

and

$$\sigma(F(z)) \subset \operatorname{clos} \mathbb{D} \quad \text{for all } z \in \mathbb{D}.$$

Find such an F if it exists.

Here $\sigma(A)$ denotes the set of eigenvalues of the square matrix A. In the case $k = 1$ this is a much-studied classical problem which has been solved by many different methods (e.g. [1, 6, 7, 8]). The case $k > 1$ appears to be much harder — none of the cited methods works. The problem has attracted a lot of attention because of its importance to the theory of H^∞ control (see [3]) and there are on the market computer packages which will often construct the desired F (see [5]). Nevertheless, from a mathematical point of view the problem cannot be regarded as definitively resolved. There is no solution of an elegance approaching that of the classical case. The numerical methods for solving it require a non-linear optimization over a parameter space of large dimension and are not guaranteed to converge.

The best results on the spectral Nevanlinna–Pick problem are due to Bercovici, Foiaş and Tannenbaum [2]. Some ingenious examples show what a subtle problem it is. In particular, even if the W_j are all diagonal, all reasonable attempts fail to decouple the problem into a series of scalar ones. The authors do obtain some positive results, but these fall short of a conclusive and easily implemented algorithm.

The purpose of this contribution is to suggest an approach based on a theorem of Schur which leads to an interpolation problem for scalar analytic functions. The theorem in questions is the following [9].

Let

$$p(z) = a_0 + a_1 z + \cdots + a_{k-1} z^{k-1} + z^k$$

and define

$$\hat{p}(z) = 1 + \bar{a}_{k-1} z + \cdots + \bar{a}_1 z^{k-1} + \bar{a}_0 z^k.$$

Then the zeros of p all lie in $\mathbb{D}$ if and only if

$$\|(p/\hat{p})(S_k)\| < 1$$

where S_k is the $k \times k$ shift matrix

$$S_k = \begin{bmatrix} 0 & 1 & 0 & \dots & 0 \\ 0 & 0 & 1 & \dots & 0 \\ . & . & . & \dots & . \\ 0 & 0 & 0 & \dots & 1 \\ 0 & 0 & 0 & \dots & 0 \end{bmatrix}$$

and the matrix norm is the usual operator norm on k-dimensional Euclidean space.

Let us restrict to the case $k = 2$. Here we find that

$$(p/\hat{p})(S_2) = \begin{bmatrix} a_0 & -\bar{a}_1 a_0 + a_1 \\ 0 & a_0 \end{bmatrix}$$

from which it is not hard to show that the zeros of p lie in $\mathbb{D}$ if and only if

$$|a_1 - \bar{a}_1 a_0| < 1 - |a_0|^2.$$

Applying this to the characteristic polynomial of 2×2 matrix A, we deduce that $\sigma(A) \subset \mathbb{D}$ if and only if

$$\left|\operatorname{tr} A - \overline{\operatorname{tr} A} \det A\right| < 1 - \left|\det A\right|^2.$$

Hence, if the spectral Nevanlinna–Pick problem with the specified data does not have a solution $F(z)$, then the functions

$$T(z) := \operatorname{tr} F(z), \qquad D(z) := \det F(z)$$

are analytic functions in $\mathbb{D}$ satisfying

$$T(z_j) = \operatorname{tr} W_j, \qquad D(z_j) = \det W_j, \qquad 1 \leqslant j \leqslant n$$
$$\left|T(z) - \overline{T(z)} D(z)\right| < 1 - \left|D(z)\right|^2, \qquad z \in \mathbb{D}.$$

Conversely, if T and D can be found satisfying these conditions then we can construct an F with the desired properties. Note that if we obtain an F which extends continuously to $\operatorname{clos}\mathbb{D}$, then in view of the subharmonicity of the spectral radius, it is sufficient to verify the spectral condition on $\mathbb{T}$. This leads to

PROBLEM 1. *Given $z_1, \dots, z_n \in \mathbb{D}$ and the complex numbers $t_1, \dots, t_n$ and $d_1, \dots, d_n$, construct if possible a pair of functions T, D in the disc algebra such that*

$$T(z_j) = t_j, \qquad D(z_j) = d_j, \qquad 1 \leqslant j \leqslant n,$$

and

$$(*) \qquad \left|T(z) - \overline{T(z)} D(z)\right| < 1 - \left|D(z)\right|^2, \qquad \text{all } z \in \mathbb{T}.$$

One approach is to look for D first, then T. It is easy enough to compute a D (when there is one) satisfying the interpolation conditions and with $\|D\|_\infty \leqslant 1$ — and indeed to obtain a parameterization of all such D. With D known, the condition (*) has a pleasing geometrical expression. For any $re^{i\theta} \in \operatorname{clos}\mathbb{D}$ let $E(re^{i\theta})$ denote the closed ellipse centered at 0, with minor axis making an angle $\frac{\theta}{2}$ with the positive θ-axis and with semi-major and semi-minor axes of lengths $1 + r$, $1 - r$ respectively. If $r = 1$, E degenerates to a line segment of length 4. Then (*) is equivalent to $T(z) \in E(D(z))$. This brings us to

PROBLEM 2. *Given $z_1, \dots, z_n \in \mathbb{D}$, $t_1, \dots, t_n \in \mathbb{C}$ and D in the disc algebra of norm $\leqslant 1$, construct if possible T in the disc algebra such that*

$$T(z_j) = t_j, \qquad 1 \leqslant j \leqslant n,$$

and

$$T(z) \in E(D(z)), \qquad \text{all } z \in \mathbb{T}.$$

The second problem can be transformed to one of a general type of frequency domain design problems arising in control and circuit theory (see "Problem OPT" in [4]). However, in the present context we are looking for a solution which is sufficiently explicit to give a lead in the solution of the $k \times k$ spectral Nevanlinna–Pick problem.

REFERENCES

1. Adamyan V. M., Arov D. Z., Krein M. G., *Analytic properties of Schmidt pairs for a Hankel operator and the Schur–Takagi problem*, Mat. Sbornik **86(128)** (1971), 34–75 (Russian); English transl. in Mat. USSR Sbornik **15** (1971), 31–73.
2. Bercovici H., Foiaş C., Tannenbaum A., *A spectral commutant lifting theorem*, Trans. Amer. Math. Soc. (to appear).
3. Doyle J. C., *Structured uncertainty in control system design*, IEEE CDC Proceedings, Fort Lauderdale, Fl., 1985.
4. Helton J. W., *Operator theory, analytic function, matrices and electrical engineering*, CMBS Series No. 68, AMS, 1987.
5. Math Works Inc., *μ-Analysis and Synthesis Toolbox for use with MATLAB*, Palo Alto, 1990.
6. Nevanlinna R., *Über beschränkte Funktionen die in gegebenen Punkten vorgeschriebene Werte annehmen*, Ann. Acad. Sci. Fenn. Ser. A **13** (1919).
7. Pick G., *Über die Beschränkungen analytischer Funktionen, welche durch vorgegebene Funktions werte bewirkt werden*, Math. Ann. **77** (1916), 7–23.
8. Sarason D., *Generalized approximation in H^∞*, Trans. Amer. Math. Soc. **127** (1967), 179–203.
9. Schur I., *Über Potenzreihen, die im Innern des Einheitskreises beschränkt sind* II, J. Reine Angew. Math. **148** (1918), 122-145.

DEPARTMENT OF MATHEMATICS
FYLDE COLLEGE
LANCASTER UNIVERSITY
LANCASTER
LA1 4YF
ENGLAND

4.2
old

SOME FUNCTION THEORETIC PROBLEMS CONNECTED WITH THE THEORY OF SPECTRAL MEASURES OF ISOMETRIC OPERATORS

V. M. ADAMYAN, D. Z. AROV, M. G. KREIN

Let V be a completely non-unitary isometric operator in a separable Hilbert space H with the defect spaces N and M:

$$V\colon H \ominus N \to H \ominus M,$$

where it is supposed for definiteness that $0 < \dim N \leqslant \dim M \leqslant \infty$. Let P_L denote the orthogonal projection of H onto a subspace L, and let $T_V = VP_{H\ominus N}$. The operator V defines in the unit disc $\mathbb{D}$ an operator-valued holomorphic function

$$\chi_V(z) = zP_N\big(I - zT_V\big)^{-1}\big|M$$

which is called the characteristic function of V.

Consider the class $\mathbb{B}(M,N)$ of all operator-valued contractive holomorphic functions in $\mathbb{D}$ taking values in the space of all bounded operators from M to N. Let $\mathbb{B}^0(M,N) = \{\chi \in \mathbb{B}(M,N) : \chi(0) = \mathbb{O}\}$. It is known that $\chi_V \in \mathbb{B}^0$ and that for every $\chi \in \mathbb{B}^0$ there exists an isometric V with the given defect spaces N and M such that $\chi_V = \chi$ (see [1]–[3]).

It is also known that in the case $\dim N = \dim M$ all unitary extensions of V not leaving H are described by the formula

$$U_\varepsilon = T_V + \varepsilon P_N$$

where ε is a unitary "parameter", $\varepsilon\colon N \to M$ ($\varepsilon^*\varepsilon = I|N$, $\varepsilon\varepsilon^* = I|M$). The spectral measure E_ε of U_ε can be determined (up to a unitary equivalence) by $\chi = \chi_V$ and ε with a help of the following formula

$$[I + \chi(z)\varepsilon\,]\,[\,I - \chi(z)\varepsilon]^{-1} = \frac{1}{2\pi}\int_{\mathbb{T}} \frac{\zeta + z}{\zeta - z}\,d\sigma_\varepsilon, \quad z \in \mathbb{D}, \tag{1}$$

where $d\sigma_\varepsilon = P_N dE_\varepsilon|N$.

The spectral measures of the minimal unitary extensions of V leaving H (now the case $\dim N \neq \dim M$ is also permitted) can be also determined by (1) where the parameter ε is already an arbitrary function in $\mathbb{B}(N,M)$. The spectral measure of U_ε is absolutely continuous if and only if the measure σ_ε in (1) is absolutely continuous with respect to the Lebesgue measure on $\mathbb{T}$.

Consider a subset $\mathbb{B}_a(M,N)$ of $\mathbb{B}^0(M,N)$ consisting of functions χ whose measure σ_ε in the Riesz–Herglotz representation

$$[I+\chi(z)\varepsilon(z)\,]\,[\,I-\chi(z)\varepsilon(z)]^{-1}=\frac{1}{2\pi}\int_{\mathbb{T}}\frac{\zeta+z}{\zeta-z}d\sigma_\varepsilon,$$

is absolutely continuous for an arbitrary choice of ε in $\mathbb{B}(N,M)$. The inclusion $\chi_V\in\mathbb{B}_a(M,N)$ is clearly equivalent to the condition that all minimal unitary extensions of V have absolutely continuous spectral measures.

PROBLEM. *Find criteria for a given χ in $\mathbb{B}^0(M,N)$ to belong to $\mathbb{B}_a(M,N)$.*

Note that for $\chi\in\mathbb{B}^0(M,N)$ the inclusion

$$(1-\|\chi\|)^{-1}\in L^1(\mathbb{T})$$

implies $\chi\in\mathbb{B}_a(M,N)$ being thus a sufficient (but not necessary) condition.

Suppose in the sequel that $\dim M<+\infty$ and let $\mathbb{B}_{\mathrm{Sz}}(M,N)$ ($\mathbb{B}^0_{\mathrm{Sz}}(M,N)$) denote the family of all χ in $\mathbb{B}(M,N)$ ($\mathbb{B}^0(M,N)$) with

$$\log\det(I-\chi^*\chi)\in L^1(\mathbb{T}). \tag{2}$$

LEMMA. *Given $\chi\in\mathbb{B}(M,N)$ the following are equivalent:*

1) *there exists an isometric operator $\varepsilon\colon N\to M$ with the corresponding measure in (1) satisfying the Szegö condition*

$$\log\det\left(\frac{d\sigma_\varepsilon}{dm}\right)\in L^1(\mathbb{T}) \tag{3}$$

2) *condition (3) holds for all isometries $\varepsilon\colon N\to M$;*
3) *$\chi\in\mathbb{B}_{\mathrm{Sz}}(M,N)$.*

If $\chi\in\mathbb{B}_{\mathrm{Sz}}(M,N)$ then the measures σ_ε are absolutely continuous for almost all ε (with respect to the invariant measure on the symmetric space of all isometries $\varepsilon\colon N\to M$).

We don't know any example of a function in $\mathbb{B}_a(M,N)$ not satisfying (2).

A more subtle sufficient condition for $\chi\in\mathbb{B}^0_{\mathrm{Sz}}(M,N)$ to belong to $\mathbb{B}_a(M,N)$ can be deduced from results of [4]–[6]. Namely, fix χ in $\mathbb{B}^0_{\mathrm{Sz}}$ and denote by P_+^{-1} and $(P_-^{-1})^*$ (unique) solutions of the factorization problem

$$I-\chi(\zeta)\chi(\zeta)^*=P_+^{-1}(\zeta)[P_+^{-1}(\zeta)]^*,\qquad \zeta\in\mathbb{T},$$
$$I-\chi(\zeta)^*\chi(\zeta)=P_-^{-1}(\zeta)[P_-^{-1}(\zeta)]^*,\qquad \zeta\in\mathbb{T},$$

in the classes of outer functions non-negative at the origin and belonging to $\mathbb{B}(N,N)$ and $\mathbb{B}^*(M,M)=\{\,h(z):h^*(1/\bar z)\in\mathbb{B}(M,M),\ |z|>1\,\}$ respectively. Let

$$q_-(\zeta)=P_-(\zeta)\chi^*(\zeta),\qquad q_+(\zeta)=P_+(\zeta)\chi^*(\zeta).$$

It follows that the values of $f_0 = q_- P_+^{-1}$ are contractions $N \to M$ a. e. on $\mathbb{T}$. Consider the Hankel operator Γ with the matrix symbol f_0. The operator Γ maps $\ell^2(N)$ into $\ell^2(M)$ and its matrix in the standard basis is $\left(\hat{f}_0(-j-k+1)\right)_{j,k\geqslant 1}$, where $\hat{f}_0$ stands for Fourier coefficients of f_0. Consider subspaces

$$N_0 = \{\, e_\xi = (\xi, 0, 0, \dots) : \xi \in N \,\}, \qquad M_0 = \{\, e_\eta = (\eta, 0, 0, \dots) : \eta \in M \,\}$$

of $\ell^2(N)$ and $\ell^2(M)$ and put

$$P_+(z,\rho)\xi = \left[\left(1 - \rho^2 \Gamma^* \Gamma\right)^{-1} N^{1/2}(\rho) e_\xi \right]_+ (z), \qquad |z| < 1,$$

$$q_+(z,\rho)\eta = \rho z \left[\Gamma^* \left(1 - \rho^2 \Gamma \Gamma^*\right)^{-1} M^{1/2}(\rho) e_\eta \right]_+ (z), \qquad |z| < 1,$$

where $\rho \in (0,1)$, $\xi \in N$, $\eta \in M$ and $N^{1/2}(\rho)$, $M^{1/2}(\rho)$ denote positive square roots of $\left[P_{N_0}\left(I - \rho^2 \Gamma^* \Gamma\right)^{-1} \mid N_0\right]^{-1}$ and $\left[P_{M_0}\left(I - \rho^2 \Gamma \Gamma^*\right)^{-1} \mid M_0\right]^{-1}$ respectively and finally

$$h_+(z) \stackrel{\text{def}}{=} \sum_{j \geqslant 1} \xi_j z^{j-1} \quad (z \in \mathbb{D}) \quad \text{for } h = \{\xi_j\}_1^\infty \in \ell^2(N).$$

It turns out that $\chi \in \mathbb{B}_a(M,N)$ provided

$$\lim_{\rho \to 1-0} P_+^{-1}(z,\rho) q_+(z,\rho) = \chi(z) \tag{4}$$

for all z in $\mathbb{D}$.

CONJECTURE. *If* $\chi \in \mathbb{B}_a(M,N) \cap \mathbb{B}^0_{\mathrm{Sz}}(M,N)$ *then* χ *satisfies* (4).

Remark. For $\chi \in \mathbb{B}^0_{\mathrm{Sz}}(M,N)$ condition (4) holds iff the following formula

$$f_\varepsilon(\zeta) = \left[q_-(\zeta) + P_-(\zeta)\varepsilon(z)\right]\left[P_+(\zeta) + q_+(\zeta)\varepsilon(\zeta)\right]^{-1}$$

establishes a one-to-one correspondence between the set of operator $(N \to M)$-valued contractive functions f_ε with the same principal part as $f_0(\zeta)$ (i.e. $\sum_{k\geqslant 1} \hat{f}_\varepsilon(-k)\zeta^{-k} = \sum_{k\geqslant 1} \hat{f}_0(-k)\zeta^{-k}$) and all functions ε in $\mathbb{B}(N,M)$.

REFERENCES

1. Livshits M. S., *On one class of linear operators in Hilbert space*, Mat. Sbornik **19 (61)** (1946), 236–260. (Russian)
2. Livshits M. S., *Isometric operators with the same defect numbers, and quasi-unitary operators*, Mat. Sbornik **26** (1950), 247–264. (Russian)
3. Sz.-Nagy B., Foiaş C., *Harmonic analysis of operators in Hilbert space*, Akad. Kiadó, Budapest, 1970.
4. Adamyan V. M., Arov D. Z., Krein M. G., *Infinite Hankel matrices and generalized Carathéodory-Fejér's and I. Schur's problems*, Funktsion. Analiz i Prilozhen. **2** (1968), no. 4, 1–17. (Russian)
5. Adamyan V. M., Arov D. Z., Krein M. G., *Infinite block-Hankel matrices and problems of extension related with them*, Izv. Akad. Nayk Arm.SSR, ser. matem. **6** (1971), 181–206. (Russian)
6. Adamyan V. M., *Non-degenerate unitary coupling of semi-unitary operators*, Funkts. Anal. i Prilozhen. **7** (1973), no. 4, 1–16. (Russian)

PROLETARSKII BULV. 12-1, 4
ODESSA-44 270044
UKRAINE

LVOVSKAYA 211, 18
ODESSA-104 270104
UKRAINE

4.3
old

THREE PROBLEMS ABOUT J-INNER MATRIX-FUNCTIONS

D. Z. Arov

1. Let $J = \begin{pmatrix} I_n & 0 \\ 0 & -I_n \end{pmatrix}$. A matrix-function (m.-f.) W meromorphic in $\mathbb{D}$ is called *J-inner* if

$$W^*(z)JW(z) \leqslant J, \quad \forall z \in \mathbb{D}, \qquad W^*(\zeta)JW(\zeta) = J \quad \text{a. e. on } \mathbb{T}.$$

Let $\mathcal{D}_+$ denote the class of m.-f. with entry functions representable as a ratio of an H^∞-function and of an outer H^∞-function. A J-inner m.-f. W is called: 1) *singular* if $W^{\pm 1} \in \mathcal{D}_+$ and 2) *regular* if there exists no nonconstant singular J-inner m.-f. W_0 such that WW_0^{-1} is J-inner.

THEOREM 1. *An arbitrary J-inner m.-f. W admits a representation $W = W_r W_s$, where W_r and W_s are respectively regular and singular J-inner m.-f.; W_r is uniquely determined by W up to a constant J-unitary right factor.*

2. The importance of the class of regular J-inner m.-f. is explained in particular by its connection with the generalized interpolation problem of Schur–Nevanlinna–Pick of finding all m.-f. S such that

$$b_1^{-1}(S - S_0)b_2^{-1} \in H_n^\infty, \qquad S \in \mathbb{B}_n, \tag{1}$$

where b_1, b_2, S_0 ($\in \mathbb{B}_n$) are given m.-f. of order n, b_1 and b_2 are inner, $\mathbb{B}_n$ denotes the set of all m.-f. of order n holomorphic and contractive in $\mathbb{D}$; H_n^p $(1 \leqslant p \leqslant +\infty)$ is the class of m.-f. of order n with entries in H^p.

Fix $z_0 \in \mathbb{D}$ with $\det b_k(z_0) \neq 0$ $(k = 1, 2)$. When S ranges over the set of solutions of problem (1), the values $S(z_0)$ fill a matrix ball. If the right and left half-radii of this ball are nondegenerate then problem (1) is called *completely indeterminate;* this definition does not depend on z_0.

Let $W = [W_{jk}]_1^2$ be an arbitrary J-inner m.-f. It has a meromorphic quasi-continuation to the exterior $\mathbb{D}_e$ of the disc $\mathbb{D}$. We denote $f^\sim(z) = f^*(1/\bar{z})$. We have [2]

$$W_{11} = b_1 p_*, \qquad W_{22} = b_2^{-1} p, \qquad S_0 \stackrel{\text{def}}{=} W_{12} W_{22}^{-1} \in \mathbb{B}_n \tag{2}$$

where b_1 and b_2 are inner m.-f., p and $p_*^\sim$ are outer m.-f., $p^{-1} \in \mathbb{B}_n$, $(p_*^\sim)^{-1} \in \mathbb{B}_n$. Singular m.-f. W are characterized by equalities $b_1 = b_2 = I_n$ in (2). The following theorem shows that it is important to establish a criterion of regularity of a J-inner m.-f.

THEOREM 2. *Let $W = [W_{jk}]_1^2$ be an arbitrary J-inner m.-f. and let b_1, b_2, S_0 be m.-f. defined in (2). Then the problem (1) with these data is completely indeterminate and the m.-f. $S = S_\varepsilon$, where*

$$S_\varepsilon = (W_{11}\varepsilon + W_{12})(W_{21}\varepsilon + W_{22})^{-1}, \qquad \varepsilon \in \mathbb{B}_n \tag{3}$$

are its solutions. The family $\{S_\varepsilon\}$ is the set of all solutions of problem (1) iff the J-inner m.-f. W is regular. For any completely indeterminate problem (1) there exists a regular J-inner m.-f. $W = [W_{jk}]_1^2$, for which formula (3) establishes a one-to-one correspondence between the set of all $\varepsilon \in \mathbb{B}_n$, and the set of all solutions of problem (1). M.-f. W may be chosen so that m.-f. b_1 and b_2 in (2) be the same as in problem (1); in this case W is defined by problem (1) up to a constant J-unitary right factor.

3. J-inner m.-f. $W = [W_{jk}]_1^2$ being arbitrary, let us consider m.-f. b_1 and b_2 corresponding to it by (2) and define a m.-f. $f_{\pm I}$,

$$f_{\pm I} = b_1^{-1}(W_{11} \pm W_{12})(W_{21} \pm W_{22})^{-1} b_2^{-1}. \tag{4}$$

It takes unitary values a.e. on $\mathbb{T}$ and $f_{\pm I} = \varphi_*^{-1}\varphi$, where $\varphi \in H_n^2$, $\varphi_*^\sim \in H_n^2$. If a m.-f. f unitary on $\mathbb{T}$ is representable in the form $f = \varphi_-^{-1}\varphi_+$, where $\varphi_+ \in H_n^2$, $\varphi_-^\sim \in H_n^2$, $\varphi_\pm$ being determined by f up to the left constant factor x with $\det x \neq 0$, then we write $\operatorname{ind} f = 0$. The following theorem holds (see [1] and Theorem 2).

THEOREM 3. *A J-inner m.-f. $W = [W_{jk}]_1^2$ is regular iff for the m.-f. $f_{\pm I}$ defined in (4) we have $\operatorname{ind} f_{\pm I} = 0$.*

COROLLARY. *For a J-inner m.-f. $W = [W_{jk}]_1^2$ to be regular it is sufficient, that*

$$\Big(1 - \|S_0(\zeta)\|\Big)^{-1} \in L^1, \qquad S_0 \overset{\text{def}}{=} W_{12}W_{22}^{-1}. \tag{5}$$

The proof of Theorems 2 and 3 is based on the results about the problem of Nehari [3–5] to which problem (1) is reduced by a substitution $f = b_1^{-1}Sb_2^{-1}$.*)

PROBLEM 1. *Find a criterion for a J-inner m.-f. W to be regular without using the notion of the index of a m.-f.*

4. It is known [6] that a product of elementary factors of Blaschke–Potapov of the 1st, 2nd, 3rd kind with the poles, respectively in $\mathbb{D}$, in $\mathbb{D}_e$ and on $\mathbb{T}$ (see [7]) is a J-inner m.-f. We have ([1])

THEOREM 4. *A J-inner m.-f. W is a product of elementary factors of the 1st and the 2nd kind iff it is regular and the m.-f. b_1 and b_2 associated with W by (2) are products of (definite) elementary factors of Blaschke–Potapov.*

Remark. Both m.-f. b_1 and b_2 in (2) are Blaschke–Potapov products iff

$$\lim_{r\uparrow 1}\int_{\mathbb{T}} \log\Big|\det\big(W_{11}(r^{-1}\zeta)W_{22}(r\zeta)\big)\Big|\,dm(\zeta) = \int_{\mathbb{T}} \log\Big|\det\big(W_{11}(\zeta)W_{22}(\zeta)\big)\Big|\,dm(\zeta). \tag{6}$$

*)see the proof in *γ-generating matrices, J-inner matrix-function and related extrapolation problems*, Theory of functions, functional analysis and appl., **51** (1989) 61–67; **52** (1989) 103–109; **53** (1989) 57–64 (Russian), English transl. in J. Sov. Math., **52** (1990) no.6, 3487–3491; **52** (1990) no.5, 3421–3425; **58** (1992) no.6, 532–537.

A corresponding condition also exists for a product of elementary factors of only the 1st (2nd) kind. In this case $b_1 = I_n$ ($b_2 = I_n$) and instead of (6) we have

$$\log\left|\det W_{11}(\infty)\right| = \int_{\mathbb{T}} \log\left|\det W_{11}(\zeta)\right| dm(\zeta), \tag{7}$$

$$\lim_{r\uparrow 1} \int_{\mathbb{T}} \log\left|\det W_{22}(r\zeta)\right| dm(\zeta) = \int_{\mathbb{T}} \log\left|\det W_{22}(\zeta)\right| dm(\zeta).$$

$$\Bigg(\log\left|\det W_{22}(0)\right| = \int_{\mathbb{T}} \log\left|\det W_{22}(\zeta)\right| dm(\zeta), \tag{8}$$

$$\lim_{r\downarrow 1} \int_{\mathbb{T}} \log\left|\det W_{11}(r\zeta)\right| dm(\zeta) = \int_{\mathbb{T}} \log\left|\det W_{11}(\zeta)\right| dm(\zeta) \Bigg).$$

COROLLARY. *Suppose condition* (5) *holds for a J-inner m.-f. $W = \left[W_{jk}\right]_1^2$. Then W is a product of elementary factors of the 1st and 2nd kind* (*only of the 1st, only of the 2nd*) *iff condition* (6) (*respectively* (7), (8)) *is valid.*

PROBLEM 2. *Find a criterion for a J-inner m.-f. to be a product of elementary factors of the 1st, 2nd and 3rd kind.*

Theorem 4 gives in fact a criterion of completeness of a simple operator in terms of its characteristic m.-f. W in the case when its eigenvalues are not on $\mathbb{T}$. The solution of Problem 2 would give a criterion without this restriction.

PROBLEM 3. *Find a criterion for a J-inner m.-f. to be a product of elementary factors of the 3rd kind.*

Let us point out that such a product is a singular m.-f. A product of elementary factors of the 1st kind arises in the "tangent" problem of Nevanlinna–Pick [8] and products of factors of the 1st and 2nd kind arise in a "bi-tangent" problem in which "tangent" data for $S(z)$ and $S^*(z)$ are given in interpolation knots z_k ($\in\mathbb{D}$). The author's attention was drawn to such a "bi-tangent" problem by B. L. Kogan. Products of elementary factors of the 3rd kind arise in the "tangent" problem which has the interpolation knots on $\mathbb{T}$. The definition and investigation of such problems is much more complicated [9, 10].

REFERENCES

1. Arov D. Z., *On the interpolation problem and Blaschke–Potapov product*, Sketch of lectures, School on theory of operators on functional spaces, Minsk, 1982, pp. 14–15. (Russian)
2. Arov D. Z., *Realization of matrix-valued functions according to Darlington*, Izv. Akad. Nauk SSSR, ser. matem. **37** (1973), no. 6, 1299–1331.
3. Adamyan V. M., Arov D. Z., Krein M. G., *Infinite Hankel matrices and generalized Carathéodory–Fejér's and I. Schur's problems*, Funktsion. Analiz i Prilozhen. **2** (1968), no. 4, 1–17. (Russian)
4. Adamyan V. M., Arov D. Z., Krein M. G., *Infinite block-Hankel matrices and problems of extension related with them*, Izv. Akad. Nauk Arm.SSR, ser. matem. **6** (1971), no. 2–3, 87–112. (Russian)
5. Adamyan V. M., *Non-degenerate unitary coupling of semi-unitary operators*, Funkts. Anal. i Prilozhen. **7** (1973), no. 4, 1–16. (Russian)
6. Arov D. Z., Simakova L. A., *On boundary values of convergent sequence of J-contractive matrix-functions*, Mat. Zametki **19** (1976), no. 4, 491–500. (Russian)
7. Potapov V. P., *Multiplicative structure of J-nonexpanding matrix-functions*, Trudy MMO **4** (1955), 125–236. (Russian)

8. Fedchina I. P., *Tangential Nevanlinna–Pick problem with multiple points*, Doklady Akad. Nauk Arm.SSR **61** (1975), no. 4, 214–218. (Russian)
9. Krein M. G., *General theorems on positive functionals*, Some questions in the theory of moments, (by Ahiezer N. I., Krein M.), Kharkov, 1938, pp. 121–150 (Russian); English transl. vol. 2, Trans. Math. Mon., AMS, 1962, pp. 124–153.
10. Melamud E. Ya., *Boundary Nevanlinna–Pick problem for J-expanding matrix-functions*, Izvestiya VUZov, Matem. **6** (1984), 36–43. (Russian)

LVOVSKAYA 211, 18
ODESSA-104 270104
UKRAINE

COMMENTARY BY THE AUTHOR

It is possible to formulate another condition for a J-inner m.-f. $W = \left[W_{jk}\right]_1^2$ to be regular. Let $\chi = W_{22}^{-1}W_{21}$, $c = (1-\chi)(1+\chi)^{-1}$ and let W be normalized by the condition $c(0) = I_n$. Let σ_W be the spectral measure of c, i.e. it is the measure that defines the function c of the Carathéodory class by the Riesz–Herglotz formula. Then the density $\sigma_a'(\mu)$ $\left(= \operatorname{Re} c(e^{i\mu})\right)$ of the absolutely continuous part of σ_a of the σ_W has right and left factorization

$$\sigma_a'(\mu) = \phi(e^{i\mu})\phi^*(e^{i\mu}) = \psi(e^{i\mu})\psi^*(e^{i\mu}),$$

where ϕ and ψ are outer m.-f. from H_n^2. We can consider the m.-f. $f_W = (\psi^*)^{-1}\phi$ that has unitary bounded values a.e. on $\partial\mathbb{D}$.

THEOREM 3A. *A J-inner m.-f. W is regular iff* 1) *σ_W is absolutely continuous and* 2) $\operatorname{ind} f_W = 0$.

For $n = 1$ this is connected with the Sarason's description [11] of the solutions of the Nehari problem in the linear-fractional form. It can be generalized for the case $n > 1$ if we consider the measure σ corresponding to $\chi = p_+^{-1}q_+$.

REFERENCE

11. Sarason D., *Exposed points in H^1*. I, The Gohberg anniversary collection, vol. II (Calgary, AB, 1988), Operator Theory: Advances and Applications, vol. 41, Birkhäuser, Basel, 1989, pp. 485–496.

4.4
old

EXTREMAL MULTIPLICATIVE REPRESENTATIONS

YU. P. GINZBURG

Let $\mathcal{C}$ be the class of entire functions W of exponential type with values in the space of all bounded operators on a separable Hilbert space and such that

$$W(0)=I,\quad W^*(\lambda)W(\lambda)\leqslant I\quad (\operatorname{Im}\lambda>0),\quad W^*(\lambda)W(\lambda)=I\quad (\operatorname{Im}\lambda=0).$$

For every $W\in\mathcal{C}$ there exists an operator-valued hermitian non-decreasing function E on $[0,l]$ ($E(0)=\mathbb{O}$, $\operatorname*{Var}_{[0,t]} E=t$) satisfying

$$W(\lambda)=\int_0^{l}\!\!\!\curvearrowleft \exp\{i\lambda\, dE(t)\}. \tag{1}$$

(see [1, 2]). Let H be the weak derivative of E. Then (1) is equivalent to

$$W(\lambda)=\int_0^{l}\!\!\!\curvearrowleft \exp\{i\lambda H(t)\, dt\}. \tag{2}$$

The function W determines H uniquely iff $I-W(\lambda)\in\mathfrak{S}_1$ ($\lambda\in\mathbb{C}$) and W, $\det W$ have the same exponential type [3]. To single out a canonical function from the family of all functions H satisfying (2) in the general case the following definition is introduced.

DEFINITION. *Let H be a weakly measurable function on $[a,b]$ ($\mathbb{O}\leqslant H$, $|H|\in L^1[a,b]$) and suppose that for every $s\in[\alpha,\beta]\subset[a,b]$ the function*

$$W_{\alpha,s}(\lambda H)\overset{\text{def}}{=}\int_\alpha^{s}\!\!\!\curvearrowleft \exp\{i\lambda H(t)\, dt\}$$

is the greatest divisor (in $\mathcal{C}$) among all divisors of $W_{\alpha,\beta}(\lambda H)$ of type $s-\alpha$. Then H is called extremal function on $[\alpha,\beta]$.

THEOREM 1. *For every W of exponential type σ in $\mathcal{C}$ there exists a unique $\widetilde{H}$ extremal on $[0,\sigma]$ and satisfying $W_{0,\sigma}(\lambda\widetilde{H})=W(\lambda)$.*

This result is a special case of a theorem proved in [2] (compare with [4]).

PROBLEM. *Find an intrinsic description of function $\widetilde{H}$ extremal on $[a,b]$ (or, at least, exhibit a large subclass of $\widetilde{H}$'s).*

The following theorem shows that the description is very likely to be of local character.

THEOREM 2 ([5], [6]). *Suppose H is extremal on $[a,b]$. Then it is extremal on any $[\alpha,\beta] \subset [a,b]$. Conversely, if for every $s \in [\alpha,\beta]$ there exists a segment $[\alpha,\beta] \subset [a,b]$ such that $s \in [\alpha,\beta]$ and H is extremal on $[\alpha,\beta]$ then H is extremal on $[a,b]$.*

CONJECTURE. *Let H be a continuous (with respect to the norm topology) operator-valued function on $[a,b]$. Then H is extremal iff all values of H are orthogonal projections.*

In the particular case $\operatorname{rank} H(t) = 1$, $t \in [a,b]$ the conjecture is true by a theorem of G. E. Kisilevskii [7] (in the form given in [8]).

Similar questions in the case when $W(1/\lambda)$ is the characteristic function of so called one-block operator have been considered in [9].

REFERENCES

1. Potapov V. P., *Multiplicative structure of J-nonexpanding matrix-functions*, Trudy MMO **4** (1955), 125–236. (Russian)
2. Ginzburg Yu. P., *Multiplicative representations and minorants of analytic operator-functions*, Funktsion. Anal. i Prilozhen. **1** (1967), no. 3, 9–23. (Russian)
3. Brodskii M. S., *Triangular and Jordan representations of linear operators*, Nauka, Moscow, 1969. (Russian)
4. Brodskii M. S., Isaev L. E., *Triangular representations of dissipative operators with resolvent of exponential type*, Doklady Akad. Nauk SSSR **188** (1969), no. 5, 971–973 (Russian); English transl. in Soviet Math. Dokl. **10** (1969).
5. Ginzburg Yu. P., *On divisors and minorants of operator-functions of bounded type*, Matem. Issledovaniya, Kishinëv **2** (1967), no. 4, 47–72. (Russian)
6. Mogilevskaya R. L., *Nonmonotone multiplicative representations of bounded analytic operator-functions*, Matem. Issledovaniya, Kishinëv **4** (1969), no. 4, 70–81. (Russian)
7. Kiselevskii G. E., *Invariant subspaces of dissipative Volterra operators with nuclear imaginary part*, Izv. Akad. Nauk SSSR, ser. matem. **32** (1968), no. 1, 3–23. (Russian)
8. Gohberg I. Ts., Krein M. G., *Theory of Volterra operators on Hilbert spaces and its applications*, Nauka, Moscow, 1967. (Russian)
9. Sakhnovich L. A., *On dissipative Volterra operators*, Mat. Sbornik **76 (118)** (1968), no. 3, 323–343. (Russian)

UL. PETRA VELIKOGO, D.21, KV.20
ODESSA, 270021
UKRAINE

4.5
old

EVALUATION OF AN INFINITE PRODUCT OF SPECIAL MATRICES

L. D. FADDEEV, N. YU. RESHETIHIN

An important role in studying the integrable models of Field Theory is played by matrix-functions of complex variable of a special form [1]. The simplest example is provided by a rational matrix:

$$L_0(z) = \frac{z+B}{z+\mu}, \tag{1}$$

where B is a matrix of size $n \times n$ and μ is a complex number. It is natural to call it the matrix Weierstrass factor for the complex plane $\mathbb{C}$ (i.e., a meromorphic function on $\mathbb{C}$ with one pole and such that $L(\infty) = 1$).

The next interesting example is given by a matrix Weierstrass factor for a strip. This function L_1 is meromorphic in the strip $\{ z \in \mathbb{C} : 0 < \operatorname{Re} z \leqslant 1 \}$ has only one pole in it and is regular at infinity, i.e.,

$$L_1(z) \to \mathcal{D}_\pm, \qquad \operatorname{Im} z \to \pm\infty,$$

where $\mathcal{D}_\pm$ are non-degenerate diagonal matrices. The boundary values of $L_1(z)$ satisfy the following relation:

$$L_1(i\lambda + 1) = A L_1(i\lambda) A^{-1}, \qquad \lambda \in \mathbb{R}, \tag{2}$$

where $A = \operatorname{diag}(1, \varepsilon, \ldots, \varepsilon^{n-1})$, $\varepsilon = \exp(\frac{2\pi i}{n})$. One can represent such a matrix-function as an infinite product of functions (1). For this purpose introduce the family of matrices

$$L^m(z) = A^m L_0(z+m) A^{-m} \tag{3}$$

and their finite product

$$L_1^N(z) = L^N(z) L^{N-1}(z) \ldots L^{-N+1}(z) L^{-N}(z). \tag{4}$$

It is easy to show that the regularized limit

$$L_1(z) = \lim_{N\to\infty}{}' \; L_1^N(z) \tag{5}$$

satisfies (2).

For $n = 1$ formulae (3)–(5) are nothing but Euler's formulae for $\sin u$, so that

$$L_1(z) = \frac{\sin \pi(z+B)}{\sin \pi(z+\mu)}.$$

PROBLEM 4.5

We calculated $L_1(z)$ for $n = 2$ in [2]. In this case $A = \operatorname{diag}(1, -1)$ and

$$B = \begin{pmatrix} S_3 & S_- \\ S_+ & -S_3 \end{pmatrix}$$

so that $\operatorname{trace}(B) = 0$.

The limit in (5) is defined as follows

$$\lim_{N\to\infty}{}' L_1^N(z) = \lim_{N\to\infty} (A^{-N} D_N L_1^N(z) D_{-N} A^N),$$

where

$$D_N = \begin{pmatrix} N^{S_3} & 0 \\ 0 & N^{-S_3} \end{pmatrix}.$$

The limit matrix $L_1(z)$ has a form

$$L_1(z) = W^{-1} \tilde{L}_1(z) W,$$

$$W = \begin{pmatrix} h(S_3)^{-1} & 0 \\ 0 & h(S_3) \end{pmatrix}, \qquad h(z) = \sqrt{\frac{\Gamma(1+R-z)\Gamma(1-R-z)}{\pi^2(z^2 - R^2)}}\, e^{-z},$$

$$\tilde{L}_1(z) = \frac{1}{\sin \pi(z+\mu)} \begin{pmatrix} \sin \pi(z + S_3), & S_- \sqrt{\dfrac{\sin^2 \pi S_3 - \sin^2 \pi R}{\pi^2(S_3^2 - R^2)}} \\ S_+ \sqrt{\dfrac{\sin^2 \pi S_3 - \sin^2 \pi R}{\pi^2(S_3^2 - R^2)}} & \sin \pi(z - S_3) \end{pmatrix},$$

where $R^2 = S_3^2 + S_+ S_-$.

We pose as a PROBLEM *the explicit calculation of the limit in* (5) *for every* n *in terms of known special functions.*

References

1. Faddeev L., *Integrable models in* $1+1$ *dimensional quantum field theory*, preprint, CEN–SACLAY, S. Ph. T. /82/76.
2. Faddeev L. D., Reshetihin N. Yu., *Hamiltonian structures for integrable fields theory models*, Teor. Mat. Fiz. **57** (1983), no. 1, 323–343. (Russian)

St.Petersburg branch of
Steklov Mathematical Institute
Fontanka 27
St. Petersburg, 191011
Russia

Department of Mathematics
Harvard University
Cambridge, Massachusetts 02138
USA

4.6
v.old

FACTORIZATION OF OPERATOR FUNCTIONS (CLASSIFICATION OF HOLOMORPHIC HILBERT SPACE BUNDLES OVER THE RIEMANNIAN SPHERE)

J. LEITERER

Let H be a Hilbert space, $L = L(H)$ the Banach space of bounded linear operators in H, and $\mathrm{GL} = \mathrm{GL}(H)$ the group of invertible operators in L. We put $\mathbb{T}_+ = \{ z \in \mathbb{C} : |z| \leqslant 1 \}$ and $\mathbb{T}_- = \{ z \in \mathbb{C} \cup \{\infty\} : 1 \leqslant |z| \leqslant \infty \}$ and denote by $\mathcal{O}(\mathbb{T}, \mathrm{GL})$, $\mathcal{O}(\mathbb{T}_+, \mathrm{GL})$, $\mathcal{O}(\mathbb{T}_-, \mathrm{GL})$ the groups of holomorphic GL-valued functions in a neighborhood of $\mathbb{T}$, $\mathbb{T}_+$, $\mathbb{T}_-$ respectively. We shall say that two functions S, T ($S, T \in \mathcal{O}(\mathbb{T}, \mathrm{GL})$) are *equivalent* if $S = A_- T A_+$ for some $A_\pm$, $A_\pm \in \mathcal{O}(\mathbb{T}_\pm, \mathrm{GL})$.

PROBLEM. *Classify the functions in $\mathcal{O}(\mathbb{T}, \mathrm{GL})$ with respect to this notion of equivalence.*

Remark. It is well-known that this problem is equivalent to the classification problem for holomorphic Hilbert space bundles over the Riemannian sphere.

1. What is known about the problem? We shall say that $\mathcal{D}$ is a diagonal function if $\mathcal{D}(z) = \sum_1^n z^{\varkappa_j} P_j$, where $\varkappa_1 < \cdots < \varkappa_n$ are integers and $P_1, \ldots, P_n$ are mutually disjoint projections in $L(H)$ such that $P_1 + \cdots + P_n = \mathbf{1}_H$; the integers $\varkappa_j$ are called the *partial indices of* $\mathcal{D}$ and the dimensions $d_j \stackrel{\mathrm{def}}{=} \dim P_j$ will be called the *dimensions* of the partial indices $\varkappa_j$. It is easily seen that the collection $\varkappa_1, \ldots, \varkappa_n$, $d_1, \ldots, d_n$ determines a diagonal function up to equivalence. For $\dim H < +\infty$ it is well-known (see, for example, [1], [2]) that every function in $\mathcal{O}(\mathbb{T}, \mathrm{GL})$ is equivalent to a diagonal function, a result that is essentially due to G. D. Birkhoff [3]. For $\dim H = \infty$ this is not true. A first counterexample was given in [4]. We represent here another counterexample: Let $H = H_1 \oplus H_2$ be a decomposition of H and $V \in L(H_1, H_2)$. Then the function defined by the block matrix

$$\begin{pmatrix} z^{-1} & 0 \\ V & z \end{pmatrix} \tag{1}$$

is equivalent to a diagonal function if and only if the operator V has a closed image in H_2, as is easily verified. However there are positive results, too:

THEOREM 1 [5]. *Let $A \in \mathcal{O}(\mathbb{T}, \mathrm{GL})$. If the values $A(z) - \mathbf{1}_H$ are compact for all z, $z \in \mathbb{T}$, then A is equivalent to a diagonal function whose non-zero partial indices have finite dimensions.*

For $A \in \mathcal{O}(\mathbb{T}, \mathrm{GL})$ we denote by W_A the Toeplitz operator defined by $W_A f = P_+(Af)$, where P_+ is the orthogonal projection from $L^2(\mathbb{T}, H)$ onto the subspace $L^2_+(\mathbb{T}, H)$ generated by the holomorphic functions on $\mathbb{T}_+$.

THEOREM 2 [4]. *A function $A \in \mathcal{O}(\mathbb{T}, \mathrm{GL})$ is equivalent to a diagonal function, whose non-zero partial indices $\varkappa_j$ have finite dimension d_j, if and only if W_A is a Fredholm operator in $L^2_+(\mathbb{T}, H)$. If the condition is fulfilled, then* $\dim \operatorname{Ker} W_A = \sum_{\varkappa_j<0} \varkappa_j d_j$ *and* $\dim \operatorname{CoKer} W_A = \sum_{\varkappa_j>0} \varkappa_j d_j$.

For further results see [4], [6], and the references in these papers.

2. A new point of view. In [6] a new simple proof was given for Theorem 1. The idea of this proof can be used to obtain some new results about general functions in $\mathcal{O}(\mathbb{T}, \mathrm{GL})$, too.

THEOREM 3. (See the proof of Lemma 1 in [6]). *Every function from $\mathcal{O}(\mathbb{T}, \mathrm{GL})$ is equivalent to a rational function of the form*

$$\sum_{j=-s}^{s} z^j T_j, \qquad T_j \in L(H).$$

Let $A \in \mathcal{O}(\mathbb{T}, \mathrm{GL})$. A couple $\varphi = (\varphi_-, \varphi_+)$ will be called a $\varkappa$-section of A if φ_-, φ_+ are holomorphic H-valued functions on $\mathbb{T}_-$, $\mathbb{T}_+$, respectively, and $z^\varkappa \varphi_-(z) = A(z)\varphi_+(z)$ for $z \in \mathbb{T}$. Then we put $\varphi(z) = \varphi_+(z)$ for $|z| \leqslant 1$ and $\varphi(z) = \varphi_-(z)$ for $1 < |z| \leqslant \infty$. For $0 \neq x \in H$ and $0 \leqslant |z| \leqslant \infty$ we denote by $\varkappa(x, z, A)$ the smallest integer $\varkappa$ such that there exists a $\varkappa$-section φ of A with $\varphi(z) = x$. From Theorem 3 it follows immediately that there are finite numbers m, $m > 1$, depending only on A and A^{-1}, such that $1 \leqslant \varkappa(x, z, A) \leqslant m$ for all z, $0 \leqslant |z| \leqslant \infty$, and $0 \neq x \in H$.

THEOREM 4. *For every function A, $A \in \mathcal{O}(\mathbb{T}, \mathrm{GL})$, there exist unique integers $\varkappa_1 < \cdots < \varkappa_n$ (the partial indices of A), unique numbers $d_1, \ldots, d_n \in \{1, 2, \ldots, \infty\}$ (the dimensions of the partial indices) and families of (not necessary closed) linear subspaces*

$$0 = M_0(z) \subsetneqq M_1(z) \subsetneqq \cdots \subsetneqq M_n(z) = H \qquad (0 \leqslant |z| \leqslant \infty)$$

such that

(*i*) *$x \in M_j(z) \setminus M_{j-1}(z)$ if and only if $\varkappa(x, z, A) = \varkappa_j$ ($j = 1, \ldots, n$; $0 \leqslant |z| \leqslant \infty$). If φ is a $\varkappa_j$-section of A and $\varphi(z) \in M_j(z_0) \setminus M_{j-1}(z_0)$ for some point z_0, then $\varphi(z) \in M_j(z) \setminus M_{j-1}(z)$ for all $0 \leqslant |z| \leqslant \infty$. If $x_1, \ldots, x_n$ are linearly independent vectors in H and, for some point z_0, φ_j are $\varkappa(x_j, z_0, A)$-sections of A with $\varphi_j(z_0) = x_j$, then the values $\varphi_1(z), \ldots, \varphi_n(z)$ are linearly independent for all $0 \leqslant |z| \leqslant \infty$.*

(*ii*) *The function A is equivalent to a diagonal function if and only if the spaces $M_j(z)$ ($0 \leqslant |z| \leqslant \infty$; $j = 1, \ldots, n$) are closed. For this it is sufficient that at least for one point z_0 the spaces $M_j(z_0)$ are closed. Further, it is sufficient that the dimensions d_j are finite with the exception of one of them.*

(*iii*) *There are Hilbert spaces $H_1, \ldots, H_{n-1}$ and holomorphic operator functions $S_j^\pm : \mathbb{T}_\pm \to L(H_j, H)$ such that $M_j(z) = \operatorname{Im} S_j^+(z)$ for $|z| \leqslant 1$, $M_j(z) = \operatorname{Im} S_j^-(z)$ for $1 < |z| \leqslant \infty$ and $z^{\varkappa_j} S_j^-(z) = A(z) S_j^+(z)$.*

(*iv*) *$d_j = \dim\big(M_j(z)/M_{j-1}(z)\big)$ for all $0 \leqslant |z| \leqslant \infty$ ($j = 1, \ldots, n$) and* $\dim \operatorname{Ker} W_A = \sum_{\varkappa_j<0} \varkappa_j d_j$, $\dim\big(L^2_+(\mathbb{T}, H)/\operatorname{clos} \operatorname{Im} W_A\big) = \sum_{\varkappa_j>0} \varkappa_j d_j$.

The proof of this theorem uses Theorem 3, the method of the proof of Lemma 2 in [6] and the open-mapping-theorem. From Theorem 4 we get a collection of invariants with respect to equivalence, the partial indices and its dimensions. However this collection does not determine the equivalence class uniquely, because, clearly, for every such collection there is a corresponding diagonal function, whereas not every function in $\mathcal{O}(\mathbb{T}, \mathrm{GL})$ is equivalent to a diagonal function.

It is easily seen that, for $\operatorname{Ker} V = \{0\}$ the function (1) has the partial indices $\varkappa_1 = 0$ and $\varkappa_2 = 1$ ($M_1(z) = H_1 \oplus \operatorname{Im} V$ for all z).

PROBLEM. *Are all functions in $\mathcal{O}(\mathbb{T}, \mathrm{GL})$ with such partial indices equivalent to a function of the form (1)?*

PROBLEM. *Can we obtain, in general, a complete classification, adding some special triangular block matrices to the diagonal functions?*

REFERENCES

1. Prössdorf S., *Einige Klassen singulärer Gleichungen*, Berlin, 1974.
2. Gohberg I. Ts., Fel'dman I. A., *Convolution equations and projection methods of its solutions*, Nauka, Moscow, 1971. (Russian)
3. Birkhoff G. D., Math. Ann. **74** (1913), 122–138.
4. Gohberg I. Ts., Leiterer J., *General theorems on factorization of operator-functions with respect to contour.* I. *Holomorphic functions*, Acta Sci. Math. **34** (1973), 103–120 (Russian);Gohberg I. Ts., Leiterer J., *General theorems on factorization of operator-functions with respect to contour.* II. *Generalization*, Acta Sci. Math. **35** (1973), 39–59. (Russian)
5. Gohberg I. Ts., *Problem of factorization of operator-functions*, Izv. Akad. Nauk SSSR, ser. matem. **28** (1964), no. 5, 1055–1082. (Russian)
6. Leiterer J., *On factorization of matrices and operator-functions*, Soobshch. Akad. Nauk Gruz.SSR **88** (1977), no. 3, 541–544. (Russian)

HUMBOLDT-UNIVERSITÄT BERLIN
FACHBEREICH MATHEMATIK
UNTER DEN LINDEN 6, PSF 1297
0-1086 BERLIN
DEUTSCHLAND

4.7

AN INVERSE PROBLEM FOR j_{qp}-INNER FUNCTIONS

D. Z. AROV, B. FRITZSCHE, B. KIRSTEIN

Let G be a simply connected domain of the extended complex plane $\mathbb{C}_0 := \mathbb{C} \cup \{\infty\}$. Then let the set $\mathcal{NM}(G)$ consist of all functions which are meromorphic in G and which can be represented as quotient of two bounded holomorphic functions in G. Let $\mathbb{D} := \{z \in \mathbb{C} : |z| < 1\}$, let $\mathbb{T} := \{z \in \mathbb{C} : |z| = 1\}$ and let $\mathbb{E} := \mathbb{C}_0 \setminus (\mathbb{D} \cup \mathbb{T})$. If $f \in \mathcal{NM}(\mathbb{D})$ (resp. $f \in \mathcal{NM}(\mathbb{E})$) and if m denotes the normalized linear Lebesgue–Borel measure on $\mathbb{T}$, then there are a Borelian subset B of $\mathbb{T}$ with $m(B) = 0$ and a Borel measurable function $\underline{f}: \mathbb{T} \to \mathbb{C}$ such that

$$\lim_{r \to 1-0} f(rz) = \underline{f}(z) \qquad (\text{resp. } \lim_{r \to 1+0} f(rz) = \underline{f}(z))$$

for all $z \in \mathbb{T} \setminus B$. The function $\underline{f}$ (which is determined m-a.e. on $\mathbb{T}$) is called the boundary function of f.

A function $f \in \mathcal{NM}(\mathbb{D})$ is said to admit a pseudocontinuation into $\mathbb{E}$, if there is a function $f^{\#} \in \mathcal{NM}(\mathbb{E})$ such that the boundary functions $\underline{f}$ and $\underline{f^{\#}}$ coincide m-a.e. on $\mathbb{T}$. Let p and q be positive integers. Then a function f, which belongs to the class $p \times q$–$\mathcal{NM}(\mathbb{D})$ of all $p \times q$ matrix-valued functions each entry function of which belongs to $\mathcal{NM}(\mathbb{D})$, is called to admit a pseudocontinuation into $\mathbb{E}$ if each of its entry functions admits a pseudocontinuation into $\mathbb{E}$.

A function $f: \mathbb{D} \to \mathbb{C}^{p \times q}$ is said to be a $p \times q$ Schur function if f is both holomorphic and contractive in $\mathbb{D}$. The symbol $\mathcal{S}_{p \times q}(\mathbb{D})$ stands for the set of all $p \times q$ Schur functions.

Obviously, the matrix $j_{pq} := \operatorname{diag}(I_p, -I_q)$ is a signature matrix, i.e. $j_{pq}{}^* = j_{pq}$ and $j_{pq}{}^2 = I_{p+q}$ hold true. A $(p+q) \times (p+q)$ matrix-valued function W is said to belong to the Potapov class $\mathcal{P}_{j_{pq}}(\mathbb{D})$ if the following three conditions are satisfied:

(*i*) W is meromorphic in the open unit disc $\mathbb{D}$.

(*ii*) $\det W$ does not identically vanish in $\mathbb{D}$.

(*iii*) For all points z of analyticity of W, the matrix $W(z)$ is j_{pq}-contractive, i.e. $W^*(z) j_{pq} W(z) \leqslant j_{pq}$ holds.

Since $\mathcal{P}_{j_{pq}}(\mathbb{D}) \subseteq (p+q) \times (p+q) - \mathcal{NM}(\mathbb{D})$ each $W \in \mathcal{P}_{j_{pq}}(\mathbb{D})$ has a boundary function $\underline{W}$ on $\mathbb{T}$. A function $W \in \mathcal{P}_{j_{pq}}(\mathbb{D})$ is called j_{pq}-inner if $\underline{W}$ satisfies $\underline{W}^*(z) j_{pq} \underline{W}(z) = j_{pq}$ for almost all $z \in \mathbb{T}$. Every j_{pq}-inner function W admits a pseudocontinuation. The function $W^{\#}$ is given by

$$W^{\#}(z) := j_{pq} \left[W^{-1}(1/\bar{z}) \right]^* j_{pq}$$

for all $z \in \mathbb{E}$ for which $1/\bar{z}$ is a point of analyticity of W^{-1}.

PROPOSITION 1. (See e. g. [1] and [7]). *Let* $W \in \mathcal{P}_{j_{pq}}(\mathbb{D})$ *be partitioned into blocks via*

$$W = \begin{pmatrix} W_{11} & W_{12} \\ W_{21} & W_{22} \end{pmatrix} \tag{1}$$

where W_{11} *is a* $p \times p$ *matrix-valued function. Then:*

(*a*) *For each point* z *of analyticity of* W_{22}, *the matrix* $W_{22}(z)$ *is non-singular.*

(*b*) $W_{11} - W_{12}W_{22}^{-1}W_{21} \in \mathcal{S}_{p\times p}(\mathbb{D})$, $W_{22}^{-1} \in \mathcal{S}_{q\times q}(\mathbb{D})$.

(*c*) *The functions* $\eta_W := W_{12}W_{22}^{-1}$ *and* $\kappa_W := W_{22}^{-1}W_{21}$ *are strictly contractive Schur functions.*

(*d*) *If* W *is* j_{pq}*-inner, then* $\det(I - \underline{\eta_W^*}\,\underline{\eta_W}) = \det(I - \underline{\kappa_W}\,\underline{\kappa_W^*})$ *and*

$$\int_{\mathbb{T}} \log\det(I - \underline{\eta_W^*}\,\underline{\eta_W})\,dm = \int_{\mathbb{T}} \log\det(I - \underline{\kappa_W}\,\underline{\kappa_W^*})\,dm > -\infty$$

Moreover, η_W *and* κ_W *admit pseudocontinuations.*

The functions η_W and κ_W are called the right canonical and the left canonical Schur function, respectively, associated with W.

Let $\mathcal{U}(p,q)$ be the set of all j_{pq}-inner functions.

PROBLEM P($\mathfrak{M}$). *Given some prescribed non-empty subset* $\mathfrak{M}$ *of* $\mathcal{U}(p,q)$.

(*a*) *Characterize all* $p \times q$ *Schur functions* f *which admit a representation* $f = \eta_W$ *with some* $W \in \mathfrak{M}$.

(*b*) *Characterize all* $q \times p$ *Schur functions* g *which admit a representation* $g = \kappa_W$ *with some* $W \in \mathfrak{M}$.

Observe that the problem can also be formulated in terms of Darlington realizations (see [1]).

There is some duality between parts (*a*) and (*b*) of the problem. Denote

$$U_{pq} := \begin{pmatrix} 0 & I_q \\ -I_p & 0 \end{pmatrix}.$$

Let $W \in (p+q)\times(p+q) - \mathcal{NM}(\mathbb{D})$ and $V \in (q+p)\times(q+p) - \mathcal{NM}(\mathbb{D})$ be such that

$$V\,U_{pq}W = U_{pq}. \tag{2}$$

Then $W \in \mathcal{U}(p,q)$ if and only if $V \in \mathcal{U}(p,q)$. Using a result due to Potapov [8], it is readily checked that (2) implies $\eta_W = \kappa_V$.

One can not expect to get an answer to Problem P($\mathfrak{M}$) for arbitrary subsets $\mathfrak{M}$ of $\mathcal{U}(p,q)$. However, for distinguished subsets $\mathfrak{M}$ this will be possible. In the case $\mathfrak{M} = \mathcal{U}(p,q)$ an answer to Problem 1 can be easily obtained from [1]:

PROPOSITION 2. *Let* $f \in \mathcal{S}_{p\times q}(\mathbb{D})$. *Then the following statements are equivalent:*

(*i*) *There exists a* $W \in \mathcal{U}(p,q)$ *such that* $\eta_W = f$.

(*ii*) *There exists a* $V \in \mathcal{U}(p,q)$ *such that* $\kappa_V = f$.

(*iii*) f *admits a pseudocontinuation and fulfills*

$$\int_{\mathbb{T}} \log\det(I - \underline{f}\,\underline{f^*})\,dm > -\infty \tag{3}$$

Subsets $\mathfrak{M}$ of $\mathcal{U}(p,q)$ which arise in the context of some types of matricial interpolation problems are of particular interest. In [7] Problem $\mathrm{P}(\mathfrak{M})$ was studied for the case that $\mathfrak{M}$ is a set of special resolvent matrices associated with a non-degenerate Schur problem. The investigations on generalized bitangential Schur–Nevanlinna–Pick Interpolation (see [2]–[5]) showed that there are two significant subclasses of $\mathcal{U}(p,q)$. To define them we need some preparations.

By the Smirnov class $\mathcal{N}_+(\mathbb{D})$ we will mean the set of all functions $f \in \mathcal{NM}(\mathbb{D})$ which are holomorphic in $\mathbb{D}$ and which fulfill

$$\int_{\mathbb{T}} \log^+ |\underline{f}(z)|\, m(dz) = \sup_{r\in[0,1)} \int_{\mathbb{T}} \log^+ |f(rz)|\, m(dz) < \infty$$

where $\log^+ x := \max(\log x, 0)$ for each $x \in [0,\infty)$.

A j_{pq}-inner function W is said to be A-singular if both W and W^{-1} belong to the set $(p+q)\times(p+q)-\mathcal{N}_+(\mathbb{D})$ of all $(p+q)\times(p+q)$ matrix-valued functions each entry function of which belongs to $\mathcal{N}_+(\mathbb{D})$.

The class of all j_{pq}-inner functions is multiplicative. If W and W_1 are j_{pq}-inner functions, then W_1 is said to be a right (resp. left) j_{pq}-inner divisor of W if WW_1^{-1} (resp. $W_1^{-1}W$) is a j_{pq}-inner function.

A j_{pq}-inner function W is called left regular (resp. right regular) if W does not contain any non-constant A-singular right (resp. left) j_{pq}-inner divisor.

If the functions V and W are connected via (2), then:

(*i*) W is A-singular if and only if V is A-singular

(*ii*) W is left regular if and only if V is right regular.

(*iii*) W is right regular if and only if V is left regular.

A-singular and regular j_{pq}-inner functions play an important role in the framework of generalized bitangential Schur–Nevanlinna–Pick interpolation (see [2]–[5]).

If $\mathcal{U}_S(p,q)$ denotes the set of all A-singular j_{pq}-inner functions, then the problem $\mathrm{P}(\mathcal{U}_S(p,q))$ has the following answer:

THEOREM 1. (See [6]). *Suppose that $f \in \mathcal{S}_{p\times q}(\mathbb{D})$ satisfies condition (iii) of Proposition 2. Then there exists a $E \in \mathcal{U}_S(p,q)$ such that $\eta_W = f$ if and only if $\underline{f}^*(I-\underline{f}\,\underline{f}^*)^{-1}$ is the boundary function of some function which belongs to the Smirnov class $q\times p$–$\mathcal{N}_+(\mathbb{D})$.*

In view of the close connection to generalized bitangential Schur–Nevanlinna–Pick interpolation, Problem $\mathrm{P}(\mathfrak{M})$ is of particular interest in the case that $\mathfrak{M}$ is the set $\mathcal{U}_{\mathrm{lr}}(p,q)$ of all left regular j_{pq}-inner functions (resp. the set $\mathcal{U}_{\mathrm{rr}}(p,q)$ of all right regular j_{pq}-inner functions). Note that a first step to attack problem $\mathrm{P}(\mathcal{U}_{\mathrm{lr}}(p,q))$ was already done:

THEOREM 2 ([4, Theorem 13]). *Suppose that $f \in \mathcal{S}_{p\times q}(\mathbb{D})$ is strictly contractive. Then the following two conditions are sufficient for the existence of some left regular j_{pq}-inner function W with $\eta_W = f$:*

(*i*) $(1-\|f\|_S)^{-1} \in L^1(m)$.

(*ii*) *There are inner functions B_1 and B_2 from $\mathcal{S}_{p\times p}(\mathbb{D})$ and $\mathcal{S}_{q\times q}(\mathbb{D})$, respectively, such that*

$$\underline{B_2}(I-\underline{f}^*\underline{f})^{-1}\underline{f}^*\underline{B_1}$$

is the boundary function of some function which belongs to the Smirnov class $q \times p - \mathcal{N}_+(\mathbb{D})$.

References

1. Arov D. Z., *Darlington realization of matrix-valued functions*, Izv. Akad. Nauk SSSR, ser. matem. **37** (1973), 1299–1331. (Russian)
2. Arov D. Z., *γ-generating matrices, J-inner matrix functions and related extrapolation problems*, deposited in Ukr. NIINTI (1986), no. 726–Uk. 86. (Russian)
3. Arov D. Z., *On regular and singular J-inner matrix-functions and related extrapolation problems*, Funkts. Anal. i Prilozh. **22** (1988), 57–59. (Russian)
4. Arov D. Z., *γ-generating matrices, J-inner matrix-functions and related extrapolation problems*, Teor. Funkts., Funktsion. Anal. i Prilozh. (Kharkov), part I, **51** (1989), 61–67; part II, **52** (1989), 103–109; part III, **53** (1990), 57–64 (Russian); English transl. in J. Soviet Math. **52** (1990), no. 6, 3487–3491; **52** (1990), no. 5, 3421–3425; **58** (1992), no. 6, 532–537.
5. Arov D. Z., *Regular J-inner matrix-functions and related continuation problems*, Linear Operators in Function Spaces (G. Arsene et al., eds.), Operator Theory: Advances and Applications, vol. 43, Birkhäuser, Basel, 1990, pp. 63–87.
6. Dubovoj V. K., Fritzsche B., Kirstein B., *Spectral association of matrix-valued Schur functions and related questions*, submitted.
7. Fritzsche B., Fuchs S. H., Kirstein B., *A Schur type matrix extension problem.* VI, Math. Nachr. (to appear).
8. Potapov V. P., *Fractional linear transformation of matrices*, Studies in the theory of operators and their applications (V. A. Marčenko, ed.), Naukova Dumka, Kiev, 1979, pp. 75–97. (Russian)

Department of Mathematics
Odessa Pedagogical Institute
270020 Odessa
Ukraine

Sektion Mathematik
Leipzig University
Augustusplatz 10
7010 Leipzig
Germany

Sektion Mathematik
Leipzig University
Augustusplatz 10
7010 Leipzig
Germany

4.8

GENERALIZED POLES OF MATRIX FUNCTIONS OF THE CLASS $\mathbf{N}_\kappa^{m\times m}$

A. DIJKSMA, H. LANGER, H. S. V. DE SNOO

1. Introduction. An $m\times m$ matrix function Q which is meromorphic (at least) on $\mathbb{C}^+\cup\mathbb{C}^-$ is said to belong to the class $\mathbf{N}_\kappa^{m\times m}$ if

(a) $Q(z)^* = Q(\bar z)$, $z\in\mathcal{D}_Q$, and

(b) the kernel $\mathrm{N}_Q(z,\zeta) := \dfrac{Q(z)-Q(\zeta)^*}{z-\bar\zeta}$, $z,\zeta\in\mathcal{D}_Q$, has κ negative squares.

Here $\mathcal{D}_Q$ denotes the domain of holomorphy of Q which is supposed to be maximal. Recall that (b) means that for arbitrary $n\in\mathbb{N}$, points $z_1,z_2,\dots,z_n\in\mathcal{D}_Q$ and vectors $x_1,x_2,\dots,x_n\in\mathbb{C}^m$ the hermitian $n\times n$ matrix $\big((x_j^*\mathrm{N}_Q(z_i,z_j)x_i)\big)_{i,j=1}^n$ has at most κ negative eigenvalues and for at least one choice of n, $z_1,z_2,\dots,z_n$, $x_1,x_2,\dots,x_n$ it has exactly κ negative eigenvalues.

For a function $Q\in\mathbf{N}_\kappa^{m\times m}$ there exists a π_κ-space Π_κ, a selfadjoint relation A in Π_κ with nonempty resolvent set $\rho(A)$ and a linear mapping Γ from $\mathbb{C}^m$ into Π_κ such that

$$Q(z) = Q(\bar\mu) + (z-\bar\mu)\Gamma^*\big(I+(z-\mu)(A-z)^{-1}\big)\Gamma, \qquad z\in\mathcal{D}_Q \tag{1}$$

where μ is an arbitrary chosen but fixed point of $\mathbb{C}^+\cap\mathcal{D}_Q$. If we impose the minimality condition

$$\Pi_\kappa = \text{c.l.s.}\Big\{\big(I+(z-\mu)(A-z)^{-1}\big)\Gamma x \;\Big|\; x\in\mathbb{C}^m,\ z\in\rho(A)\Big\},$$

which we always can and will do in the sequel, then the relation A is uniquely determined up to π-unitary equivalence.

In the following we shall suppose that the relation A is even an operator (see also Remark 1 below). Then according to a theorem of L. S. Pontrjagin, A has a κ-dimensional nonpositive invariant subspace $\mathcal{L}_-$ and, in particular, at least one eigenvalue with a nonpositive eigenvector f, that is, $[f,f]\leqslant 0$. It follows easily from the representation (1) that the resolvent set of A coincides with $\mathcal{D}_Q$ and that the isolated eigenvalues of A are the poles of Q.

Let λ_0 be an isolated eigenvalue of A. Then λ_0 is of finite algebraic multiplicity and a basis of the algebraic eigenspace consisting of Jordan chains of A at λ_0 can be chosen. On the other hand, with a pole λ_0 of the $m\times m$ matrix function Q there are associated pole functions $g(z)$ of Q. For the convenience of the reader we recall the definition of a pole function. To this end, we denote by $H(\lambda_0;\mathbb{C}^m)$ the set of all m-vector functions which are holomorphic at λ_0 and by $\mathcal{U}(\lambda_0)$ ($\mathring{\mathcal{U}}(\lambda_0)$) an arbitrary (deleted) neighborhood of λ_0. Then if Q has a pole at λ_0 the function $g\in H(\lambda_0;\mathbb{C}^m)$ is called a *pole function* of Q at λ_0 of order $k\in\mathbb{N}$, if

(i) $g(\lambda_0)\neq 0$

(*ii*) $g(z) = Q(z)f(z)$, $z \in \mathring{\mathcal{U}}(\lambda_0)$, where $f \in H(\lambda_0; \mathbb{C}^m)$ has a zero of order k at λ_0, and

(*iii*) if $f_1 \in H(\lambda_0; \mathbb{C}^m)$ also satisfies $g(z) = Q(z)f_1(z)$, $z \in \mathring{\mathcal{U}}(\lambda_0)$, then the order of the zero of f_1 at λ_0 is at most k.

It is known (see, e.g., [B], [DLS]) that if λ_0 is a pole of $Q \in \mathbf{N}_\kappa^{m\times m}$, there is a bijective correspondence between the Jordan chains of A of length k at λ_0 and (equivalence classes of) pole functions of Q of order k at λ_0; it is such that if $g_0, g_1, \ldots, g_{k-1}$ and $\hat{g}_0, \hat{g}_1, \ldots, \hat{g}_{k-1}$ are Jordan chains of A at λ_0 and g, $\hat{g}$ are the corresponding pole functions: $Q(z)f(z) = g(z)$, $Q(z)\hat{f}(z) = \hat{g}(z)$, then

$$[g_p, \hat{g}_q] = \lim_{z,\zeta\to\lambda_0} D_z^p D_\zeta^q \hat{f}(\zeta)^* \mathrm{N}_Q(z,\zeta) f(z),$$

where $D_z^p = (1/p!)(d/dz)^p$. In particular, a canonical system of pole functions (see [GS]) can be chosen which determines completely the structure and inner product on the algebraic eigenspace of A at λ_0.

2. Open problem. Let Q and A be as in Section 1. An eigenvalue λ_0 $(= \bar{\lambda}_0)$ of A which is not isolated in the spectrum $\sigma(A)$ of A is called a *generalized pole* of function Q. It is known (see [BL], [DLS]) that a generalized pole λ_0 of $Q \in \mathbf{N}_\kappa^{m\times m}$ can be characterized as follows: There exists an m-vector function f defined and holomorphic on $\mathcal{U}(\lambda_0) \cap \mathbb{C}^+$ such that

(α) $f(z) \to 0$ if $z \hat{\to} \lambda_0$,

(β) $Q(z)f(z)$ has a finite nonzero limit if $z \hat{\to} \lambda_0$, and

(γ) $f(z)^* \mathrm{N}_Q(z,z) f(z)$ has a finite limit if $z \hat{\to} \lambda_0$.

Here $z \hat{\to} \lambda_0$ means that z tends to λ_0 nontangetially. So it seems natural to call the function g: $g(z) := Q(z)f(z)$, which is holomorphic in $\mathcal{U}(\ell_0) \cap \mathbb{C}^+$, a *pole function* of Q at λ_0.

PROBLEM. *Characterize the algebraic eigenspace of Q at the nonisolated eigenvalue λ_0 by means of the pole functions of Q at λ_0.*

E.g., if $g_0, g_1, \ldots, g_{k-1}$ is a Jordan chain of A at λ_0, then the matrix

$$\left([g_p, g_q]\right)_{p,q=0}^{k-1} \tag{2}$$

is a Hankel matrix $[g_p, g_q] = c_{p-q}$ and there exists a p_0, $k-1 \leqslant p_0 \leqslant 2k-1$, such that $c_r = 0$ for $0 \leqslant r \leqslant p_0 - 1$. With the Jordan chain there should be associated a pole function g of Q at λ_0 such that k, p_0 and the inner product can be determined from the boundary behavior of g at λ_0.

Partial results are the following:

1. *With each pole function g of Q at λ_0, $g(z) = Q(z)f(z)$, where f has the properties (α)–(γ) above, there is associated an eigenvector g_0 of A at λ_0 such that the limit in (γ) is $[g_0, g_0]$; see* [BL], [DLS].

2. *If* $\ker(A-\lambda_0)$ *is nondegenerate then the number of linearly independent equivalence classes of pole functions of* Q *al* λ_0 *coincides with* $\dim\ker(A-\lambda_0)$, *and if* $g^1, g^2, \ldots, g^n$ *is a basis of* $\ker(A-\lambda_0)$ *there exists a basis of pole functions*

$$g^1(z) = Q(z)f^1(z),\ g^2(z) = Q(z)f^2(z), \ldots, g^n(z) = Q(z)f^n(z)$$

of Q *at* λ_0 *such that*

$$[g^i, g^j] = \lim_{z,\zeta \hat{\to} \lambda_0} f^j(\zeta)^* \mathrm{N}_Q(z,\zeta) f^i(z).$$

Therefore the above problem remains open only for eigenvalues with a degenerate kernel; evidently,these eigenvalues do all belong to $\sigma(A|\mathcal{L}_-)$ where $\mathcal{L}_-$ is a κ-dimensional nonpositive invariant subspace of A.

A similar result as in case 2 holds if the algebraic eigenspace of A at λ_0 is nondegenerate: Then the whole space can be split into the orthogonal sum of this algebraic eigenspace, where λ_0 is an isolated spectral point of the restriction of A, and a complementary subspace where $A-\lambda_0$ is injective.

If $m = 1$, a generalized pole λ_0 of the (scalar) function Q, such that to A at λ_0 there corresponds a nonpositive eigenvector, is a generalized zero of negative type of the function $\hat{Q} := -Q^{-1}$ in the sense of [L]. In this case the function $g(z) \equiv -1$ can serve as a pole function, $f(z) = Q(z)$, $\dim\ker(A-\lambda_0) = 1$ and the number κ_0 of nonpositive eigenvalues of the matrix (2), where $g_0, g_1, \ldots, g_{k-1}$ is the maximal Jordan chain of A at λ_0, is characterized as follows:

$$\lim_{z \hat{\to} \lambda_0} (z-\lambda_0)^{-(2\kappa_0-1)} f(z) \text{ exists and is finite and nonpositive, and}$$

$$\lim_{z \hat{\to} \lambda_0} (z-\lambda_0)^{-(2\kappa_0+1)} f(z) \text{ exists and is positive } (\leqslant \infty).$$

3. Closing remarks.

Remark 1. If A in (1) is not an operator but a relation there is a decomposition of the π_κ-space Π_κ:

$$\Pi_\kappa = \Pi^1_{\kappa_1} \oplus \Pi^2_{\kappa_2}, \tag{3}$$

which reduces A and is such that $A_1 := A|_{\Pi^1_{\kappa_1}}$ is a selfadjoint operator and $\sigma(A_1)$ contains all the eigenvalues of A in the finite complex plane with a nonpositive eigenvector. The decomposition (3) corresponds to a decomposition $Q(z) = Q_1(z) + Q_2(z)$ of the function Q, where, e.g.,

$$Q_1(z) = Q(\bar{\mu}) + (z-\bar{\mu})\Gamma_1^*(I + (z-\mu)(A_1-z)^{-1})\Gamma_1, \qquad z \in \mathcal{D}_{Q_1},$$

with $\Gamma_1 := P_1\Gamma$, P_1 is the orthogonal projection of Π_κ onto $\Pi^1_{\kappa_1}$ corresponding to (3). Then the generalized poles of Q, which are eigenvalues of A with a nonpositive eigenvector, are the generalized poles of Q_1, and there is an easy connection between the corresponding functions.

Remark 2. If A in (1) is not an operator and the singular part of A contains a nonpositive vector, the above problem arises also for the "generalized pole ∞".

Remark 3. For a given function $Q \in \mathbf{N}_\kappa^{m\times m}$ the corresponding space Π_κ can be realized as a reproducing kernel space $\mathfrak{P}(Q)$ of m-vector functions which are meromorphic in $\mathbb{C}^+ \cup \mathbb{C}^-$; see, e.g., [DLS]. It seems to be natural to expect that the pole functions of Q at λ_0 are in direct connection with those elements of Π_κ which represent the elements of the algebraic eigenspace of A at λ_0.

Remark 4. A solution to the above problem would be useful in the study of boundary eigenvalue problems with eigenvalue depending boundary conditions (see, e.g., [DLS]) as in those problems the underlying operators are often described by holomorphic functions, which are sometimes called Tichmarsh–Weyl coefficients.

References

[B] Borogovac M., *Zeros of meromorphic matrix functions of the class* $\mathbf{N}_\kappa^{n\times n}$, Math. Nachr. (to appear).

[BL] Borogovac M., Langer H., *A characterization of generalized zeros of negative type of matrix functions of class* $\mathbf{N}_\kappa^{n\times n}$, Operator theory: Adv. Appl. **28** (1988), 17–26.

[DLS] Dijksma A., Langer H., de Snoo H. S. V., *Eigenvalues and pole functions of Hamiltonian systems with eigenvalue depending boundary conditions*, submitted for publication.

[GS] Gohberg I. C., Sigal E. I., *An operator generalization of the logarithmic residue theorem and the theorem of Rouché*, Mat. Sb. **84** (1971), 607–629 (Russian); English transl. in Math. USSR–Sb. **13** (1971), 603–625.

[L] Langer H., *A characterization of generalized zeros of negative type of functions of the class* $\mathbf{N}_\kappa$, Operator Theory: Adv. Appl. **17** (1986), 201–212.

Department of Mathematics
University of Groningen
Postbox 800
9700 AV Groningen
The Netherlands

Department of Mathematics
University of Regensburg
Postbox 397
8400 Regensburg
Germany

Department of Mathematics
University of Groningen
Postbox 800
9700 AV Groningen
The Netherlands

4.9

LOCAL AND GLOBAL ANALYTIC EQUIVALENCE OF ANALYTIC OPERATOR FUNCTIONS

I. Gohberg, M. A. Kaashoek, L. Rodman

Let $\mathcal{B}$ be a Banach space, $\mathcal{L}(\mathcal{B})$ the algebra of all linear bounded operators on $\mathcal{B}$, Ω a domain in the complex plane. Two operator functions $T\colon \Omega \to \mathcal{L}(\mathcal{B})$, $S\colon \Omega \to \mathcal{L}(\mathcal{B})$, analytic on Ω, are said to be *globally analytically equivalent,* if there exist analytic operator functions $E\colon \Omega \to \mathcal{L}(\mathcal{B})$, $F\colon \Omega \to \mathcal{L}(\mathcal{B})$ whose values are invertible operators such that

$$T(z) = E(z)S(z)F(z), \qquad z \in \Omega.$$

The functions S and T are called *locally analytic equivalent* if for every $z_0 \in \Omega$ there exists an open neighborhood U of z_0 in Ω and analytic operator functions $E_{z_0}: U \to \mathcal{L}(\mathcal{B})$, $F_{z_0}: U \to \mathcal{L}(\mathcal{B})$ whose values are invertible operators such that

$$T(z) = E_{z_0}(z)S(z)F_{z_0}(z), \qquad z \in \Omega.$$

In operator theory the concepts of global and local equivalence have been studied and employed in [3], [7], [8] and [5], [6].

In general, local analytic equivalence does not imply global analytic equivalence. A rather involved example to this effect is given in [10]; in this example, $\mathcal{B}$ is a separable Hilbert space, the domain Ω is simply connected, the locally analytically equivalent operator functions S and T are such that $C(z)T(z) = S(z)$, $z \in \Omega$, for some continuous function $C(\cdot)$ on Ω with invertible operator values, and nevertheless T and S are not globally analytically equivalent. Another example of this kind is given in [6]. In the latter example, $\mathcal{B}$ is a Banach space for which the group of invertible linear bounded operators on $\mathcal{B}$ is not connected, $\Omega = \{ z \in \mathbb{C} : 1 < |z| < 2 \}$, $S(z)$, $z \in \Omega$, is an analytic projection valued function which is locally analytically equivalent to a constant projection T_0; nevertheless, for any Banach space Z the extended functions

$$\begin{pmatrix} S(z) & 0 \\ 0 & I \end{pmatrix} \quad \text{and} \quad \begin{pmatrix} T_0 & 0 \\ 0 & I \end{pmatrix}$$

(where I is the identity operator on Z) are not globally analytically equivalent.

On the other hand, for certain classes of operator valued functions analytic equivalence of T and S does imply global analytic equivalence. For example, this is the case if the values of T (and therefore also the values of S) are Fredholm operators, as follows from more general results proved in [10].

A general problem (still largely unsolved) of characterization of local and global analytic equivalence was posed in [4]. In a series of papers [5], [6] and [9] global analytic equivalence for certain classes of operator functions was studied from the point of view of linearizations (see also [11] for an exposition of some of this material in the framework of operator polynomials). Recall that a linear function $zA + B$ is called a *linearization* of an analytic operator function $S(z)$, $z \in \Omega$ if for some Banach space Z the extended function $\begin{pmatrix} S(z) & 0 \\ 0 & I_z \end{pmatrix}$ is globally analytically equivalent to $zA + B$ (on Ω).

It is of interest to find additional classes of operator functions (besides Fredholm operator valued ones) for which local analytic equivalence implies global analytic equivalence. Of particular interest, also in view of the linearization approach mentioned above, are operator functions of the form $zI - A$, where A is a constant operator. It was proved in [9] that if $zI - A_1$ is globally analytically equivalent to $zI - A_2$ on domain Ω that contains the spectra $\sigma(A_1)$ and $\sigma(A_2)$, then the operators A_1 and A_2 are similar. Thus, the following problem occurs:

PROBLEM 1. *Given $A_1, A_2 \in \mathcal{L}(\mathcal{B})$ and a domain $\Omega \supset \sigma(A_1) \cup \sigma(A_2)$. Prove that $zI - A_1$ is locally analytically equivalent to $zI - A_2$ on Ω if and only if A_1 and A_2 are similar, or provide a counterexample showing that this statement is false.*

Again, some particular cases are known. It is proved in [1] that if A_1 and A_2 are spectral operators (in the sense of Dunford [2]) on a Hilbert space, then the local analytic equivalence of $zI - A_1$ and $zI - A_2$ on a domain containing $\sigma(A_1) \cup \sigma(A_2)$ implies the similarity of A_1 and A_2.

We remark that Problem 1 makes sense and is of interest in more general framework of analytic functions whose values are elements in a Banach algebra with identity.

REFERENCES

1. Apostol C., *On a spectral equivalence of operators*, Integral Equations and Operator Theory **32** (1988), 15–35.
2. Dunford N., *Spectral operators*, Pacific J. of Math. **4** (1954), 321–354.
3. Gohberg I., *On some questions of spectral theory of finite-meromorphic operator-function*, Izv. Arm. Akad. Nauk **6 (2–3)** (1971), 160–181. (Russian)
4. Gohberg I., *Problems on equivalence and on factorization*, Informal note of a lecture given at the University of Maryland, College park, MD, February 5, 1975.
5. Gohberg I., Kaashoek M. A., Lay D., *Spectral classification of operators and operator functions*, Bull. Amer. Math. Soc. **82** (1976), 587–589.
6. Gohberg I., Kaashoek M. A., Lay D., *Equivalence, linearization and decomposition of holomorphic operator functions*, J. Funct. Analysis **28** (1978), 102–144.
7. Gohberg I., Sigal E. I., *An operator generalization of the logarithmic residue theorem and Rouché's theorem*, Mat. Sb. **84 (126)** (1971), 607–629 (Russian); English transl. in Math. USSR Sb. **13** (1971), 603–625.
8. Gohberg I., Sigal E. I., *Global factorization of a meromorphic operator-function and some of its applications*, Mat. Issled. **6** (1971), no. 1, 63–82. (Russian)
9. Kaashoek M. A., van der Mee C., Rodman L., *Analytic operator functions with compact spectrum. I. Spectral nodes, linearization and equivalence*, Integral Equations and Operator Theory **4** (1981), 504–547.
10. Leiterer J., *Local and global equivalence of meromorphic operator functions.* I, Math. Nachr. **83** (1978), 7–29; II, Math. Nachr. **84** (1978), 145–170.
11. Rodman L., *An introduction to operator polynomials*, OT 38, Birkhäuser Verlag, Basel, 1989.

RAYMOND AND BEVERLY SACKLER
FACULTY OF EXACT SCIENCES
SCHOOL OF MATH. SCIENCES
TEL-AVIV UNIVERSITY
RAMAT-AVIV 69978, ISRAEL

FAC. WISKUNDE EN INF.
VRIJE UNIVERSITEIT
DE BOELELAAN 1081A
1081 HV AMSTERDAM
THE NETHERLANDS

DEPT. OF MATHEMATICS
COLLEGE OF WILLIAM AND MARY
WILLIAMSBURG
VA 23185
U.S.A.

4.10

LIFTINGS OF VECTOR-VALUED ANALYTIC FUNCTIONS

M. PUTINAR

In various contexts of complex analysis and operator theory the following question appears.

OPEN PROBLEM 1. *Let D be a polydisk of $\mathbb{C}^n$ and let $u(z)\colon E \to F$ be an analytic map of continuous, linear operators between Fréchet spaces. Assume that u induces an onto map at the level of sheaves: $\mathcal{O}\hat{\otimes}E \to \mathcal{O}\hat{\otimes}F$. Does then u induce an onto map at the level of global sections: $\mathcal{O}(D)\hat{\otimes}E \to \mathcal{O}\hat{\otimes}F$?*

For E and F Banach spaces the answer is affirmative, see [1] and [5].

A positive solution to the previous problem would imply that a Fréchet analytic sheaf on Stein space is acyclic whenever it admits locally Fréchet free resolutions to the left, see [3]. This class of analytic sheaves (called Fréchet quasi-coherent) has appeared recently as a technical tool in several places, see [2], [3], and [4].

For analytic complexes of Banach spaces on complex manifolds it is known that the pointwise exactness implies the exactness at the level of sheaves, see [5]. The proof of the same assertion on singular spaces seems to face additional difficulties. More exactly, let

$$E.(x)\colon 0 \to E_k \overset{u_k(x)}{\longrightarrow} E_{k-1} \to \cdots \to E_2 \overset{u_2(x)}{\longrightarrow} E_1 \overset{u_1(x)}{\longrightarrow} E_0 \to 0$$

be a bounded complex of Banach spaces whose boundaries are continuous linear operators which depend analytically on $x \in X$, where X is a reduced analytic space.

OPEN PROBLEM 2. *Assume that the complex $E.(x)$ is pointwise exact on X. Is there the complex of sheaves $\mathcal{O}_X\hat{\otimes}E$ still exact?*

The problem is of course local, so one may assume that X is a closed analytic subset of a polydisk $D \subset \mathbb{C}^n$. The answer would be affirmative if one can extend $E.(x)$ to an analytic complex on D. This is possible for instance if $E_0, E_1, \dots, E_k$ are Hilbert spaces or ℓ_1-spaces. In particular the assertion is true for the complexes of finite dimensional vector spaces.

REFERENCES

1. Leiterer J., *Banach coherent analytic sheaves*, Math. Nachr. **85** (1978), 91–109.
2. O'Brian N. R., Toledo D., Tong Y. L. L., *A Grothendieck–Riemann–Roch formula for maps of complex manifolds*, Math. Ann. **271** (1985), 493–526.
3. Putinar M., *Une caractérisation des modules analytiques quasi-cohérents*, C. R. Acad. Sci. Paris **309** (1989), no. I, 881–884.
4. Ramis J. L., Ruget G., *Résidus et dualité*, Invent. Math. **26** (1974), 89–131.
5. Taylor J. L., *Analytic functional calculus for several commuting operators*, Acta Math. **125** (1970), 1–38.

INSTITUTE OF MATHEMATICS
P. O. BOX 1–764
BUCHAREST RO 70–700
ROMANIA

AND

DEPARTMENT OF MATHEMATICS
UNIVERSITY OF KANSAS
LAWRENCE, KANSAS 66045
U.S.A.

S.4.11
v.old

A PROBLEM ON OPERATOR VALUED BOUNDED ANALYTIC FUNCTIONS

B. SZŐKEFALVI-NAGY

Let $\mathcal{D}$, $\mathcal{D}^*$ be two Hilbert spaces and $\mathcal{B}(\mathcal{D}, \mathcal{D}^*)$ the space of all bounded linear operators mapping $\mathcal{D}$ into $\mathcal{D}^*$. The following result was proved in [1].

THEOREM. *Suppose θ is a bounded $\mathcal{B}(\mathcal{D}, \mathcal{D}^*)$-valued function analytic in the unit disc $\mathbb{D}$. The following assertions are equivalent:*

(a) *there exists a bounded $\mathcal{B}(\mathcal{D}^*, \mathcal{D})$-valued function Ω analytic in $\mathbb{D}$ and satisfying*

$$\Omega(\lambda)\theta(\lambda) = I_{\mathcal{D}} \qquad (\lambda \in \mathbb{D}) \tag{1}$$

(b) *the kernel function K_ε:*

$$K_\varepsilon(\lambda, \mu) \stackrel{\text{def}}{=} \frac{\theta^*(\mu)\theta(\lambda) - \varepsilon^2 I_{\mathcal{D}}}{1 - \bar{\mu}\lambda} \qquad (\lambda, \mu \in \mathbb{D})$$

is positive definite, i.e.

$$\sum_{k=1}^{n}\sum_{h=1}^{n}(K_\varepsilon(\lambda_k, \lambda_h)d_h, d_k) \geqslant 0 \tag{2}$$

for any finite systems $\{\lambda_1, \ldots, \lambda_n\}$, $\{d_1, \ldots, d_n\}$, where $\lambda_j \in \mathbb{D}$, $d_j \in \mathcal{D}$.

Condition (1) obviously implies that

$$\|\theta(\lambda)d\| \geqslant \varepsilon\|d\| \qquad (|\lambda| < 1). \tag{3}$$

The QUESTION is *whether (3) implies (2) with the same ε or at least with some, possibly different, positive constant.*

In the special case when $\dim \mathcal{D} = 1$ and $\dim \mathcal{D}^* < \infty$ the equivalence of (1) and (3), and thus the equivalence of (2) and (3), follows from the Corona Theorem of L. Carleson, cf. [2]. A proof of the equivalence of (2) and (3) in the general case, and possibly with operator theoretic arguments, would be an important achievement.

REFERENCES

1. Szőkefalvi-Nagy B., Foiaş C., *On contractions similar to isometries and Toeplitz operators*, Ann. Acad. Scient. Fennicae, Ser. A. I. Mathematica **2** (1976), 553–564.
2. Arveson W., *Interpolation problems in nest algebras*, J. Func. Anal. **20** (1975), 208–233.

BOLYAI INST. OF MATH.
ARADI VÉRTANÚK TERE 1
6720 SZEGED
HUNGARY

Commentary

This interesting question has been considered in several publications. Using some refinements of T. Wolff's corona argument, V. A. Tolokonnikov [3] (see also [4], p. 101) and M. Rosenblum [7] proved (independently) that (3) $\Longrightarrow$ (1) if $\dim \mathcal{D} \left(\stackrel{\text{def}}{=} d \right) = 1$. Moreover, Tolokonnikov obtained an estimate of the solution Ω. This estimate (in somewhat simplified form) looks as follows:

$$c_1(\varepsilon) \leqslant \min\{ 20 \left(\log \frac{e}{\varepsilon} \right)^{3/2} \frac{1}{\varepsilon^2}, \quad (1 - 46(1-\varepsilon))^{-1} \} \qquad (0 < \varepsilon < 1),$$

where

$$c_d(\varepsilon) \stackrel{\text{def}}{=} \sup_{\theta,\, \varepsilon^2 I \leqslant \theta^* \theta \leqslant I} \; \inf_{\Omega,\, \Omega\theta = I} \|\Omega\|_\infty, \qquad d = \dim \mathcal{D}.$$

For small values of ε a better estimate is due to Uchiyama [6]

$$c_1(\varepsilon) = O\left(\varepsilon^{-2} \log \frac{1}{\varepsilon} \right);$$

V. I. Vasyunin has shown that

$$c_1(\varepsilon) \leqslant \frac{1}{\varepsilon} + \frac{1}{\varepsilon^2} \left(4\sqrt{e \log \frac{1}{\varepsilon}} + 3e\sqrt{6} \log \frac{1}{\varepsilon} \right).$$

V. I. Vasyunin (see [5]) has proved that (3) $\Longrightarrow$ (1) if $d < +\infty$ (with the estimate $c_d(\varepsilon) \leqslant \sqrt{d} c_1(\varepsilon^d)$).

S. R. Treil has shown in [9, 10] that (3) does not imply (1) for $d = \infty$. Therefore the question arises to find another local condition that would be sufficient for existence of Ω satisfying (1).

If $\mathcal{D} \subset \mathcal{D}^*$ and $\dim \mathcal{D} < +\infty$ then assertions (a) and (b) in the Problem are equivalent to the possibility "to enlarge" θ to a *square* matrix $\tilde{\theta}$ analytic in $\mathbb{D}$ and satisfying $\theta = \tilde{\theta} | \mathcal{D}$, $\sup_{\lambda \in \mathbb{D}} \|\tilde{\theta}(\lambda)\| < \infty$, $\sup_{\lambda \in \mathbb{D}} \|(\tilde{\theta}(\lambda))^{-1}\| < \infty$, (see [5]).

There exists a connection between the corona theorem (for $\dim \mathcal{D} = 1$) and the left invertibility of the vector Toeplitz operator T_{θ^*} ([1], see also [8]).

References

3. Tolokonnikov V. A., *Estimates in the Carleson Corona theorem and finitely generated ideals of the algebra H^∞*, Funktsion. Anal. i Prilozhen. **14** (1980), no. 4, 85–86 (Russian); English transl. in Funct. Anal. and its Appl. **14** (1980), no. 4.
4. Nikolskii N. K., *Treatise on the Shift Operator*, Nauka, Moscow, 1980 (Russian); English transl. Springer-Verlag, 1986.
5. Tolokonnikov V. A., *Estimates in the Carleson Corona theorem, ideals of the algebra H^∞, a problem of Sz.-Nagy*, Zapiski Nauchn. Semin. LOMI **113** (1981), 178–198 (Russian); English transl. in Soviet Math. J. **22** (1983), 1814–1828.
6. Uchiyama A., *Corona theorems for countably many functions and estimates for their solutions*, Preprint, University of California at Los Angeles, 1981.
7. Rosenblum M., *A corona theorem for countably many functions*, Integral Equat. and Operator Theory **3** (1980), no. 2, 125–137.

8. Schubert C. F., *The corona theorem as an operator theorem*, Proc. Amer. Math. Soc. **69** (1978), no. 1, 73–76.
9. Treil S. R., *Angles between coinvariant subspaces and an operator-valued corona problem. A question of Szőkefalvi-Nagy*, Doklady Akad. Nauk SSSR **302** (1988), 1063–1067 (Russian); English transl. in Soviet Math. Doklady **38** (1989), 394–399.
10. Treil S. R., *Geometric methods in spectral theory of vector-valued functions: some recent results*, Toeplitz Operator and Spectral Function Theory (Nikolskii N. K., ed.), Operator Theory: Adv. and Appl., vol. 42, Birkhäuser Verlag, Basel, 1989, pp. 209–280.

Chapter 5

GENERAL OPERATOR THEORY

Edited by

Peter Rosenthal
Department of Mathematics
University of Toronto
Toronto, Ontario, M5S 1A1
Canada

INTRODUCTION

Operator theory began with the study of integral and differential operators. There is still much research concerning concrete operators which arise in various analytic contexts, such as Toeplitz and Hankel operators; questions about such operators appear in other chapters of this problem book.

This chapter, however, concerns the study of unitarily invariant properties of bounded linear operators, often subject to given restrictive hypotheses. Normal operators, for example, are quite well understood (although there still remain some interesting questions). There is also much known about compact operators (but there is still nothing approaching a structure theorem analogous, for example, to the Jordan canonical form for operators of finite rank). For general operators there is a huge body of knowledge, but it remains fragmentary; there is clearly a great deal remaining to be discovered.

Stating and disseminating research problems has always provided useful direction to mathematical research, from Fermat's theorem, through Hilbert's famous list of problems at the beginning of this century, to, in our era and for our subject, Halmos' "Ten problems in Hilbert space" [7] (see [8] for a progress report).

One operator theory problem that has stimulated the production of many beautiful theorems is the invariant subspace problem, which is the best-known and (in my view) most interesting unsolved problem in operator theory. There have been many affirmative results under additional hypotheses (see [10] and [2]), including the remarkable breakthroughs by Lomonosov [9] and Brown [4] (also note the extension in [5]—see [3]). In addition, there are counterexamples for operators on Banach spaces ([6], [11], [1]), even on ℓ^1 [12]. But the general problem remains elusive.

My guess is that there will be a counterexample found, probably before the end of this century. Counterexamples to any of the related reductive algebra, transitive algebra or hyperinvariant subspace problems (see [10] for statements of these problems) would also be of great interest. Regardless of the disposition of the invariant subspace problem itself, there will undoubtedly be much additional research into existence of invariant subspaces in particular cases, into the structure of the invariant subspaces of particular operators and into reflexive operator algebras.

We can't all solve the invariant subspace problem, and even the mathematician who is destined to solve it will likely want to work on other problems in the interim, at least occasionally. Hence what follows.

This chapter contains a number of interesting problems about various aspects of the theory of bounded linear operators on Hilbert space. Some of the problems are brand new, some go back to the 1984 edition of this problem book (a reasonable amount of inquiry suggests that they remain unsolved), and a couple are slightly older. But none of these problems has received so much attention that it can be said with assurance that the problem is very difficult; many are probably tractable.

The problem titles are quite descriptive, and a quick perusal of the brief note about a problem will give you much more information than could be conveyed in this introduction, so it is suggested that you begin by skimming the chapter.

INTRODUCTION

It has been said that "A mathematician is someone with a solution looking for a problem". I hope that some of your solutions are applicable to some of the problems which follow.

Acknowledgements: I am very grateful to Pat Broughton, Nadia Cavaliere and Karin Smith for secretarial help, to Shlomo Rosenoer for several translations, and to David McIntosh for editorial and TEXnical assistance.

REFERENCES

1. Beauzamy B., *Un operateur sans sous-espace invariant non-trivial: simplification de l'exemple de P. Enflo*, Integral Equations and Operator Theory **8** (1985), 314–384.
2. Beauzamy B., *Introduction to Operator Theory and Invariant Subspaces*, North-Holland, Amsterdam, 1988.
3. Bercovici H., *Notes on invariant subspaces*, Bull. Amer. Math. Soc. (N.S.) **23** (1990), 1–33.
4. Brown S., *Invariant subspaces for subnormal operators*, Integral Equations and Operator Theory **1** (1978), 310–333.
5. Brown S., Chevreau B., Pearcy C., *On the structure of contraction operators II*, J. Funct. Anal. **76** (1988), 30–55.
6. Enflo P., *On the invariant subspace problem in Banach spaces*, Acta Math. **158** (1987), 213–313.
7. Halmos P. R., *Ten problems in Hilbert space*, Bull. A.M.S. **76** (1970), 887–933.
8. Halmos P. R., *Ten years in Hilbert space*, Integral Equations and Operator Theory **2** (1979), 393–428.
9. Lomonosov V., *Invariant subspaces for operators commuting with compact operators*, Funkts. Anal. i Prilozh. **7** (1973), 55–56 (Russian); English transl. in Funct. Anal. Appl. **7** (1973), 213–214.
10. Radjavi H., Rosenthal P., *Invariant Subspaces*, Springer-Verlag, Berlin, 1973.
11. Read C. J., *A solution to the invariant subspace problem*, Bull. Lond. Math. Soc. **16** (1984), 337–401.
12. Read C. J., *A short proof concerning the invariant subspace problem*, J. Lond. Math. Soc. **34** (1986), 335–348.

5.1

PERTURBATION OF EIGENVALUES FOR NORMAL OPERATORS

JOHN A. HOLBROOK

This article is related to the problems treated by Chandler Davis in item 4.32 of LNM 1043.† The intervening years have shed light on some of those problems, and have suggested additional research directions. For example, Davis discussed the evidence known at the time for the following conjecture: if A and B are normal matrices, then

$$\min_{\sigma} \max_{k} |\alpha_k - \beta_{\sigma(k)}| \leqslant \|A - B\|. \tag{1}$$

Here the norm on the RHS is the operator norm, while the α_k and β_k on the LHS are the eigenvalues of A and B respectively and σ denotes an arbitrary permutation. The LHS of (1) is the natural definition of 'spectral distance' between A and B; henceforth we denote it by $sd(A, B)$. The conjecture has a long history (see Mirsky [17], for example), but it is now known to be false (see Holbrook [14] for details). There remains the QUESTION: *what constant needs to be inserted on the RHS of* (1)*?*

To be more precise, let

$$c_n = \max\left\{ \frac{sd(A,B)}{\|A - B\|} : A \text{ and } B \text{ are normal } n \times n \text{ matrices} \right\}. \tag{2}$$

It is easy to show that $c_2 = 1$ but the examples of [14] imply that $c_3 \geqslant 1.016$; clearly $c_{n+1} \geqslant c_n$. Bhatia, Davis, Koosis, and McIntosh (see [5] and [6]) were able to obtain an upper bound for the c_n that is independent of n. They show that all the c_n are bounded by a constant somewhat less than 3. It seems very likely that the true values of c_n are smaller, especially for moderate n.

Much of the fascination with (1) stems from the many interesting conditions under which it is true. Conditions on the spectral geometry go back to the beginning of the subject, and continue to be discovered. Weyl [19] noted that (1) holds for self-adjoint A and B, i.e. when the spectra lie in the real line. In fact, it follows from the methods of Bhatia [3] that (1) holds whenever $\sigma(A)$ (the spectrum of A) lies in a (straight) line segment L_A, $\sigma(B)$ also lies in a segment L_B, and L_A is parallel to L_B. On the other hand, Sunder [18] shows that (1) also holds if L_A and L_B are at right angles. Recently Farid [12] has established that (1) holds for L_A and L_B at any angle, provided that the line segments do not cross. Somewhat analogous results are known for spectra lying on circular arcs. Bhatia and Davis [4] showed that (1) holds for unitary A and B; in terms of spectral geometry this corresponds to the condition that the spectra $\sigma(A)$ and $\sigma(B)$ lie on a common circle. Bhatia and Holbrook [7] extends this result to the case of concentric circles: if $\sigma(A)$ and $\sigma(B)$ lie on circles C_A and C_B respectively, then (1)

†See S.5.18 for an abridged form of the paper. – Ed.

holds, provided that C_A and C_B are concentric. The methods of [19], [3], [18], [12], [4], and [7] are strikingly diverse, but it is tempting to think that the results on spectral geometry may be linked by appropriate spectral mapping principles.

In [14] there is a characterization of extremal pairs A,B, minimizing $\|A - B\|$ for a given spectral geometry, but it is assumed there that the eigenvalues are not collinear. In view of the results summarized above for the case where the spectra lie on line segments, it may be helpful to find the extreme pairs in that case also. We might then hope to discover whether the result of Farid [12] is best possible, i.e. whether (1) may fail if L_A and L_B cross.

In [3] Bhatia introduced the normal path inequality, and Bhatia and Holbrook develop this technique further in [7], [8], and [9]. One form of this inequality is the following:

$$sd(A,B) \leqslant \min\{\ \|\gamma\| : \gamma \text{ is a normal path from } A \text{ to } B\ \}. \tag{3}$$

Here $\|\gamma\|$ is the arclength (with respect to the operator norm) of the continuous path γ among the normal matrices, with endpoints at A and B. For normal matrices of a given size $n \times n$, let w_n denote the largest value of the RHS of (3) over all A and B such that $\|A - B\| = 1$. Thus w_n is a geometric property of the set of normals of size $n \times n$, related to its maximum 'warp' with respect to the (non-Euclidean) operator norm metric. It is a curious fact that $w_2 = 1$ (see [7]), but (3) shows that $c_n \leqslant w_n$, so that $c_n > 1$ for all $n \geqslant 3$. The problem of evaluating the w_n or estimating them from above seems interesting in itself, and success may even yield better bounds for spectral variation, better for example than the $c_n < 3$ obtained by Bhatia et al. in [5] and [6].

In [5] and [6] Bhatia et al. obtain bounds on eigenvalue perturbation by first studying the perturbation of spectral subspaces. While this approach does not always lead to optimal estimates of the c_n, it raises interesting related questions. The thesis of McEachin [16] includes recent work on subspace perturbation.

Those wishing to explore infinite-dimensional versions of the eigenvalue perturbation problem could start by consulting Azoff and Davis [1], Davidson [10], [11], and the thesis of Marcoux [15], where the program is extended to the estimation of the distance between unitary orbits of weighted shifts.

Another aspect of the problem arises when we consider matrix norms other than the operator norm. For example, we may compare the Schatten p-norm $\|A - B\|_p$ between two normal matrices with the appropriate p-spectral distance:

$$sd_p(A,B) = \min_{\sigma} \|\alpha - \sigma(\beta)\|_p, \tag{4}$$

where α and β now denote the vectors of eigenvalues and $\sigma(\beta)$ denotes the vector obtained by reordering the components of β via a permutation σ. The classic result of Hoffman and Wielandt [13] says that, when $p = 2$,

$$sd_p(A,B) \leqslant \|A - B\|_p, \tag{5}$$

for all normal matrices A and B. In [18] Sunder points out that (5) does not persist for $p < 2$, while the examples in [14] imply that (5) may fail when p is sufficiently large. Bhatia has suggested that more powerful methods may reveal that the Hoffman-Wielandt case ($p = 2$) is the only case where the spectral inequality holds (with constant 1) for all normals.

Finally we note that the monograph [2] by Bhatia is a valuable reference for many of the results mentioned here, apart from the most recent ones.

REFERENCES

1. Azoff E., Davis C., *On the distance between unitary orbits of self-adjoint operators*, Acta Sci. Math. (Szeged) **47** (1984), 419–439.
2. Bhatia R., *Perturbation Bounds for Matrix Eigenvalues*, Pitman Research Notes in Mathematics No. 162, Longman Scientific & Technical, Essex UK, 1987.
3. Bhatia R., *Analysis of spectral variation and some inequalities*, Trans. Amer. Math. Soc. **272** (1982), 323–331.
4. Bhatia R., Davis C., *A bound for the spectral variation of a unitary operator*, Linear Mult. Alg. **15** (1984), 71–76.
5. Bhatia R., Davis C., Koosis P., *An extremal problem in Fourier analysis with applications to operator theory*, J. Funct. Anal. **82** (1989), 138–150.
6. Bhatia R., Davis C., McIntosh A., *Perturbation of spectral subspaces and solution of linear operator equations*, Linear Algebra Appl. **52** (1983), 45–67.
7. Bhatia R., Holbrook J., *Short normal paths and spectral variation*, Proc. Amer. Math. Soc. **94** (1985), 377–382.
8. Bhatia R., Holbrook J., *Unitary invariance and spectral variation*, Linear Alg. Appl. **95** (1987), 43–68.
9. Bhatia R., Holbrook J., *A softer, stronger Lidskii theorem*, Proc. Indian Acad. Sci. (Math. Sci.) **99** (1989), 75–83.
10. Davidson K., *The distance between unitary orbits of normal operators*, Acta Sci. Math. (Szeged) **50** (1986), 213–223.
11. Davidson K., *The distance between unitary orbits of normal elements in the Calkin algebra*, Proc. Roy. Soc. Edinburgh **99A** (1984), 35–43.
12. Farid F., *The spectral variation for two matrices with spectra on two intersecting lines*, Linear Algebra Appl. (to appear).
13. Hoffman A. J., Wielandt H. W., *The variation of the spectrum of a normal matrix*, Duke Math. J. **20** (1953), 37–39.
14. Holbrook J., *Spectral variation for normal matrices*, Linear Algebra Appl. (to appear).
15. Marcoux L., *On the distance between unitary orbits of weighted shifts*, thesis, University of Waterloo, 1988.
16. McEachin R., *Analysis of an inequality concerning perturbation of self-adjoint operators*, thesis, University of Illinois at Urbana-Champaign, 1990.
17. Mirsky L., *Symmetric gauge functions and unitarily invariant norms*, Quarterly J. Math. Oxford Ser. 2 **11** (1960), 50–59.
18. S. Sunder V., *Distance between normal operators*, Proc. Amer. Math. Soc. **84** (1982), 483–484.
19. Weyl H., *Das asymptotische Verteilungsgesetz der Eigenwerte linearer partieller Differentialgleichungen*, Math. Ann. **71** (1912), 441–479.

UNIVERSITY OF GUELPH
GUELPH, ONTARIO
CANADA, N1G 2W1

5.2

UNSOLVED PROBLEMS ON REDUCIBILITY OF SEMIGROUPS OF OPERATORS

HEYDAR RADJAVI

Let $\mathcal{S}$ be a semigroup of bounded operators on the complex Hilbert space $\mathcal{H}$. (By a semigroup we mean a collection closed under multiplication.) We say $\mathcal{S}$ is *reducible* if its members have a common invariant (closed) subspace other than $\{0\}$ and $\mathcal{H}$. The invariant subspace problem for a general Hilbert-space operator, which is still unsettled, is: is a singly generated semigroup reducible? Of course, if $\mathcal{S}$ is a commutative semigroups of compact operators, then it is reducible by Lomonosov's theorem [8]; observe that the non-trivial case of this assertion is the one in which every member of $\mathcal{S}$ is quasinilpotent. The following problem was posed in [9] and is still unsolved.

PROBLEM 1. *Is a semigroup of compact quasinilpotent operators reducible?*

If "semigroup" is replaced by "algebra" in this problem, the answer is yes by Lomonosov's theorem. If the semigroup contains a nonzero member of some Schatten class $\mathfrak{S}_p$, the answer is yes [9]. In particular, if $\mathcal{H}$ is finite-dimensional, this is Levitzki's theorem [7]. In the case of infinite dimensions, a question more general than Problem 1 is: if the spectral radius is submultiplicative on a semigroup $\mathcal{S}$ of compact operators, is $\mathcal{S}$ reducible? The only hard case of this is Problem 1, to which it can be reduced: a semigroup with at least one non-quasinilpotent compact member on which the spectral radius is submultiplicative is reducible [10].

Without the compactness hypothesis, Problem 1 has a negative answer [3,4]. On the other hand, if a semigroup of quasinilpotent operators contains a nonzero compact operator its reducibility problem is equivalent to Problem 1. For we can consider the (semigroup) ideal of the semigroup generated by that compact operator and use the fact that if a nonzero ideal of $\mathcal{S}$ is reducible, then so is $\mathcal{S}$.

As observed in [9], if Problem 1 has an affirmative answer, then a stronger assertion is true: every such semigroup $\mathcal{S}$ is then *triangularizable*. This means the existence of a collection $\mathcal{C}$ of invariant subspaces for $\mathcal{S}$ such that $\mathcal{C}$ is a maximal chain of subspaces of $\mathcal{H}$. Since the algebra generated by a semigroup $\mathcal{S}$ is simply the linear span of $\mathcal{S}$, and since the triangularizability of $\mathcal{S}$ is equivalent to that of the algebra generated by $\mathcal{S}$, the preceding remarks show that the following problem has an affirmative answer if Problem 1 does [9].

PROBLEM 2. *If every member of the semigroup generated by the compact operators A and B is quasinilpotent, is $A + B$ quasinilpotent?*

Hypotheses other than compactness sometimes yield reducibility for a semigroup. One such condition is "positivity" for matrix entries or for kernels of integral operators. Let $\mathcal{H} = \mathcal{L}^2(\mathcal{X}, \mu)$, where $\mathcal{X}$ is a locally compact Hausdorff Lindeloff space and μ is a

σ-finite, regular Borel measure. By an integral operator on $L^2(\mathcal{X}, \mu)$ we mean a bounded operator T with a kernel K_T such that

$$(Tf)(x) = \int K_T(x,y)\, f(y)\, d\mu(y)$$

for $f \in L^2(\mathcal{X}, \mu)$. Such an operator need not be compact [5]. It follows from much more general, elegant results of Ando [1] and Krieger [6] that a quasinilpotent operator T with non-negative kernel K_T has a non-trivial invariant subspace M of special kind: there is a measurable set E of $\mathcal{X}$ such that

$$M = \{f : f = 0 \quad \text{a.e. on} \quad E\}.$$

If a collection of integral operators has an invariant subspace of this form (with $\mu(E)\mu(\mathcal{X} \backslash E) \neq 0$) we say it is (not just reducible, but) *decomposable.*

PROBLEM 3. *Is a semigroup of quasinilpotent integral operators with non-negative kernels decomposable?*

Only partial answers seem to be known to this problem and to the corresponding triangularizability problem. In [2] Problem 3 is answered affirmatively for special cases where certain semicontinuity assumptions are made on the kernels. No such hypotheses are needed, of course, if $\mathcal{S}$ is singly generated, as noted above.

REFERENCES

1. Ando T., *Positive operators in semi-ordered linear spaces*, J. Fac. Sci. Hokkaido Univ. Ser. I. **13** (1957), 214–228.
2. Choi M.-D., Nordgren E., Radjavi H., Rosenthal P., Zhong Y., *Triangularizing semigroups of quasinilpotents with non-negative entries*, Indiana Univ. Math. J. (to appear).
3. Hadwin D., Nordgren E., Radjabalipour M., Radjavi H., Rosenthal P., *A nil algebra of bounded operators on Hilbert space with semisimple norm closure*, Int. Equat. Oper. Th. **9** (1986), 739–743.
4. Hadwin D., Nordgren E., Radjabalipour M., Radjavi H., Rosenthal P., *On simultaneous triangularization of collections of operators*, Houston Math. J. (to appear).
5. Halmos P. R., Sunder V. S., *Bounded Integral Operators on L^2 Spaces*, Springer-Verlag, New York, 1978.
6. Krieger H. J., *Beiträge zur Theorie positiven Operatoren*, Schriffenriehe der Institute für Math: Reihe A, Heff 6, Akademie-Verlag, Berlin, 1969.
7. Levitzki J., *Über nilpotente Unterringe*, Math. Ann. **105** (1931), 620–627.
8. Lomonosov V., *Invariant subspaces for operators commuting with compact operators*, Funct. Anal. and Appl. **7** (1973), 213–214.
9. Nordgren E., Radjavi H., Rosenthal P., *Triangularizing semigroups of compact operators*, Indiana Univ. Math. J. **33** (1984), 271–275.
10. Radjavi H., *On reducibility of semigroups of compact operators*, Indiana Univ. Math. J. **39** (1990), 499–515.

DALHOUSIE UNIVERSITY
HALIFAX, NOVA SCOTIA
CANADA B3H 3J5

5.3

COMPACT OPERATORS AND MASAS

KENNETH R. DAVIDSON

There is an old and interesting question asking for a lattice characterization of those CSL's (commutative subspace lattices) $\mathcal{L}$ for which the compact operators in Alg($\mathcal{L}$) are weak* dense in the algebra (cf. [1, 2]). We wish to ask two related questions.

1. *Is the WOT closed algebra generated by a compact operator and a masa always reflexive?*
2. *Is there a non-nilpotent compact operator K and a masa $\mathcal{M}$ with the property that for all $\varepsilon > 0$, there are invariant projections E_n for K in $\mathcal{M}$ such that $\bigvee_{n\geqslant 1} E_n = I$ and $\|E_n K\| < \varepsilon$ for all $n \geqslant 1$?*

Both of these questions are asking, in different ways, what restrictions can be placed on the invariant subspaces of a compact operator. If K is Hilbert-Schmidt, it is straightforward using known results to show that 1. has a positive answer and 2. has a negative one. However, there is a recent example of Haydon [3] of a compact CSL which has lots of 'small' projections spanning to I. If this algebra contains compact operators, it would give a positive answer to question 2.

REFERENCES

1. Davidson K. R., *Problems in reflexive algebras*, Proc. GPOTS meeting 1987, Rocky Mountain Math. J. **20** (1990), 317–330.
2. Davidson K. R., Pitts D. R., *Compactness and complete distributivity for commutative subspace lattices*, J. London Math. Soc. **42** (1990), 147–159.
3. Haydon R., *A compact lattice of projections*, preprint, 1990.

PURE MATH. DEPT.
UNIVERSITY OF WATERLOO
WATERLOO, ONTARIO N2L-3G1
CANADA

5.4

DIFFERENTIATION AND TRANSLATION INVARIANT SUBSPACES OF HILBERT SPACES OF ENTIRE FUNCTIONS

Aharon Atzmon

Let $\mathcal{H}$ be a Hilbert space of entire functions with reproducing kernel. It is easy to verify that the existence of a reproducing kernel in $\mathcal{H}$ is equivalent to the condition that convergence in $\mathcal{H}$ implies uniform convergence on compact subsets of $\mathbb{C}$. (The non-trivial direction is proved by an application of the uniform boundedness principle). Assume in addition that $\mathcal{H}$ is invariant under differentiation. Then by the closed graph theorem the differentiation operator D is continuous on $\mathcal{H}$.

Problem 1. *Let $\mathcal{H}$ be a Hilbert space with the above properties. Does the differentiation operator D on $\mathcal{H}$ have a proper invariant subspace?*

As shown in [1], this problem is equivalent to the invariant subspace problem for Hilbert space operators.

If f is a function on $\mathbb{C}$ then, for every $z \in \mathbb{C}$, we shall denote by f_z the function on $\mathbb{C}$ defined by $f_z(\zeta) = f(\zeta + z)$, $\zeta \in \mathbb{C}$. A Hilbert space $\mathcal{H}$ is called translation invariant if $f \in \mathcal{H}$ implies that $f_z \in \mathcal{H}$ for every $z \in \mathbb{C}$. If in addition for every $f \in \mathcal{H}$ the mapping $z \to f_z$ from $\mathbb{C}$ into $\mathcal{H}$ is continuous, we shall say that the translation group acts continuously on $\mathcal{H}$.

As shown in [1] a Hilbert space $\mathcal{H}$ with reproducing kernel is differentiation invariant if and only if the translation group acts continuously on $\mathcal{H}$, and moreover, in this case a closed subspace of $\mathcal{H}$ is differentiation invariant if and only if it is translation invariant. Thus an equivalent formulation of Problem 1 (hence of the invariant subspace problem for Hilbert space operators) is the following problem in Harmonic analysis:

Problem 2. *Let $\mathcal{H}$ be a Hilbert space of entire functions with reproducing kernel on which the translation group acts continuously. Does $\mathcal{H}$ have a proper translation invariant subspace?*

In view of this it is of interest to consider the existence of proper translation invariant subspaces of special classes of Hilbert spaces of entire functions which satisfy the conditions of Problem 2. One such class is obtained as follows. Let w be a positive continuous function on $\mathbb{C}$ and denote by $\mathcal{H}_w$ the vector space of all entire functions f for which the norm

$$\|f\|_w = \left(\int_{\mathbb{C}} \left|\frac{f(\zeta)}{w(\zeta)}\right|^2 dm(\zeta)\right)^{\frac{1}{2}}$$

(where m denotes area measure on $\mathbb{C}$) is finite. It is well-known that with respect to this norm $\mathcal{H}_w$ is a Hilbert space with reproducing kernel. Assume further that the weight function w satisfies the condition

$$(\dagger) \qquad \sup\left\{\frac{w(\zeta+z)}{w(\zeta)} : \zeta \in \mathbb{C}, \quad |z| < 1\right\} < \infty.$$

It is easy to verify that this condition implies that the translation group acts continuously on $\mathcal{H}_w$.

PROBLEM 3. *Let w be a continuous function on $\mathbb{C}$ which satisfies condition (†), and assume that $\mathcal{H}_w$ is infinite dimensional. Does $\mathcal{H}_w$ have a proper translation invariant subspace?*

A particular example of weights for which the answer to Problem 3 is not clear is given by the weights $w_\alpha(z) = \exp(r^\alpha \cos \alpha\theta)$, $z = re^{i\theta} \in \mathbb{C}$, $|\theta| \leqslant \pi$, where $1/2 < \alpha < 1$. It can be shown that for $0 < \alpha \leqslant 1/2$ the corresponding spaces have proper translation invariant subspaces.

L. de Branges [5] introduced a class of Hilbert spaces of entire functions as follows. For every entire function E which satisfies the inequality $|E(\bar{z})| < |E(z)|$ for $\operatorname{Im} z > 0$, let $\mathcal{H}(E)$ denote the vector space of all entire functions F such that

$$\|F\|_E^2 = \int_{-\infty}^{\infty} \left|F(t)/E(t)\right|^2 dt < \infty$$

and such that the functions F/E and F^*/E (where $F^*(z) = \overline{F}(\bar{z})$, $z \in \mathbb{C}$) are of bounded type and nonpositive mean type in the upper half plane.

De Branges proved that with the above norm $\mathcal{H}(E)$ is a Hilbert space, which contains nonzero functions, with reproducing kernel which can be expressed in terms of E.

PROBLEM 4. *What are the additional conditions on the entire function E which imply that the translation group acts continuously on the Hilbert space $\mathcal{H}(E)$?*

It can be shown that a necessary condition for the translation group to act continuously on $\mathcal{H}(E)$ is that E is of finite exponential type. We conjecture that a sufficient condition is that E is of finite exponential type and satisfies the condition

$$\sup\left\{\left|\frac{E(t+x)}{E(t)}\right|,\ t \in \mathbb{R}\right\} < \infty$$

for every $x \in \mathbb{R}$.

PROBLEM 5. *Let E be an entire function such that the translation group acts continuously on the Hilbert space $\mathcal{H}(E)$. Does this space have a proper translation invariant subspace?*

In [2], [3] and [4] we have constructed nuclear Fréchet spaces of entire functions on which the differentiation operator and the translation group act continuously, and which have no proper differentiation and no proper translation in invariant subspaces.

REFERENCES

1. Atzmon A., *A model for operators with cyclic adjoint*, Integral Equations and Operator Theory **10** (1987), 153–163.
2. Atzmon A., *Irreducible representations of abelian groups*, Harmonic Analysis, Lecture Notes in Math. **1359** (1988), 83–92.
3. Atzmon A., *Spaces of entire functions with no proper translation invariant subspace*, Commutative Harmonic Analysis, Contemporary Math. **91** (1989), 1–8.

4. Atzmon A., *Fréchet spaces of entire functions with transitive differentiation*, Journal d'Analyse, Szolem Mandelbrojt Commemorative volume (1992) (to appear).
5. De Branges L., *Hilbert Spaces of Entire Functions*, Prentice-Hall, Englewood Cliffs, 1968.

Raymond and Beverly Sackler
Faculty of Exact Science
School of Mathematical Sciences
Tel Aviv University
Tel Aviv 69978, Israel

5.5

SPECTRUM ASSIGNMENT PROBLEMS

LEIBA RODMAN, ILYA SPITKOVSKY

Let A and B be $n \times m$ matrices, respectively (with complex entries). An interesting question (motivated also by important applications in modern control theory) is: Find all possible spectra (or, more generally, all possible Jordan forms) of matrices of the form $A + BF$, where F is a variable $m \times n$ matrix. This problem is called the *pole assignment* problem. Recently, a complete solution was given in [8, 9], in terms of the invariants of the Kronecker canonical form of the pencil $[\lambda I - A,\ B]$.

The corresponding problem, called the *spectrum assignment* problem, for A and B operators acting in an infinite dimensional Hilbert space was studied in [1,3,6,4,5]. Denote by $L(H,G)$ the set of linear bounded operators $H \to G$, where H and G are infinite dimensional Hilbert spaces.

THEOREM 1. *Let $A \in L(H,H)$, $B \in L(G,H)$. Assume that the operator*

$$(1) \qquad [B, AB, \ldots, A^{p-1}B] \in L(G^p, H)$$

is right invertible for some integer p. Then for every compact $\Lambda \subset \mathbb{C}$ there exists $F \in L(H,G)$ such that $\sigma(A + BF) = \Lambda$.

Note that the right invertibility of (1) is equivalent to the right invertibility of the operator $[\lambda I - A,\ B] \in L(H \oplus G, H)$ for every $\lambda \in \mathbb{C}$ [3,6].

In [2] this result was extended as follows (providing thereby a partial answer to Problem 7.2.3 in [5]).

THEOREM 2. *Let $A \in L(H,H)$, $B \in L(G,H)$ be such that*

$$(2) \qquad Im[B, AB, \ldots, A^{p-1}B] = Im[B, AB, \ldots, A^{p}B]$$

and the linear set (2) is closed and distinct from H, for some integer $p \geqslant 1$. If $Im\,B$ is not finite dimensional, then for every compact set $\Lambda \subset \mathbb{C}$ such that

$$\Lambda \supseteq \{\, \lambda \in \mathbb{C} \mid [\lambda I - A,\ B] \text{ is not right invertible} \,\}$$

there exists $F \in L(H,G)$ for which

$$\sigma(A + BF) = \Lambda.$$

PROBLEM 1 (reformulated Problem 7.2.2 in [5]). *Study the fine spectral structure of operators of the form $A + BF$, where $A \in L(H,H)$, $B \in L(G,H)$ are given and $F \in L(H,G)$ is variable, in the spirit of the finite dimensional result.*

Results in this direction (concerning the semi-Fredholm properties of $A + BF$) are obtained in [2].

PROBLEM 2. *Characterize $\sigma(A+BF)$ for certain classes of pairs of operators (A,B) not satisfying the hypotheses of Theorem 3. For example, let A be compact (with infinitely many non-zero eigenvalues) and* $\operatorname{rank} B = 1$.

The second part of Problem 2 can be generalized:

PROBLEM 3. *Let $A \in L(H,H)$ be a compact operator with infinitely many non-zero eigenvalues. Describe the spectra, and, more generally, Jordan structure of non-zero eigenvalues, of operators of the form $A + C$, where $C \in L(H,H)$ is variable and* $\operatorname{rank} C = 1$.

In connection with Problem 3 note that the finite dimensional analog of this problem is solved completely in [7].

One can specialize Problem 2 in another direction, for multiplication operators:

PROBLEM 4. *Let H be a suitable Hilbert space of analytic functions, A be the multiplication operator by $a(z)$, and B the rank 1 operator $(Bf)(z) = f(z_0)\psi(z)$, where z_0 and $\psi(z) \in H$ are fixed. Characterize the possible spectral properties of $A + BF$, where F is variable.*

One observation in this direction [2]: Let $H = H^2(\mathbb{D})$ be the standard Hardy space on the open unit disc $\mathbb{D}$, A be the multiplication by z, and define B by $(Bf)(z) = f(0)$. Then an operator $F : H \to \mathbb{C}$ such that $\sigma(A+BF) = \Lambda$ exists if and only if $\overline{\mathbb{D}} \subset \Lambda$, the set $\Lambda \setminus \overline{\mathbb{D}} = \{z_k\}$ is not more than countable, and

$$\sum_k (|z_k| - 1) < \infty.$$

We remark that it is not known whether Theorems 2 and 3 are valid for Banach space operators.

Finally, we note that in applications (control of systems with infinite number of degrees of freedom) the spectrum assignment problems with unbounded self-adjoint operator A and finite rank operator B (possibly also unbounded) are especially important.

REFERENCES

1. Eckstein G., *Exact controllability and spectrum assignment*, Operator Theory: Advances and Applications **2** (1981), 81–94.
2. Gurvits L., Rodman L., Spitkovsky I., *Spectrum assignment for Hilbert space operators*, preprint.
3. Kaashoek M. A., van der Mee C. V. M., Rodman L., *Analytic operator functions with compact spectrum III. Hilbert space case: inverse problem and applications*, J. of Operator Theory **10** (1983), 219–250.
4. Rodman L., *On exact controllability of operators*, Rocky Mountain J. of Math. **20** (1990), 549–560.

5. Rodman L., *An Introduction to Operator Polynomials*, OT: Adv. Appl., vol. 38, Birkhäuser Verlag, Basel, Boston, 1989.
6. Takahashi K., *Exact controllability and spectrum assignment*, J. of Math. Anal. and Appl. **104** (1984), 537–545.
7. Thompson R. C., *Invariant factors under rank one perturbations*, Canadian J. Math. **XXXII** (1980), 240–245.
8. Zaballa I., *Matrices with prescribed rows and invariant factors*, Linear Algebra and its Applications **87** (1987), 113–146.
9. Zaballa I., *Interlacing and majorization in invariant factor assignment problems*, Linear Algebra and its Applications **121** (1989), 409–421.

Department of Mathematics
The College of William and Mary
Williamsburg, Virginia 23187-8795
U.S.A.

Department of Mathematics
The College of William and Mary
Williamsburg, Virginia 23187-8795
U.S.A.

5.6
old

WHAT IS A FINITE OPERATOR?

DOMINGO A. HERRERO

An operator A acting on a complex separable Hilbert space $\mathcal{H}$ is called *finite* if the identity operator 1 is perpendicular to the range of the inner derivation δ_A induced by A; that is,

$$\inf \|\delta_A(X) - 1\| = \|1\| = 1$$

where $\delta_A(X) = AX - XA$ and X runs over the algebra $\mathcal{L}(\mathcal{H})$ of all (bounded linear) operators acting on $\mathcal{H}$.

The notion of finite operator was introduced by J. P. Williams in [7], where he proved the following result.

THEOREM 1. *These are equivalent conditions on an operator A:*

(i) $\inf_{X \in \mathcal{L}(\mathcal{H})} \|\delta_A(X) - 1\| = 1$.

(ii) 0 *belongs to the closure of the numerical range of* $\delta_A(X)$ *for each* X *in* $\mathcal{L}(\mathcal{H})$.

(iii) *There exists* $f \in \mathcal{L}(\mathcal{H})^*$ *such that* $f(1) = 1 = \|f\|$ *and*

$$\ker f \supset \operatorname{ran} \delta_A.$$

The origin of the term *finite* is this: if $A \subset \mathcal{L}(\mathcal{H})$ has a finite dimensional reducing subspace $\mathcal{M} \neq \{0\}$, $\{e_j\}_{j=1}^m (m = \dim \mathcal{M})$ is an orthonormal basis of $\mathcal{M}$, and $f(B) = \frac{1}{m} \sum_{j=1}^{m} \langle Be_j, e_j \rangle$, then f satisfies the conditions of Theorem 1 (iii) and therefore A is a finite operator.

Thus, if $R_n = \{T \in \mathcal{L}(\mathcal{H}) : T \text{ has a reducing subspace of dimension } n\}$ $(n = 1, 2, \dots)$, then $R = \bigcup_{n=1}^{\infty} R_n$ is a subset of the family (Fin) of all finite operators; furthermore, since (Fin) is closed in $\mathcal{L}(\mathcal{H})$ [7], $R^- \subset$ (Fin).

CONJECTURE 1 (J. P. Williams [7]). (Fin) $= R^-$.

As we have observed above, if $A \in R$, then it is possible to construct f as in Theorem 1 (iii) such that f does not vanish identically on $\mathcal{K}(\mathcal{H})$, the ideal of all compact operators. J. H. Anderson proved in [1] (Theorem 10.10 and its proof) that this fact can actually be used to characterize R as follows: $A \in R$ if and only if there exists $f \in \mathcal{L}(\mathcal{H})^*$ such that $f(1) = 1 = \|f\|$, and $\ker f \supset \operatorname{ran} \delta_A$, but $\ker f \not\supset \mathcal{K}(\mathcal{H})$; furthermore, if $A \in (\text{Fin}) \setminus R$, then every f as in Theorem 1 (iii) is necessarily a *singular* functional (i.e., $\ker f \supset \mathcal{K}(\mathcal{H})$). This is true, in particular, if $A \in R^- \setminus R$).

Since (Fin) $\supset R^-$, it is plain that (Fin) contains all *quasidiagonal operators* (in the sense of Halmos; see [4]). Moreover, (Fin) contains every operator of the form

$$(\dagger) \qquad A = T \oplus B + K,$$

where $T \in \mathcal{L}(\mathcal{H}_0)$ (for some finite or infinite dimensional subspace $\mathcal{H}_0$ of infinite codimension), $B \cong \bigoplus_{n=1}^{\infty} B_n$ for a suitable uniformly bounded sequence $\{B_n\}_{n=1}^{\infty}$ of operators acting on finite dimensional subspaces (i.e., B is a *block-diagonal* operator [4]) and K is compact.

Let R_0 denote the family of all operators of the form (†). It is apparent that

$$R_0 \subset R^- \subset (\mathrm{Fin})$$

and

$$R_0' \stackrel{\text{def}}{=} R_0 \backslash R \subset R' \stackrel{\text{def}}{=} R^- \backslash R \subset (\mathrm{Fin})' \stackrel{\text{def}}{=} (\mathrm{Fin}) \backslash R.$$

CONJECTURE 2 (D. A. Herrero). $(\mathrm{Fin})' = R_0'$; *moreover, if* $A \in (\mathrm{Fin})'$ *then there exists* $K \in \mathcal{K}(\mathcal{H})$ *and* Q *quasidiagonal such that*

$$A - K \cong A \oplus (Q \oplus Q \oplus Q \oplus \cdots). \tag{‡}$$

Several remarks are relevant here:

(1) J. W. Bunce has obtained several other equivalencies, in addition to the three given by Theorem 1 (see [2]).

(2) R^- properly includes $\bigcup_{n \geqslant 1}(R_n^-)$ [3, p. 262], [5, Example 11].

(3) In [1, Corollary 10.8], J. H. Anderson proved that every unilateral weighted shift is a finite operator, by exhibiting a functional f satisfying the conditions of Theorem 1 (iii). Recently, D. A. Herrero proved that all the (unilateral or bilateral) weighted shift operators belong to R^- [5].

(4) Suppose that $A \in R^-$; then there exists a sequence $\{P_n\}_{n=1}^{\infty}$ of non-zero finite rank orthogonal projections such that $\|AP_n - P_nA\| \to 0$ $(n \to \infty)$. Passing, if necessary, to a subsequence we can directly assume that $P_n \to H$ (weakly, as $n \to \infty$) for some hermitian operator H, $0 \leqslant H \leqslant 1$. It is easily seen that A commutes with H. If H has a non-zero finite rank spectral projection, then $A \in R$. If either $H = 0$ or $H = 1$, then it is not difficult to show that $A \in R_0$.

(5) According to a well-known result of D. Voiculescu, $A - K \cong A \oplus (Q \oplus Q \oplus Q \oplus \cdots)$ (as in (‡)) if and only if the C^*-algebra generated by $\pi(A)$ in the Calkin algebra $\mathcal{L}(\mathcal{H})/\mathcal{K}(\mathcal{H})$ admits a $*$-representation ρ such that $\rho\circ\pi(A) \cong Q$ [6]. Suppose that $A \in (\mathrm{Fin})'$. *Is it possible to use the singular functionals f provided by Theorem 1 (iii) and a Gelfand–Naimark–Segal type construction in order to construct a $*$-representation ρ with the desired properties (i.e. so that $\rho\circ\pi(A)$ is quasidiagonal)?*

(6) *Is it possible, at least, to show that the existence of such a singular functional implies that, for each $\varepsilon > 0$, A admits a finite rank perturbation F_ε, with $\|F_\varepsilon\| < \varepsilon$, such that $A - F_\varepsilon \cong A_\varepsilon \oplus A_\varepsilon'$, where A_ε acts on a non-zero finite dimensional subspace?* (An affirmative answer to this last question implies that $(\mathrm{Fin})' = R_0'$).

(7) Any partial answer to the above questions will also shed some light on several interesting problems related to quasidiagonal operators.

REFERENCES

1. Anderson J. H., *Derivations, commutators and essential numerical range*, Dissertation, Indiana University, 1971.
2. Bunce J. W., *Finite operators and amenable C^*-algebras*, Proc. Amer. Math. Soc. **56** (1976), 145–151.
3. Bunce J. W., Deddens J. A., *C^*-algebras generated by weighted shifts*, Indiana Univ. Math. J. **23** (1973), 257–271.
4. Halmos P. R., *Ten problems in Hilbert space*, Bull. Amer. Math. Soc. **76** (1970), 887–933.
5. Herrero D. A., *On quasidiagonal weighted shifts and approximation of operators*, Indiana Univ. Math. J. **33** (1984), 549–571.
6. Voiculescu D., *A non-commutative Weyl-von Neumann theorem*, Rev. Roum. Math. Pures et Appl. **21** (1976), 97–113.
7. Williams J. P., *Finite operators*, Proc. Amer. Math. Soc. **26** (1970), 129–136.

CHAPTER EDITOR'S NOTE

Domingo Herrero died on April 13, 1991, in his early fifties. We lost a leading operator-theorist and a remarkably joyous and loving person; Domingo enhanced the lives of all those who had the good fortune to come into contact with him.

5.7

MULTIPLICATIVE COMMUTATOR AND PRODUCT OF INVOLUTIONS

P. Y. WU

1. Multiplicative commutator. An operator T acting on a complex separable Hilbert space $\mathcal{H}$ is called a *multiplicative commutator* if $T = ABA^{-1}B^{-1}$ for some invertible operators A and B on $\mathcal{H}$. This is in contrast to another notion of commutator: the *additive* ones, operators of the form $AB - BA$. A characterization of this latter class of operators has long been known: for finite-dimensional $\mathcal{H}$, T is an additive commutator if and only if it has trace 0 [7]; for $\mathcal{H}$ infinite-dimensional, T is such a commutator if and only if it is not of the form $\lambda I + K$, where λ is a nonzero scalar and K is a compact operator [1]. However, the corresponding question for multiplicative commutators (on infinite-dimensional spaces) is unanswered to this day. A. Brown and C. Pearcy, after their successful settling of the question for additive commutators, launched an attack on the multiplicative ones and obtained some partial results [2]. Based on their findings, they proposed the following

CONJECTURE 1. *T is a multiplicative commutator if and only if T is invertible and not of the form $\lambda I + K$, where λ is a scalar with $|\lambda| \neq 1$ and K is compact.*

The necessity is easy to prove and already in [2]. It is also shown there that the sufficiency is true if T is normal or $T = \lambda + K$, where $|\lambda| = 1$ and K is of finite rank.

Note that, for finite-dimensional $\mathcal{H}$, multiplicative commutators are exactly operators with determinant 1 [8]. In [3], there are some results concerning unitary ones.

2. Product of involutions. An operator T is an *involution* if $T^2 = I$. One question of interest in this area is to determine which operator is expressible as the product of (finitely many) involutions. If $\mathcal{H}$ is finite-dimensional, then this is answered in [5]: T is such a product if and only if its determinant is ± 1, and, in this case, four involutions would suffice. For infinite-dimensional $\mathcal{H}$, the class of such products is much larger. In fact, it was shown in [6] that this class coincides with that of all invertible operators and, moreover, each such operator is the product of at most seven involutions. The question whether seven is optimal remains open. In a recent research, undertaken together with J.-H. Wang [9], we discovered a close connection between such products and multiplicative commutators. Using techniques similar to those in [2], we are able to prove the following:

(1) If $T = \lambda I + K$, where λ is a scalar and K is compact, and T is the product of four involutions, then $|\lambda| = 1$.
(2) A normal invertible operator is the product of four involutions if and only if it is not of the form $\lambda I + K$, where $|\lambda| \neq 1$ and K is compact.
(3) If $T = \lambda I + K$, where $|\lambda| = 1$ and K is of finite rank, then T is the product of four involutions.

Based on these observations, it seems quite plausible that the classes of products of four involutions and multiplicative commutators coincide. This we propose as

CONJECTURE 2. *T is the product of four involutions if and only if it is a multiplicative commutator.*

As for the minimal number of involutions required to form products for invertible operators, we are able to show, borrowing techniques from [4], that every invertible operator is the product of six involutions. This number can be further reduced to five for invertible normal operators. Thus, in general, we would expect the following to be true:

CONJECTURE 3. *Every invertible operator is the product of five involutions.*

Two remarks are in order. Firstly, if Conjectures 1 and 2 hold, then Conjecture 3 follows by an easy argument. Secondly, for nonseparable $\mathcal{H}$, we should replace compact operators in all the above formulations by operators in the (unique) maximal ideal of the algebra of operators on $\mathcal{H}$.

REFERENCES

1. Brown A., Pearcy C., *Structure of commutators of operators*, Ann. Math. **82** (1965), 112–127.
2. Brown A., Pearcy C., *Multiplicative commutators of operators*, Canad. J. Math. **18** (1966), 737–749.
3. Fan K., *Some remarks on commutators of matrices*, Arch. Math. **5** (1954), 102–107.
4. Fong C. K., Sourour A. R., *The group generated by unipotent operators*, Proc. Amer. Math. Soc. **97** (1986), 453–458.
5. Gustafson W. H., Halmos P. R., Radjavi H., *Products of involutions*, Linear Algebra Appl. **13** (1976), 157–162.
6. Radjavi H., *The group generated by involutions*, Proc. Roy. Irish Acad. **81A** (1981), 9–12.
7. Shoda K., *Einige Sätze über Matrizen*, Jap. J. Math. **13** (1936), 361–365.
8. Shoda K., *Über den Kommutator der Matrizen*, J. Math. Soc. Jap. **3** (1951), 78–81.
9. Wang J.-H., *Decomposition of operators into quadratic types*, Ph.D. dissertation, National Chiao Tung University, 1991.

DEPARTMENT OF APPLIED MATHEMATICS
NATIONAL CHIAO TUNG UNIVERSITY
HSINCHU, TAIWAN
REPUBLIC OF CHINA

5.8

INVARIANT SUBSPACES OF MULTIPLICATION BY THE INDEPENDENT VARIABLE ON A RIEMANN SURFACE

D. V. YAKUBOVICH

Let R be a compact, not necessarily connected Riemann surface with boundary ∂R and $\Pi = \Pi_R$ a projection of $\overline{R} = R \cup \partial R$ to $\mathbb{C}$ interpreted as an independent variable. The general question is to study the structure of invariant subspaces of the multiplication operator $M_R f = \Pi_R f$ acting in the Smirnov space $E^p(R)$, $1 \leqslant p < \infty$ (one can replace $E^p(R)$ with the Hardy class $H^p(R)$). The investigation of invariant subspaces of Toeplitz operators with good analytic and rational non-analytic symbols reduces to this problem [1, 2]; see also [3, 4]. The spectral multiplicity of M_R for a wide class of Riemann surfaces R is calculated in [1].

Put $\gamma = \Pi(\partial R)$. For $\lambda \in \mathbb{C} \backslash \gamma$, the valence $\kappa(\lambda)$ of the projection Π is defined as the number of pre-images of λ on R counted with their multiplicities. Consider the following "absence of holes" condition:

(H) Each bounded connected component Ω of $\mathbb{C} \backslash \gamma$ has a common boundary arc with a component Ω' of $\mathbb{C} \backslash \gamma$ such that $\kappa(\Omega') < \kappa(\Omega)$.

As shown in [2], if some natural requirements on the "regularity" of (R, Π) (sufficient smoothness of γ, uniform boundedness of κ, etc.) and the condition (H) are fulfilled, then the study of the lattice of invariant subspaces $\mathrm{Lat}(M_\Pi)$ reduces to the following special case of "two-sheeted" Riemann surface: $R_0 = \widetilde{A} \cup \widetilde{B}$, $\widetilde{A} \cap \widetilde{B} = \emptyset$, and the projection Π maps $\widetilde{A}$ and $\widetilde{B}$ homeomorphically onto two simply connected domains A, B with piecewise smooth boundaries intersecting in two points. We can identify $E^p(R_0)$ with $E^p(A) \oplus E^p(B)$.

Suppose $\mathcal{E} \in \mathrm{Lat}(M_{R_0})$ and V is a "good" subset of R_0. The subspace

$$\mathrm{tr}_V \mathcal{E} = \mathrm{clos}_{E^p(V)} \{ f|V \colon f \in \mathcal{E} \}$$

of $E^p(V)$ is called *the trace of* $\mathcal{E}$ *on* V. Obviously, $\mathrm{tr}_V \mathcal{E}$ is an invariant subspace of the multiplication operator M_V by $\Pi|V$ acting on $E^p(V)$. Put $\widetilde{C} = \Pi^{-1}(A \cap B)$. *The standard invariant subspace* $\mathrm{St}(\mathcal{E})$ ($\in \mathrm{Lat}(M_{R_0})$) *corresponding to* $\mathcal{E}$ is

$$\mathrm{St}(\mathcal{E}) = \{ f \in E^p(R_0) \colon f|V \in \mathrm{tr}_V \mathcal{E}, \quad V = \widetilde{A}, \widetilde{B}, \widetilde{C} \}.$$

Note that the invariant subspaces of $E^p(V)$ for $V = \widetilde{A}, \widetilde{B}, \widetilde{C}$ are described easily with the help of the Beurling–Lax–Halmos theorem. We have $\mathrm{St}(\mathcal{E}) \supset \mathcal{E}$.

THEOREM 1 [2]. *For every invariant subspace* $\mathcal{E}$ *of* $E^p(R_0)$ *there exists a measurable function* ρ *on* $(\partial B) \cap A$ *such that*

$$\mathcal{E} = \{ f \in \mathrm{St}(\mathcal{E}) \colon f\rho \in L^1_{\mathrm{weak}}((\partial B) \cap A) \}. \tag{1}$$

Here $L^1_{\text{weak}}(\Gamma) = \{\varphi \in L^{1/2}(\Gamma)\colon tm_1\{|\varphi| > t\} \to 0,\ t \to +\infty\}$ for an arc Γ, where m_1 is the arc length measure. It follows that every invariant subspace $\mathcal{E}$ in $\operatorname{Lat}(M_R)$ is determined by three inner functions $\Theta_{\widetilde{A}}, \Theta_{\widetilde{B}}, \Theta_{\widetilde{C}}$ arising in the description of the traces $\operatorname{tr}_{\widetilde{A}}\mathcal{E}, \operatorname{tr}_{\widetilde{B}}\mathcal{E}, \operatorname{tr}_{\widetilde{C}}\mathcal{E}$, and by a measurable function ρ. The conditions analogous to those in [5, Theorem 4] ensure $\mathcal{E} = \operatorname{St}(\mathcal{E})$. In particular, $\operatorname{St}(\mathcal{E}) = E^p(R_0) \implies \mathcal{E} = E^p(\mathcal{R}_0)$ (this fact is contained in [1]). In these cases we can put $\rho \equiv 1$ in (1). In general, however, the connection between $\Theta_{\widetilde{A}}, \Theta_{\widetilde{B}}, \Theta_{\widetilde{C}}$ and possible ρ remains mysterious.

PROBLEM 1. *Find a description of the lattice of invariant subspaces in $E^p(R_0)$ with a complete characterization of the functional parameters involved.*

We formulate also two weaker questions related to this problem. For any subspace $\mathcal{E}$ of M_R we can define the rank function $r_{\mathcal{E}}$ on $\mathbb{C}\backslash\gamma$ by $r_{\mathcal{E}}(\lambda) = \dim\operatorname{Ker}((M_R - \lambda|\mathcal{E})^*)$. It is easy to see that $r_{\mathcal{E}}$ is locally constant on $\mathbb{C}\backslash\gamma$ and $r \leqslant \kappa$.

PROBLEM 1′. *Let $\mathcal{E} \in \operatorname{Lat}(M_{R_0})$. Does the equality $r_{\mathcal{E}}(\lambda) = 2$, $\lambda \in A \cap B$, imply that $\mathcal{E} = \operatorname{St}(\mathcal{E})$?*

The examples of non-standard subspaces contained in [2] have the form $\mathcal{E} = \mathcal{E}_f \overset{\text{def}}{=} \operatorname{span}\{M^n_{R_0}f\colon n \geqslant 0\}$, where $f \in E^p(R_0)$ (so that $r_{\mathcal{E}}(\lambda) \equiv 1$ in $A \cap B$). They suggest the following

PROBLEM 1″. *Suppose $f \in E^2(R_0)$, $\partial A \cap \partial B = \{\xi, \xi'\}$ and the functions $f \circ (\Pi|\widetilde{A})^{-1}$, $f \circ (\Pi|\widetilde{B})^{-1}$ are analytic in ξ'. Are the invariant subspaces $\mathcal{E}$ of M_{R_0} satisfying $\operatorname{St}(\mathcal{E}_f) \supset \mathcal{E} \supset \mathcal{E}_f$ linearly ordered by inclusion?*

The case when the hole absence condition (H) is violated is in fact much more difficult. However, the invariant subspaces in E^2 of an annulus which can be viewed as a simplest Riemann surface with holes have been described completely by D. Hitt and D. Sarason [6, 7]. Using the technique of traces of subspaces, their result extends to the case of finitely connected domain with analytic inner boundaries [8]. For simplicity we formulate this result for a twice-connected domain $R = D \setminus \overline{V}$, where $\overline{V} \subset D$, and the domains D and V are simply connected. For an invariant subspace $\mathcal{E}$ of the multiplication operator $M_z f = zf$ on $E^2(R)$, we define a function $\chi_{\mathcal{E}} \in L^\infty(\partial V)$ from the equations $\mathcal{H} = \operatorname{clos}_{L^2(\partial V)}\{f|\partial V\colon f \in \mathcal{E}\} = \overline{\chi}_{\mathcal{E}} \cdot E^2(\partial V)$, $|\chi_{\mathcal{E}}| \equiv 1$ (we put $\chi_{\mathcal{E}} \equiv 0$ if $\mathcal{H} = L^2(\partial V)$). We denote by $\operatorname{GCD}(\mathcal{E})$ the great common divisor of inner parts of functions in an invariant subspace $\mathcal{E}$ of M_z in $E^2(R)$ (see [6]). The classification of invariant subspaces of M_z in $E^2(R)$ reduces to the study of those $\mathcal{E}$ in $\operatorname{Lat}(M_z)$ satisfying $\operatorname{GCD}(\mathcal{E}) = 1$. In fact, each invariant subspace $\mathcal{E}$ has a form $\mathcal{E} = \varphi \cdot \mathcal{E}'$, where $\varphi = \operatorname{GCD}(\mathcal{E})$, $\mathcal{E}' \in \operatorname{Lat}(M_z)$ and $\operatorname{GCD}(\mathcal{E}') = 1$. Let τ be a conformal mapping of V onto the unit disk.

THEOREM 2 [8]. *Suppose ∂V is an analytic curve, and $\mathcal{E}$ is an invariant subspace of M_z in $E^2(R)$ with $\operatorname{GCD}(\mathcal{E}) = 1$. If $\chi_{\mathcal{E}} \neq 0$, then there exist outer $g \in E^2(V)$, inner Θ in V and $m \in \mathbb{Z}$ with the property $|g|^2 = \operatorname{Re}(\tau\Theta v) + 1$ a.e. on ∂V for some $v \in E^{1/2}(V) \cap L^1_{\text{weak}}(\partial V)$ such that*

$$\mathcal{E} = \{x\colon \chi \cdot x|\partial V \in E^2(V),\ |xg^{-1}| \in L^2(\Gamma)\}, \tag{2}$$

where $\chi = \tau^m \Theta g/\bar{g}$, $|\chi| = 1$ a.e. on ∂V, and $\chi = \chi_{\mathcal{E}}$.

Conversely, for every g, Θ, m as above the formula (2) *defines a closed invariant subspace $\mathcal{E}$ such that* $\mathrm{GCD}(\mathcal{E}) = 1$ *and* $\chi_{\mathcal{E}} = \tau^m \Theta g/\bar{g}$.

If $\chi_{\mathcal{E}} \equiv 0$ *on* ∂V, *then* $\mathcal{E} = E^2(R)$.

We remark that the possibility that $\chi_{\mathcal{E}}$ may be zero on some components of the inner boundary of a multiply connected domain is missing in the formulations of Theorems 2 and 3 in [8]. In this case, all the conditions on the functions in the invariant subspace under consideration corresponding to these components must be removed.

The proof of this theorem relies heavily on the existence of an analytic continuation of τ to a neighbourhood of ∂V.

PROBLEM 2. *Does Theorem 2 hold true for non-analytic boundary ∂V, in particular, for finitely smooth ∂V?*

One can ask also similar questions about $E^p(R)$, $1 \leqslant p < \infty$. Partial results are contained in [8, 9]. Of course, an efficient description of invariant subspaces of M_R for general Riemann surfaces (R, Π) is desirable. Note that it is unclear how to calculate the spectral multiplicity of operators $M_R|\mathcal{E}$, where $\mathcal{E} \in \mathrm{Lat}(M_R)$.

REFERENCES

1. Solomyak B. M., Volberg A. L., *Multiplicity of analytic Toeplitz operators*, Operator Theory: Advances and Appl. **42** (1989), 87–192.
2. Yakubovich D. V., *Riemann surface models of Toeplitz operators*, Operator Theory: Advances and Appl. **42** (1989), 305–415.
3. Yakubovich D. V., *On the spectral theory of Toeplitz operators with smooth symbols*, Algebra i Analiz **3** (1991), no. 4, 208–226 (Russian); English transl. in St. Petersburg Math. J.
4. Yakubovich D. V., *Spectral multiplicity of a class of Toeplitz operators with smooth symbol*, Report No.6, Inst. Mittag–Leffler, 1990/1991.
5. Yakubovich D. V., *Multiplication operators on special Riemann surfaces as models of Toeplitz operators*, Dokl. Akad. Nauk SSSR **302** (1988), no. 5, 1068–1072; English translation in Soviet Math. Dokl. **38** (1989), no. 2, 400–404.
6. Hitt D., *Invariant subspaces of H^2 of an annulus*, Pacif. J. Math. **134** (1988), no. 1, 101–120.
7. Sarason D., *Nearly invariant subspaces of the backward shift*, Operator Theory: Advances and Appl. **35** (1988), 481–494.
8. Yakubovich D. V., *Invariant subspaces of multiplication by z in E^p of a multiply connected domain*, Zapiski nauchn. semin. LOMI **178** (1989), 166–183. (Russian)
9. Hitt D., *The structure of invariant subspaces of H^p of an annulus*, preprint, 1989.

DEPT. OF MATHEMATICS AND MECHANICS
ST.-PETERSBURG STATE UNIVERSITY
BIBLIOTECHNAYA PL. 2.
STARYI PETERHOF, ST.-PETERSBURG, 198904
RUSSIA

5.9
old

MAXIMAL NON-NEGATIVE INVARIANT SUBSPACES OF $\mathcal{J}$-DISSIPATIVE OPERATORS

T. YA. AZIZOV, I. S. IOHVIDOV

Let $\mathcal{H}$ be a $\mathcal{J}$-space (Krein space); i.e., a Hilbert space with an inner product (x, y) and indefinite $\mathcal{J}$-form $[x, y] = (\mathcal{J}x, y)$, $\mathcal{J} = \mathcal{J}^* = \mathcal{J}^{-1}$ (for more detailed information see, for instance, [1] or [2]). A subspace L is called *non-negative* if $[x, x] \geqslant 0$ for $x \in L$, and *maximal non-negative* if it is non-negative and has no proper non-negative extensions.

A linear operator $\mathcal{A}$ on $\mathcal{H}$ with a domain $D_{\mathcal{A}}$ is called *dissipative* (*$\mathcal{J}$-dissipative*) if $\mathrm{Im}(\mathcal{A}x, x) \geqslant 0$ ($\mathrm{Im}[\mathcal{A}x, x] \geqslant 0$) for all $x \in D_{\mathcal{A}}$. Such an $\mathcal{A}$ is called *maximal dissipative* (*maximal $\mathcal{J}$-dissipative*) if it has no proper dissipative ($\mathcal{J}$-dissipative) extensions.

PROBLEM 1. *Does there exist a maximal non-negative invariant subspace for any bounded $\mathcal{J}$-dissipative operator $\mathcal{A}$ with $D_{\mathcal{A}} = \mathcal{H}$?*

This problem has a positive solution if $\mathcal{A}$ is a *uniformly $\mathcal{J}$-dissipative* operator i.e. there is a constant $\gamma > 0$ such that $\mathrm{Im}[\mathcal{A}x, x] \geqslant \gamma\|x\|^2$. In that case $\sigma(\mathcal{A}) \cap \mathbb{R} = \emptyset$ and hence the Riesz projection generated by the set $\sigma(\mathcal{A}) \cap \mathbb{C}^+$ gives us the desired subspace. Note that $\mathcal{J}$-dissipativity of $\mathcal{A}$ implies the uniform dissipativity of $\mathcal{A}_\varepsilon = \mathcal{A} + i\varepsilon\mathcal{J}$ ($\varepsilon > 0$). As $\mathcal{A}_\varepsilon$ possesses a maximal non-negative invariant subspace, it is natural to use the "passage to the limit" for $\varepsilon \to 0$. Such a passage—M. G. Krein's method (see [2])—leads to a positive solution of Problem 1 if $(\mathrm{I} + \mathcal{J})\mathcal{A}(\mathrm{I} - \mathcal{J}) \in \mathfrak{S}_\infty$. In the general case Problem 1 has not yet been solved and therefore subclasses of operators for which it has a positive solution are being considered on the one hand, and on the other hand attempts are being made to construct counterexamples.

THEOREM. *If $\widehat{\mathcal{H}} = \mathcal{H} \oplus \mathcal{H}$ is a $\widehat{\mathcal{J}}$-space, $\mathcal{J} = \begin{pmatrix} 0 & \mathrm{I} \\ \mathrm{I} & 0 \end{pmatrix}$, then $\widehat{\mathcal{A}} = \begin{pmatrix} 0 & -\mathrm{I} \\ \mathcal{A} & 0 \end{pmatrix}$ is a continuous $\widehat{\mathcal{J}}$-dissipative operator if and only if $\mathcal{A}$ is a continuous dissipative operator in $\mathcal{H}$; in that case $\widehat{\mathcal{A}}$ has a maximal non-negative (with respect to the $\widehat{\mathcal{J}}$-form) invariant subspace.*

Proof. One verifies immediately that a maximal non-negative subspace $\mathcal{L}$ of $\widehat{\mathcal{H}}$ is invariant under $\widehat{\mathcal{A}}$ iff it is the graph of an operator $(-i\mathcal{K})$ where $\mathcal{K}$ is dissipative in $\mathcal{H}$ and $\mathcal{K}^2 = \mathcal{A}$ $\left(\mathcal{L} = \bigvee\{\langle x, -i\mathcal{K}x\rangle\}_{x\in\mathcal{H}}\right)$. Such an operator $\mathcal{K}$ does exist and is bounded, by the theorem of Matsaev–Palant [3]. □

Matsaev–Palant's result about the square root of dissipative operator was developed by H. Langer in [4]. There it was proved in particular that each maximal dissipative operator possesses a maximal dissipative square root. This result allows one to omit the requirement of the continuity of $\mathcal{A}$ in the above Theorem and replace it by the maximal dissipativity condition.

Problem 2. *Let* $\widehat{\mathcal{H}} = \mathcal{H} \oplus \mathcal{H}$, $\widehat{\mathcal{J}} = \begin{pmatrix} 0 & \mathcal{J} \\ \mathcal{J} & 0 \end{pmatrix}$, $\mathcal{J} = \mathcal{J}^* = \mathcal{J}^{-1}$, $\widehat{\mathcal{A}} = \begin{pmatrix} 0 & -\mathrm{I} \\ \mathcal{A} & 0 \end{pmatrix}$, *where* $\mathcal{A}$ *is a continuous* $\mathcal{J}$*-dissipative operator in* $\mathcal{H}$*. Does there exist a maximal (in* $\widehat{\mathcal{H}}$*) non-negative subspace invariant under the* $\widehat{\mathcal{J}}$*-dissipative operator* $\widehat{\mathcal{A}}$*?*

References

1. Bognar J., *Indefinite Inner Product Spaces*, Springer–Verlag, 1974.
2. Azizov T. Ya., Iohvidov I. S., *Linear operators on spaces with indefinite metric and their applications*, Itogi nauki i tekhniki (Surveys of science and technology), Series "Mathematical Analysis", vol. 17, VINITI, Moscow, 1979, pp. 105–207. (Russian)
3. Matsaev W. I., Palant Yu. A., *On powers of a bounded dissipative operator*, Ukrainskiĭ Matematicheskiĭ Zhournal (Ukrainian Mathematical Journal) **14** (1962), 329–337. (Russian)
4. Langer H., *Über die Wurzeln eines maximalen dissipativen Operators*, Acta Math. **XIII** (1962), no. 3–4, 415–424.

A/JA 9, Voronezh-68,
394068,
Russia

5.10
v.old

ARE MULTIPLICATION AND SHIFT UNIFORMLY ALGEBRAICALLY APPROXIMATE?

A. M. VERSHIK

0. New dimensions. A family $\mathcal{A} = \{A_\omega : \omega \in \Omega\}$ of bounded operators on Hilbert space $\mathcal{H}$ is called *uniformly algebraically approximable* or (briefly) *approximable* if for every positive ε and for every $\omega \in \Omega$ there exists an operator $A_{\omega,\varepsilon}$ such that

a) $\sup\limits_{\omega\in\Omega} \|A_{\omega,\varepsilon} - A_\omega\| < \varepsilon$;

b) the $*$-algebra (i.e. the algebra containing B^* together with B) spanned by $\{A_{\omega,\varepsilon}\}_{\omega\in\Omega}$ is finite-dimensional; it is denoted $\mathcal{A}_\varepsilon$†.

In particular, an operator A is called *approximable* if the family $\{A\}$ is approximable. In this case $\{\operatorname{Re} A, \operatorname{Im} A\}$ is approximable also. Given an approximable family $\mathcal{A}$ and $\varepsilon > 0$ let $\mathcal{A}_\varepsilon^0$ denote the algebra of the least dimension $\dim \mathcal{A}_\varepsilon^0$ among algebras satisfying a) and b). The function $\varepsilon \to \mathcal{H}(\varepsilon, \mathcal{A}) \stackrel{\text{def}}{=} \log_2 \dim \mathcal{A}_\varepsilon^0$ is called the *entropy growth* of $\mathcal{A}$.

1. The main problem. To obtain convenient criteria for a family of operators (in particular, for a single non-self-adjoint operator) to be approximable, and to develop a functional calculus for approximable families. Some concrete analytic problems are given in section 5.

2. Known approximable families. The first is $\{A\}$ with $A = A^*$. Indeed, let $A_\varepsilon = \sum\limits_{i=1}^{n} \lambda_i (P_{\lambda_i} - P_{\lambda_{i-1}})$ where $\{\lambda_i\}_{i=1}^n$ forms an ε-net for the spectrum of A and $\{P_\lambda\}$ is the spectral measure of A. In this case $\mathcal{H}(\varepsilon, A)$ coincides with the usual ε-entropy of $\operatorname{spec} A$ considered as a compact subset in $\mathbb{R}$.

Let $\mathcal{A} = \{A_1, \dots, A_n\}$ be a family of commuting self-adjoint operators. It is clearly approximable with $A_{i,\varepsilon}$ defined analogously. The entropy $\mathcal{H}(\varepsilon, \mathcal{A})$ is again the ε-entropy of the joint spectrum in $\mathbb{R}^n$.

The same holds for a finite family of commuting normal operators.

Next let $\mathcal{A}$ be a finite or compact family of compact operators. Then the operators A_ε can be chosen to have finite rank and therefore $\mathcal{A}$ is approximable.

Given an approximable family $\mathcal{A}'$ and a finite collection $B_1, \dots, B_n$ of compact operators consider the family $\mathcal{A} = \mathcal{A}' \bigcup \{B_1, \dots, B_n\}$. Then $\mathcal{A}$ is approximable. In particular, any operator with compact imaginary part is approximable.

Let $\mathcal{H} = \int_X \oplus \mathcal{H}_x \, dx$ and let $\dim \mathcal{H}_x = n$. Then $\mathcal{A} = \{A_i \stackrel{\text{def}}{=} \int_X \oplus A_i^x \, dx : i = 1, \dots, n\}$ is approximable.

†We do not require the identity in $\mathcal{A}_\varepsilon$ to be the identity operator on $\mathcal{H}$ in order to include consideration of compact operators. If the identity of $\mathcal{A}_\varepsilon$ is the identity operator I on $\mathcal{H}$ then $\mathcal{A}_\varepsilon$ does not contain compact operators and defines a decomposition $\mathcal{H} = \mathcal{H}_1^\varepsilon \otimes \mathcal{H}_2^\varepsilon$ with $\dim \mathcal{H}_2^\varepsilon < \infty$ and $A_{\omega,\varepsilon} = \{I_1 \otimes a_{\omega,\varepsilon} : a_{\omega,\varepsilon} \in \mathcal{L}(\mathcal{H}_2^\varepsilon),\ \omega \in \Omega\}$. In general it is convenient to consider all algebras in the Calkin algebra. (See [1] for definitions of the theory of C^*-algebras).

Consider $\mathcal{A} = \{P_1, P_2\}$, P_i being an orthogonal projection, $i = 1, 2$. This is a special case of the previous example because there exists a decomposition $\mathcal{H} = \int_X \oplus \mathbb{C}^2\, dx$ such that $P_i = \int_X \oplus P_i^x\, dx$ (see [2] for example).

The unilateral shift U is approximable. If $\mathcal{B}$ denotes the C^*-algebra generated by U then it contains the ideal $\mathcal{LC}(\mathcal{H})$ of all compact operators and $\mathcal{B}/\mathcal{LC}(\mathcal{H}) = C(\mathbb{T})$ (cf., e.g., [3]). If follows that U is approximable in the Calkin algebra.

It was actually proved in [4] that any finite family of commuting quasi-nilpotent operators is approximable (see [5]).

3. Known non-approximable families.

THEOREM. *If a family $\mathcal{A} = \{U_i, i = 1, \dots, n\}$ of unitary operators is approximable then the C^*-algebra generated by $\{U_i\}$ is amenable; in particular, if $\mathcal{A}$ is a group algebra then the group is amenable.*

See for example [6] for the proof.

COROLLARY. *If $\mathcal{A} = \{P_i\}_{i=1}^n$, $n > 2$ is a family of orthogonal projections in general position then $\mathcal{A}$ is not approximable.*

Indeed, set $U_i = 2P_i - I$. Then $U_i = U_i^* = U_i^{-1}$ and G is a free product of n copies of $\mathbb{Z}_2$ which cannot be amenable for $n > 2$.

Therefore a family of two (or more) unitary or self-adjoint operators picked at random cannot be approximable in general. This implies that a single randomly chosen non-self-adjoint operator is not approximable either. The property of approximability imposes some restrictions on the structure of invariant subspaces (see the footnote † to Section 1).

Consider a family $\mathcal{A}_n = \{U_1, \dots, U_n\}$ of partial isometries bound by the relation $\sum_{i=1}^{n} U_i U_i^* = I$, $n \geqslant 2$. Then $\mathcal{A}$ is not approximable [7] although the algebra generated by $\mathcal{A}$ is amenable [8].

Any algebra generated by an approximable family, being a subalgebra of an inductive limit of C^*-algebras of type I, is amenable as a C^*-algebra [7]. However, the class of such algebras is narrower than the class of all amenable algebras. If an approximable family generates a factor in $\mathcal{H}$ then it is clearly hyperfinite [6]. All that gives necessary conditions of approximability.

4. Justification of the problem. Many families of operators arising in the scope of a single analytic problem turn out to be approximable, apparently because the operators simultaneously considered in applications cannot be "too non-commutative" (see [9], [10], problems of the perturbation theory, of representations of some non-commutative groups, etc.). Besides, approximable families are the simplest non-commutative families after the finite-dimensional ones.

On the other hand, an approximable family admits a developed functional calculus based on the usual routine of standard matrix theory. Indeed, functions of non-commutative elements belonging to an approximable family can be defined as the uniform limits of corresponding functions of matrices. Therefore it looks plausible that a well-defined functional calculus, as well as symbols, various models and canonical forms, can be defined for such a family. This in turn can be applied to the study of lattices of invariant subspaces, etc. In particular, if A is an approximable non-self-adjoint operator whose spectrum contains at least two points then, apparently, it can be proved that A has a non-trivial invariant subspace.

It is known that the weak approximation, which holds for any finite family, is not sufficient to develop a substantial functional calculus for non-commuting operators. However, it is possible to consider other intermediate (between and uniform and weak) notions of approximation (see, for instance, the definition of pseudo-finite family in [6]).

5. More concrete problems. Our topic can be very clearly expressed by the following questions.

a) Let G be a locally compact abelian group with the dual group $\widehat{G}$. Given $g \in G$ and $\phi \in \widehat{G}$ consider operators $Uf(x) = f(gx)$ and $Vf = \phi f$ on $L^2_m(G)$. For example, for $G = \mathbb{Z}$ let

$$(Uf)_n = f_{n+1}, \qquad (Vf)_n = e^{i\alpha n} f_n, \quad \alpha \in \mathbb{T}, \qquad f = \{f_n\} \in \ell^2,$$

and for $G = \mathbb{T}$

$$Uf(e^{i\theta}) = f(e^{i(\theta+\alpha)}), \qquad Vf(e^{i\theta}) = e^{i\theta} f(e^{i\theta}), \qquad f \in L^2_m(\mathbb{T}),$$

and finally for $G = \mathbb{R}$

$$Uf(x) = f(x+r), \qquad Vf(x) = e^{itx} f(x), \qquad f \in L^2(\mathbb{R}),\ t, r \in \mathbb{R}\,.$$

PROBLEM. *Is the pair $\{U, V\}$ approximable?*

The answer to this question requires a detailed (useful for its own sake) investigation of the Hilbert space geometry of spectral subspaces of these operators. One of the approaches reduces the problem to the following. Consider a partition of $\mathbb{T} = \bigcup_{i=1}^n \ell_i$ by a finite number of arcs ℓ_i. Then $L^2(\mathbb{T}) = \sum_{i=1}^n \oplus L^2_{\ell_i}$.

Let $\mathcal{H} = \mathcal{H}_{\alpha,\varepsilon}, \varepsilon > 0$ be the subspace of $L^2(\mathbb{T})$ consisting of functions whose Fourier coefficients may differ from zero only for integers n satisfying $|\{n\alpha\}| < \varepsilon$. Here $\{x\}$ stands for the fractional part of x and α is irrational. *What is the mutual position of subspaces $\mathcal{H}_{\alpha,\varepsilon}$ and $L^2_{\ell_i}$ in $L^2_m(\mathbb{T})$, i.e., what are their stationary angles, the mutual products of the orthogonal projections, etc.?* Since U and V satisfy $VUV^{-1}U^{-1} = e^{i\alpha}I$ (Heisenberg equation), the above question can be reformulated as follows. *Is it possible to solve this equation approximately in matrices with any prescribed accuracy in the norm topology?*

The shift U can be replaced by a more general dynamical system with invariant measure (X, T, μ). Then $U_T f(x) = f(Tx)$ and $V_\phi f = \phi f$, $\phi \in L^\infty(X)$, $f \in L^2_\mu(X)$. Whether $\{U_T, V_\phi\}$ is approximable or not depends essentially on properties (and not only spectral ones) of the dynamical system. The author knows of no literature on the subject. Note that numerous approximation procedures existing in Ergodic Theory are useless here because it can be easily shown that the restriction of the uniform operator topology to the group of unitary operators generated by the dynamical system induces the discrete topology on the group.

Note also that if the answer is positive, some singular integral operators as well as the operators of Bishop-Halmos type [11] would turn out to be approximable which would lead to the direct proofs of the existence of invariant subspaces (see section 4).

b) Let A be a contraction on $\mathcal{H}$. *Are there convenient criteria for A to be approximable expressed in terms of its unitary dilation or characteristic function?*

c) Let

$$Af(x) = \int_X K(x,y)f(y)\,d\mu(y), \qquad f \in L^2_\mu(X).$$

Find approximability criteria in terms of K.
Non-negative kernels $K \geqslant 0$ are especially interesting.

d) *For what countable solvable groups G of rank 2 does the regular unitary representation of G in $\ell^2(G)$ generate approximable families? For what general locally compact groups does this hold?*

REFERENCES‡

1. Dixmier J., *Les C^*-algèbres et leurs représentations*, Gauthier-Villard, Paris, 1969.
2. Halmos P., *Two subspaces*, Trans. Amer. Math. Soc. **144** (1969), 381–389.
3. Coburn L., *C^*-algebras, generated by semigroups of isometries*, Trans. Amer. Math. Soc. **137** (1969), 211–217.
4. Apostol C., *On the norm-closure of nilpotents. III*, Rev. Roum. Math. Pures Appl. **21** (1976), no. 2, 143–153.
5. Apostol C., Foiaş C., Voiculescu D., *On strongly reductive algebras*, Rev. Roum. Math. Pures Appl. **21** (1976), no. 6, 611–633.
6. Vershik A. M., *Countable groups which are close to finite groups*, Appendix to the Russian translation of Greenleaf's "Invariant means on topological groups", Moscow, "Mir", 1973 (Russian);Vershik A. M., *Amenability and approximation of infinite groups*, Selecta Math. Sov. **2** (1982), no. 4, 311–330.
7. Rosenberg J., *Amenability of cross products of C^*-algebras*, Commun. Math. Phys. **57** (1977), no. 2, 187–191.
8. Arzumanyan V. A., Vershik A. M., *Factor-representations of a crossed product of a commutative C^*-algebra and a semi-group of its endomorphisms*, Doklady AN SSSR **238** (1978), no. 3, 511–516. (Russian)
9. Sz-Nagy B., Foiaş C., *Harmonic Analysis of Operators on Hilbert Space*, Amsterdam-Budapest, 1970.
10. Gohberg I. Tz., Kreĭn M. G., *Theory of Volterra Operators on Hilbert Space and its Applications*, Nauka, Moscow, 1967. (Russian)
11. Davie A., *Invariant subspaces for Bichop's operator*, Bull. London Math. Soc. (1974), no. 6, 343–348.

ST. PETERSBURG BRANCH OF
STEKLOV MATHEMATICAL INSTITUTE
FONTANKA 27
ST. PETERSBURG, 191011
RUSSIA

‡M. I. Zaharevich turned my attention to [4] and A. A. Lodkin to [7].

Commentary by the Author (1984)

During recent years, considerable progress in the field discussed in this paper has been made, and new problems have arisen. We list the most important facts.

A C^*-algebra will be called an AF-algebra if it is generated by an inductive limit of finite-dimensional C^*-algebras. C^*-subalgebras of AF-algebras will be called AFI-algebras. A family of operators generating an AFI-algebra is called approximable. The problem was to find conditions of approximability for a family of operators or one (non-self-adjoint) operator and to give quantitative characteristics of the corresponding AF-algebras, etc.

1. In [12] a positive answer to Question a), Section 5 was actually given. Namely, the approximability problem is solved for the pair of unitaries U, V

$$(Uf)(\zeta) = f(\zeta\alpha), \qquad (Vf)(\zeta) = \zeta f(\zeta),$$

$f \in L^2(\mathbb{T})$, ζ, $\alpha \in \mathbb{T}$. This is the simplest of the non-trivial cases. In [12] the authors made use of the fact that these operators are the only (up to equivalence) solutions of the Heisenberg equation: $UVU^{-1}V^{-1} = \alpha I$.

2. In [13] and [14] the approximability of an arbitrary dynamical system (i.e., of the pair (U_T, M_ϕ), where $(U_T f)(x) = f(Tx)$, $f \in L^2(X, \mu)$, T being an automorphism of (X, μ), $M_\phi f = \phi f$, $\phi \in L^\infty$) is proved. The result is based on a new approximation technique developed for the purposes of ergodic theory ("adic realization" of automorphisms, Markov compacta). Later in [15] conditions on a topological dynamical system (i.e., on a homeomorphism of a compactum) were found under which the skew product $C(X)\widetilde{\otimes}\ell^1(\mathbb{Z})$ is an AFI-algebra.

3. These results in turn allowed the description in some cases of an important algebraic invariant of algebras generated by dynamical systems, the K-functor (cf. [12, 16, 17]). This makes it possible to apply K-theory in ergodic theory.

Nevertheless, we still have neither general approximability criteria, nor complete information on approximable non-self-adjoint operators. Since many important families turned out to be approximable, the questions on construction of functional calculus, estimates of the norms of powers, resolvents etc. are of great importance. Let us mention the following.

Some concrete questions.

A. *Is the group algebra of a discrete amenable algebra approximable?*

B. *Is true that an arbitrary approximable operator has a non-trivial invariant subspace?*

C. *What might the K-functor (as an ordered group) of an AFI-algebra look like?*

D. *How are the properties of a dynamical system related to the entropy growth (i.e. the growth of dimensions of finite-dimensional subalgebras of an AF-algebra which contains the algebra generated by the dynamical system)?*

References

12. Pimsner M., Voiculescu D., *Imbedding the irrational rotation C^*-algebras into an AF-algebra*, J. Oper. Theory **4** (1980), 201–210.

13. Vershik A. M., *Uniform algebraic approximation of shift and multiplication operators*, Dokl. Akad. Nauk SSSR **259** (1981), no. 3, 526–529 (Russian); English transl. in Soviet Math. Dokl. **24** (1981), no. 1, 97–100.
14. Vershik A. M., *A theorem on Markov periodic approximation in ergodic theory*, Zapiski nauchnych seminarov LOMI **115** (1982), 72–82. (Russian)
15. Pimsner M., *Imbedding the compact dynamical system*, preprint No. 44, INCREST, 1982.
16. Connes A., *An analogue of the Thom isomorphism for crossed products of a C^*-algebra by an action of* $\mathbb{R}$, Adv. in Math. **39** (1981), 31–55.
17. Effros E. G., *Dimensions and C^*-algebras*, C.B.M.S. Region Conf. Series no. 46, AMS, Providence, 1981.

5.11
old

A PROBLEM ON EXTREMAL SIMILARITIES

DOUGLAS N. CLARK

Let T be a Hilbert space operator which is similar to an isometry S:

$$T = L^{-1}SL. \tag{1}$$

Two interesting quantities associated with T are

$$k(T) = \inf\{ k : \|p(T)\| \leqslant k\|p\|_\infty \},$$

where $\| \|_\infty$ is the sup over $\mathbb{D}$ of the polynomial p, and

$$d(T) = \inf\{ \|L\|\|L^{-1}\| : (1) \text{ holds} \}.$$

The quantity $k(T)$ is called the k-norm of T and $d(T)$ is related to the distortion coefficient; Holbrook [3].

From (1), it follows that, for any polynomial p,

$$\|p(T)\| \leqslant \|L\|\|L^{-1}\|\|p\|_\infty,$$

and therefore

$$k(T) \leqslant d(T). \tag{2}$$

Since estimates on $k(T)$ yield information about functional calculus and spectral sets, this gives one reason why $d(T)$ is interesting. Another reason is the frequent occurrence of the quantity $\|L\|\|L^{-1}\|$ (see, for example, [2, p. 248]).

One way to try to compute $d(T)$ is to characterize L satisfying (1) and

$$\|L\|\|L^{-1}\| = d(T). \tag{3}$$

Suppose L satisfies (1) *and*

$$\|Lx\| = \lim_{n\to\infty} \|T^n x\| \tag{4}$$

for all x; then is L a similarity satisfying (3)*?*

The answer is "yes" in two (extreme) cases: that in which $\|L\| = 1$ and that in which $\|L^{-1}\| = 1$. In fact, in the latter case $\|L\| = d(T) = k(T)$, by (2) and by

$$\begin{aligned} k(T)\|x\| &\geqslant \|T^n x\| \geqslant \|Lx\| - \varepsilon \geqslant \|L\|\|x\| - 2\varepsilon \\ &\geqslant \|L\|\|L^{-1}\|\|x\| - 2\varepsilon \geqslant d(T)\|x\| - 2\varepsilon, \end{aligned}$$

if n and x are chosen correctly.

If L satisfies (1), (4) and $\|L\| \leqslant 1$, the proof of (3) is similar and uses the fact that, in this case

$$d(T) = 1/\left[\inf_x \lim_{n\to\infty} \|T^n x\|\right].$$

See [1] and [4].

The inequality $\|L\| \leqslant 1$ holds, in particular, when T is a contraction. In that case, strict inequality holds in (2).

In [1], this author studied $k(T)$ and $d(T)$ in connection with some recent results on similarity of Toeplitz operators. One result was that, in most cases, the similarity satisfying (1) and (4) could be computed explicitly.

REFERENCES

1. Clark D. N., *Toeplitz operators and k-spectral sets*, Indiana Univ. Math. J. **33** (1984), 127–141.
2. Cowen M. J., Douglas R. G., *Complex geometry and operator theory*, Acta Math. **141** (1978), 187–261.
3. Holbrook J. A. R., *Distortion coefficients for cryptocontractions*, Linear Algebra Appl. **18** (1977), 229–256.
4. Sz.-Nagy B., Foiaş C., *On contractions similar to isometries and Toeplitz operators*, Ann. Acad. Sci. Fenn. Ser. A.I. **2** (1976), 553–564.

DEPARTMENT OF MATHEMATICS
UNIVERSITY OF GEORGIA
ATHENS, GEORGIA 30602
U.S.A.

5.12
v.old

ESTIMATES OF OPERATOR POLYNOMIALS ON THE SCHATTEN – VON NEUMANN CLASSES

V. V. PELLER

Precise estimates of functions of operators is an essential part of general spectral theory of operators. In the case of Hilbert space one of the best known and most important inequalities of this type is von Neumann's inequality (cf. [1]):

$$\|\phi(T)\| \leqslant \max\{\,|\phi(\zeta)| : \ |\zeta| \leqslant 1\,\}$$

for any contraction T (i.e., $\|T\| \leqslant 1$) on a Hilbert space and for any complex polynomial ϕ ($\mathcal{P}_A$ is the set of such polynomials). For a Banach space X an explicit calculation of the norm

$$|\!|\!|\phi|\!|\!|_X \stackrel{\text{def}}{=} \sup\{\,\|\phi(T)\| : \ T \text{ is a contraction on } X\,\}, \qquad \phi \in \mathcal{P}_A,$$

would be an analogue of von Neumann's inequality.

In 1966 V. I. Matsaev conjectured that for infinite-dimensional L^p spaces, $|\!|\!|\cdot|\!|\!|_{L_p}$ coincides with the p-multiplier norm. This means that for any contraction T on L^p, $\|\phi(T)\|_{L^p} \leqslant |\phi|_p \stackrel{\text{def}}{=} \|\phi(S)\|_{\ell^p}$, where S is the shift operator on ℓ^p: $S(\lambda_0, \lambda_1, \dots) = (0, \lambda_0, \lambda_1, \dots)$. This inequality was proved in [2] for absolute contractions T (i.e. $\|T\|_{L^1} \leqslant 1$, $\|T\|_{L^\infty} \leqslant 1$) and in [3], [5] (independently) for operators T having a contractive majorant (for dominated contractions); i.e. for such T that there exists a positive contraction $\widetilde{T}$ on L^p satisfying $|Tf| \leqslant \widetilde{T}|f|$ a.e. for $f \in L^p$ (see also the survey article [4]).

Let $\mathfrak{S}_p = \mathfrak{S}_p(\mathcal{H})$ be the space of compact operators a on an infinite dimensional Hilbert space $\mathcal{H}$ with $\|a\|_{\mathfrak{S}_p} \stackrel{\text{def}}{=} \left(\operatorname{trace}(a^*a)^{p/2}\right)^{1/p} < \infty$. The dual of $\mathfrak{S}_p$, $1 < p < \infty$, can be identified with $\mathfrak{S}_{p'}$ with respect to the duality $(a, b) = \operatorname{trace} ab$, $a \in \mathfrak{S}_p$, $b \in \mathfrak{S}_{p'}$. We are interested in the $\mathfrak{S}_p$-version of von Neumann's inequality. Let $\mathcal{K} = \ell^2(\mathcal{H})$ and s be the shift operator on K $\left(s(x_0, x_1, \dots) = (0, x_0, x_1, \dots),\ x_i \in \mathcal{H}\right)$, and s^* be the adjoint operator. The operator S on $\mathfrak{S}_p(\mathcal{K})$ is defined by $Sa = sas^*$, $a \in \mathfrak{S}_p(\mathcal{K})$. This operator seems to play a role similar to that of the shift on ℓ^p. Let us introduce the notation $|\phi|_{\mathfrak{S}_p} \stackrel{\text{def}}{=} \|\phi(S)\|_{\mathfrak{S}_p}$.

CONJECTURE 1. $|\!|\!|\phi|\!|\!|_{\mathfrak{S}_p} = |\phi|_{\mathfrak{S}_p}$. *In other words for any contraction T on $\mathfrak{S}_p$*

$$\|\phi(T)\|_{\mathfrak{S}_p} \leqslant |\phi|_{\mathfrak{S}_p}, \qquad \phi \in \mathcal{P}_A. \tag{1}$$

The conjecture is true for $p = 1, 2, \infty$.

PROPOSITION 1. *Inequality (1) holds for isometries (not necessarily invertible) on the space $\mathfrak{S}_p$.*

Proof. This uses an idea due to A. K. Kitover. Consider an operator $W\colon \mathfrak{S}_p(\mathcal{H}) \to \mathfrak{S}_p(\mathcal{K})$ defined by $Wa = (a, rTa, r^2T^2a, \dots)$, where $0 < r < 1$. Then $Wa \in \mathfrak{S}_p(\mathcal{K})$ for any $a \in \mathfrak{S}_p(\mathcal{H})$. Put $S^*a = s^*as$. It is easy to see that $\phi(S^*)W = W\phi(rT)$, $\phi \in \mathcal{P}_A$, so

$$\|W\phi(rT)a\|_{\mathfrak{S}_p} \leqslant \|\phi(S^*)\|_{\mathfrak{S}_p}\|Wa\|_{\mathfrak{S}_p},$$

$$\|Wa\|^p_{\mathfrak{S}_p} = \sum_{n\geqslant 0} r^{np}\|T^na\|^p_{\mathfrak{S}_p} = \|a\|^p_{\mathfrak{S}_p}\sum_{n\geqslant 0} r^{np},$$

$$\|W\phi(rT)a\|^p_{\mathfrak{S}_p} = \|\phi(rT)a\|^p_{\mathfrak{S}_p}\sum_{n\geqslant 0} r^{np}.$$

Therefore $\|\phi(r(T))\|_{\mathfrak{S}_p} \leqslant \|\phi(S^*)\|_{\mathfrak{S}_p} = \|\phi(S)\|_{\mathfrak{S}_{p'}} = |\phi|_{\mathfrak{S}_{p'}}$. Making $r \to 1$ we obtain $\|\phi(T)\|_{\mathfrak{S}_p} \leqslant |\phi|_{\mathfrak{S}_{p'}}$. It is not difficult to show that $|\phi|_{\mathfrak{S}_p} = |\phi|_{\mathfrak{S}_{p'}}$. □

CONJECTURE 2. $|\phi|_{\mathfrak{S}_p} = |\phi|_p$, $\phi \in \mathcal{P}_A$.

Let us show that $|\phi|_{\mathfrak{S}_p} \geqslant |\phi|_p$, $\phi \in \mathcal{P}_A$. Let $a = \sum_{j\geqslant 0} \lambda_j(\cdot, S^je)S^je \in \mathfrak{S}_p(\mathcal{K})$, where $e = (x, 0, 0, \dots) \in \mathcal{K}$. Then $\|\phi(S)a\|_{\mathfrak{S}_p} = \|\phi(S)(\lambda_0, \lambda_1, \dots)\|_{\ell^p}$, so $|\phi|_{\mathfrak{S}_p} \geqslant |\phi|_p$. □

Perhaps it is possible to prove (1) for certain classes of contractions T (or even for any contraction T) applying one of the following two methods used by the author for L^p-spaces.

DEFINITION.

1. A net of operators $\{T_\alpha\}$ on a Banach space X is said to converge to an operator T in the *pw*-topology if $\lim_\alpha (T^n_\alpha x, y) = (Tx, y)$, $x \in X$, $y \in X^*$, $n \geqslant 0$.
2. Let J be an isometric imbedding of $\mathfrak{S}_p$ into $\mathfrak{S}_p$, P be the norm one projection from $\mathfrak{S}_p$ onto JP (it does exist [6]) and T be an operator on $\mathfrak{S}_p$. An operator U on $\mathfrak{S}_p$ is called a dilation of T if $PU^nJ = JT^n$, $n \geqslant 0$.

In view of Proposition 1, an operator T on $\mathfrak{S}_p$ has to satisfy (1) if it can be approximated by isometries in the *pw*-topology or it has an isometric dilation on $\mathfrak{S}_p$. Thus one should describe operators on $\mathfrak{S}_p$, having an isometric dilation on $\mathfrak{S}_p$, and the closure of the set of isometries on $\mathfrak{S}_p$ in the *pw*-topology. For this perhaps it would be useful to apply a known description of the set of $\mathfrak{S}_p$-isometries U (cf. [6]). It is known that each contraction on a Hilbert space has a unitary dilation on a Hilbert space [1] and can be approximated by unitary operators in the *pq*-topology (cf. [4]). The set of operators on $L^p[0,1]$, $p \neq 2$, having a unitary dilation on an L^p-space coincides with the closure in the *pw*-topology of the set of unitary operators and coincides with the set of operators having a contractive majorant (cf. [3], [4]). (Earlier for positive contractions the existence of unitary dilations was established in [7]).

QUESTIONS. *Is it true that*

1) *Any $\mathfrak{S}_p$-contraction has an isometric dilation?*
2) *Any absolute $\mathfrak{S}_p$-contraction (i.e. contraction on $\mathfrak{S}_1$ and on $\mathfrak{S}_\infty$) has an isometric dilation?*

3) pw-clos U *coincides with the set of all contractions on* $\mathfrak{S}_p$?
4) pw-clos U *contains the set of absolute contractions on* $\mathfrak{S}_p$?

The affirmative answer to 1) and 3) would imply the validity of Conjecture 1. If Conjecture 2 is also valid then this would imply the validity of V. I. Matsaev's conjecture because ℓ^p can be isometrically imbedded into $\mathfrak{S}_p$ in such a way that there exists a contractive projection onto its image. In conclusion let us indicate a class of operators satisfying (1).

PROPOSITION 2. *Let* a, b *be contractions on* $\mathcal{H}$, $Tc = acb$. *Then the operator* T *on* $\mathfrak{S}_p$ *has an isometric dilation and can be approximated by isometries.*

This follows from the fact that a and b have a unitary dilation on a Hilbert space and can be pw-approximated by unitary operators on $\mathcal{H}$.

REFERENCES

1. Sz.-Nagy B., Foiaş C., *Harmonic Analysis of Operators on Hilbert Space*, North Holland-Akademia Kiado, Amsterdam-Budapest, 1970.
2. Peller V. V., *Analog of J. von Neumann's inequality for L^p space*, Dokl. Akad. Nauk SSSR **231** (1976), no. 3, 539–542. (Russian)
3. Peller V. V., *L'inégalité de von Neumann, la dilation isométrique et l'approximation par isométries dans L^p*, C. R. Acad. Sci. Paris **287** (1978), no. 5 A, 311–314.
4. Peller V. V., *Analog of von Neumann's inequality, isometric dilation of contractions and approximation by isometries in the space of measurable functions*, Trudy MIAN **155** (1981), 103–150. (Russian)
5. Coifman R. R., Rochberg R., Weiss G., *Applications of transference: The L^p version of von Neumann's inequality and the Littlewood-Paley-Stein theory*, Proc. Conf. Math. Res. Inst. Oberwolfach, Intern. ser. Numer. Math., vol. 40, Birkhäuser, Basel, 1978, pp. 53–63.
6. Arazy J., Friedman J., *The isometries of $C_p^{n,m}$ into C_p*, Isr. J. Math. **26** (1977), no. 2, 151–165.
7. Akcoglu M. A., Sucheston L., *Dilations of positive contractions on L^p-spaces*, Canad. Math. Bull. **20** (1977), no. 3, 285–292.

STEKLOV MATHEMATICAL INSTITUTE
ST. PETERSBURG BRANCH
FONTANKA 27
ST. PETERSBURG, 191011
RUSSIA

AND

DEPARTMENT OF MATHEMATICS
KANSAS STATE UNIVERSITY
MANHATTAN
KANSAS, 66506
USA

5.13
old

A QUESTION IN CONNECTION WITH MATSAEV'S CONJECTURE

A. K. KITOVER

This problem is closely related to the preceding one [1], where definitions of all notions used here can be found. I propose to consider Matsaev's conjecture for the operator T on the 2-dimensional space ℓ_2^p, $1 < p < 2$, defined by the matrix

$$\frac{1}{\sqrt[p]{2}}\begin{pmatrix} 1 & -1 \\ 1 & 1 \end{pmatrix}$$

in the standard basis of ℓ_2^p. The contraction T is of interest because it has some resemblance with unitary operators on 2-dimensional Hilbert space and, as is well-known, the abundance of unitary operators plays a decisive role in the proof of von Neumann's inequality. Moreover T has no contractive majorant and so the results of Akcoglu and Peller cannot be applied to it. Thus it seems plausible that the validity of Matsaev's conjecture in the general case depends essentially on the answer to the following

QUESTION. *Is Matsaev's conjecture true for T?*

REFERENCE

1. Peller V. V., *Estimates of operator polynomials on the Schatten–von Neumann classes*, This "Collection", Problem 5.12.

2395, 43D AVENUE, APT 1,
SAN FRANCISCO,
USA

5.14
old

ON THE CONNECTION BETWEEN THE INDICES OF AN OPERATOR MATRIX AND ITS DETERMINANT

I. A. FEL'DMAN, A. S. MARKUS

Let $\mathcal{H}$ be a Hilbert space, and $\mathcal{L}(\mathcal{H})$ be the algebra of all linear bounded operators in $\mathcal{H}$. An operator $A \in \mathcal{L}(\mathcal{H})$ is called a Fredholm operator if $\dim \operatorname{Ker} A < \infty$ and $\dim(\operatorname{Ran} A)^{\perp} < \infty$. The number $\operatorname{ind} A = \dim \operatorname{Ker} A - \dim(\operatorname{Ran} A)^{\perp}$ is called the index of A.

If $\mathcal{H}^n$ is the orthogonal sum of n copies of the space $\mathcal{H}$, then any operator $\mathcal{A} \in \mathcal{L}(\mathcal{H}^n)$ can be represented in the form of an operator matrix $\{A_{jk}\}_{j,k=1}^{n}$ $(A_{jk} \in \mathcal{L}(\mathcal{H}))$.

Set $[A, B] = AB - BA$. Let $\mathcal{A} = \{A_{jk}\}$ and suppose all commutators $[A_{jk}, A_{j'k'}]$ to be compact. Define the determinant $\det \mathcal{A}$ in the usual way. The order of the factors A_{jk} in each term is of no importance in this connection, since various possible results differ from each other by compact summand. N. Ja. Krupnik showed ([1], see also [2], p. 195) that $\mathcal{A}$ is a Fredholm operator if and only if $\det \mathcal{A}$ is. On the other hand it is known that under these conditions the equality

$$\operatorname{ind} \mathcal{A} = \operatorname{ind} \det \mathcal{A} \tag{1}$$

does not hold in general (see the example below).

In [3] it was stated that the equality (1) holds if the condition of compactness of commutators is replaced by the condition of their nuclearity.

The question of preciseness of conditions $[A_{jk}, A_{j'k'}] \in \mathfrak{S}_1$ arises naturally.

CONJECTURE 1. *Let $\mathfrak{S}$ be any symmetrically-normed ideal* (see [4]) *of the algebra $\mathcal{L}(\mathcal{H})$, different from $\mathfrak{S}_1$. There exists a Fredholm operator $\mathcal{A} = \{A_{jk}\}$, such that $[A_{jk}, A_{j'k'}] \in \mathfrak{S}$ but* (1) *does not hold.*

The weaker conjecture given below is also of some interest.

CONJECTURE 2. *For any $p > 1$ there exists a Fredholm operator $\mathcal{A} = \{A_{jk}\}$ such that $[A_{jk}, A_{j'k'}] \in \mathfrak{S}_p$ but* (1) *does not hold.*

Note that in the example below Conjecture 2 is confirmed only for $p > 2$.

EXAMPLE. Let $\mathcal{H} = L^2(S^2)$ where S^2 is the two-dimensional sphere. There exist singular integral operators $A_{jk} \in \mathcal{L}(\mathcal{H})$ such that

$$\mathcal{A} = \begin{pmatrix} A_{11} & A_{12} \\ A_{21} & A_{22} \end{pmatrix}$$

is a Fredholm operator and $\operatorname{ind} \mathcal{A} = 1$ ([5], Ch. XIV, §4). As $\operatorname{ind} \det \mathcal{A} = 0$ ([5], Ch. XIII, Theorem 3.2) equality (1) does not hold. It can be assumed that the symbols of the operators A_{jk} are infinitely smooth (for $\xi \neq 0$) and therefore the commutators $[A_{jk}, A_{j'k'}]$

map $L^2(S^2)$ into $W_1^2(S^2)$ ([6], Theorem 3). Hence $s_m([A_{jk}, A_{j'k'}]) = O(m^{-1/2})$ (see e.g. [7]).

We note in conclusion that some conditions sufficient for the validity of (1) have been found in [8–10]. In these papers, as well as in [1], [3], operators on Banach spaces are considered.

REFERENCES

1. Krupnik N. Ya., *On normal solvability and index of singular integral equations*, Uchyonye Zapiski Kishinevskogo universiteta (Notices of the University of Kishinev) **82** (1965), 3–7. (Russian)
2. Gohberg I. Tz., Feldman I. A., *Convolution Equation and its Solution by Projection Methods*, "Nauka", Moscow, 1971. (Russian)
3. Markus A. S., Feldman I. A., *On the index of operator matrix*, Functional Analysis and Its Applications **11** (1977), no. 2, 83–84. (Russian)
4. Gohberg I. Tz., Krein M. G., *Introduction to the Theory of Non-self-adjoint Operators on Hilbert Space*, "Nauka", Moscow, 1965. (Russian)
5. Michlin S. G., Prössdorf S., *Singuläre Integral-operatoren*, Akademie-Verlag, Berlin, 1980.
6. Seeley R.T., *Singular integrals on compact manifolds*, Amer. J. Math. **81** (1959), 658–690.
7. Paraska V. I., *On asymptotic behaviour of eigenvalues and singular numbers of linear operators which increase the order of smoothness*, Matematičeskiĭ Sbornik **68 (110)** (1965), 623–631. (Russian)
8. Krupnik N. Ya., *Certain general problems of the theory of one-dimensional singular operators with matrix coefficients*, Non-self-adjoint Operators, Publ. House Shtiintsa, Kishinev, 1976, pp. 91–112. (Russian)
9. Krupnik N. Ya., *Conditions of existence of a n-symbol and a sufficient set of n-dimensional representations of a Banach algebra*, Linear Operators, Publ. House Shtiintsa, Kishinev, 1980, pp. 84–97. (Russian)
10. Vasilevskiĭ N. L., Truhilio R., *On theory of OP-operators in operator matrix algebras*, Linear Operators, Shtiintsa, Kishinev, 1980, pp. 3–15. (Russian)

DEPT. OF MATH. AND COMP. SCI.,
BEN GURION UNIV. OF NEGEV,
P.O. BOX 653, BEER-SHEVA, 84105,
ISRAEL

DEPT. OF MATH. AND COMP. SCI.,
BEN GURION UNIV. OF NEGEV,
P.O. BOX 653, BEER-SHEVA, 84105,
ISRAEL

5.15
old

SOME PROBLEMS ON COMPACT OPERATORS WITH POWER-LIKE BEHAVIOUR OF SINGULAR NUMBERS

M. SH. BIRMAN, M. Z. SOLOMYAK

Classes of compact operators with power-like behaviour of eigenvalues and singular numbers arise quite naturally in studying spectral asymptotics for differential and pseudodifferential operators. Presented are three problems related to the theory of such classes.

Let $\mathcal{B}$ be the algebra of all bounded operators on a Hilbert space $\mathcal{H}$. Given A in the ideal $\mathcal{C}$ of all compact operators in $\mathcal{B}$ define $s_n(A)$, $n = 1, 2, \ldots$, the singular numbers of A. For $0 < p < \infty$ let

$$\Sigma_p = \{ A \in \mathcal{C} : \ \|A\|_p \stackrel{\text{def}}{=} \sup_n \left(n^{1/p} \cdot s_n(A)\right) < \infty \},$$
$$\Sigma_p^0 = \{ A \in \Sigma_p : \ s_n(A) = o(n^{-1/p}) \}.$$

See [1–4] for details concerning Σ_p-spaces.

While studying spectral asymptotics the main interest is focused not on the spaces Σ_p, Σ_p^0 themselves, but on the quotient spaces

$$\sigma_p = \Sigma_p / \Sigma_p^0.$$

The spaces σ_p, $0 < p < \infty$ (for details see [5]) are complete and non-separable with respect to the quasi-norm $|a|_p = \limsup\{n^{1/p} s_n(A)\}$, $a = A + \Sigma_p^0$. The natural limit case of σ_p-spaces is the Calkin algebra $\sigma_\infty = \mathcal{B}/\mathcal{C}$.

The multiplication of operators induces the multiplication of elements $a \in \sigma_p$, $b \in \sigma_q$, $0 < p, q \leqslant \infty$. The product belongs to the class σ_r, $r^{-1} = p^{-1} + q^{-1}$. Taking adjoints of operators induces the involution $a \mapsto a^*$ in σ_p-spaces. So one can consider commuting classes, self-adjoint classes, normal classes, etc.

PROBLEM 1. *Let $a \in \sigma_p$, $a^* a = a a^*$. Is there a normal operator in the class a?*

It is known (see [6]), that the answer is negative if $p = \infty$. This is due to the fact that in the σ_∞-space there is the Index, i.e., a nontrivial homomorphism of the group of invertible elements of the algebra σ_∞ onto the group $\mathbb{Z}$, as well as to the fact that the spectrum of an element $a \in \sigma_\infty$ can separate the complex plane. These two circumstances do not occur if $p < \infty$.

An analogous question on self-adjoint classes has an affirmative answer (and it is trivial): if $a \in \sigma_p$, $a^* = a$, then for an arbitrary $A \in a$ the operator $\frac{1}{2}(A + A^*)$ is self-adjoint and belongs to the class a.

Closely related to PROBLEM 1 are the following two problems.

PROBLEM 2. *Let* $a \in \sigma_p$, $b \in \sigma_q$, $ab = ba$. *Are there commuting operators* $A \in a$, $B \in b$?

PROBLEM 3. *Let* $a = a^* \in \sigma_p$, $b = b^* \in \sigma_q$, $ab = ba$. *Are there self-adjoint commuting operators* $A \in a$, $B \in b$?

Problem 1 and Problem 3 in the case $q = p$ are evidently equivalent. On the other hand, a positive answer to Problem 2 does not automatically yield a positive answer to Problem 1.

REFERENCES

1. Gohberg I. Tz., Kreĭn M. G., *Introduction to the Theory of Linear Non-self-adjoint Operators*, Nauka Publ. House, Moscow, 1965. (Russian)
2. Birman M. Sh., Solomyak M. Z., *Estimates of s-numbers of integral operator*, Uspekhi Mat. Nauk **XXXII** (1977), no. 1 (193), 17–84. (Russian)
3. Simon B., *Trace ideals and their applications*, London Math. Soc. Lect. Note Series, vol. 35, Cambridge Univ. Press, 1979.
4. Triebel H., *Interpolation Theory. Function Spaces. Differential Operators*, Berlin, 1978.
5. Birman M. Sh., Solomyak M. Z., *Compact operators with power-like asymptotic behaviour of s-numbers*, Zapiski nauchn. semin. LOMI **126** (1983), 21–30. (Russian)
6. Brown L., Douglas R., Fillmore P., *Unitary equivalence modulo the compact operators and extensions of C^*-algebras*, Lecture Notes in Math. **345** (1973), 58–128.

DEPT. OF PHYSICS
ST.-PETERSBURG STATE UNIVERSITY
BIBLIOTECHNAYA PL. 2.
STARYI PETERHOF, ST.-PETERSBURG, 198904
RUSSIA

DEPARTMENT OF THEORETICAL MATHEMATICS
WEIZMANN INSTITUTE OF SCIENCE
REHOVOT 76100
ISRAEL

TWO PROBLEMS ABOUT THE OPERATORS $b(X)a(\mathcal{D})$

M. Z. SOLOMYAK

For a function u on $\mathbb{R}^d, d \geqslant 1$ we put by definition

$$b(X)a(\mathcal{D})\colon u(x) \mapsto b(X)\Phi^{-1}_{\xi\to x}\big(a(\xi)\Phi_{x\to\xi}u\big), \tag{1}$$

where Φ is the Fourier operator. We consider (1) as an operator in $L_2(\mathbb{R}^d)$. Operators of this type appear in many applications. A number of papers was devoted to the problem of estimating singular numbers of operators (1); see for instance [BS], [C], [S] and especially [BKS] where a summary of the up-to-date results in this field is presented.

The most principal results concern estimates of $b(X)a(\mathcal{D})$ in the von Neumann-Schatten classes $\mathfrak{S}_p$, $0 < p < \infty$, and in their weak analogs $\mathfrak{S}_{p,w}$. Recall that an operator T acting in a Hilbert space H belongs to $\mathfrak{S}_{p,w}$, $0 < p < \infty$, if and only if the following quantity (quasi-norm) is finite:

$$\|T\|_{p,w} = \sup_n \big(n^{1/p}s_n(T)\big).$$

$\mathfrak{S}_{p,w}$ is a natural analogue of the classical "weak" spaces of functions $L_{p,w}$ or of numerical sequences $\ell_{p,w}$.

THEOREM 1 ([C]). *If $a \in L_{p,w}(\mathbb{R}^d)$ and $b \in L_p(\mathbb{R}^d)$ for some $p \in (2,\infty)$, then*

$$\|b(X)a(\mathcal{D})\|_{p,w} \leqslant C(p)\|a\|_{L_{p,w}}\|b\|_{L_p}. \tag{2}$$

To formulate the estimates in $\mathfrak{S}_{p,w}$ for $p < 2$, we need specific function spaces. With a function $f \in L_{2,loc}(\mathbb{R}^d)$ we associate the numerical sequence $v(f) = \{v_n(f)\}$, $n \in \mathbb{Z}^d$, where $v_n^2(f) = \int_{Q_n} |f|^2\,dx$, $Q_n = n + (0,1)^d$. By definition, $f \in G_p(\mathbb{R}^d)$ if and only if $v(f) \in \ell_p$ and $f \in G_{p,w}(\mathbb{R}^d)$ if and only if $v(f) \in \ell_{p,w}$. The spaces G_p, $G_{p,w}$ are equipped with the natural (quasi)-norms $\|f\|_{G_p} = \|v(f)\|_{\ell_p}$, $\|f\|_{G_{p,w}} = \|v(f)\|_{\ell_{p,w}}$.

THEOREM 2. ([S] for $1 \leqslant p < 2$, [BKS] for all $p \in (0,2)$). *If $a \in G_{p,w}(\mathbb{R}^d)$ and $b \in G_p(\mathbb{R}^d)$ for $0 < p < 2$ then*

$$\|b(X)a(\mathcal{D})\|_{p,w} \leqslant C(d,p)\|a\|_{G_{p,w}}\|b\|_{G_{p,w}}. \tag{3}$$

Both the above Theorems are sharp in the following sense: there exist specific functions $a \in L_{p,w}$ $(p > 2)$ or $a \in G_{p,w}$ $(p < 2)$ such that the condition $b \in L_p$ $(b \in G_p)$ is necessary and sufficient for the operator (1) to belong to $\mathfrak{S}_{p,w}$ (see the discussion in [BKS] and in the papers mentioned there).

Comparing Theorems 1, 2 we see that the value $p = 2$ is dropped. It seems to be natural: indeed, the functional spaces $L_{p,w}$ and $G_{p,w}$ are of "different nature" and it is not quite clear what must be the "right" functional space for the case $p = 2$.

So, our first problem reads as follows:

PROBLEM 1. *To find a sharp estimate of the operator* (1) *in* $\mathfrak{S}_{2,\infty}$.

Estimates of $b(X)a(\mathcal{D})$ in $\mathfrak{S}_p$ read as follows:
We have

$$\|b(X)a(\mathcal{D})\|_p \leqslant (2\pi)^{-d/p}\|a\|_{L_p}\|b\|_{L_p}, \qquad 2 \leqslant p < \infty \tag{4}$$

(this is a straightforward consequence of Riesz–Torin Theorem for $\mathfrak{S}_p$-spaces, see [S]), and

$$\|b(X)a(\mathcal{D})\|_p \leqslant C(d,p)\|a\|_{G_p}\|b\|_{G_p}, \qquad 0 < p \leqslant 2,$$

(here the proof is rather complicated; see [BS]).

Here the value $p = 2$ is not dropped. This matches with the evident fact that $G_2 = L_2$.

Obviously, we have the equality in (4) for $p = 2$. It follows that the condition $a, b \in L_2$ is not only sufficient but also necessary for membership of $b(X)a(\mathcal{D})$ ($\neq 0$) in $\mathfrak{S}_2$. It is quite trivial; but the following statement ([S]) looks rather unexpected.

THEOREM 3. *Let* $0 \neq b(X)a(\mathcal{D}) \in \mathfrak{S}_1$. *Then* $a, b \in G_1$.

In fact, the proof given in [S] can be easily extended to the case $1 \leqslant p < 2$; but this proof fails for the "non-normed case" $p < 1$. So we state our second Problem.

PROBLEM 2. *Does the assumption*

$$0 \neq b(X)a(\mathcal{D}) \in \mathfrak{S}_p, \qquad 0 < p < 1,$$

imply that $a, b \in G_p$?

Note that there is no statement similar to Theorem 3 but for the case $p > 2$: the condition $a, b \in L_p$ is *a fortiori* not necessary for the membership of $b(X)a(\mathcal{D})$ in $\mathfrak{S}_p$.

REFERENCES

[BKS] Birman M., Karadzhov G., Solomyak M., *Boundedness conditions and spectrum estimates for the operators $b(X)a(\mathcal{D})$ and their analogs*, Adv. in Sov. Math. (M. Birman, ed.), vol. 7, AMS, Providence RI 1991.

[BS] Birman M. S., Solomyak M. Z., *Estimates for the singular numbers of integral operators*, Uspekhi Mat. Nauk **32** (1977), no. 1 (193), 17–84 (Russian); English transl. in Russian Math. Surveys **32** (1977), no. 1, 15–89.

[C] Cwikel M., *Weak type estimates for singular values and the number of bound states of Schrödinger operators*, Ann. Mat. **106** (1977), 93–100.

[S] Simon B., *Trace Ideals and their Applications*, Cambridge Univ. Press, Cambridge, 1979.

DEPARTMENT OF THEORETICAL MATHEMATICS
WEIZMANN INSTITUTE OF SCIENCE
REHOVOT 76100
ISRAEL

5.17
old

BOUNDEDNESS OF CONTINUUM EIGENFUNCTIONS AND THEIR RELATION TO SPECTRAL PROBLEMS

BARRY SIMON

We will describe a set of problems for matrices acting on $\ell^2(\mathbb{Z})$. There are analogous problems for $\ell^2(\mathbb{Z}^\nu)$ and for suitable elliptic operators on $L^2(\mathbb{R}^\nu)$. Let A be a bounded self-adjoint operator on $\ell^2(\mathbb{Z})$ whose matrix elements obey $a_{ij} \equiv (\delta_i, A\delta_j) = 0$ if $|i-j| \geqslant K$. A fundamental result asserts the existence of a measure $d\rho(E)$, a function $n(E)$ taking values $0, 1, \ldots, \infty$ (infinity allowed) with $n(E) \geqslant 1$ $d\rho$-a.e. E and $n(E) = 0$ if $E \notin \operatorname{supp} \rho$ and for each E, $n(E)$ linearly independent sequences $u_\alpha(E; n)$; $\alpha = 1, \ldots, n(E)$ (not necessarily in ℓ^2) so that

(a) $|u_\alpha(E;n)| \leqslant C(1+|n|)$;
(b) $\sum a_{ij} u_\alpha(E;j) = E\, u_\alpha(E;i)$;
(c) Let $\mathcal{H}' = L^2(\mathbb{R}; \mathbb{C}^{n(E)}; d\rho)$, i.e. functions, f, on $\mathbb{R}$ with $f(E)$ having values in $\mathbb{C}^{n(E)}$ (where $\mathbb{C}^\infty = \ell^2$) and let C_0 denote sequences in $\ell^2(\mathbb{Z})$ of compact support. Define U taking C_0 into $\mathcal{H}'$ by $(Ug)_\alpha(E) = \sum_m \overline{u_\alpha(E;m)} g(m)$. Then U extends to a unitary map of $\ell^2(\mathbb{Z})$ onto $\mathcal{H}'$;
(d) $U(Ag) = E(Ug)$.

These continuum eigenfunction expansions are called BGK expansions in [1] in honor of the work of Berezanskii, Browder, Gårding, Gel'fand and Kac, who developed them in the context of elliptic operators. See [1, 2, 3] for proofs. These expansions don't really contain much more information than the spectral theorem. The most significant additional information concerns the boundedness properties of u; see [4, 5] for applications.

Actually, the general proofs show that $(1+|n|)$ in part (a) can be replaced by $(1+|n|)^\alpha$ for any $\alpha > 1/2$. Indeed, one shows that for any $g \in \ell^2$, one can arrange that for $(d\rho)$-a.e. E $g(\cdot)u_\alpha(E,\cdot) \in \ell^2$. If one could arrange a set, S, of good E's where $gu \in \ell^2$ *for all* $g \in \ell^2$ with $\rho(\mathbb{R} \setminus S) = 0$, then on S, $u \in \ell^\infty$. This leaves open:

QUESTION 1. *Is it true that for $(d\rho)$-a.e. E, each $u_\alpha(E,\cdot)$ is bounded?*

There is a celebrated counterexample of Maslov [6] to the boundedness in the one dimensional elliptic case. As explained in [1], Maslov's analysis is wrong, and it is not clear whether his example has bounded u's a.e. We believe the answer to Question 1 (and all other yes/no questions below) is affirmative, but for what we have to say below, a weaker result would suffice:

QUESTION 2. *Is it at least true that for $d\rho$-a.e. E and all α : $\frac{1}{2N+1} \sum_{|n| \leqslant N} |u_\alpha(E,n)|^2$ is bounded?*

QUESTION 3. *Is it true that*

$$\lim_{n\to\infty} \frac{1}{2N+1} \sum_{|n|\leqslant N} \left|u_\alpha(E,n)\right|^2 \equiv k(\alpha, E)$$

exists?

The $\overline{\lim}$ we will denote by $\bar{k}(\alpha, E)$.

Given a subset M, of $\{(E,\alpha) : E \in \mathbb{R}, \alpha \leqslant N\}$ we define

$$(P(M)g)(n) = \sum_\alpha \int_{\{E:(E,\alpha)\in M\}} u_\alpha(E,n)(Ug)_\alpha(E)d\rho(E)$$

where a suitable limit in mean may need to be taken. Define

$$\begin{aligned} M_1 &= \{\,(E,\alpha) : u_\alpha(E,\cdot) \in \ell^2\,\} \\ M_2 &= \{\,(E,\alpha) : \bar{k}(\alpha,E) = 0 \text{ but } (E,\alpha) \notin M_1\,\} \\ M_3 &= \{\,(E,\alpha) : \bar{k}(\alpha,E) \neq 0\,\}. \end{aligned}$$

Obviously, $P(M_1)$ is the projection onto the point spectrum of A.

QUESTION 4. *Is it true that $P(M_2)$ is the projection onto the singular continuous space of A and $P(M_3)$ the projection onto the absolutely continuous spectrum of A?*

Among other things this result would imply that in the Jacobi case (where the number K of the third sentence in this note is 2), the singular spectrum is simple.

In higher dimensions, one can see situations where A separates (i.e. $\ell^2(\mathbb{Z}^\nu) = \ell^2(\mathbb{Z}^{\nu_1}) \otimes \ell^2(\mathbb{Z}^{\nu_2})$ and $A = A_1 \otimes I + I \otimes A_2$) where A has a. c. spectrum with eigenfunctions decaying in ν_2 dimensions but of plane wave form in the remaining ν_1-dimensions. One can also imagine a. c. spectrum from combining singular spectrum for A_1 and A_2. In either case $k = 0$ for lots of continuum a. c. eigenfunctions.

QUESTION 5. *Is there a sensible (i.e. not obviously false) version of Question 4 in the multidimensional case?*

There are examples [7] of cases where A has only point spectrum but there is an eigenfunction with $k(\alpha, E) > 0$ (since it occurs on a set of ρ-measure zero, it is not a counterexample to a positive answer to a Question 4). Does the second part of Question 4 have a positive converse?

QUESTION 6. *Is it true that if $Au = Eu$ has a bounded eigenfunction with $\bar{k} > 0$ for a set, Q, of E's of positive Lebesgue measure, then A has some a. c. spectrum on Q?*

QUESTION 7. *What is the proper analog of Question 6 for singular continuous spectrum?*

REFERENCES

1. Simon B., *Schrödinger semigroups*, Bull. Amer. Math. Soc. **7** (1982), 447–526.
2. Berezanskii Yu. M., *Expansions in eigenfunctions of selfadjoint operators*, "Naukova Dumka", Kiev, 1965 (Russian); English transl. in *Math. Monogr.*, vol. 17, Amer. Math. Soc., Providence, 1968.
3. Kovalenko V. F., Semenov Yu. A., *Some questions on expansions in generalized eigenfunctions of a Schrödinger operator with strongly singular potentials*, Uspekhi Mat. Nauk **33** (1978), no. 4, 107–140 (Russian); English transl. in Russian Math. Surveys **33** (1978), 119–157.
4. Pastur L., *Spectral properties of disordered systems in onebody approximation*, Comm. Math. Phys. **75** (1980), 179.
5. Avron J., Simon B., *Singular continuous spectrum for a class of almost periodic Jacobi matrices*, Bull. Amer. Math. Soc. **6** (1982), 81–86.
6. Maslov V. P., *On the asymptotic behaviour of eigenfunctions of the Schrödinger equation which are distributions*, Uspekhi Mat. Nauk **16** (1961), no. 4, 253–254. (Russian)
7. Simon B., Spencer T., *Trace class perturbations and the absence of absolutely continuous spectra*, Comm. Math. Phys. **125** (1989), 113–125.

DEPARTMENT OF MATHEMATICS AND PHYSICS
CALIFORNIA INSTITUTE OF TECHNOLOGY
PASADENA, CALIFORNIA 91125
U.S.A.

S.5.18
old

PERTURBATION OF SPECTRUM OF NORMAL OPERATORS AND OF COMMUTING TUPLES

CHANDLER DAVIS

Recent progress has somewhat clarified the subject of perturbation of spectrum of normal operators and of K-tuples of commuting self-adjoints.

$\mathcal{H}$ will be a Hilbert space of n dimensions. The spectral resolution of a normal operator A will be written $A = \sum_{j=1}^{n} \alpha_j u_j u_j^*$, here the u_j are orthonormal eigenvectors, with eigenvalues α_j corresponding; and the notation x^*, for any $x \in \mathcal{H}$, denotes the linear functional corresponding to x. Similarly for normal B let us write $B = \sum_{j=1}^{n} \beta_j v_j v_j^*$. As the distance δ between $\sigma(A)$ and $\sigma(B)$ let us use

$$\delta = \min_{\pi} \max_{j} |\alpha_j - \beta_{\pi j}|, \tag{1}$$

the minimum being over all permutations of $\{1, 2, \dots, n\}$.

PROBLEM 1. *Find the best constant c such that, for all normal A and B,*

$$\delta \leqslant c\|A - B\|. \tag{2}$$

The problem of K-tuples of commuting self-adjoints may be more important, but so far seems less tractable. I will use the following notation. If $A^{(1)}, A^{(2)}, \dots, A^{(k)}$ are self-adjoint and commute, then for orthonormal $u_1, \dots, u_n$ and corresponding real $\alpha_j^{(J)}$ we have $A^{(J)} = \sum_{j=1}^{n} \alpha_j^{(J)} u_j u_j^*$; I will let $\tilde{A}$ denote the operator-matrix of one column whose J-th entry is $A^{(J)}$, so that $\tilde{A} = \sum_{j=1}^{n} \tilde{\alpha}_j u_j u_j^*$, and we may speak of $\tilde{\alpha}_j \in \mathbb{R}^k$ as the j-th eigenvalue of $\tilde{A}$. (As an operator from $\mathcal{H}$ to $\mathcal{H}^k$, it does not have eigenvalues in the usual sense.) Similarly $\tilde{B} = \sum_{j=1}^{n} \tilde{\beta}_j v_j v_j^*$. As above, the distance will be

$$\delta = \min_{\pi} \max_{j} \|\tilde{\alpha}_j - \tilde{\beta}_{\pi j}\|,$$

the minimum being over all permutations, and the norm being that of $\mathbb{R}^k$.

PROBLEM 2. *Find the best constant c_k such that, for all K-tuples $\tilde{A}$ and $\tilde{B}$, $\delta \leqslant c_k\|\tilde{A} - \tilde{B}\|$.*

DEPARTMENT OF MATHEMATICS
UNIVERSITY OF TORONTO
TORONTO M5S 1A1, ONTARIO
CANADA

COMMENTARY BY CHAPTER'S EDITOR

The solutions and new related problems are described by John Holbrook in Problem 5.1 above.

S.5.19
old

COMPOSITION OF INTEGRATION AND SUBSTITUTION

YU. I. LYUBICH

For ϕ a continuous function mapping the closed unit interval into itself and vanishing at 0, the operator I_ϕ is defined on $C[0,1]$ by the formula

$$(I_\phi f)(x) = \int_0^{\phi(x)} f(t)\, dt.$$

It is not hard to see that I_ϕ is quasinilpotent if $\phi(x) \leqslant x$ for all x. The problem asked if the converse is true.

COMMENTARY BY CHAPTER'S EDITOR

The answer is affirmative, and the same result holds if I_ϕ is regarded as an operator on L_p. This was shown independently by Whitely [1] and by Yusun [2]; the latter paper also contains an interesting result on triangularizing certain quasinilpotent integral operators with non-negative kernels.

REFERENCES

1. Whitely R., *The spectrum of a Volterra composition operator*, Int. Equat. Oper. Theory **10** (1987), 146–149.
2. Tong Yusun, *Quasinilpotent integral operators*, Acta Math. Sinica **32** (1989), 727–735. (Chinese)

DEPARTMENT OF MATHEMATICS
TECHNION
HAIFA 32000
ISRAEL

Chapter 6

PERTURBATION THEORY
SCATTERING THEORY

Edited by

M. Sh. Birman
Department of Physics
St. Petersburg State University
Staryi Peterhof, St. Petersburg, 198904
Russia

INTRODUCTION

A considerable part of Operator Theory can be attributed, almost without stretching the truth, to the spectral perturbation theory. At the same time, it is rather the method than the essence of the question that the name "perturbation theory" hints to. Therefore, one could equally well attribute to it nothing at all. The present small chapter did not exist in the preceding edition. Now it is distinguished mainly in order to achieve a greater unity in other parts. Consequently, the problems constituting Chapter 6 are fairly diverse. Seven problems from the total number of 12 are "old", two of them are solved. A number of papers are connected with Problems 6.4 and S.6.11 (and some of them are very recent). All this is reflected by Commentaries. No advance appears to have been made in Problems 6.1, 6.6, 6.9. But according to the editors' opinion they have not lost their topicality and hence are reproduced here. No system should be sought to trace in the order of the problems in the chapter. We note only that the topics of Problem 6.5 and 6.4 (the former one is new) are related to each other. Problems 6.8, 6.9, 6.10 pertain to Scattering Theory proper, and problems 6.7, S.6.11 are connected with this theory ideologically.

6.1
old

PERTURBATION THEORY AND INVARIANT SUBSPACES

L. DE BRANGES

If $\mathbf{C}$ is a given coefficient Hilbert space, let $\mathbf{C}(z)$ be the Hilbert space [1] of square summable power series $f(z) = \sum a_n z^n$ with coefficients in $\mathbf{C}$

$$\|f(z)\|^2_{\mathbf{C}(z)} = \sum |a_n|^2.$$

If $B(z)$ is a power series whose coefficients are operators on $\mathbf{C}$ and which represents a function which is bounded by one in the unit disc, then multiplication by $B(z)$ is contractive in $\mathbf{C}(z)$. Consider the range $\mathcal{M}(B)$ of multiplication by $B(z)$ in $\mathbf{C}(z)$ in the unique norm such that multiplication by $B(z)$ is a partial isometry of $\mathbf{C}(z)$ onto $\mathcal{M}(B)$. Define $\mathcal{H}(B)$ to be the complementary space to $\mathcal{M}(B)$ in $\mathbf{C}(z)$. Then the difference-quotient transformation $f(z)$ into $[f(z) - f(0)]/z$ in $\mathcal{H}(B)$ is a canonical model of contractive transformations in Hilbert space which has been characterized [2] as a conjugate isometric node with transfer function $B(z)$.

If $\varphi(z)$ is a power series whose coefficients are operators on $\mathbf{C}$ and which represents a function with positive real part in the unit disc, then

$$B(z) = [1 - \varphi(z)]/[1 + \varphi(z)]$$

is a power series which represents a function which is bounded by one in the unit disc. Define $\mathcal{L}(\varphi)$ to be the unique Hilbert space of power series with coefficients in $\mathbf{C}$ such that multiplication by $1 + B(z)$ is an isometry of $\mathcal{L}(\varphi)$ onto $\mathcal{H}(B)$. Then the difference-quotient transformation has an isometric adjoint in $\mathcal{L}(\varphi)$.

The overlapping space $\mathcal{L}$ of $\mathcal{H}(B)$ is the set of elements $f(z)$ of $\mathbf{C}(z)$ such that $B(z)f(z)$ belongs to $\mathcal{H}(B)$ in the norm

$$\|f(z)\|^2_{\mathcal{L}} = \|f(z)\|^2_{\mathbf{C}(z)} + \|B(z)f(z)\|^2_{\mathcal{H}(B)}.$$

The overlapping space $\mathcal{L}$ is isometrically equal to a space $\mathcal{L}(\Theta)$.

A fundamental theorem of perturbation theory [3] states that a partially isometric transformation exists of $\mathcal{L}(\varphi)$ into $\mathcal{L}(\Theta)$ which commutes with the difference-quotient transformation. The transformation is a computation of the wave-limit. The wave-limit is isometric on the square summable elements of $\mathcal{L}(\varphi)$ and annihilates the orthogonal complement of the square summable elements of $\mathcal{L}(\varphi)$. If T denotes the adjoint in $\mathbf{C}(z)$ of multiplication by $B(z)$ as a transformation in $\mathbf{C}(z)$, then the wave-limit agrees with $1 + T$ on square summable elements of $\mathcal{L}(\varphi)$. A fundamental problem is to determine the range of the wave-limit in $\mathcal{L}(\Theta)$. It is known [4] that the range can be a proper subspace of $\mathcal{L}(\Theta)$. The orthogonal complement of the range of the wave-limit in $\mathcal{L}(\Theta)$ is the overlapping space of a space $\mathcal{H}(C)$ such that $B(z) = A(z)C(z)$ for a space $\mathcal{H}(A)$ which is contained isometrically in $\mathcal{H}(B)$.

CONJECTURE. *The range of the wave-limit contains every element of $\mathcal{L}(\Theta)$ if the self-adjoint part of the operator $\varphi(0)$ is of Matsaev class.*

The Matsaev class seems a reasonable candidate because the existence of invariant subspaces is known for contractive transformations T such that $1 - T^*T$ is of Matsaev class. Invariant subspaces exist which cleave the spectrum of the transformation. An integral representation of the transformation exists in terms of invariant subspaces [5]. For reasons of quasi-analyticity, such results do not hold for any larger class of completely continuous operators.

Some recent improvements in the spectral theory of nonunitary transformations link the Matsaev class to the theory of overlapping spaces [6].

REFERENCES

1. De Branges L., *Square Summable Power Series*, Addison–Wesley (to appear).
2. De Branges L., *The model theory for contractive transformations*, Proceedings of the Symposium on the Mathematical Theory of Networks and Systems in Beersheva, Springer–Verlag (to appear).
3. De Branges L., Shulman L., *Perturbations of unitary transformations*, J. Math. Anal. Appl. **23** (1968), 294–326.
4. De Branges L., *Perturbation theory*, J. Math. Anal. Appl. **57** (1977), 393–415.
5. Gohberg I. C., Krein M. G., *The Theory of Volterra Operators in Hilbert Space and its Applications*, Nauka, Moscow, 1967 (Russian); English transl. Translations of Mathematical Monographs, vol. 24, Amer. Math. Soc., 1970.
6. De Branges L., *The expansion theorem for Hilbert spaces of analytic functions*, Topics in Operator Theory (Rehovot, 1983), Birkhäuser, Basel-Boston, Mass., 1984, pp. 75–107.

DEPARTMENT OF MATHEMATICS
PURDUE UNIVERSITY
WEST LAFAYETTE, INDIANA 47901
USA

6.2

QUASIDIAGONALITY AND THE MACAEV IDEAL

DAN VOICULESCU

We denote by $\mathfrak{S}_\omega$ the normed ideal of compact operators on Hilbert space defined by the norm $|T|_\infty^- = \sum_{j\in\mathbb{N}} \lambda_j j^{-1}$ where $\lambda_1 \geqslant \lambda_2 \geqslant \ldots$ are the eigenvalues of $(T^*T)^{1/2}$. This ideal contains all Schatten–von Neumann classes $\mathfrak{S}_p$ $(1 \leqslant p < \infty)$. It corresponds to the double index $(\infty, 1)$ on the Lorentz scale and is known as the Matsaev ideal ([1]).

In [4] we proceed that for every normed ideal $(\mathcal{J}, |\cdot|_{\mathcal{J}})$ of compact operators such that $\mathcal{J} \subset \mathfrak{S}_\omega$, $\mathcal{J} \neq \mathfrak{S}_\omega$ and any bounded operator T there are finite rank operators $A_k \uparrow I$, $0 \leqslant A_k \leqslant I$ such that $\lim_{k\to\infty} \big|[A_k, T]\big|_{\mathcal{J}} = 0$.

If $\mathcal{J}$ is replaced by $\mathfrak{S}_\omega$, there are operators for which the above property does not hold ([5]). This kind of situation is also related to the entropy of dynamical systems ([6]).

An operator T, such that there exist finite rank projections $P_k \uparrow I$, $P_k = P_k^* = P_k^2$, so that $\lim_{k\to\infty} \big|[P_k, T]\big|_{\mathcal{J}} = 0$ is called *$\mathcal{J}$-quasidiagonal* (see [3] for this refinement of Halmos' notion of quasidiagonality [2]). If $\mathcal{J}$ is the ideal $\mathcal{K}$ of compact operators, an operator quasidiagonal relative to $\mathcal{K}$ is called *quasidiagonal*.

PROBLEM. *Let $\mathcal{J}$ be a normed ideal such that $\mathcal{J} \supset \mathfrak{S}_\omega$, $\mathcal{J} \neq \mathfrak{S}_\omega$. Does it follow that every quasidiagonal operator T is also $\mathcal{J}$-quasidiagonal?*

REFERENCES

1. Gohberg I. T., Krein M. G., *Introduction to the theory of non-selfadjoint operators*, Moscow, 1965. (Russian)
2. Halmos P. R., *Ten problems in Hilbert space*, Bull. Amer. Math. Soc. **76** (1970), 887–933.
3. Voiculescu D., *Some extensions of quasitriangularity*, Rev. Roumaine Math. Pures Appl. **18** (1973), 1303–1320.
4. Voiculescu D., *On the existence of quasicentral approximate units relative to normed ideals*, Part I, J. Functional Analysis **90** (1990), no. 1, 1–36.
5. Voiculescu D., *A note on quasidiagonal operators*, Operator Theory: Advances and Applications **32** (1988), Birkhäuser Verlag, 265–274.
6. Voiculescu D., *Entropy of dynamical systems and perturbations of Hilbert space operators*, preprint I.H.E.S., 1990.

DEPARTMENT OF MATHEMATICS
UNIVERSITY OF CALIFORNIA
BERKELEY, CA 94720
U.S.A.

A QUESTION OF POLYNOMIAL APPROXIMATION ARISING IN CONNECTION WITH THE LACUNAE OF THE SPECTRUM OF HILL'S EQUATION

H. P. McKean

Let $Q = -\frac{d^2}{dx^2} + q(x)$ be a Hill's operator with $q \in C_1^\infty$, the class of real infinitely differentiable functions of period 1. The spectrum determined by the periodic and anti-periodic solutions $Qf = \lambda f$, $0 \leqslant x < 1$ comprise a simple (periodic) ground state λ_0 followed by separated pairs $\lambda_{2n-1} < \lambda_{2n}$, $n = 1, 2, \ldots$ of alternately anti-periodic and periodic eigenvalues increasing to $+\infty$, the equality or inequality signifying the dimensionality ($= 1$ or 2) of the eigenspace; see [1]. The intervals $[\lambda_{2n-1}, \lambda_{2n}]$, $n = 1, 2, \ldots$ are the *lacunae* of the spectrum of Q in $L^2(\mathbb{R})$. Hochstadt [2] proved that the infinite differentiability of q is reflected in the rapid vanishing of the lengths l_n of the lacunae $[\lambda_{2n-1}, \lambda_{2n}]$ as $n \uparrow +\infty$. Trubowitz [3] proved that the real analyticity of q is equivalent to $l_n \leqslant ae^{-bn}$, $n \uparrow \infty$. A comparison between l_n and $\widehat{q}(n) = \int_0^1 e^{2\pi inx} q(x)\, dx$ springs to mind. Interest in sharpening these results arises in connection with the following geometrical problem.

Let M, $M \subset C_1^\infty$, be the class of functions giving rise to a fixed periodic and anti-periodic spectrum $\lambda_0 < \lambda_1 \leqslant \lambda_2 < \lambda_3 \leqslant \lambda_4 < \ldots$ and let g, $g \leqslant \infty$, be the number of pairs of simple eigenvalues $\lambda_{2n-1} < \lambda_{2n}$. M is a compact g-dimensional torus identifiable as the real part of the Jacobi variety of the hyperelliptic curve of genus g, $g \leqslant \infty$, with branch points over the real spectrum, augmented by the point at ∞; see [4] and [5] for $g < \infty$, and [6] for $g = \infty$. M admits a family of transitive commuting (iso-spectral) flows expressible in Hamiltonian form as $\frac{\partial q}{\partial t} = Xq$ with $Xq = (\operatorname{grad} \lambda)'$, prime signifying $\frac{d}{dx}$, in which λ is a simple eigenvalue of Q and $\operatorname{grad} \lambda$ is the functional gradient $\partial\lambda/\partial q(x) =$ the square of the normalized eigenfunction $[f(x)]^2$. The *local* flows: $\partial q/\partial t = X_1 q = \partial q/\partial x$ (translation), $X_2 q = 3q\frac{\partial q}{\partial x} - \frac{1}{2}\frac{\partial^3 q}{\partial x^3}$ (Korteweg – de Vries), etc. are more familiar. The latter belong to the span of the former, but can be expressed in an independent fashion $[X_m q = (\operatorname{grad} H_m)']$ via the rule $X_m q = \left(qD + Dq - \frac{1}{2}D^3\right) \operatorname{grad} H_{m-1}$ starting from $H_{-1} = 1$; for example,

$$H_0 = \int_0^1 q\, dx, \qquad H_1 = \int_0^1 \frac{q^2}{2}\, dx, \qquad H_2 = \int_0^1 \left(\frac{q'^2}{4} + \frac{q^3}{2}\right) dx.$$

THE GEOMETRICAL QUESTION *is to decide if the local vector fields X_1, X_2, etc. span the tangent space of M at each point.* This is always the case if $g < \infty$; see [5] or [7]. McKean–Trubowitz [6] make the question precise for $g = \infty$ and prove the following necessary and sufficient condition. Let F be the space of sequences $f(\lambda_{2n})$, $n = 1, 2, 3, \ldots$, with quadratic form $\|f\|^2 = \sum_{n \geqslant 1} l_n |f(\lambda_{2n})|^2 < \infty$. Let P be the subspace $f(\lambda_{2n}) = p(\lambda_{2n})$, $n = 1, 2, 3, \ldots$, with p a polynomial. *Then the spanning of the local* *vector fields takes place if and only if P spans F.* The condition is met if q is real analytic ($l_n < ae^{-bn}$, $n \uparrow +\infty$). It is known that $\lambda_{2n} \sim n^2\pi^2 + c_0 + c_1 n^{-2} + c_2 n^{-4} + \dots$ ($n \uparrow +\infty$), permitting the application of a result of Koosis [8] in the case of purely simple spectrum to verify that the spanning takes place in that circumstance only if $\sum_{n\geqslant 1} \frac{\log l_n}{n^2} = -\infty$. Contrariwise, the spanning *cannot take place* if q vanishes on an interval; in fact, if the local vector fields span the tangent space, then the associated gradients $\partial H/\partial q$ span the normal space of M, and the two together (tangent and normal) fill up the whole of the ambient space, taken to be $L^2(0,1)$, which is impossible since X_q and $\partial H/\partial q$ are universal polynomials in q, q', q'', etc. without constant term and consequently vanish on the same interval as q. *It seems likely that the spanning becomes critical in the vicinity of quasi-analytic q.* The same questions arise for Q on the line with $q \in C^\infty_\downarrow$, the class of infinitely differentiable functions of rapid decay at $\pm\infty$. The rate of vanishing of the lacunae is replaced by the rate of decay of the reflection coefficient $s_{12}(k)$ of Faddeev [9], e.g., $q \in C^\infty_\downarrow$ is reflected in $s_{12} \in C^\infty_\downarrow$, while the analyticity of q in a horizontal strip is reflected in $|s_{12}(k)| < ae^{-b|k|}$, $k \to \pm\infty$; see [10]. The torus M is now replaced by g-dimensional ($g \leqslant \infty$) cylinder specified by fixing $|s_{12}|$ and finite number of bound states (negative simple eigenvalues) $-k_n^2$ ($n = 1, \dots, g$), and the vector fields $X_q = (\operatorname{grad} H)'$ determined from $H = |s_{12}(k)|$, $k \in \mathbb{R}$, or from $H = -k_n^2$, ($n = 1, \dots, g$) presumably span the tangent space; see [11] for preliminary information. The local vector fields $X_1 q = q'$, $X_2 q = 3qq' - \frac{1}{2}q'''$ operate as before, and the question is the same as before: *do they span the tangent space of M?* The technical clarification of the question is necessary part of any discussion.

References

1. Magnus W., Winkler, *Hill's Equation*, Interscience–Wiley, New York, 1966.
2. Hochstadt H., *Function theoretic properties of the discriminant of Hill's equation*, Math. Zeit **82** (1963), 237–242.
3. Trubowitz E., *The inverse problem for periodic potentials*, Comm. Pure Appl. Math. **30** (1977), 321–337.
4. Dubrovin B. A., Novikov S. P., *A periodic problem for the Korteweg–de Vries and Sturm–Liouville equation. Their connection with algebraic geometry*, Doklady Akad. Nauk SSSR **219** (1974), no. 3, 531–534. (Russian)
5. McKean H. P., van Moerbeke P., *The spectrum of Hill's equation*, Invent. Math. **30** (1975), 217–274.
6. McKean H. P., Trubowitz E., *Hill's operator and hyperelliptic function theory in the presence of infinitely many branch points*, Comm. Pure Appl. Math. **29** (1976), 143–226.
7. Lax P., *Periodic solutions of the K dV equation*, Comm. Pure Appl. Math. **28** (1975), 141–188.
8. Koosis P., *Weighted polynomial approximation on arithmetic progressions of integrals or points*, Acta Math. **116** (1966), 223–277.
9. Faddeev L. D., *Properties of the S-matrix of the one-dimensional Schrödinger equation*, Trudy Mat. Inst. Steklov AN SSSR **73** (1964), 314–336. (Russian)
10. Deift P., Trubowitz E., *Inverse scattering on the line*, Comm. Pure Appl. Math. **32** (1979), no. 2, 121–251.
11. McKean H. P., *Theta functions, solutions, and singular curves*, (Proc. Conf., Park City, Utah., 1977, 237–254), Dekker, New York, 1979, Lecture Notes in Pure and Appl. Math., 48.

New York University
Courant Institute of Mathematical Sciences
251 Mercer Street
New York, N. Y. 10012
U.S.A.

6.4
old

WHEN ARE DIFFERENTIABLE FUNCTIONS DIFFERENTIABLE?

HAROLD WIDOM

If $f\colon \mathbb{R} \to \mathbb{R}$ is continuous and A is a C^*-algebra then there is defined by the usual functional calculus a mapping $f_A\colon x \mapsto f(x)$ from the linear space of hermitian elements of A into itself.

What is a necessary and sufficient condition on f that for all A the function f_A is differentiable everywhere?

Taking $A = \mathbb{C}$ shows that f must be differentiable. In fact:

(1) *If f_A is differentiable for all A then $f \in C^1(\mathbb{R})$.*

Proof. Let A be the algebra of bounded functions on an interval $[a, b]$. The differentiability of f_A at a function x asserts that for every ε there is a δ such that for any function h with $\|h\| < \delta$

$$\|f(x+h) - f(x) - f'_A(x) \cdot h\| \leqslant \varepsilon \cdot \|h\|.$$

This shows immediately that $f'_A(x)$ must be mapping $h \to (f' \circ x)h$. Let $s_0, t_0 \in [a, b]$ satisfy $|s_0 - t_0| < \delta$ and take $x(t)$ to be the identity function t and $h(t)$ the constant function $s_0 - t_0$. Then

$$\|h\| = |s_0 - t_0| < \delta$$

and $f(x+h) - f(x) - f'_A(x)h$ is equal at $t = t_0$ to

$$f(s_0) - f(t_0) - f'(t_0)(s_0 - t_0).$$

Thus $|f(s_0) - f(t_0) - f'(t_0)(s_0 - t_0)| \leqslant \varepsilon |s_0 - t_0|$. Interchanging s_0 and t_0, adding, and dividing by $|s_0 - t_0|$ give

$$|f'(s_0) - f'(t_0)| \leqslant 2\varepsilon. \quad \square$$

It is even easier to show that if $f \in C^1$ and A is commutative then f_A is differentiable. For general A all I know is this:

(2) *If in a neighborhood of each point of $\mathbb{R}$ the function f is equal to a function whose derivative has Fourier transform belonging to $L^1(\mathbb{R})$ then f_A is differentiable for all A.*

Proof. Of course "each point" in the assumption on f can be replaced by "each compact set" and since the differentiability of f_A at x depends on the values of f in an arbitrary neighborhood of the interval $\left[-\|x\|, \|x\|\right]$ we may assume f itself has derivative whose Fourier transform belongs to $L^1(\mathbb{R})$.

Let x and h be hermitian. From the identity

$$\frac{d}{ds}e^{is(x+h)}e^{-isx} = ie^{is(x+h)}he^{-isx}$$

we obtain upon integrating with respect to s over $[0,t]$ and right multiplying by e^{itx}

$$e^{it(x+h)} = e^{itx} + i\int_0^t e^{is(x+h)}he^{i(t-s)x}\,ds.$$

Applying the Fourier inversion formula gives

$$\begin{aligned} f(x+h) - f(x) & \\ &= \int_{-\infty}^{+\infty} i\hat{f}(t)\,dt \int_0^t e^{is(x+h)}he^{i(t-s)x}\,ds \\ &= \int_{-\infty}^{+\infty} i\hat{f}(t)\,dt \int_0^t e^{isx}he^{i(t-s)x}\,ds + \int_{-\infty}^{+\infty} i\hat{f}(t)\,dt \int_0^t \left[e^{is(x+h)} - e^{isx}\right]he^{i(t-s)x}\,ds \\ &= \quad \mathrm{I} \quad + \quad \mathrm{II}\quad , \end{aligned}$$

let us say. The inner integral in I has norm at most $|t|\cdot\|h\|$ and so (since $t\hat{f}(t) \in L^1(\mathbb{R})$) the double integral makes sense and represents a continuous linear function of h. In fact it will define $f'_A(x)h$. To show this it suffices to show that the double integral II has norm $o(\|h\|)$ as $\|h\| \to 0$. But the norm of the inner integral in II is $o(\|h\|)$ for each t and is at most $2|t|\,\|h\|$ for all t and so the conclusion follows from the dominated convergence theorem.

PROBLEM 1. *Fill the gap between (1) and (2). In particular, is $f \in C^1$ a sufficient condition for the differentiability of f_A for all A?*

Here is a concrete example. Let A be the algebra of bounded operators on $L^2(a,b)$. If x is M_t, multiplication by the identity function, and h is the integral operator with kernel $K(s,t)$, then formally $f'_A(x)h$ is the integral operator with kernel

$$(*) \qquad K(s,t)\,\frac{f(s)-f(t)}{s-t}.$$

(This is easily checked by a direct computation if f is a polynomial.) Hence we have a concrete analogue of Problem 1:

PROBLEM 2. *Find a necessary and sufficient condition on f that whenever $K(s,t)$ is the kernel of a bounded operator on $L^2(a,b)$ then so also is the kernel $(*)$.*

NATURAL SCIENCES DIV.
UNIVERSITY OF CALIFORNIA AT SANTA CRUZ
SANTA CRUZ, CA 95064
USA

EDITORS' NOTE

Both problems 1 and 2 were extensively investigated by M. Š. Birman and M. Z. Solomyak within the very general scope of their theory of double operator integrals ([1], [2] and references therein; see also previous papers [3], [4]). They obtained a series of sharp sufficient conditions mentioned in Problem 2 and also sharp sufficient conditions for f to be differentiable on the set of all selfadjoint operators (Birman and Solomyak considered the Gâteaux differentiation but their techniques actually gives the existence of the Fréchet differential). Let us cite some results.

Suppose that $[a,b] \subset (0,T)$ and f can be extended from $[a,b]$ as a T-periodic function with Fourier series $\sum_{k=-\infty}^{\infty} \hat{f}(k)e^{\frac{2\pi i}{T}kt}$. Put $R_n(t) = \sum_{|k|\geqslant n} \hat{f}(k)e^{\frac{2\pi i}{T}kt}$. If there exists a sequence $\{\varepsilon_n\}_{n\geqslant 1}$ of positive numbers with $\sum \varepsilon_n < +\infty$ such that

$$\sup_{0\leqslant t\leqslant T} \sum_{n\geqslant 1} \varepsilon_n^{-1} |R_n(t)|^2 < +\infty,$$

then the kernel $(*)$ defines a bounded operator on $L^2(a,b)$ whenever $K(s,t)$ does. In particular this is the case if

$$\sum_{n=1}^{\infty} \|R_n\|_\infty < +\infty. \tag{1}$$

Condition (1) is satisfied e.g. if f' belongs to the Hölder class Λ_α with a positive (arbitrary small) α or if f' has absolutely convergent Fourier series.

If f is defined on the whole real line and $f \,\big|\, [a,b]$ satisfies the above condition for any $a, b \in \mathbb{R}$ then f is differentiable on the set of all selfadjoint operators.

The Birman–Solomyak theory encompasses many other related problems (e.g. for unbounded selfadjoint operators and for the differentiation with respect to an operator ideal). In particular they considered Problem 2 in a more general setting, namely replacing the quotient $\frac{f(s)-f(t)}{s-t}$ by a function $h(s,t)$. They reformulated this general problem as follows: for which $h(s,t)$ is $\varphi(s)\psi(t)h(s,t)$ the kernel of a nuclear operator $T_{\varphi,\psi}$ for any $\varphi, \psi \in L^2$ with $\|T_{\varphi,\psi}\|_{\mathfrak{S}_1} \leqslant \text{const}\, \|\varphi\|_{L^2}\|\psi\|_{L^2}$? This equivalence leads (via V. V. Peller's criterion [5] of nuclearity of Hankel operators) to a *necessary condition* for f to satisfy the requirements of Problems 2 and 1. Indeed, putting $\varphi = \psi \equiv 1$ we see that $\frac{f(s)-f(t)}{s-t}$ should be the kernel of a nuclear operator. It follows from [5], [6] that *this is the case iff f belongs to the Besov class $B_1^1[a,b]$ for any $a, b \in \mathbb{R}$. So the condition $f \in C^1$ is not sufficient in both Problems 1 and 2.*

Let us mention also an earlier paper by Yu. B. Farforovskaya [7] where explicit examples of *selfadjoint operators A_n, B_n with spectra in $[0,1]$ and of functions f_n are constructed such that $\|A_n - B_n\| \to 0$, $|f_n(x) - f_n(y)| \leqslant |x-y|$ and $\frac{\|f_n(A_n)-f_n(B_n)\|}{\|A_n-B_n\|} \to \infty$.* Note that the existence of such sequences $\{A_n\}$, $\{B_n\}$, $\{f_n\}$ follows also from the above mentioned Peller's results.

References

1. Birman M. Sh., Solomyak M. Z., *Notes on the function of spectral shift*, Zap. Nauchn. Semin. LOMI **27** (1972), 33–46. (Russian)
2. Birman M. Sh., Solomyak M. Z., *Double Stieltjes operator integrals*, III, Problems of Math. Physics (1973), no. 6, 27–53. (Russian)
3. Krein M. G., *Some new studies in the theory of perturbations of the self-adjoint operators*, First Math. Summer School, I, Kiev, 1964, pp. 103–187. (Russian)
4. Daletskii Yu. L., Krein S. G., *Integration and differentiation of functions of Hermitian operators and applications to the theory of perturbations*, Trudy Sem. Funkts. Anal., I, Voronezh, 1956, pp. 81–105. (Russian)
5. Peller V. V., *Hankel operators of class $\mathfrak{S}_p$ and their applications (rational approximation, Gaussian processes, the problem of majorization of operators)*, Matem. Sbornik **113** (1980), no. 4, 539–581. (Russian)
6. Peller V. V., *Vectorial Hankel operators, commutators and related operators of the Schatten–von Neumann class $\mathfrak{S}_p$*, Integr. Equat. and Oper. Theory **5** (1982), no. 2, 244–272.
7. Farforovskaya Yu. B., *An estimate of the norm $| f(B) - f(A) |$ for selfadjoint operators A and B*, Zapiski nauchn. sem. LOMI **56** (1976), 143–162 (Russian); English transl. in J. Soviet Math. **14** (1980), no. 2, 1133–1149.

Commentary by V. V. Peller

It is explained in the Note above that the problem under consideration essentially reduces to the problem of the description of the functions f for which the function $\check{f}$,

$$\check{f} = \frac{f(s) - f(t)}{s - t},$$

is a *Schur multiplier*, i.e. $\check{f}(s,t)\varphi(s)\psi(t)$ is the kernel of a nuclear integral operator for any φ, $\psi \in L^2(I)$ for any interval I.

It is technically more convenient to consider the same problem for functions f on the unit circle $\mathbb{T}$. The reasoning mentioned in the Note leads via the nuclearity criterion for Hankel operators to the fact that if $\check{f}$ is a Schur multiplier, then f belongs to the Besov space B_1^1 (see [8]). Using more sophisticated arguments, Peller [8] also showed that the nuclearity criterion allows one to obtain a stronger necessary condition. Let h_f be the harmonic extension of f to the unit disc. If $\check{f}$ is a Schur multiplier, then the measure

$$\left|(\nabla^2 h_f)(x,y)\right| dx\, dy$$

is a Carleson measure on the unit disc (see [8], [9]).

In [8] a sharp sufficient condition in order that $\check{f}$ be a Schur multiplier was found. It was shown that if f belongs to the Besov class $B_{\infty 1}^1$, then $\check{f}$ is a Schur multiplier.

In Peller's paper [10] similar results were obtained for functions f on the real line, where both necessary conditions and sufficient conditions were found for the function $\check{f}$ on $\mathbb{R} \times \mathbb{R}$ to be a Schur multiplier. This allows one to consider the case of unbounded operators.

Note that similar results for Schur multipliers of Schatten – von Neumann classes $\mathfrak{S}_p$, $0 < p < 1$, were obtained in [11].

Recently Arazy, Barton, and Friedman [12] improved the sufficient condition $f \in B_{\infty 1}^1$ for functions f on the unit circle. They showed that if f is analytic in the unit disc $\mathbb{D}$ and

$$\sup_{\zeta \in \mathbb{T}} \int_{\mathbb{D}} \frac{1 - |z|^2}{|1 - \bar{z}\zeta|^2} \, |f''(z)| \, dx\, dy < \infty$$

then $\check{f}$ is a Schur multiplier.

Let us mention some more problems in operator theory which lead to the problem of determining when $\check{f}$ is a Schur multiplier.

1. Let A and B be self-adjoint operators such that $A - B$ is nuclear. M. G. Krein [13] associated with each such pair a function ξ on $\mathbb{R}$ (the spectral shift) and proved that for sufficiently smooth functions f the following formula holds

$$\text{trace}(\varphi(A) - \varphi(B)) = \int_{\mathbb{R}} \varphi'(t)\xi(t)\,dt \tag{1}$$

(see [1], where the relationships between (1) and the problem under consideration are discussed in detail). The right-hand side of (1) is well defined for any Lipschitz function f. However Farforovskaya proved [14] that (1) cannot be true for any Lipschitz f. It follows from the results of [8] that (1) is valid if $f \in B^1_{\infty 1}(\mathbb{R})$ and the above necessary conditions on f are also necessary for the validity of (1). In particular there are functions f in $C^1(\mathbb{R})$ with compact support for which (1) is wrong.

2. In [15] Birman and Solomyak considered the following problem. Let $\mathcal{H}$ and $\mathcal{K}$ be Hilbert spaces, A a self-adjoint operator on $\mathcal{H}$, B a self-adjoint operator on $\mathcal{K}$, and J a bounded operator from $\mathcal{K}$ to $\mathcal{H}$. It is shown in [15] that if $\check{f}$ is a Schur multiplier, then under certain natural restrictions on an operator ideal $\mathfrak{S}$

$$\|f(A)J - Jf(B)\|_{\mathfrak{S}} \leqslant \text{const}\|AJ - JB\|_{\mathfrak{S}}.$$

The case considered in Widom's problem corresponds to the situation $\mathcal{H} = \mathcal{K}$, $J = I$. Another interesting case is $\mathcal{H} = \mathcal{K}$, $B = A$ which allows one to estimate the norm of the commutator $[f(A), J]$ in terms of the norm $[A, J]$.

3. Let A and B be bounded almost commuting self-adjoint operators on Hilbert space, i.e. the operator $AB - BA$ is nuclear. The problem is to construct a functional calculus on a class $\mathcal{C}$ of functions in two real variables which satisfies the following properties:

(i) $(\alpha_1\varphi_1 + \alpha_2\varphi_2)(A, B) = \alpha_1\varphi_1(A, B) + \alpha_2\varphi_2(A, B)$, $\varphi_1, \varphi_2 \in \mathcal{C}$, $\alpha_1, \alpha_2 \in \mathbb{C}$;
(ii) $(\varphi\psi)(A, B) - \varphi(A, B)\psi(A, B) \in \mathfrak{S}_1$, $\varphi, \psi \in \mathcal{C}$;
(iii) if $\varphi(s, t) = f(s)$, then $\varphi(A, B) = f(A)$;
if $\varphi(s, t) = g(t)$, then $\varphi(A, B) = g(B)$.

It is easy to construct such a functional calculus on the class of polynomials. If $\varphi(s, t) = \sum \varphi_{nk} s^n t^k$, one can put

$$\varphi(A, B) = \sum \varphi_{nk} A^n B^k.$$

Pincus, Carey, and Helton and Howe (see [16], [17], [18]) proved that for the above polynomial calculus the following trace formula holds

$$\text{trace}\big(\varphi(A, B)\psi(A, B) - \psi(A, B)\varphi(A, B)\big) = \frac{1}{2\pi i} \iint_{\mathbb{C}} \left(\frac{\partial\varphi}{\partial s}\frac{\partial\psi}{\partial t} - \frac{\partial\varphi}{\partial t}\frac{\partial\psi}{\partial s} \right) g(s, t)\,ds\,dt, \tag{2}$$

where g is the so-called Pincus principal function which is associated with any pair (A, B) of almost commuting self-adjoint operators, g has compact support, and $g \in L^1(\mathbb{C})$. As in the case of the trace formula (1), the right-hand side is well-defined for any Lipschitz φ and ψ. It was shown by Peller [9] that the above necessary condition on f in order that $\check{f}$ be a Schur multiplier imply that it is impossible to construct a functional calculus on the class of continuously differentiable functions that satisfies (i)–(iii). Using the results by Birman and Solomyak [15] mentioned above, Peller [9] constructed a functional calculus satisfying (i)–(iii) and (2) on a big class of functions in two real variables.

4. In [19] Johnson and Williams considered the following problem. Let A be a self-adjoint operator on a Hilbert space H (in fact they considered a more general case when A is normal). The problem was to describe the operators B on H such that the range of the derivation Δ_B,

$$\Delta_B T \stackrel{\text{def}}{=} BT - TB, \qquad T \text{ bounded on } \mathcal{H},$$

is contained in the range of Δ_A. Surprisingly their results lead to the above problem about Schur multipliers. Roughly speaking they proved that the range of Δ_B is contained in the range of Δ_A if and only if $B = f(A)$, where f is a function such that $\check{f}$ is a Schur multiplier.

References

8. Peller V. V., *Hankel operators in the perturbation theory of unitary and selfadjoint operators*, Funktsional. Anal. i ego Prilozhen. **19** (1985), no. 2, 37–51. (Russian)
9. Peller V. V., *Functional calculus for a pair of almost commuting selfadjoint operators*, Report no. 14, Institut Mittag-Leffler, 1990/1991.
10. Peller V. V., *Hankel operators in the perturbation theory of unbounded selfadjoint operators*, Analysis and Partial Differential Equations. A collection of papers dedicated to Misha Cotlar, Marcel Dekker, Inc., 1990, pp. 529–544.
11. Peller V. V., *For which f does $A - B \in \mathfrak{S}_p$?*, Operator Theory: Advances and Applications **24** (1987), Birkhäuser Verlag, 289–294.
12. Arazy J., Barton T. J., Friedman Y., *Operator differentiable functions*, Integral Equations Operator Theory **13** (1990), 461–487.
13. Krein M. G., *On some investigations in the perturbation theory of selfadjoint operators*, The First Summer School, Kiev, 1964, pp. 103–187. (Russian)
14. Farforovskaya Yu. B., *An example of a Lipschitz function of a selfadjoint operator giving a nonnuclear increment under a nuclear perturbation*, Zapiski Nauchn. Semin. LOMI **30** (1972), 146–153. (Russian)
15. Birman M. S., Solomyak M. Z., *Operator integration, perturbations, and commutators*, Zapiski Nauchn. Semin. LOMI **170** (1989), 34–66. (Russian)
16. Pincus J., *Commutators and systems of singular integral equations*, I, Acta Math. **121** (1968), 219–249.
17. Helton J. W., Howe R., *Integral operators, commutators, traces, index, and homology*, Lecture Notes in Math. **345** (1973), 141–209.
18. Carey R. W., Pincus J. D., Indiana Univ. Math. J. **23** (1974), 1031–1042.
19. Johnson B. E., Williams J. P., *The range of a normal derivation*, Pacific J. Math. **58**, 105–122.

Steklov Mathematical Institute
St. Petersburg branch
Fontanka 27
St. Petersburg, 191011
Russia

and

Departement of Mathematics
Kansas State University
Manhattan
Kansas, 66506
USA

6.5

SPECTRAL SHIFT FUNCTION AND DOUBLE OPERATOR INTEGRALS

M. SH. BIRMAN

Questions to be discussed are related to works [1], [2] and to Problem 6.4. Let $\mathfrak{H}$ be a separable Hilbert space, $\mathfrak{S}_1$ the ideal of trace class operators on $\mathfrak{H}$. Put $V = V^* \in \mathfrak{S}_1$, $H_s = H_0 + sV$, $H_0 = H_0^*$, $s \in \mathbb{R}$. It is well known (M. G. Krein [3], [4]), that for a suitable function class of φ the following trace formula holds:

$$\text{(1)} \qquad \operatorname{Tr}(\varphi(H_1) - \varphi(H_0)) = \int_{\mathbb{R}} \varphi'(\lambda)\xi(\lambda)\, d\lambda, \qquad \xi = \bar{\xi} \in L_1(\mathbb{R}),$$

where ξ is the so called *spectral shift function* (s.s.f.) for the couple (H_1, H_0). Equality (1) is deduced in [3], [4] from the following representation of logarithm of the determinant of perturbation:

$$\log\det\big((H_1 - zI)(H_0 - zI)^{-1}\big) = \int_{\mathbb{R}} \frac{\xi(\lambda)\, d\lambda}{\lambda - z}, \qquad \operatorname{Im} z \neq 0.$$

Another possible direction to investigate s.s.f. was sketched in [2] (there and in [5] possibilities to enlarge the function class of φ, satisfying (1), are discussed). Let $E_s(\cdot)$ be the spectral measure of H_s. Then the following double operator integral (d.o.i.) representation holds:

$$\text{(2)} \qquad \varphi(H_1) - \varphi(H_0) = \int_{\mathbb{R}}\int_{\mathbb{R}} \frac{\varphi(\mu) - \varphi(\lambda)}{\mu - \lambda}\, dE_1(\mu)\, V\, dE_0(\lambda).$$

For general information on d.o.i. we refer to [1]. It is shown there (see also [6]) that (2) remains true and $\varphi(H_1) - \varphi(H_0) \in \mathfrak{S}_1$ whenever d.o.i. in the right hand side of (2) defines a continuous linear transformation (in the variable V) on the class $\mathfrak{S}_1$. Moreover, for a suitable function class of φ the following representations are valid (see [2]):

$$\frac{d\varphi(H_s)}{ds} = \int_{\mathbb{R}}\int_{\mathbb{R}} \frac{\varphi(\mu) - \varphi(\lambda)}{\mu - \lambda}\, dE_s(\mu)\, V\, dE_s(\lambda),$$

$$\text{(3)} \qquad \operatorname{Tr}\frac{d\varphi(H_s)}{ds} = \int_{\mathbb{R}} \varphi'(\lambda)\, d(\operatorname{Tr} E_s(\lambda)V)$$

(the derivative is understood with respect to the norm in $\mathfrak{S}_1$). It follows from (3) that

$$\text{(4)} \qquad \operatorname{Tr}(\varphi(H_1) - \varphi(H_0)) = \int_{\mathbb{R}} \varphi'(\lambda)\, d\Xi(\lambda),$$

where the Borel measure Ξ is defined by

$$\Xi(\delta) = \int_0^1 \operatorname{Tr}\bigl(E_s(\delta)V\bigr)\, ds, \qquad \delta \subset \mathbb{R}. \tag{5}$$

Comparing (4), (5) with (1) shows that for any interval $\delta \subset \mathbb{R}$

$$\int_\delta \xi(\lambda)\, d\lambda = \Xi(\delta), \tag{6}$$

and, therefore, Ξ is absolutely continuous. Relations (5), (6) are useful for investigation of s.s.f. But, for the time being, they do not lead to an independent approach to construction of s.s.f., because no direct deduction of absolute continuity of Ξ from (5) is available.

QUESTIONS.

1) *How can absolute continuity of Ξ be extracted from (5)?*
2) *Is it possible to deduce trace formula (1) from (2)?*

REFERENCES

1. Birman M. Sh., Solomyak M. Z., *Double Stieltjes operator integrals*, III, Problems of Math. Physics (1973), no. 6, 27–53. (Russian)
2. Birman M. Sh., Solomyak M. Z., *Notes on the function of spectral shift*, Zap. Nauchn. Semin. LOMI **27** (1972), 33–46. (Russian)
3. Krein M. G., *On the trace formula in the theory of perturbations*, Matem. Sbornik **33** (1953), no. 3, 597–626. (Russian)
4. Krein M. G., *Some new studies in the theory of perturbations of the self-adjoint operators*, First Math. Summer School, I, Kiev, 1964, pp. 103–187. (Russian)
5. Peller V. V., *Hankel operators in the perturbation theory of unitary and selfadjoint operators*, Funktsional. Anal. i Prilozhen. **19** (1985), no. 2, 37–51. (Russian)
6. Birman M. S., Solomyak M. Z., *Operator integration, perturbations, and commutators*, Zapiski Nauchn. Semin. LOMI **170** (1989), 34–66. (Russian)

DEPT. OF PHYSICS
ST. PETERSBURG STATE UNIVERSITY
STARYI PETERHOF, ST. PETERSBURG, 198904
RUSSIA

6.6
old

RE-EXPANSION OPERATORS AS OBJECTS OF SPECTRAL ANALYSIS

M. Sh. Birman

1. Notations. L^2_s and L^2_a are subspaces of even and odd functions in $L^2(\mathbb{R})$; Φ is the Fourier transform; Σ is the multiplication by $\operatorname{sign} x$ on $L^2(\mathbb{R})$; $K \stackrel{\text{def}}{=} \Phi^*\Sigma\Phi$ is the Hilbert transform; $Y \stackrel{\text{def}}{=} K\Sigma = \Phi^*\Sigma\Phi\Sigma$. Let $\mathcal{J}_s$ ($\mathcal{J}_a$) be the following unitary mapping from L^2_s (L^2_a) onto $L^2(\mathbb{R}_+)$:

$$u \mapsto \sqrt{2}\,u \mid \mathbb{R}_+.$$

Φ_s and Φ_a are the Fourier cosine and sine transforms on $L^2(\mathbb{R}_+)$; $\Pi \stackrel{\text{def}}{=} \Phi_a^*\Phi_s$, $M \stackrel{\text{def}}{=} i\Pi$. Let σ denote the multiplication by $\sigma_n = \operatorname{sign}(n+\frac{1}{2})$ on $\ell^2(\mathbb{Z})$. Integrals with singular kernels are understood in the sense of principal value.

2. Re-expansion operators appear quite often in scattering theory. Namely, the wave operators for a pair of self-adjoint operators H_0, H

$$W_\pm(H, H_0) = \operatorname*{s\text{-lim}}_{t\to\pm\infty} \exp(itH)\exp(-itH_0)$$

can be obtained as follows. A given function is expanded with respect to the eigenfunctions of H_0 and then the inverse transform using the eigenfunctions of H is taken. Let for example, H_s and H_a be the operators $-\frac{d^2}{dx^2}$ on $L^2(\mathbb{R}_+)$ with the domains defined by $u'(0) = 0$ and $u(0) = 0$ respectively. Then $W_\pm(H_a, H_s) = \pm M$. Indeed, let

$$g_0 = \Phi_s f_0, \qquad g_1 = \Phi_a f_1,$$
$$\exp(-itH_s)f_0 = u_0(t), \qquad \exp(-itH_a)f_1 = u_1(t).$$

Then

$$u_0(t)(x) = \sqrt{\frac{2}{\pi}} \int_{\mathbb{R}_+} e^{-ik^2t} g_0(k)\cos kx\, dk\,,$$

$$u_1(t)(x) = \sqrt{\frac{2}{\pi}} \int_{\mathbb{R}_+} e^{-ik^2t} g_1(k)\sin kx\, dk\,.$$

Simple calculations by the stationary phase method show that $u_0(t) \sim u_1(t)$ when $t \to \pm\infty$ provided $g_1 = \pm i g_0$.

The re-expansion operator Π arises in the polar decomposition of $A = -i\frac{d}{dx}$ on $L^2(\mathbb{R}_+)$ with the boundary condition $u(0) = 0$, namely, $A = M^*|A|$ and $A^* = M|A^*|$.

Let us verify the first equality. Since $|A| = (A^*A)^{1/2} = H_a^{1/2}$, we have (using the above notation $g_1 = \Phi_a f_1$)

$$(|A|f_1)(x) = \sqrt{\frac{2}{\pi}} \int\limits_{\mathbb{R}_+} k \sin kx \, g_1(k)\, dk,$$

$$(Af_1)(x) = i\sqrt{\frac{2}{\pi}} \int\limits_{\mathbb{R}_+} k \cos kx \, g_1(k)\, dk,$$

hence

$$A = -i\Phi_s^*\Phi_a|A| = M^*|A|.$$

Concrete re-expansion operators are apparently interesting from the analytical point of view and as the objects of spectral analysis. In this connection (see Sect. 5) we propose some problems. But at first (in Sect. 3, 4) we use re-expansion operators as specimens for observations.

The author thanks N. K. Nikol'skii and M. Z. Solomyak, whose remarks are incorporated into the text.

3. Put $Y = Y_s \oplus Y_a$, in accordance with the decomposition $L^2(\mathbb{R}) = L_s^2 \oplus L_a^2$. It is easy to see that

$$Y_s = \mathcal{J}_s^* M \mathcal{J}_s, \qquad Y_a = \mathcal{J}_a^* M^* \mathcal{J}_a,$$

hence

$$(Mu)(x) = \frac{1}{\pi i} \int\limits_{\mathbb{R}_+} \frac{2xu(t)}{t^2 - x^2}\, dt. \tag{1}$$

The operator V defined by

$$Vu(\sigma) = e^{\sigma/2} u(e^{\sigma})$$

maps isometrically $L^2(\mathbb{R}_+)$ onto $L^2(\mathbb{R})$. Clearly, VMV^* coincides with the convolution operator $f \to f * \varphi$, where

$$\varphi(s) = -\frac{1}{\pi i}\frac{e^{s/2}}{\sin hs}.$$

Taking the Fourier transform, we obtain*

$$M = V^*\Phi^* E \Phi V, \tag{2}$$

where E is the multiplication by the function

$$\varepsilon(\tau) = \frac{\sin h\, \pi\tau + i}{\cos h\, \pi\tau}, \qquad \tau \in \mathbb{R} \tag{3}$$

on $L^2(\mathbb{R})$. It follows from (2) and (3) that the spectrum of Π (respectively of Y) is absolutely continuous and fills out the semi-circle $\mathbb{T} \cap \{ z \in \mathbb{C}\colon \operatorname{Re} z \geqslant 0 \}$ (respectively $\mathbb{T}$). These imply that Y *is unitarily equivalent to the shift operator* on $\ell^2(\mathbb{Z})$. Note that

$$|\Phi_a - \Phi_s| = |I - \Pi| = \sqrt{2}$$

and that the equality $\Phi_a u = \Phi_s u$ is impossible for $u \in L^2(\mathbb{R}_+) \setminus \{0\}$, though it holds for $u_0(t) = t^{-1/2}$ and in fact $\Phi_a u_0 = \Phi_s u_0 = u_0$.

*The spectral decomposition of M can be also deduced from [1] (see Ch. IX) but the proof given here is more direct and simple.

4. A re-expansion operator on $L^2(\Delta)$, $\Delta = (-\pi, \pi)$, with analogous properties appears in connection with the system

$$\left\{\frac{1}{\sqrt{2\pi}} \exp i(n + \frac{1}{2})t\right\}, \qquad n \in \mathbb{Z},$$

(but not with the usual trigonometric system). Let $\widetilde{\Phi}\colon L^2(\Delta) \to L^2(\mathbb{Z})$ be the Fourier transform corresponding to this system. Let $\Delta_+ = (0, \pi)$, $\widetilde{\Phi}_s$ and $\widetilde{\Phi}_a$ be maps of $L^2(\Delta_+)$ onto $\ell^2(\mathbb{Z}_+)$ corresponding to the systems

$$\left\{\sqrt{\frac{2}{\pi}} \sin(n + \frac{1}{2})t\right\} \quad \text{and} \quad \left\{\sqrt{\frac{2}{\pi}} \cos(n + \frac{1}{2})t\right\}.$$

Further put

$$\widetilde{\Pi} = \widetilde{\Phi}_a^* \widetilde{\Phi}_s, \quad \widetilde{M} = i\widetilde{\Pi}, \quad \widetilde{K} = \widetilde{\Phi}^* \sigma \widetilde{\Phi}.$$

The sense of the following notations is clear by analogy with Sect. 3. The operator $\widetilde{K}$ acts as follows

$$(4) \qquad (\widetilde{K}u)(t) = \frac{1}{2\pi i} \int\limits_{\Delta} \frac{u(s)\, ds}{\sin \frac{s-t}{2}}.$$

Changing variables by the formulae

$$f = Gu\,, \quad f(y) = \sqrt{2} \cos \frac{s}{2}\, u(s)\,, \quad y = \tan \frac{s}{2}\,,$$

we reduce $\widetilde{K}$ to the Hilbert transform: $\widetilde{K} = G^* K G$. This implies that $\widetilde{Y} \overset{\text{def}}{=} \widetilde{K}\widetilde{\Sigma}$ can be written in the form $\widetilde{Y} = G^* Y G$. Further, decompose $\widetilde{Y}$ into two parts (even and odd): $\widetilde{Y} = \widetilde{Y}_s \oplus \widetilde{Y}_a$. Then $\widetilde{Y}_s = G_s^* Y_s G_s$, where $G_s = G|\widetilde{L}_s^2$. Since $\widetilde{Y}_s = \widetilde{\mathcal{J}}_s^* \widetilde{M} \widetilde{\mathcal{J}}_s$ and $Y_s = \mathcal{J}_s^* M \mathcal{J}_s$, we obtain a unitary equivalence of $\widetilde{M}$ and M, namely

$$\widetilde{M} = (\widetilde{\mathcal{J}}_s G_s^* \mathcal{J}_s^*) M (\mathcal{J}_s G_s \widetilde{\mathcal{J}}_s^*) = G_+^* M G_+\,,$$

where $G_+ = G|L^2(\Delta_+)$. Note that $\widetilde{M}$ describes the non-trivial part of the scattering matrix for the diffraction on a semi-infinite screen (this is shown in [2], where a unitary equivalence of $\widetilde{M}$ and of the multiplication by function (3) is presented in an explicit form). Another (and a more elementary) situation where $\widetilde{M}$ appears is the following. Let $\mathcal{B}_0$ (resp. $\mathcal{B}_\pi$) be $-i\frac{d}{dx}$ on $L^2(\Delta_+)$ with the boundary condition $u(0) = 0$ (resp. $u(\pi) = 0$). Then $\mathcal{B}_0 = \widetilde{M}^* |\mathcal{B}_0|$, $\mathcal{B}_\pi = \widetilde{M} |\mathcal{B}_\pi|$. To write $\widetilde{Y}$ in a matrix form we note that the operator $\widetilde{\Phi}\widetilde{Y}\widetilde{\Phi}^*$ $(= \sigma\widetilde{\Phi}\widetilde{\Sigma}\widetilde{\Phi}^*)$ on $\ell^2(\mathbb{Z})$ has the following bilinear form

$$\frac{2i}{\pi} \sum_{\substack{n,m \in \mathbb{Z} \\ n \not\equiv m \pmod 2}} \frac{\sigma_n a_n \bar{b}_m}{n - m}.$$

5. Problems.

1) The equality $\Pi = -iM = -i\mathcal{J}_s Y_s \mathcal{J}_s^*$ implies that Π is bounded on $L^p(\mathbb{R}_+)$, $1 < p < \infty$. *What is the norm of $I - \Pi$?* (If $p = 2$ see (4)). It is not excluded that the answer can be extracted from the results of [1], Ch. IX.

2) Multi-dimensional analogues of the operator Y can be described in the following way. Let λ, ν be unimodular functions on $\mathbb{R}^m$ satisfying $\lambda(rx) = \lambda(x)$, $\nu(rx) = \nu(x)$ for $r > 0$. Let L, N be multiplications by these functions on $L^2(\mathbb{R}^m)$. If Φ_m is the Fourier transform on $L^2(\mathbb{R}^m)$ then $Y \stackrel{\text{def}}{=} \Phi_m^* L \Phi_m N$ is a unitary operator. *It would be of interest to investigate its spectral properties.* It might be reasonable to impose some additional conditions on λ and ν (e.g. some symmetry conditions).

3) Let q be an even positive function on $\mathbb{R}$ and $\{p_n\}$ be the orthogonal family of polynomials in $L^2(-a, a; q)$, $0 < a \leqslant \infty$. Then an analogue of the operator $\widetilde{\Pi}$ appears in $L^2(0, a; q)$, namely, the re-expansion operator from the even polynomials $\{p_{2n}\}$ to the odd ones $\{p_{2n+1}\}$. *It would be of interest to investigate its spectral properties.*

4) Consider the following systems in $L^2(\Delta_+)$:

$$\left\{\sqrt{\frac{2}{\pi}}\,\sin nt\right\}, \quad \left\{\sqrt{\frac{2}{\pi}}\,\cos nt\right\}, \qquad n \in \mathbb{N}.$$

The second system has defect 1. Let P be the re-expansion operator "from sines to cosines". This is a semi-unitary operator on $L^2(\Delta_+)$ with defect indices $(1, 0)$. *Is it completely nonunitary? In other words is the orthogonal system $\{P^n \mathbf{1} : n \geqslant 0\}$ complete in $L^2(\Delta_+)$?*

5) The operator Y is connected with the harmonic conjugation. *What does the theory of invariant subspaces of the shift mean in terms of Y? What is the role of zeros and poles of the function* (3) *in this connection?*

References

1. Gohberg I. C., Krupnik N. Ya., *One-dimensional linear singular integral equations*, "Stiinca", Kishinёv, 1973 (Russian); German transl. *Einführung in die Theorie der eindimensionalen singulären Integral-operatoren*, Birkhäuser Verlag, 1979.
2. Il′in E. M., *Scattering characteristics of a problem of diffraction by a wedge and by a screen*, Zapiski nauchn. sem. LOMI **107** (1982), 193–197 (Russian); English transl. in J. Soviet Math. **36** (1987), no. 3, 417–419.

Dept. of Physics
St. Petersburg State University
Staryi Peterhof, St. Petersburg, 198904
Russia

6.7

POINTWISE CONVERGENCE TO THE INITIAL DATA FOR SOLUTIONS TO SCHRÖDINGER TYPE EVOLUTION EQUATIONS

M. BEN-ARTZI AND A. DEVINATZ

Let $H = -d^2/dx^2$, considered as a self-adjoint operator in $L^2(\mathbb{R})$, and let $u(t,x) = (\exp itH)u_0(x)$, $u_0 \in L^2(\mathbb{R})$ be a solution to the Schrödinger initial value problem $i\partial u/\partial t = Hu$, $u(0,x) = u_0(x)$. In 1979 L. Carleson [5] showed that if $u_0 \in H^{1/4}(\mathbb{R})$ then

$$\lim_{t\to 0} u(t,x) = u_0(x), \qquad \text{a.e. in } x. \tag{1}$$

It was subsequently shown by Dahlberg and Kenig [6] that the result does not hold for $u_0 \in H^s(\mathbb{R})$ for $s < 1/4$.

Since then there have been various attempts to obtain similar results for solutions to n-dimensional initial value problems of Schrödinger type. For $H = -\Delta$, considered as self-adjoint operator in $L^2(\mathbb{R}^n)$, $n \geqslant 2$, the correct exponent s appears to be unknown. The best result to date seems to be $u_0 \in H^s(\mathbb{R}^n)$, $s > 1/2$, see [2], [7], [8] and [9]. In a recent paper J. Bourgain [3] has shown that (1) is true for $u_0 \in H^\rho(\mathbb{R}^2)$ for some $\rho < 1/2$.

In [2] the proposers considered this problem for a wide class of operators of the form $H = P(D)$, where $P(\xi)$ is a suitable symbol. For these symbols we obtained the result that the a.e. convergence to the initial function was true for $u_0 \in H^s(\mathbb{R}^n)$, $s > 1/2$. This convergence was an immediate consequence of certain global estimates which were obtained for solutions of the corresponding Schrödinger type initial value problem. (See also [4], [6], [8], and [9]). The methods used by the proposers were based on the proposers' approach to limiting absorption principles exposed in [1].

The methods used in the just mentioned papers do not seem to yield any information on the best possible exponent for which (1) is true. Hence the question is the following:

Taking $u_0 \in H^s(\mathbb{R}^n)$, what is the best possible exponent s for (1) to hold when $H = -\Delta$, or when $-\Delta$ is replaced by a more general self-adjoint operator in $L^2(\mathbb{R}^n)$?

REFERENCES

1. Ben-Artzi M., Devinatz A., *The limiting absorption principle for partial differential operators*, Memoirs Amer. Math. Soc. **66** (1987), no. 364.
2. Ben-Artzi M., Devinatz A., *Local smoothing and convergence properties of Schrödinger type equations*, J. Functional Analysis (to appear).
3. Bourgain J., *A remark on Schrödinger operators*, Israel J. Math (to appear).
4. Constantin P., Saut J.-C., *Local smoothing properties of dispersive equations*, J. Amer. Math. Soc. **1** (1988), 413–439.
5. Carleson L., *Some analytical problems related to statistical mechanics*, Lecture Notes in Math. **779** (1979), 5–45.

6. Dahlberg B. E. J., Kenig C. E., *A note on almost everywhere behavior of solutions to the Schrödinger equation*, Lecture Notes in Math. **908** (1981), 205–209.
7. Kenig C. E., Ponce G., Vega L., *Oscillatory integrals and regularity of dispersive equations*, Indiana U. Math. J. **40** (1991), no. 1, 33-69.
8. Sjölin P., *Regularity of solutions to the Schrödinger equation*, Duke Math. J. **55** (1987), 699–715.
9. Vega L., *Schrödinger equations: Pointwise convergence to the initial data*, Proc. Amer. Math. Soc. **102** (1988), 874–878.

INSTITUTE OF MATHEMATICS
HEBREW UNIVERSITY
JERUSALEM 91904
ISRAEL

DEPARTMENT OF MATHEMATICS
NORTHWESTERN UNIVERSITY
EVANSTON, ILLINOIS 60208
U.S.A.

6.8

ENERGY ESTIMATES FOR LIMITING RESOLVENTS

M. BEN-ARTZI AND A. DEVINATZ

The function $u(t,x) = \exp(it\Delta)u_0(x)$ is a solution to the Schrödinger initial value problem $i\partial u/\partial t = -\Delta u$, $u(0,x) = u_0(x) \in L^2(\mathbb{R}^n)$. In [4] the proposers showed that if we write $u_0 = u_1 + u_2$, where u_1 and u_2 are square integrable, and the Fourier transform, $\hat{u}_1(\xi)$, vanishes for $|\xi| \leqslant R$ and $\hat{u}_2(\xi)$ vanishes for $|\xi| \geqslant R$, $R > 0$, then if $u_1(t,x) = \exp(it\Delta)u_1(x)$ and $u_2(t,x) = \exp(it\Delta)u_2(x)$, it follows that $u_1(t,x)$ is analytic in $\mathbb{R} \times \mathbb{R}^n$ and for every multi-index α, $sup_{t,x}|D^\alpha u_1(t,x)| \leqslant C_\alpha \|u_0\|_{L^2}$, while $(1+|x|^2)^{-1/2}u_2(t,x) \in L^2(\mathbb{R}_t; H^{1/2}(\mathbb{R}^n_x))$. These are, of course, local smoothness result in x for $u(t,x)$, a.e. $t \in \mathbb{R}$. (See also [5], [7], [8] and [9]).

The result for $u_1(t,x)$ is trivial, while the result for $u_2(t,x)$ is based on the proposers' approach to limiting absorption principles in [3]. It is obtained from an inequality which is proved making essential use of "energy estimates" for the limiting resolvents $R_0^\pm(\lambda) = \lim_{\varepsilon\to 0+}(-\Delta - \lambda \pm i\varepsilon)^{-1}$. In the case under consideration these estimates are

$$(1) \qquad \|R_0^\pm(\lambda)\|_{L^{2,s},L^{2,-s}} \leqslant C\lambda^{-1/2}, \qquad \lambda > 0, \qquad s > 1/2,$$

where $L^{2,s} = \{\, f : \|f\|_s^2 = \int_{\mathbb{R}^n}(1+|x|^2)^s|f(x)|^2\,dx < \infty \,\}$.

I. High Energy Estimates. The problem is to try to reproduce the above results for Schrödinger operators with variable potentials; i.e. replacing $-\Delta$ by $H = -\Delta + V$, where V is a symmetric short range potential. Assuming that H has no eigenvalues, in order to get the above results it is enough to get "high energy" estimates of the for (1); i.e. for λ sufficiently large, for the limiting resolvents $R^\pm(\lambda)$ connected with H. Results for a limited class of short range potentials have been obtained in [2] and [6].

One possible approach to this problem is to consider the wave operators $W_\pm = s\text{-}\lim_{t\to\pm\infty} \exp(itH)\exp(it\Delta)$. As is well known (assuming H has no eigenvalues) these operators are unitary on $L^2(\mathbb{R}^n)$ and give a unitary equivalence between H and $-\Delta$. Thus for any $\varepsilon > 0$,

$$(2) \qquad R(\lambda \pm i\varepsilon) = W_\pm R_0(\lambda \pm i\varepsilon)W_\pm^*.$$

If we knew that $W_\pm$ could be extended as bounded operators from $L^{2,-s}$ to $L^{2,-s}$, $s > 1/2$, we could let $\varepsilon \to 0$ in (2) and thus use the estimates for $R_0^\pm(\lambda)$ to obtain the same estimates for $R^\pm(\lambda)$.

Thus we are led to two specific questions:

1. Can estimates of the form (1) be obtained for $R^\pm(\lambda)$ for high energy values λ?
2. In the absence of eigenvalues for H, can the wave operators be extended as bounded operators from $L^{2,-s}$ to $L^{2,-s}$, $s > 1/2$?

II. Low Energy Estimates. The question is whether the estimates (1) can be obtained for the limiting resolvents $R^{\pm}(\lambda)$ connected with $H = -\Delta + V$ for small values of the energy variable λ? This appears to be unknown even for real $V \in C_0^\infty(\mathbb{R}^n)$. In the one dimensional case this type of result has been obtained in [1]. Of course a positive answer to question 2 above would give such an estimate for all $\lambda > 0$.

REFERENCES

1. Ben-Artzi M., Dermenjian Y. and Guillot J.-C., *Acoustic waves in stratified fluids: A spectral theory*, Comm. in Partial Differential Equations **14** (1989), no. 4, 479–517.
2. Ben-Artzi M., *Global estimates for the Schrödinger equation*, J. Functional Analysis (to appear).
3. Ben-Artzi M., Devinatz A., *The limiting absorption principle for partial differential operators*, Memoirs Amer. Math. Soc. **66** (1987), no. 364.
4. Ben-Artzi M., Devinatz A., *Local smoothing and convergence properties of Schrödinger type equations*, J. Functional Analysis (to appear).
5. Constantin P., Saut J.-C., *Local smoothing properties of dispersive equations*, J. Amer. Math. Soc. **1** (1988), 413–439.
6. Constantin P., Saut J.-C., *Local smoothing properties of Schrödinger equations*, Indiana U. Math. J. **38** (1989), 791–810.
7. Kenig C. E., Ponce G., Vega L., *Oscillatory integrals and regularity of dispersive equations*, Indiana U. Math. J. **40** (1991), no. 1, 33–69.
8. Sjölin P., *Regularity of solutions to the Schrödinger equation*, Duke Math. J. **55** (1987), 699–715.
9. Vega L., *Schrödinger equations: Pointwise convergence to the initial data*, Proc. Amer. Math. Soc. **102** (1988), 874–878.

INSTITUTE OF MATHEMATICS
HEBREW UNIVERSITY
JERUSALEM 91904
ISRAEL

DEPARTMENT OF MATHEMATICS
NORTHWESTERN UNIVERSITY
EVANSTON, ILLINOIS 60208
U.S.A.

CHAPTER EDITOR'S NOTE

Concerning (1) for large $\lambda > 0$ see [10].

REFERENCE

10. Agmon S., *Spectral properties of Schrödinger operators and scattering theory*, Ann. Scuola Norm. Sup. Pisa II **2** (1975), 151–218.

6.9
old

SCATTERING THEORY FOR COULOMB TYPE PROBLEMS

L. A. SAKHNOVICH

1. Let self-adjoint operators A and A_0 act in a Hilbert space H and suppose the spectrum of A_0 is absolutely continuous. Suppose further that there exists a unitary operator function $W_0(t)$ satisfying the conditions: $W_0(t)W_0(\tau) = W_0(\tau)W_0(t)$, $W_0(t)A_0 = A_0W_0(t)$, $\operatorname{s-lim}_{t\to\infty} W_0^*(t+\tau)W_0(t) = E$ and there exist limits

$$\operatorname{s-lim}_{t\to\pm\infty} \exp(iAt)\exp(-iA_0t)W_0(t) = U_\pm(A, A_0), \tag{1}$$

$$\operatorname{s-lim}_{t\to\pm\infty} W_0^{-1}(t)\exp(iA_0t)\exp(-iAt)P = U_\pm(A_0, A), \tag{2}$$

where P is the orthogonal projection onto the absolutely continuous part of A. For the generalized wave operators $U_\pm$ [1–5] the equality $AU_\pm(A, A_0) = U_\pm(A, A_0)A_0$ holds. The factor $W_0(t)$ is not uniquely defined. The factor $\widetilde{W_0}(t) = W_0(t)V_\pm$ $(t \gtrless 0)$ can be used too when obvious requirements to $V_\pm$ are fulfilled. Due to this ambiguity the naturally looking definition

$$S(A, A_0) = U_+^*(A, A_0)U_-(A, A_0) \tag{3}$$

of the scattering operator becomes senseless since $\widetilde{S}(A, A_0) = V_+^*S(A, A_0)V_-$ is in fact an arbitrary unitary operator commuting with A_0.

PROBLEM 1. *Find physically motivated normalization of $W_0(t)$ when $t \to \pm\infty$, removing the non-uniqueness in definition of the scattering operator.*

This problem has been solved for the scattering with the Coulomb main part, i.e. when

$$Af = -\frac{d^2f}{dr^2} + \left[\frac{l(l+1)}{r^2} - \frac{2z}{r} + q(r)\right]f, \qquad A_0f = -\frac{d^2f}{dr^2} + \frac{l(l+1)}{r^2}f, \tag{4}$$

where $0 < r < \infty$, $l > 0$. For the system (4) the factor $W_0(t)$ is of the form [1–2]: $W_0(t) = \exp\left[i(\operatorname{sign} t)\ln|t|\, z/\sqrt{A_0}\,\right]$. In [2] it is proved that with such $W_0(t)$ the scattering operator (3) coincides with the results of the stationary scattering theory. This fact suggests that the normalization of $W_0(t)$ is physically reasonable. A similar problem has been solved for the Dirac equations with the Coulomb main part [5]. Consider the system

$$Af = -\frac{d^2f}{dx^2} + q(x)f, \qquad A_0f = -\frac{d^2f}{dx^2}, \qquad -\infty < x < \infty, \tag{5}$$

where $q(x) \approx \alpha_\pm/x$, $x \to \pm\infty$.

PROBLEM 2. *Find $W_0(t)$ and solve Problem 1 for the system (5).*

When $\alpha_+ = -\alpha_-$ the system (5) can be reduced to the Coulomb case, when $\alpha_+ = \alpha_-$ it is considered in [2]. The case $\alpha_- = 0$, $\alpha_+ \neq 0$ is of importance in a number of physical problems [6–7].

2. The Coulomb interaction of n particles is described by the system

$$A = -\sum_{k=1}^{n} \Delta_k/2m_k + \sum_{1\leqslant k<l\leqslant n} z_{k,l}/r_{k,l}, \qquad A_0 = -\sum_{k=1}^{n} \Delta_k/2m_k\,, \tag{6}$$

where r_k is the radius-vector of the k-th particle, $r_{k,l} = |r_k - r_l|$. Taking bound states into account leads to the transition from A_0 to the extended operator $\widetilde{A}_0$ [8]. The operators $\widetilde{A}_0$ and $\widetilde{W}_0(t)$ for the system (6) have been constructed in [9]. The construction is effective when $n = 3$. In [9] the existence of $U_\pm(A, \widetilde{A}_0)$ is proved.

PROBLEM 3. *Prove the existence of $U_\pm(\widetilde{A}_0, A)$ for the system (6), i.e. the completeness of the corresponding wave operators.*

3. We shall consider the non-standard inverse

PROBLEM 4. *Let operator A_0 and generalized scattering operator S be known. Recover the operator A and the corresponding $W_0(t)$, find classes where the problem has one and only one solution.*

Consider a model example, namely

$$A_0 f = xf(x)\,, \qquad Af = xf(x) + i\left[\int_a^x f(t)p(t)\,dt - \int_x^b f(t)p(t)\,dt\right]p(x)\,, \tag{7}$$

where $p(x)$ belongs to the Hölder class and $p(x) > 0$, $x \in (a, b)$. There is an effective solution of the "direct" problem (the construction of $W_0(t)$ and of S) for the system (7) [2]:

$$W_0(t)f = f(x)\exp\left[ip^2(x)\ln|t|\right], \quad Sf = f(x)s(x)\,, \quad s(x) = \psi_-(x)/\psi_+(x)\,, \tag{8}$$

where $\psi_\pm = \lim_{y\to\pm 0} \psi(x + iy)$ and

$$\psi(z) = \frac{\Phi(z) + \Phi^{-1}(z)}{2}\,, \qquad \Phi(z) = \exp\left[\frac{i}{2}\int_a^b \frac{p^2(u)}{u - z}\,du\right]. \tag{9}$$

Suppose in addition that

$$\int_a^b \left[p^2(t)/(t-a)\right] dt < \pi\,, \qquad \int_a^b \left[p^2(t)/(b-t)\right] dt < \pi\,.$$

Then operator A has no discrete spectrum. The formulae (8), (9) give an effective solution of Problem 4 for system (7) by the factorization method, i.e. $\Phi(z)$ and $p(x)$ are found by $S(x)$ and hence A and $W_0(t)$. Some problems with non-local potentials and the case (5) with $\alpha_+ = \alpha_-$ can be reduced to the system (7).

4. Now we come to *stationary inverse problems.* When $z > 0$ the discrete spectrum of the operator A (see (4)) is described by the Ritz half-empiric formula

$$\lambda_n = -z^2/\left[n + l + \delta_l + \varkappa(n,l)\right]^2, \qquad \varkappa(n,l) \to 0, \quad n \to \infty;$$

its proof is given in [10]. The number δ_l is called a quantum defect of the discrete spectrum. The same number δ_l serves as a deviation measure of the operator (4) from the case of hydrogen nucleus $[q(r) = 0]$.

PROBLEM 5. *Find a method to recover the potential $q(r)$ from δ_l $(l = 0, 1, 2, \dots)$.*

Here the representation of δ_l by solutions of the Schrödinger equation [10] and the transformation operator [11] can be useful. The definition of the quantum defect δ_k $(k = \pm 1, \pm 2, \dots)$ is introduced in [12] for the Dirac radial equation too

$$\frac{d}{dr}\psi \begin{bmatrix} 0 & 1 \\ -1 & 0 \end{bmatrix} - \psi H(r) - \lambda\psi = 0, \qquad \psi = [\psi_1, \psi_2],$$

$$H(r) = \begin{bmatrix} \frac{2z}{r} + q(r) - m & -\frac{k}{r} \\ -\frac{k}{r} & \frac{2z}{r} + q(r) + m \end{bmatrix}. \tag{10}$$

So Problem 5 can be formulated for the case (10) too. Note that the classic inverse problem for the Dirac equation has not yet been solved even when the Coulomb member is missing. The systems of the Dirac type have been thoroughly investigated when (see (10))

$$H(r) = \begin{bmatrix} V(r) - m & W(r) \\ W(r) & V(r) + m \end{bmatrix},$$

where $V(r)$, $W(r)$ are real functions from $L(0, \infty)$. The peculiarity of the system (10) is defined by the fact that the element $W(r)$ is known $[W(r) = -K/r \,\overline{\in}\, L(0, \infty)]$. It is therefore perhaps not necessary to use the scattering (or spectrum) data over the whole energy interval $-\infty < \lambda < \infty$. We come to a peculiar half-inverse problem both for the Coulomb and non-Coulomb case.

PROBLEM 6. *Let z, m, $W(r) = -k/r$ $(k = \pm 1, \pm 2, \dots)$ be known. Reconstruct the element $V(r) = 2z/r + q(r)$ by the scattering (or spectrum) data, belonging to the energy interval $0 \leqslant \lambda < \infty$.*

The model half-inverse problem has been solved in [13].

REFERENCES

1. Dollard J., *Asymptotic convergence and Coulomb interaction,* J. Math. Phys. **5** (1964), 729–738.
2. Sakhnovich L. A., *Generalized wave operators,* Matem. Sb. **81** (1970), no. 2, 209–227. (Russian)
3. Sakhnovich L. A., *Generalized wave operators and regularization of the perturbation expansion,* Teoret. Mat. Fiz. **2** (1970), no. 1, 80–86. (Russian)
4. Buslaev V. S., Matveev V. B., *Wave operators for the Schrödinger equation with slowly decreasing potential,* Teoret. Mat. Fiz. **2** (1970), 367–376. (Russian)
5. Sakhnovich L. A., *The invariance principle for generalized wave operators,* Funkts. Anal. i Prilozh. **5** (1971), no. 1, 61–68. (Russian)
6. Tunel effects in the solids, "Mir", Moscow, 1973. (Russian)

7. Brodskii A. M., Gurevich A. Yu., *Theory of electron emission from metals*, 1973. (Russian)
8. Faddeev L. D., *Mathematical Questions in the quantum theory of scattering for a system of three particles*, Trudy Matem. Inst. Steklov AN SSSR **69** (1963). (Russian)
9. Sakhnovich L. A., *On taking into account all scattering channels in the N-body problem with Coulomb interaction*, Teoret. Mat. Fiz. **13** (1972), no. 3, 421–427. (Russian)
10. Sakhnovich L. A., *On the Ritz formula and the quantum defects of the spectrum for a radial Schrödinger equation problem with Coulomb interaction*, Izv. Akad. Nauk SSSR, ser. matem. **30** (1966), no. 6, 1297–1310. (Russian)
11. Kostenko N. M., *A transformation operator*, Izv. Vyssh. Uchebn. Zav. Matematika **9** (1977), 43–47. (Russian)
12. Sakhnovich L. A., *On properties of the discrete and continuous spectra of the radial Dirac equation*, Doklady Akad. Nauk SSSR **185** (1969), no. 1, 61–64. (Russian)
13. Sakhnovich L. A., *A semi-inverse problem*, Uspekhi Matem. Nauk **18** (1963), no. 3, 199–206. (Russian)

PR. DOBROVOLSKOGO 154, 199
ODESSA 270111
UKRAINE

6.10

UNIFICATION OF THE TRACE-CLASS AND SMOOTH APPROACHES IN SCATTERING THEORY

D. R. YAFAEV

Let H_0 and H be self-adjoint operators in a Hilbert space $\mathcal{H}$. The main problem of the mathematical scattering theory (ST) is the proof of the existence of the strong limits (of the wave operators)

$$W_{\pm}(H,H_0) = \operatorname*{s-lim}_{t\to\pm\infty} \exp(iHt)\exp(-iH_0t)\,P_0,$$

where P_0 is the orthogonal projection on the absolutely continuous subspace of H_0. The operator $W_{\pm}(H,H_0)$ is called complete if its range coincides with the absolutely continuous subspace of H. This is equivalent to the existence of $W_{\pm}(H,H_0)$.

There are two essentially different approaches in ST. The first of them (trace-class—see e.g. [1]) does not make any assumptions about the "unperturbed" operator H_0. Its basic result is the Kato–Rosenblum theorem which guarantees the existence of $W_{\pm}(H,H_0)$ if the perturbation $V = H - H_0$ belongs to the trace-class $\mathfrak{S}_1$. According to the Weyl–von Neumann–Kuroda theorem this condition cannot be relaxed in terms of operator ideals. The second, smooth, approach relies on a certain regularity of the eigenfunction expansion of the operator H_0. The perturbation V is supposed to be in some sense smooth with respect to H_0. There are different versions of this notion. For example, in the Friedrichs model [2] V is an integral operator with smooth kernel in the spectral representation of H_0. The assumptions of trace-class and smooth ST are quite different. Thus the following problem naturally arises.

PROBLEM 1. *To develop a theory unifying the results of trace-class and smooth approaches.*

This problem admits of course different interpretations but becomes unambiguously posed in the context of applications, especially to differential operators (see Problem 2 below). In the abstract theory there exists a general scheme (see e.g. [3]) containing formally both approaches. This scheme is formulated in the framework of the so-called stationary method in which wave operators are constructed in terms of the resolvents of the operators H_0 and H (instead of their unitary groups). The results of [3] require assumptions on the resolvents of both operators H_0 and H. This is not convenient for verification. Here we shall give more explicit result of such type which is closer in spirit to the perturbation theory. Let $\mathfrak{S}_r$ be the ideal of compact operators T such that the sequence of eigenvalues of $|T|$ belongs to the space ℓ_r. Suppose that $V = G^*\mathcal{V}G$, where $\mathcal{V} = \mathcal{V}^*$. For simplicity we assume that $\mathcal{V}$ and G are bounded and that the null-space of G is trivial. The proof of the following result can be found in [4].

THEOREM 1. *Suppose that the operator $B(z) = G(H_0 - z)^{-1}G^* \in \mathfrak{S}_r$, $\operatorname{Im} z \neq 0$, for some $r < \infty$ and that $B(z)$ has angular boundary values in $\mathfrak{S}_r$ as $z \to \lambda \pm i0$ for almost all $\lambda \in \mathbb{R}$. Then the wave operators $W_\pm(H, H_0)$ exist. (The wave operators $W_\pm(H_0, H)$ also exist so that $W_\pm(H, H_0)$ are complete.)*

This assertion resembles the following typical result (see e.g. [5]) of the smooth ST.

THEOREM 2. *Suppose that $B(z)$ is compact for $\operatorname{Im} z \neq 0$ and $B(z)$ depends in the operator norm continuously on the parameter z as it approaches an open set of full measure in $\mathbb{R}$. Then conclusions of Theorem 1 hold.*

Theorem 2 looks like a limit case of Theorem 1 corresponding to $r = \infty$. Compared to it Theorem 2 requires more with respect to smoothness of $B(z)$ in z but the operator topology in Theorem 2 is weaker than that in Theorem 1. On the other hand, Theorem 1 contains a typical result of the trace-class ST. In fact, as shown in [6], for an arbitrary self-adjoint operator A and arbitrary $T \in \mathfrak{S}_2$ the operator-function $T(A - z)^{-1}T^*$ has angular boundary values in $\mathfrak{S}_2$ (more precisely, in $\mathfrak{S}_r$ for every $r > 1$ but not in $\mathfrak{S}_1$, according to the results of [7]) as $z \to \lambda \pm i0$ for almost all $\lambda \in \mathbb{R}$. It follows that assumptions of Theorem 1 are fulfilled if $GE_0(X) \in \mathfrak{S}_2$ (here $E_0(\cdot)$ is the spectral family of H_0) for any bounded interval $X \subset \mathbb{R}$ and $G(|H_0| + I)^{-1/2} \in \mathfrak{S}_{2r}$ for some $r \in [2, \infty)$. Thus Theorem 1 seems to establish a bridge between trace-class and smooth results.

However, applications to differential operators show that this unification is not quite satisfactory. Suppose now that $\mathcal{H} = L_2(\mathbb{R}^d)$, $H_0 = -\Delta + p(x)$, $H = H_0 + q(x)$ where real bounded functions p and q satisfy the bounds $p = O(|x|^{-\alpha})$, $q = O(|x|^{-\beta})$ as $|x| \to \infty$. Trace-class ST shows (see e.g. [8]) that the wave operators $W_\pm(H, H_0)$ exist (and are complete) if p is an arbitrary bounded function (i.e. $\alpha = 0$) and $\beta > d$. Smooth ST establishes such results only under very special assumptions about p:

1) $p = 0$;
2) p is short-range, i.e. $\alpha > 1$;
3) p is long-range, i.e. $p = O(|x|^{-\varepsilon})$ and $|x|\,|\partial p/\partial |x|| = O(|x|^{-\varepsilon})$ for some $\varepsilon > 0$;
4) p is periodic;
5) p is a potential energy of a many-particle quantum system (with, possibly, long-range pair potentials).

On the other hand, in all these cases it requires only that $\beta > 1$ in the bound for q. Compared to these results, Theorem 1 does not give us new conditions of the existence of wave operators for the pair under consideration. In particular, it leaves open the following

PROBLEM 2. *Let $d > 1$. Do wave operators $W_\pm(H, H_0)$ exist for arbitrary $p \in L_\infty(\mathbb{R}^d)$ and q satisfying the bound $q = O(|x|^{-\beta})$ with $\beta > 1$ only?*

Note that in case of the positive solution of Problem 2 wave operators are automatically complete under its assumptions. We conjecture, on the contrary, that Problem 2 has a negative solution. Moreover, we expect that the absolutely continuous part of the spectrum is no longer stable in the situation under consideration.

This conjecture could be related to the results on the spectrum of the Schrödinger operator $H_0 = -\Delta + p$ with a "generic" bounded potential p. Indeed, as shown in [9], in case $d = 1$ the spectrum of H_0 is pure point and hence does not have an absolutely continuous part. Trace-class ST establishes that the latter result formulated in probabilistic terms has a certain stability with respect to perturbations $q = O(|x|^{-\beta})$, $\beta > 1$. It can be conjectured that for $d \geqslant 2$ the structure of the spectrum of the Schrödinger operator with a bounded potential p can be so wild that for short-range perturbations q a new absolutely continuous part can appear or disappear. Nevertheless this is not possible if $q = O(|x|^{-\beta})$ with $\beta > d$.

REFERENCES

1. Ahiezer N. I., Glasman I. M., *Theory of Linear Operators in Hilbert Space*, "Nauka", Moscow, 1966 (Russian); English transl. Ungar, New York, vol. 1, 1961, vol. 2, 1963.
2. Faddeev L. D., *On the Friedrichs model in the perturbation theory of the continuous spectrum*, Trudy Matem. Inst. Steklov AN SSSR **73** (1964), 292–313. (Russian)
3. Birman M. Sh., Yafaev D. R., *A general scheme in the stationary scattering theory*, Problems of Math. Phys., LGU **12** (1987), 89–117. (Russian)
4. Yafaev D. R., *Mathematical Scattering Theory. V.1. General theory*, Amer. Math. Soc., 1992.
5. Kuroda S. T., *Scattering theory for differential operators*, J. Math. Soc. Japan **25** (1973), 75–104, 222–234.
6. Birman M. Sh., Entina S. B., *Stationary approach in the abstract scattering theory*, Izv. AN USSR **31** (1967), 401–430. (Russian)
7. Naboko S. N., *On the boundary problems of analytic operator-functions with positive imaginary part*, Zapiski nauchn. semin. LOMI **157** (1987), 55–69. (Russian)
8. Birman M. Sh., Yafaev D. R., *On the trace method in the theory of potential scattering*, Zapiski nauchn. semin. LOMI **171** (1989), 12–35. (Russian)
9. Gol′dshtein I. Ya., Molchanov S. A., Pastur L. A., *A typical one-dimensional Schrödinger operator has a pure point spectrum*, Funkts. Anal.ı Prilozh. **11** (1977), no. 1, 1–10 (Russian); English transl. in Funct. Anal. Appl. **11** (1977), no. 1, 1–8.

STEKLOV MATHEMATICAL INSTITUTE
ST. PETERSBURG BRANCH
FONTANKA 27
ST. PETERSBURG, 191011
RUSSIA

AND

UNIVERSITÉ DE NANTES
DEPARTEMENT DE MATHÉMATIQUES
2 RUE DE LA HOUSSINIERE
44072 NANTES CEDEX 03
FRANCE

S.6.11
v.old

ZERO SETS OF OPERATOR FUNCTIONS WITH A POSITIVE IMAGINARY PART

L. D. FADDEEV AND B. S. PAVLOV

Let E be a separable Hilbert space, M be a function analytic in the unit disk $\mathbb{D}$, taking values in the space of bounded operators on E and continuous up to the boundary of $\mathbb{D}$. Suppose also that

$$M(\zeta) = I + C(\zeta), \qquad \zeta \in \operatorname{clos}\mathbb{D},$$

where $C(\zeta)$ is a compact operator on E. We also assume that the following properties are satisfied:

(A) For a modulus of continuity ω the inequality

$$|M(\zeta) - M(\zeta')| \leqslant \omega(|\zeta - \zeta'|)$$

holds with $\zeta, \zeta' \in \mathbb{D}$.

(B) M has a positive imaginary part in $\mathbb{D}$:

$$\operatorname{Im} M(\zeta) \overset{\text{def}}{=} \frac{M(\zeta) - M(\zeta)^*}{2i} > 0, \qquad \zeta \in \mathbb{D}.$$

A point ζ in $\operatorname{clos}\mathbb{D}$ will be called a *root* of M if

$$\inf_{\|e\|=1} \|M(\zeta)e\| = 0.$$

Since $I - M(\zeta)$ is compact, for any root ζ there exists $e \in E$ such that $M(\zeta)e = 0$. It is not hard to verify that the roots of a function with a positive imaginary part can lie only on $\mathbb{T}$. Denote the set of all roots of M by Λ and let $m\Lambda_\delta$ be the Lebesgue measure of its δ–neighborhood in $\mathbb{T}$.

CONJECTURE 1. *Under hypotheses* (A), (B) *the inequality* $m\Lambda_\delta \leqslant C\omega(\delta)$ *holds for a positive constant* C.

It seems to be natural to weaken hypothesis (A) and replace it by the following one taking into account the behavior of M only near set Λ.

(A′) $|M^{-1}(z)|^{-1} \leqslant \omega\big(\operatorname{dist}(z, \Lambda)\big)$.

CONJECTURE 2. *Under hypotheses* (A′), (B) *the inequality*

$$m\Lambda_\delta \leqslant C\omega(\delta)$$

holds for a positive constant C.

References

1. Faddev L. D., *On the Friedrichs model in perturbation theory*, Trudy Matem. Inst. Steklov AN SSSR **30** (1964), 33–75. (Russian)
2. Pavlov B. S. and Petras S. V., *On singular spectrum of a weakly perturbed multiplication operator*, Funkts. Anal. i Prilozh. **4** (1970), no. 2, 54–61. (Russian)
3. Pavlov B. S., *A uniqueness theorem for functions with a positive imaginary part*, Problems of Math. Physics, No. 4: Spectral Theory. Wave Processes, Leningr. State Univ., Leningrad, 1970, pp. 118–124. (Russian)

Steklov Mathematical Institute
St. Petersburg branch
Fontanka 27
St. Petersburg 191011
Russia

Department of Physics
St. Petersburg University
Petrodvoretz
St. Petersburg 198904
Russia

Commentary by S. N. Naboko

Conjecture 1 was proved in [4, 5] under the additional condition $C(\xi) \in \mathfrak{S}_1$, $\xi \in \mathbb{D}$, and later in paper [6] it was proved in general case. The proof of Conjecture 1 allows (cf. [4, 5]) to describe the structure of the singular spectrum of selfadjoint Friedrichs model (see the discussion after the statement of Conjecture 2 in the previous edition of this Problems Book). It is well known that just these spectral problems caused Conjectures 1, 2 (see also papers [7, 8] that deal with further generalizations of results presented in [4, 5]).

Conjecture 2 in general case is wrong [9] (an analogous result was received independently in [6]). Nevertheless Conjecture 2 turns out to be true under an additional condition $C(\xi) \in \mathfrak{S}_1$, $\xi \in \mathbb{D}$ [4], or if the modulus of continuity $\omega(\delta)$ satisfies the condition $\varliminf_{\delta \to 0} \delta/\omega(\delta) > 0$ [6]. Note that we could not replace the nuclearity of $C(\xi)$, $\xi \in \mathbb{D}$, by a weaker condition $C(\xi) \in \mathfrak{S}_p$, $\xi \in \mathbb{D}$, $p > 1$, keeping Conjecture 2 to be true [6, 9].

Closely connected with Conjectures 1, 2, detailed investigations of the structure of the zero-sets (or singularities) of operator-valued functions with positive imaginary part (so-called R-functions) with the property $C(\xi) \in \mathfrak{S}_p$, $\xi \in \mathbb{D}$, $p > 1$, were carrying out in [6, 9, 10, 11]. These results show that the structure of the zero-sets (singularities) essentially depends on not only modulus of continuity ω (as in Conjectures 1, 2), but on the value of the number $p > 1$ as well. For example, in the case $\omega(t) = t^\alpha$, $0 < \alpha < 1$, the Hausdorff dimension of the zero-set $\mathfrak{S}_p$-valued R-function does not exceed $(1-\alpha)p$. This estimate is sharp and could not be improved [6, 8, 10]. The most difficult case of trace-class-valued R-functions ($p = 1$) was considered in [11].

References

4. Naboko S. N., *Uniqueness theorems for operator-valued functions with a positive imaginary part and the singular spectrum in the selfadjoint Friedrichs model*, Dokl. Akad. Nauk SSSR **275** (1984), no. 6, 1310–1313. (Russian)

5. Naboko S. N., *Uniqueness theorems for operator-valued functions with positive imaginary part, and the singular spectrum in the selfadjoint Friedrichs model*, Arkiv för Matem. **25** (1987), no. 1, 115–140.
6. Mikityuk Ya. V., *Uniqueness theorems for analytic operator functions with a non-negative imaginary part*, Funktsion. Anal. i Prilozhen. **22** (1988), no. 1, 73–74. (Russian)
7. Mikityuk Ya. V., *On the singular spectrum of selfadjoint operators*, Dokl. Akad. Nauk SSSR **303** (1988), no. 1, 33–36. (Russian)
8. Yakovlev S. I., *On perturbations of the singular spectrum in the selfadjoint Friedrichs model*, Vestnik LGU, ser.I (1990), no. 1, 117–119. (Russian)
9. Naboko S. N., *On the structure of zeros of operator functions with a positive imaginary part of the classes* $\mathfrak{S}_p$, Dokl. Akad. Nauk SSSR **295** (1987), no. 3, 538–541. (Russian)
10. Naboko S. N., *Uniqueness theorems for operator R-functions of classes* $R_0(\mathfrak{S}_p)$, $p > 1$, Problemy Matem. Fiziki **13** (1991), 169–191. (Russian)
11. Naboko S. N., *On the structure of singularities of operator functions with a positive imaginary part*, Funktsion. Anal. i Prilozhen. **25** (1991), no. 4, 1–13. (Russian)

UL. ORDZHONIKIDZE 37–1, 5
ST. PETERSBURG 196234
RUSSIA

S.6.12
old

POINT SPECTRUM OF PERTURBATIONS OF UNITARY OPERATORS

N. G. MAKAROV

Let U be a unitary operator with purely singular spectrum and let an operator K be of trace class.

QUESTION. *Can the point spectrum of the perturbed operator $U + K$ be uncountable?*

If it is not assumed that the spectrum of U is singular the answer is YES. A necessary and sufficient condition for a subset of $\mathbb{T}$ to be the point spectrum of some trace class perturbation of some (arbitrary) unitary operator was given in [1]: such a subset must be a countable union of Carleson sets (for the definition see, e.g., Problem 14.3 of this "Collection"). A version of the reasoning in [1] allows to reduce our QUESTION to a question of function theory.

PROPOSITION. *Let E be a subset of $\mathbb{T}$. The following are equivalent.*

(1) *E is the point spectrum of some trace class perturbation of a unitary operator with singular spectrum.*

(2) *There exist two distinct inner functions Θ_1 and Θ_2 such that*

$$E = \left\{ \zeta \in \mathbb{T} \colon \frac{\Theta_1(z) - \Theta_2(z)}{z - \zeta} \in L^2(\mathbb{T}) \right\}.$$

Note that if one of the inner functions is constant then the latter set is countable since it is the point spectrum of a unitary operator, namely of a rank one perturbation of a restricted shift (cf. [2]).

REFERENCES

1. Makarov N. G., *Unitary point spectrum of almost unitary operators*, Zapiski nauchn. sem. LOMI **126** (1983), 143–149 (Russian); English transl. in J. Soviet Math. **27** (1984), no. 1, 2517–2520.
2. Clark D., *One dimensional perturbations of restricted shifts*, J. Analyse Math. **25** (1972), 169–191.

STEKLOV MATHEMATICAL INSTITUTE
ST. PETERSBURG BRANCH
FONTANKA 27
ST. PETERSBURG, 191011
RUSSIA

AND

DEPARTEMENT OF MATHEMATICS
CALIFORNIA INSTITUTE OF TECHNOLOGY
253-37 PASADENA
CALIFORNIA 91125
USA

COMMENTARY BY THE AUTHOR

A positive answer is given by the author in [3].

REFERENCE

3. Makarov N. G., *One-dimensional perturbations of singular unitary operators*, Acta Sci. Math. **52** (1988), 459–463.

Chapter 7

HANKEL AND TOEPLITZ OPERATORS

Edited by

Jaak Peetre
Matematiska institutionen
Box 6701
S-113 85 Stockholm
Sweden

INTRODUCTION

The subject matter of this Chapter—the theory of "Ha-plitz operators" in its classical setting—has been described in the first edition of this book (see Introduction to Chapter 5 there). Also some of the applications to other disciplines in both pure and applied mathematics were mentioned.

Most of the old problems have been reprinted in this issue (with slight editorial changes) in the problem form, or as solutions collected at the end of the chapter. Only two very old ones (5.1 and 5.4) have been omitted. In addition, there are two newcomers from an other chapter (4.18 and 4.24, now reissued as 7.13 and 7.2 respectively).

In the past 5 or 6 years a new development has taken place, which has entirely reshaped the subject and considerably widened its basis. This is also reflected by several of the new problems presented now.

Let us briefly describe what this new development is about.

In the classical situation a Hankel or a Toeplitz operator is often made to operate on the standard Hardy space $H^2(\mathbb{T})$, its range being in $H^2(\mathbb{T})$ (in the Toeplitz case) or its orthogonal complement $H^2_-(\mathbb{T})$ in $L^2(\mathbb{T}, dm)$ (in the Hankel case). The group $\mathrm{SU}(1,1)$ of 2×2 complex matrices $g = \begin{pmatrix} a & b \\ c & d \end{pmatrix}$ subject to the conditions $ad - bc = 1$, $\bar{d} = a$, $\bar{c} = b$ acts on the complex plane via the formula $z \mapsto g(z) = \dfrac{az+b}{cz+d}$, leaving the unit disk $\mathbb{D}$ invariant, and this action extends to a unitary action on the Hilbert space $L^2(\mathbb{T}, dm)$ given by the formula $f(z) \mapsto f(g(z))(g'(z))^{\frac{1}{2}}$, leaving $H^2(\mathbb{T})$ invariant. The basic observation is now that the group intertwines with Ha-plitz operators: If we conjugate such an operator with a group operator, we get another operator of the same type but with a different symbol. This again suggests several generalizations:

1° Consider more general operators with the same intertwining property. This leads to consider (at least) Hankel operators of higher "weight".

2° Replace $H^2(\mathbb{T})$ by a weighted Bergman space $A^{\alpha,2}(\mathbb{D})$ $(\alpha > -1)$: $f \in A^{\alpha,2}(\mathbb{D})$ if and only if f is an analytic function on $\mathbb{D}$ which is square integrable against the measure $\dfrac{\alpha+1}{\pi}(1-|z|^2)^\alpha$. The group $\mathrm{SU}(1,1)$ acts on this space via $f(z) \mapsto f(g(z))(g'(z))^{\frac{\alpha+2}{2}}$. As a limiting case $(\alpha \to -1)$ we get back the Hardy space. As another limiting case $(\alpha \to \infty)$ we obtain the Fock (or Bargmann-Segal) space (entire functions square integrable with respect to a Gaussian measure); we have first to write the definition for a disc $\mathbb{D}_R$ of radius R and then let R and α tend to infinity in such a way that the ratio α/R^2 stays finite. Then the rôle of $\mathrm{SU}(1,1)$ is played by the Heisenberg group.

3° Finally, one can give up the disk as the "base space" and consider more general domains, even in higher dimensions. This includes, in particular, the case of general bounded symmetric domains. It is interesting to note that in this way one gets an unexpected contact with physics (quantization, Berezin transform, coherent states, Feynman integral etc.).

The point is that with each of the cases 2° and 3° there goes along a natural definition of "Ha-plitz" operators.

4° One can also give up the requirement of homogeneity, completely or partially. This leads, in particular, to similar problems on planar domains of arbitrary connectivity or more generally Riemann surfaces and on strictly pseudo-convex domains in $\mathbb{C}^n$ etc.

Acknowledgement. The Chapter Editor would like to express his, and the mathematical community's thanks to Karin Lindberg of the Mittag-Leffler Institute, to Barbro Fernström of Stockholm University, and to his wife Eila Jansson for invaluable secretarial aid.

7.1
old

QUASINILPOTENT HANKEL OPERATORS

S. C. POWER

Hankel operators possess little algebraic structure. This fact handicaps attempts to elucidate their spectral theory. The following sample problem is of untested depth and has some interesting function theoretic and operator theoretic connections.

PROBLEM. *Does there exist a non-zero quasinilpotent Hankel operator?*

A Hankel operator A on H^2 is one whose representing matrix is of the form $(a_{i+j})_{i,j=0}^{\infty}$, with respect to the standard orthonormal basis. A well known theorem of Nehari shows that we may represent A as $A = S_\varphi = PJM_\varphi|H^2$ where P is the orthogonal projection of L^2 onto H^2, J is the unitary operator defined by $(Jf)(z) = f(\bar{z})$, for f in L^2, and M_φ denotes multiplication by a function φ in L^∞. The symbol function φ and the defining sequence a_n are connected by $\hat{\varphi}(-n) = a_n$, $n = 0, 1, 2, \ldots$. The following observation appears to be new and provides a little evidence against existence.

PROPOSITION. *There does not exist a non zero nilpotent Hankel operator.*

Proof. Suppose $A \neq 0$ and is nilpotent. Then ker A is a non-zero invariant subspace for the unilateral shift U, since $AU = U^*A$. By Beurling's theorem this subspace is of the form uH^2 for some non constant inner function u. Thus, with the representation above, we have $S_{\varphi u} = 0$ and hence $\widehat{\varphi u}(n) = 0$ for $n \geqslant 1$. So the symbol function φ may be written in the factored form $\varphi = z\bar{u}h$ for some h in H^∞, and we may assume (by cancellation) that u and h possess no common inner divisors. The operator $S_{z\bar{u}}$ is a partial isometry with support space $K = H^2 \ominus uH^2$ and final space $K^\dagger = H^2 \ominus u^\dagger H^2$, where $g^\dagger(z) = \overline{g(\bar{z})}$. By the hypothesizes nilpotence of $S_{z\bar{u}h} = S_{z\bar{u}M_h}$ it follows that for some non-zero function f in $K^\dagger$, hf belongs to uH^2. Hence u divides f, and $f = uf_1$ with f_1 in H^2. Since f belongs to $K^\dagger$ we have $P(\bar{u}^\dagger uf_1) = 0$. This says that the Toeplitz operator $T_{\bar{u}^\dagger u}$ has non-trivial kernel. But $\ker T^*_{\bar{u}^\dagger u} = \ker T_{u^\dagger \bar{u}} = \ker T_{(u\bar{u}^\dagger)^\dagger} = (\ker T_{\bar{u}^\dagger u})^\dagger$, and we have a contradiction of Coburn's alternative: Either the kernel or the co-kernel of a non zero Toeplitz operator is trivial. □

Function theory. The evidence for existence is perhaps stronger. There are many compact non-selfadjoint Hankel operators, so perhaps a non-zero one can be found which has no non-zero eigenvalues. A little manipulation reveals that λ is an eigenvalue for S_φ if and only if there is a non-zero function f in H^2 (the eigenvector) and a function g in zH^2 such that

$$\varphi(z) = \frac{\lambda f(\bar{z}) + g(z)}{f(z)}. \tag{1}$$

Since continuous functions induce compact Hankel operators it would be sufficient then to find a continuous function φ which fails to be representable in this way for every $\lambda \neq 0$.

Whilst the singular numbers of a Hankel operator A (the eigenvalues of $(A^*A)^{1/2}$) have been successfully characterized (see for example [3, Chapter 5]), less seems to be known about eigenvalues.

Operator theory. It is natural to examine (1) when the symbol can be factored as $\varphi = z\bar{u}h$ (cf. the proof above) with u an interpolating Blaschke product. The corresponding Hankel operators and function theory are tractable in certain senses (see [1], [2, Part 2] and [3, Chapter 4]), partly because the functions $(1-|\alpha_i|^2)^{1/2}(1-\bar{\alpha}_i z)^{-1}$, where $\alpha_1, \alpha_2, \ldots$ are the zeros of u, form a Riesz basis for $H^2 \ominus uH^2$. It turns out that S_φ is compact if $h(\alpha_1), h(\alpha_2), \ldots$ is a null sequence. A quasinilpotent compact Hankel operator of this kind will exist if and only if the following problem for operators on ℓ^2 can be solved.

PROBLEM. *Construct an interpolating sequence α_i and a compact diagonal operator D so that the equation $DXx = \lambda x$ admits no proper solutions x in ℓ^2 when $\lambda \neq 0$.* Here X is the bounded (!) operator on ℓ^2 associated with $\alpha_1, \alpha_2, \ldots$ determined by representing matrix

$$x_{i,j} = \frac{(1-|\alpha_i|^2)^{1/2}(1-|\alpha_j|^2)^{1/2}}{1-\alpha_i\alpha_j}.$$

References

1. Clark D. N., *On interpolating sequences and the theory of Hankel and Toeplitz matrices*, J. Functional Anal. **5** (1970), 247–258.
2. Khruščëv S. V., Nikol'skiĭ N. K., Pavlov B. S., *Unconditional bases of exponentials and of reproducing kernels*, Complex analysis and spectral theory. Seminar, Leningrad 1979/80, Lect. Notes Math., vol. 864, Springer-Verlag, Berlin–Heidelberg–New York, 1981, pp. 214–335.
3. Power S. C., *Hankel Operators on Hilbert Space*, Research Notes in Mathematics 64, Pitman, London, 1982.

DEPARTMENT OF MATHEMATICS
UNIVERSITY OF LANCASTER
BAILRIGG, LANCASTER LAI 4YW
ENGLAND

Editor's note

The *first* PROBLEM (existence) is solved by A. Megretskii [4]: the operator S_ϕ with $\phi(z) = i + \sum_{n\geqslant 1} 2^{-n} z^{-2^n}$ is a compact quasinilpotent Hankel operator with $\ker S_\phi = \{0\}$. Moreover, in [4] the spectrum of all S_ϕ's with lacunary Fourier decomposition $\phi(z) = \sum_{n\geqslant 0} a_n z^{-2^n}$, $\{a_n\} \in \ell^2$, is computed in terms of a "branched dynamical system". Some *new open problems* are raised.

Reference

4. Megretskii A. V., *A quasinilpotent Hankel operator*, Algebra i Analiz **2** (1990), no. 4, 201–212 (Russian); English transl. in Leningrad Math. J. **2** (1991), no. 4.

7.2
old

ESTIMATES OF FUNCTIONS OF HILBERT SPACE OPERATORS, SIMILARITY TO A CONTRACTION AND RELATED FUNCTION ALGEBRAS

V. V. PELLER

For a given class of operators on Banach spaces we can consider the problem to estimate norms of functions of these operators. Sometimes, dealing with operators on Hilbert space, we can obtain sharper estimates than in the case of an arbitrary Banach space. A remarkable example of such a phenomenon is the following J. von Neumann's inequality:

$$\|\varphi(T)\| \leqslant \|\varphi\|_\infty = \sup_{|\zeta|<1} |\varphi(\zeta)|$$

for any contraction T (i.e. $\|T\| \leqslant 1$) on Hilbert space and for any complex polynomial φ (denote by $\mathcal{P}_A$ the set of all complex polynomials).

We consider here some other classes of operators.

Operators with the growth of powers of order $\alpha, \alpha \geqslant 0$**.** This class consists of operators satisfying $\|T^n\| \leqslant c(1+n)^\alpha$, $n \geqslant 0$. Clearly, for any such operator on a Banach space we have

$$\|\varphi(T)\| \leqslant c\|\varphi\|_{\mathcal{F}\ell^1_{(\alpha)}} \stackrel{\text{def}}{=} c\sum_{n\geqslant 0} |\hat{\varphi}(n)|(1+n)^\alpha, \quad \varphi \in \mathcal{P}_A. \tag{1}$$

It is easy to see (cf. [1]) that the fact that inequality (1) cannot be improved for Hilbert space operators is equivalent to the fact that $\mathcal{F}\ell^1_{(\alpha)}$ is *an operator algebra* (with respect to the pointwise multiplication), i.e. it is isomorphic to a subalgebra of the algebra of bounded operators on Hilbert space. It was proved by N. Th. Varapoulos [2] that this is the case if $\alpha > 1/2$ and so for $\alpha > 1/2$ (1) cannot be improved. It follows from [2], [3], [4] that $\mathcal{F}\ell^1_{(\alpha)}$ is not an operator algebra for $\alpha \leqslant 1/2$ and so (1) can be improved in this case. Now the PROBLEM is to *find sharp estimates of*

$$\|\varphi(T)\| \quad \textit{for } T \textit{ satisfying} \quad \|T^n\| \leqslant c(1+n)^\alpha.$$

Note that some estimates improving (1) are obtained in [1]. This general problem apparently is very difficult. Let us consider the most interesting case $\alpha = 0$.

Power bounded operators. We mean operators on Hilbert space satisfying $\|T^n\| \leqslant c$, $n \geqslant 0$. It is well-known (see ref. in [1]) that such operators are not necessarily *polynomially bounded* (i. e. $\|\varphi(T)\| \leqslant \text{const}\,\|\varphi\|_{L^\infty}, \varphi \in \mathcal{P}_A$). In [1] the following estimates of polynomials of power bounded operators are obtained

$$\|\varphi(T)\| \leqslant \text{const}\,\|\varphi\|_{\mathcal{L}} \leqslant \text{const}\,\|\varphi\|_{\text{VMO}_A \hat{\otimes} H^1} \leqslant \text{const}\,\|\varphi\|_{B^0_{\infty 1}}. \tag{2}$$

Here

$$\|\varphi\| = \inf\Big\{ \|\{\gamma_{mk}\}_{m,k\geqslant 0}\|_{\ell^1\check{\otimes}\ell^1} : \sum_{m+k=n} \gamma_{mk} = \hat{\varphi}(n), n \geqslant 0 \Big\},$$

where $\ell^1\check{\otimes}\ell^1$ is the injective tensor product;

$$\|\varphi\|_{\mathrm{VMO}_A\hat{\otimes}H^1} = \inf\Big\{ \sum_{m\geqslant 0} \|f_m\|_{\mathrm{VMO}_A}\|g_m\|_{H^1} : \sum_{m\geqslant 0} f_m * g_m = \varphi \Big\},$$

where $\mathrm{VMO}_A = \{f = \sum_{n\geqslant 0} \hat{f}(n)z^n : \hat{f}(n) = \hat{g}(n), n \geqslant 0, \text{ for some } g \in C(\mathbb{T})\}$;

$$\|\varphi\|_{B^0_{\infty 1}} = |\varphi(0)| + \int_0^1 \|\varphi_r'\|_\infty \, dr, \ \varphi_r(z) \stackrel{\text{def}}{=} \varphi(rz),$$

is the norm of φ in the Besov space $B^0_{\infty 1}$.

The fact that the inequalities in (2) are precise is equivalent to the fact that the sets $\mathcal{L}$, $\mathrm{VMO}_A\hat{\otimes}H^1$, $B^0_{\infty 1}$ form operator algebras with respect to the pointwise multiplication. For $B^0_{\infty 1}$ this is not the case [1]. For $\mathcal{L}$, $\mathrm{VMO}_A\hat{\otimes}H^1$ *the question is open.* It is even *unknown whether $\mathcal{L}$ forms a Banach algebra.*

If $\mathrm{VMO}_A \hat{\otimes} H^1$ is an operator algebra then the norms $\|\cdot\|_{\mathcal{L}}$ and $\|\cdot\|_{\mathrm{VMO}_A\hat{\otimes}H^1}$ are equivalent. The question of whether this is the case can be reformulated in the following way.

Let $\mathcal{M}H^1$ be the set of all *Fourier multipliers of* H^1, i.e.

$$\mathcal{M}H^1 = \{f : g \in H^1 \implies f * g \in H^1\}.$$

Let V^2 be the set of matrices $\{\alpha_{nk}\}_{n,k\geqslant 0}$ such that

$$\sup_{N>0} \|\{\alpha_{nk}\}_{0\leqslant n,k\leqslant N}\|_{\ell^\infty\hat{\otimes}\ell^\infty} < \infty,$$

where $\ell^\infty \hat{\otimes} \ell^\infty$ is the projective tensor product.

QUESTION 1. *Is it true that*

$$\psi \in \mathcal{M}H^1 \iff \Gamma_\psi \in V^2?$$

Recall that $\Gamma_\psi = \{\hat{\psi}(n+k)\}_{n,k\geqslant 0}$ is the Hankel matrix. It is easy to show (see [1]) that $\Gamma_\psi \in V^2 \implies \psi \in \mathcal{M}H^1$.

Similarly we can define the spaces V^M of tensors $\{\alpha_{n_1\ldots n_m}\}_{n_j\geqslant 0}$ and the Hankel tensor $\Gamma^M_\psi = \{\hat{\psi}(n_1 + \cdots + n_M)\}_{n_j\geqslant 0}$.

QUESTION 2. *Is it true that*

$$\psi \in \mathcal{M}H^1 \implies \Gamma^M_\psi \in V^M \quad \textit{and} \quad \log\|\Gamma^M_\psi\|_{V^M} \leqslant \textit{const}\cdot M?$$

If Question 2 has a positive answer then $\mathrm{VMO}_A \hat{\otimes} H^1$ is an operator algebra (see [1]) and so is $\mathcal{L}$ and the estimates $|\varphi(T)| \leqslant \text{const}\,\|\varphi\|_{\mathcal{L}} \leqslant \text{const}\,\|\varphi\|_{\mathrm{VMO}_A\hat{\otimes}H^1}$ cannot be improved.

QUESTION 3. *Is it true that*

$$\Gamma_\psi \in V^2 \implies \Gamma_\psi^M V^M \quad \text{and} \quad \log \|\Gamma_\psi^M\|_{V^M} \leqslant \text{const} \cdot M?$$

An affirmative answer would imply that $\mathcal{L}$ is an operator algebra and the estimate $|\varphi(T)| \leqslant \text{const}\, \|\varphi\|_{\mathcal{L}}$ is the best possible ([1]). Moreover in this case the estimate is attained on the Davie's example (see [1]) of power bounded non polynomially bounded operator.

Similarity to a contraction. Here we touch the well-known problem (see e.g. [5]) of *whether each polynomially bounded operator T on Hilbert space is similar to a contraction* (i.e. whether there exists an invertible operator V such that $\|VTV^{-1}\| \leqslant 1$).

In [1] we consider operators R_f on $\ell^2 \oplus \ell^2$ defined by

$$R_f = \begin{pmatrix} S^* & \Gamma_f \\ \mathbf{0} & S \end{pmatrix},$$

where S is the shift operator on ℓ^2. It was proved in [1] that R_f is power bounded if and only if f belongs to the Zygmund class Λ_1, i. e. $|f''(\zeta)| \leqslant \text{const}\, \dfrac{1}{1-|\zeta|}$, $|\zeta| < 1$. It was also shown in [1] that among R_f there are many power bounded operators, non polynomially bounded. It seems reasonable to try to construct a counterexample to the problem stated above on the class of the operators R_f. It is very easy to calculate the function of R_f. Namely,

$$\varphi(R_f) = \begin{pmatrix} \varphi(S^*) & \Gamma_{\varphi'(S^*)f} \\ \mathbf{0} & \varphi(S) \end{pmatrix}.$$

THEOREM. *If $f' \in BMO_A$ then R_f is polynomially bounded.*

(Recall that

$$\text{BMO}_A = \Big\{ f = \sum_{n \geqslant 0} \hat{f}(n) z^n : \hat{f}(n) = \hat{g}(n), n \geqslant 0, \text{ for some } g \text{ in } L^\infty \Big\}.)$$

Proof. By Nehari's theorem (see [6]) $|\Gamma_{\varphi'(S^*)f}| \asymp \|\varphi'(S^*)f\|_{\text{BMO}_A}$. We have

$$\|\varphi'(S^*)f\|_{\text{BMO}_A} = \sup_{\|g\|_{H^1} \leqslant 1} |(\varphi'(S^*)f, g)| = \sup_{\|g\|_{H^1} \leqslant 1} |(f, \varphi' g)|.$$

Now $\varphi' g = (\varphi g)' - \varphi g'$. To finish the proof we use the fact that

$$F \in H^1 \iff \int_{-\pi}^{\pi} \left(\int_0^1 |F'(re^{i\theta})|^2 (1-r)\, dr \right)^{\frac{1}{2}} d\theta$$

(see [7]).

It follows that

$$\int_{-\pi}^{\pi} \left(\int_0^1 |\varphi' g(re^{i\theta})|^2 (1-r)\, dr \right)^{\frac{1}{2}} d\theta < \infty$$

and

$$|(f, \varphi' g)| \leqslant \text{const}\, (\|f'\|_{\text{BMO}} + |f(0)|) \int_{-\pi}^{\pi} \left(\int_0^1 |\varphi' g(re^{i\theta})|^2 (1-r)\, dr \right)^{\frac{1}{2}} d\theta. \quad \square$$

QUESTION 4. *Is it true that if R_f is polynomially bounded then $f' \in BMO_A$?*

This question is related to a question of R. Rochberg [8] concerning Hankel operators.

QUESTION 5. *Does there exist f with $f' \in BMO_A$ such that R_f is similar to no contraction?*

Operators with the growth of resolvents of order α. We consider here the operators satisfying

$$\|R(\lambda, T)\| \leqslant \frac{c}{(|\lambda| - 1)^\alpha}, \quad |\lambda| > 1, \alpha \geqslant 1.$$

It is not difficult to prove that for any such operator on a Banach space we have

$$\|\varphi(T)\| \leqslant \text{const}\, \|\varphi\|_{B_1^\alpha}, \tag{3}$$

where the Besov space B_1^α consists of the functions f satisfying

$$\iint_{\{|z|<1\}} |f^{(n)}(z)|(1 - |z|)^{n-\alpha-1}\, dxdy < \infty,$$

n being an integer greater than α.

Inequality (3) is the best possible on the class of all Banach spaces. (It is enough to consider multiplication by z on B_1^α.) The fact that (3) is the best possible for Hilbert space operators is equivalent to the fact that the algebra B_1^α is an operator algebra.

Consider the following operators on a commutative Banach algebra A

$$\mathcal{H}_\varphi \colon A \to A^*, \quad (\mathcal{H}_\varphi x, y) \stackrel{\text{def}}{=} (\varphi, xy), \ \varphi \in A^*.$$

It is proved by A. M. Tonge [9] that if all operators $\mathcal{H}_\varphi$ are 2-absolutely summing then A is an operator algebra and by P. Charpentier [4] that if A is an operator algebra then all operators $\mathcal{H}_\varphi$ can be factored through Hilbert space.

In the case of B_1^α the operators $\mathcal{H}_\varphi$ look as follows

$$\mathcal{H}_\varphi \colon B_1^\alpha \to B_\infty^{-\alpha}, \quad \mathcal{H}_\varphi x = \mathbb{P}_+ \varphi \tilde{x}, \quad \varphi \in B_\infty^{-\alpha},$$

where

$$B_\infty^{-\alpha} = \{\, f : |f(\zeta)| \leqslant \text{const}\,(1 - |\zeta|)^{-\alpha},\ |\zeta| < 1\,\},\ \tilde{x}(\zeta) = x(\bar{\zeta}),\ \zeta \in \mathbb{T},\ \mathbb{P}_+ f = \sum_{n \geqslant 0} \hat{f}(n) z^n.$$

The space B_1^α is isomorphic to ℓ^1 (see [10]). It follows from Grothendiek's theorem (see [11]) that operators on ℓ^1 factored through Hilbert space are 2-absolutely summing.

QUESTION 6. *Is it true that for any $\varphi \in B_\infty^{-\alpha}$ the operator $\mathcal{H}_\varphi \colon B_1^\alpha \to B_\infty^{-\alpha}$ is 2-absolutely summing?*

The answer is positive if and only if B_1^α is an operator algebra which is equivalent to the fact that (3) cannot be improved. If the answer is negative, *find estimates sharper than* (3). Even the QUESTION *of estimating* $\|T^n\|$ *is not yet solved.* It follows from (3) that $\|T^n\| \leqslant c(1+n)^\alpha$. This question for $\alpha = 1$ was considered in [12]. The best example I know is due to S. N. Naboko (unpublished), and another one is due to J. A. van Casteren [13]. They constructed for any $\beta < 1/2$ weighted shifts T such that

$$\|R(\lambda, t)\| \leqslant \frac{c}{|\lambda| - 1} \quad \text{and} \quad \|T^n\| \geqslant c(1+n)^\beta.$$

References

1. Peller V. V., *Estimates of functions of power bounded operators on Hilbert spaces*, J. Oper. Theory **7** (1982), 341–372.
2. Varopoulos N. Th., *Some remarks on Q-algebras*, Ann. Inst. Fourier (Grenoble) **22** (1972), 1–11.
3. Varopoulos N. Th., *Sur les quotiens des algèbres uniformes*, C. R. Acad. Sci. Paris **274** (1972), 1344–1346.
4. Charpentier P., *Q-algèbres et produits tensoriels topologiques*, Thèse, Orsay, 1973.
5. Halmos P., *Ten problems in Hilbert space*, Bull. Am. Math. Soc. **76** (1970), 887–933.
6. Sarason D., *Function theory on the unit circle*, Notes for lectures at Virginia Polytechnic Inst. and State Univ., Blacksburg, Va., 1978.
7. Fefferman Ch., Stein E. M., *H^p spaces of several variables*, Acta Math. **129** (1972), 137–193.
8. Rochberg R., *A Hankel type operator arising in deformation theory*, Proc. Sympos. Pure. Math. **35** (1979), no. 1, 457–458.
9. Tonge A. M., *Banach algebra and absolutely summing operators*, Math. Proc. Camb. Philos. Soc. **80** (1976), 465–473.
10. Lindenstrauss J., Pełczyński A., *Contribution to the theory of classical Banach spaces*, J. Funct. Anal. **8** (1971), 225–249.
11. Lindenstrauss J., Pełczyński A., *Absolutely summing operators in $\mathcal{L}_p$-spaces and their applications*, Studia Math. **29** (1968), 275–326.
12. Shields A. L., *On Möbius bounded operators*, Acta Sci. Math. **40** (1978), 371–374.
13. van Casteren J. A., *Operators similar to unitary and self-adjoint ones*, Pacif. J. Math. **104** (1983), 241–255.

Steklov Mathematical Institute
St. Petersburg branch
Fontanka 27
St. Petersburg, 191011
Russia

and

Department of Mathematics
Kansas State University
Manhattan
Kansas, 66506
USA

Editors' note

A recent publication [14] could be of interest.

Reference

14. Paulsen V. I., *Toward a theory of K-spectral sets*, Surveys of some recent results in operator theory (J. Conway and B. Morrel, ed.), Pitnam Res. Math. Series, no. 171, Longman, UK, 1988, pp. 221–240.

7.3

SINGULAR VALUES OF HANKEL OPERATORS

SVANTE JANSON

1. Distribution of singular values. Consider a Hankel operator $H_{\bar{f}}$ with a conjugate analytic symbol $\bar{f}$ on the Hardy space H^2, and let s_n, $n = 0, 1, \ldots$, be its singular values. Peller [7, 8] proved the basic result that for every p, $0 < p < \infty$,

$$\|f\|_{B_p} \asymp \left(\sum_0^\infty s_n^p\right)^{1/p} \tag{1}$$

where B_p is the Besov space $B_p^{1/p,p}$. Although this result is very elegant, it is sometimes too crude to consider the ℓ^p-norms only, and more refined statements about the distributions of $\{s_n\}$ would be interesting.

For example, suppose that $1 < p < \infty$ and define $\varphi_f(z) = (1 - |z|^2)f'(z)$. Then $\|f\|_{B_p} \asymp \|\varphi_f\|_{L^p(d\lambda)}$, where $d\lambda$ is the invariant measure $(1-|z|^2)^{-2}\,dm$ on the unit disc, so (1) reads

$$\|\varphi_f\|_{L^p(d\lambda)} \asymp \|(s_n)\|_{\ell^p}. \tag{2}$$

This suggests looking for relations between the decreasing rearrangement φ_f^* of φ_f (with respect to $d\lambda$) and the function $s(t) = s_{[t]}$ (note that φ_f^* and $s(t)$ both are decreasing functions on $(0,\infty)$), or equivalently, between the corresponding distribution functions $\lambda\{z : \varphi_f(z) > t\}$ and $\#\{n : s_n > t\}$. Simple equivalence of these functions can not hold, since (2) fails for $p = 1$ or ∞, but one might hope for integral estimates.

Some such estimates may be derived from (2). For example, if $1 < p_0 < p_1 < \infty$, it follows, using the Adamyan-Arov-Kreĭn theorem [5], that

$$K(t, \varphi_f; L^{p_0}(d\lambda), L^{p_1}(d\lambda)) \leqslant C\,K(t, s_n; \ell^{p_0}, \ell^{p_1}),\ t > 0$$

and thus by Holmstedt's formula [3], replacing t by t^δ with $\delta = 1/p_0 - 1/p_1$,

(3)

$$\left(\int_0^t \varphi_f^*(u)^{p_0}\,du\right)^{1/p_0} + t^\delta\left(\int_t^\infty (\varphi_f^*(u))^{p_1}\,du\right)^{1/p_1} \leqslant C\left(\int_0^t s(u)^{p_0}\,du\right)^{1/p_0} + C\,t^\delta\left(\int_t^\infty s(u)^{p_1}\,du\right)^{1/p_1}$$

which implies

$$\varphi_f^*(t) \leqslant C\left(\frac{1}{t}\int_0^t s(u)^{p_0}\,du\right)^{1/p_0} + C\left(\frac{1}{t}\int_t^\infty s(u)^{p_1}\,du\right)^{1/p_1}. \tag{4}$$

Note that, conversely, (4) implies, by Hardy's inequality, (2) and (1) for $p_0 < p < p_1$.

It should be possible to obtain sharper results of this type (and conversely, estimating s_n using φ_f^*); for example end point results corresponding to $p_0 \to 1$ and $p_1 \to \infty$ in (4).

The same problem may be considered for other Hankel operators, for example on the Bergman space where (1) is valid also for $p = \infty$ and the situation ought to be somewhat simpler.

The methods of Rochberg and Semmes [9, 10] are probably useful.

2. Lower bounds. Consider now a big Hankel operator $H_{\bar{f}}$ with non-constant conjugate analytic symbol on the Bergman space on the unit disc. It was shown by Arazy, Fisher and Peetre [1] that $H_{\bar{f}}$ may belong to the Schatten ideal $\mathfrak{S}_p$, $p > 1$, but not to $\mathfrak{S}_1$; more precisely, see Nowak [6], it belongs to $\mathfrak{S}_{1\infty}$ when f is sufficiently smooth but never to $\mathfrak{S}_{1q}$, $q < \infty$. Presumably, this can be sharpened further.

Conjecture 1. *$H_{\bar{f}}$ never belongs to $\mathfrak{S}_{1\infty}^{\infty}$, i.e.*

$$\limsup_{n\to\infty} ns_n > 0.$$

It seems likely that more is true: (This can be proved when f is sufficiently smooth, but is not known in general.)

Conjecture 2.

$$\liminf_{n\to\infty} ns_n > 0.$$

Perhaps even more is true.

Problem. *Does $\lim_{n\to\infty} ns_n$ always exist (possibly infinite)?*

The conjectures and problem may also be considered for other situations where there is a cut-off, for example for the Bergman or Hardy space on the unit ball in $\mathbb{C}^N$, $N \geqslant 2$, cf. [2] and [4], replacing ns_n by $n^{1/2N}s_n$. For the Bergman space, Conjecture 1 holds by [2], Remark 7.2.

References

1. Arazy J., Fisher S., Peetre J., *Hankel operators on weighted Bergman spaces*, Amer. J. Math. **110** (1988), 989–1053.
2. Arazy J., Fisher S., Janson S., Peetre J., *Membership of Hankel operators on the ball in unitary ideals*, Proc. London Math. Soc. **43** (1991), 485–508.
3. Bergh J., Löfström J., *Interpolation Spaces*, Springer-Verlag, 1976.
4. Rochberg R., Feldman M., *Singular value estimates for commutators and Hankel operators on the unit ball and the Heisenberg group*, Analysis and PDE, A collection of papers dedicated to Mischa Cotlar (Cora Sadosky, ed.), Marcel Dekker Inc., 1990, pp. 121–160.
5. Nikol'skiĭ N. K., *Treatise on the shift operator*, Springer-Verlag, Berlin-Heidelberg-New York-Tokyo, 1986.
6. Nowak K., *Weak type estimates for singular values of commutators on weighted Bergman spaces*, Indiana Univ. Math. J. (to appear).
7. Peller V. V., *Hankel operators of class $\mathfrak{S}_p$ and their applications*, Mat. Sb. **113** (1980), 538–581 (Russian); English transl. in Math. USSR, Sb. **41** (1982), 443–479.

8. Peller V. V., *A description of Hankel operators of class* $\mathfrak{S}_p$ *for* $p > 0$*, an investigation of the rate of rational approximation, and other applications*, Mat. Sb. **122** (1983), 481–510 (Russian); English transl. in Math. USSR, Sb. **50** (1985), 465–494.
9. Rochberg R., Semmes S., *Nearly weakly orthonormal sequences, singular value estimates and Calderón-Zygmund operators*, J. Funct. Anal. **86** (1989), 237–306.
10. Rochberg R., Semmes S., *End point results for estimates of singular values of integral operators.*, Contribution to operator theory and its applications (Gohberg et al, eds.), OT 35, Operator Theory: Advances and Applications, Birkhäuser, Basel-Boston-Berlin, 1988, pp. 217–231.

Matematiska Institutionen
Thunbersvägen 3
S-752 38 Uppsala
Sweden

7.4

THREE QUESTIONS ABOUT HANKEL OPERATORS AND THEIR GENERALIZATIONS

RICHARD ROCHBERG

1. Kronecker's Theorem for generalized Hankel operators. Kronecker's classical result [K] gives a complete description of finite rank Hankel operators. Analogous results are known for Hankel operators on Bergman and Fock spaces [JPR]. However, for the generalized Hankel operators on the Fock space introduced in [JPW] the question is open. One formulation is this. Given a positive integer, N, we want to find all entire functions $h, b, f_1, \ldots f_N, g_1, \ldots g_N$ which solve the functional equation

$$h(x-y)b(x+y) = \sum_{i=1}^{N} f_i(x)g_i(y). \tag{1$_N$}$$

(In fact we are only interested in functions which are $O(\exp c|z|^2)$ at infinity for some c.) x and y vary over $\mathbb{C}^n$ but the case $N = 1$ is open. Looking at [JPW] we find solutions

$$h(z) = e^{\alpha z^2} P_1(z)$$
$$b(z) = e^{\alpha z^2} P_2(z)$$

where α is in $\mathbb{C}$ and P_1 and P_2 satisfy constant coefficient differential equations (i.e. the P_i are linear combinations of exponentials with appropriate degenerate cases).

It is not hard to see that, if $N = 1$, all solutions of $(1)_1$ are of this form (and, in fact P_1 and P_2 must be single terms). However the full story is more complicated. Lee Rubel pointed out to me that, for $N = 2$, one can find solutions of $(1)_2$ using elliptic functions. For instance, if we use the notation of Chapter I of [M] then, for the Jacobi theta functions θ_{00} and θ_{11}, writing $c = \theta_{00}^{-1}(0)$, we have

$$\theta_{00}(x+y)\theta_{00}(x-y) = c\,\theta_{00}^2(x)\theta_{00}^2(y) + c\,\theta_{11}^2(x)\theta_{11}^2(y).$$

Page 22 of [M] gives a dozen variations on this. (As a consequence, the choice $h = b =$ the Weierstrass σ-function is also a solution to $(1)_2$.) Riemann theta functions of m variable give analogous solutions of $(1)_N$ for $N = 2^m$ (see Example xi, Section 286, [B]).

Finally, in [GR] there is an explicit PDE for functions which have the form on the right hand side on $(1)_N$. Although the equation is quite complicated and highly nonlinear, even when $N = 2$, so is the PDE satisfied by the σ-function. (Recall that $(-\log\sigma)''$ is the Weierstrass $\wp$-function.)

At this point I don't even have a good conjecture. It would be quite interesting to know if the full story could be told with exponentials and theta functions or, on the other hand, what else is going on.

2. The Theorem of Adamyan, Arov, and Kreĭn. Let H_b be a Hankel operator on the Hardy space and let

$$s_n(H_b) = \text{distance}(H_b, \text{ operators of rank} \leqslant n)$$
$$\tilde{s}_n(H_b) = \text{distance}(H_b, \text{ Hankel operators of rank} \leqslant n).$$

Trivially $s_n \leqslant \tilde{s}_n$. The theorem of Adamyan, Arov, and Kreĭn [AAK] is the converse estimate

$$\tilde{s}_n \leqslant s_n. \tag{2}$$

It is unknown if the analog of (2) holds for various classes of generalized Hankel operators, for instance the Hankel operators on the Bergman and Fock space studied in [JPR].

The proof of (2) uses the refined function theory of the Hardy space (inner-outer factorization, etc.). Without such tools, perhaps (2) is too much to hope for. However it would still be very interesting to know if there are constants c and d so that

$$\tilde{s}_{cn} \leqslant d s_n. \tag{2-weak}$$

Estimates such a (2) show up in various places; one interesting recent example involves the interpolation theory of the Schatten-von Neumann ideals, $\mathfrak{S}_p$, of compact operators. Pisier [P] has obtained rather precise results on the interpolation theoretic structure of $\mathfrak{S}_p \cap$ (upper triangular matrices) as a subscale on the $\mathfrak{S}_p$'s. Analogous results involving Hankel operators instead of upper triangular matrices can be obtained using (2). Similar results for generalized Hankel operators would follow from (2-weak) [J].

3. Unitary equivalence of Hankel operators. Suppose H_{b_1} and H_{b_2} are two unitarily equivalent Hankel operators on the Hardy space. What can be said about the relationship of b_1 and b_2? This question was mentioned to me a couple of years ago by Stephen Semmes but it may be much older. The question is even interesting (especially interesting?) if b_1 and b_2 are rational and (equivalently) the H's have finite rank.

References

[AAK] Adamyan V. M., Arov D. Z., Kreĭn M. G., *Analytic properties of Schmidt pairs for a Hankel operator and the generalized Schur-Takagi problem*, Mat. Sb., Nov. Ser. **86** (1971), 34–75 (Russian); English transl. in Math. USSR Sbornik **15** (1971), 31–73.

[B] Baker H. F., *Abel's Theorem and the Allied Theory Including the Theory of Theta Functions*, Cambridge University Press, Cambridge, 1897.

[GR] Gauchman H., Rubel L., *Sums of products of functions of x times functions of y*, Linear Algebra Appl. **125** (1989), 19–63.

[J] Janson S., manuscript in preparation 1991.

[JPR] Janson S., Peetre J., Rochberg R., *Hankel forms and the Fock space*, Rev. Mat. Iberoamer. **3** (1987), 61–138.

[JPW] Janson S., Peetre J., Wallstén R., *A new look on Hankel forms over Fock space*, Studia Math. **94** (1989), 33–41.

[K] Kronecker L., *Zur Theorie der Elimination einer Variablen aus zwei algebraischen Gleichungen*, Monatsber. Königl. Preuss. Akad. Wiss. (1881), 535–600.
[M] Mumford D., *Tata Lectures on Theta* I, II, Birkhäuser, Basel, 1983.
[P] Pisier G., *Interpolation between H^p spaces and non-commutative generalizations* I, preprint, 1990.

Department of Mathematics
Box 1146
Washington University
St. Louis, MO 63130
USA

Editors' note

Recent paper [MPT] contains a solution of the following problem close to Problem 3 of 7.4: to characterize self-adjoint operators which are unitarily equivalent to a Hankel (self-adjoint) operator on the Hardy space.

Reference

[MPT] Megretskii A., Peller V., Treil S., *The inverse spectral problem for self-adjoint Hankel operators*, Acta Math. (to appear).

7.5

HANKEL OPERATORS ON THE BERGMAN SPACE OF A STRICTLY PSEUDOCONVEX DOMAIN

KEHE ZHU

Let Ω be a bounded domain in $\mathbb{C}^n$ with normalized volume measure dv. Let P be the orthogonal projection (called the Bergman projection) from $L^2(\Omega, dv)$ onto the closed subspace $L^2_a(\Omega)$ (called the Bergman space) consisting of holomorphic functions. For a function f in $L^2(\Omega, dv)$ the Hankel operator H_f is the (possibly unbounded) operator with domain $L^2_a(\Omega)$ and range $L^2(\Omega, dv)$ defined by $H_f g = (I - P)(fg)$, where I is the identity operator. The problem (or rather the project) here is to describe operator theoretic properties of H_f in terms of function theoretic properties of f, and vice versa.

Let β be the Bergman distance function on Ω and let

$$D(z,r) = \{\, w \in \Omega : \beta(z,w) < r \,\}$$

for $z \in \Omega$ and $r > 0$. Denote the dv-integral average of an function f over $D(z,r)$ by $\widehat{f}_r(z)$. The mean oscillation of f over $D(z,r)$ is then defined by

$$MO_r(f)(z) = \left[\frac{1}{v(D(z,r))} \int_{D(z,r)} \left| f(w) - \widehat{f}_r(z) \right|^2 dv(w) \right]^{\frac{1}{2}} .$$

CONJECTURE 1. *Suppose Ω is a strictly pseudoconvex domain in $\mathbb{C}^n$ and f is a function in $L^2(\Omega, dv)$. Then for any $r, p > 0$ we have*

1) *H_f and $H_{\bar{f}}$ are both bounded if and only if $MO_r(f)$ is bounded on Ω;*
2) *H_f and $H_{\bar{f}}$ are both compact if and only if $MO_r(f)(z) \to 0$ as z approaches the boundary of Ω;*
3) *H_f and $H_{\bar{f}}$ are both in the Schatten class S_p if and only if $MO_r(f)$ belongs to $L^p(\Omega, d\lambda)$, where $d\lambda(z) = K(z,z)\, dv(z)$ and $K(z,w)$ is the Bergman kernel of Ω.*

1) and 2) above are proved in [2] for any bounded symmetric domain in $\mathbb{C}^n$. In particular, they hold for the open unit ball in $\mathbb{C}^n$ which is the most special strictly pseudoconvex domain. We also mention that 1) and 2) are proved in [3] for plane domains with finite connectivity and Jordan boundary. When Ω is a strictly convex domain in $\mathbb{C}^n$, partial results regarding questions 1) and 2) are obtained in [4].

Question 3) above is still open even for the open unit disk. However, we do know that (in the case of a bounded symmetric domain) the "only if" part in 3) is true when $p \geqslant 2$. This follows easily from the results in [5], [6].

The Berezin transform of a function f in $L^1(\Omega, dv)$ is the function $\widetilde{f}$ on Ω defined by

$$\widetilde{f}(z) = K(z,z)^{-1} \int_\Omega f(w)|K(z,w)|^2 \, dv(w).$$

For f in $L^2(\Omega, dv)$ we let $MO(f)$ denote the function on Ω defined by

$$MO(f)(z) = \left[\widetilde{|f|^2}(z) - |\widetilde{f}(z)|^2\right]^{1/2}.$$

The function $MO(f)$ describes (in an implicit but invariant way) the mean oscillation of the function f in the Bergman metric.

CONJECTURE 2. *Suppose Ω is a strictly pseudoconvex domain in $\mathbb{C}^n$ and f is a function in $L^2(\Omega, dv)$. For any $p > 0$ we have*

1) *H_f and $H_{\bar{f}}$ are both bounded if and only if $MO(f)$ is bounded on Ω;*
2) *H_f and $H_{\bar{f}}$ are both compact if and only if $MO(f)(z) \to 0$ as z approaches the boundary of Ω;*
3) *H_f and $H_{\bar{f}}$ are both in the Schatten class S_p if and only if $MO(f)$ belongs to $L^p(\Omega, d\lambda)$.*

Again 1) and 2) are proved in [2] for bounded symmetric domains and in [3] for plane domains with finite connectivity and Jordan boundary. When $p \geqslant 2$, 3) is proved in [6] for the open unit ball and in [5] for bounded symmetric domains. When $0 < p \leqslant 2$, the "if" part of 3) is known for bounded symmetric domains. [4] contains partial results about 1) and 2) in the case of a strictly convex domain.

The above conjectures are essentially about Hankel operators with real symbols. We also formulate these problems for Hankel operators with (conjugate) holomorphic symbols, a case which is probably easier to prove.

CONJECTURE 3. *Suppose Ω is a strictly pseudoconvex domain in $\mathbb{C}^n$ and f is a function in $L^2_a(\Omega)$. For large p (determine exactly how large) we have*

1) *$H_{\bar{f}}$ is bounded if and only if f belongs to the Bloch space of Ω;*
2) *$H_{\bar{f}}$ is compact if and only if f belongs to the little Bloch space of Ω;*
3) *$H_{\bar{f}}$ is in the Schatten class S_p if and only if f belongs to certain Besov space.*

This conjecture has been settled for all bounded symmetric domains. See [1], [2], [6]. 1) and 2) are proved in [4] for strictly convex domains and in [3] for plane domains with finite connectivity and Jordan boundary. It is interesting to note that 3) is trivial for bounded symmetric domains with rank > 1 (like the polydisk in $\mathbb{C}^n$), because in this case the only compact Hankel operator with conjugate holomorphic symbol is the zero operator [2].

REFERENCES

1. Arazy J., Fisher S., Janson S., Peetre J., *Membership of Hankel operators on the ball in unitary ideals*, J. London Math. Soc. **43** (1991), 485–508.
2. Békollé D., Berger C., Coburn L., Zhu K., *BMO in the Bergman metric on bounded symmetric domains*, J. Funct. Anal. **93** (1990), 310–320.

3. Li H., *BMO and Hankel operators on multiply connected domains*, preprint, 1990.
4. Li H., *BMO and Hankel operators on strictly convex domains*, preprint, 1990.
5. Zheng D., *Schatten class Hankel operators on the Bergman space of bounded symmetric domains*, Integral Equations Oper. Theory **13** (1990), 442–459.
6. Zhu K., *Schatten class Hankel operators on the Bergman space of the unit ball*, Am. J. Math. **113** (1991), 147–167.

DEPARTMENT OF MATHEMATICS
STATE UNIVERSITY OF NEW YORK
ALBANY, NY 12222
USA

7.6

SOME PROBLEMS ABOUT THE BOUNDEDNESS AND COMPACTNESS FOR SOME HANKEL-TYPE OPERATORS AND PSEUDODIFFERENTIAL OPERATORS

QIHONG FAN, LIZHONG PENG

In this note, we will describe two kinds of operators. Their Hilbert – Schmidt norm can be characterized easily, but it is difficult to characterize their boundedness and compactness.

Let $D = \{ (x,y), x^2 + y^2 < 1 \}$ be the unit disk, $\mathrm{PW}(D) = \{ \mathcal{F}^{-1}(\chi_D f), f \in L^2(D) \}$ the Paley–Wiener space in D, and P the projection defined by $(Pf)\hat{}(\xi) = \chi_D(\xi)\hat{f}(\xi)$. In [1] C. Fefferman proved that the operator P is bounded only in L^2. It is not a standard Calderón–Zygmund operator. It is natural to ask:

PROBLEM 1. *Find the necessary and sufficient conditions for $b(x)$ such that the commutator $[b, P]$ is bounded or compact.*

This problem is closely related to the boundedness and compactness of some Hankel-type operators. In [3] R. Rochberg proposed a definition of Toeplitz and Hankel operators on $\mathrm{PW}(D)$ with symbol $b(x)$. They are defined by $T_b(f) = P(bf)$ and $H_b(f) = P(b\bar{f})$ for $f \in \mathrm{PW}(D)$ respectively. Because $\mathrm{PW}(D)$ is preserved when taking complex conjugates, these two operators on $\mathrm{PW}(D)$ are unitary equivalent. It is easy to characterize the Hilbert – Schmidt norm of T_b. In fact (see [2]),

$$\|T_b\|_{HS} \sim \|b\|_{B_2^{3/4,2}(2D)}.$$

The condition for boundedness and compactness are yet unknown.

PROBLEM 2. *Find the necessary and sufficient conditions for the symbol $b(x)$ such that the operators T_b and H_b are bounded or compact.*

The next problem is on pseudodifferential operators. It is related to the study of the Schrödinger operator with discrete spectrum. Let $\sigma(x,\xi) \in C^\infty(\mathbb{R}^{2n})$ be a symbol, and define the corresponding operator $\sigma(x, D)$ by

$$\sigma(x, D)f(x) = \int e^{2\pi i x\xi} \sigma(x,\xi)\hat{f}(\xi)\, d\xi.$$

For the boundedness of this type of operators, there is the well-known Calderón–Vaillancourt theorem. It says that if $|D_x^\alpha D_\xi^\beta \sigma(x,\xi)| \leqslant C_{\alpha\beta}$, when $|\alpha|, |\beta| \leqslant N$, where N is a constant, then $\sigma(x, D)$ is bounded. If

$$\lim_{R\to\infty} \sup_{|(x,\xi)|\geqslant R} |D_x^\alpha D_\xi^\beta \sigma(x,\xi)| = 0 \quad \text{for any } |\alpha|, |\beta| \leqslant N,$$

then the operator $\sigma(x, D)$ is compact (see [4]). It is known that these conditions are far from necessary. In fact, if the symbol $\sigma(x, \xi) \in L^2(\mathbb{R}^{2n})$, then the corresponding operator $\sigma(x, D)$ is a Hilbert – Schmidt operator and the Hilbert – Schmidt norm of $\sigma(x, D)$ is equivalent to the $L^2(\mathbb{R}^{2n})$ norm of the symbol $\sigma(x, \xi)$. From this one can construct a symbol $\sigma(x, \xi)$, such that $\sigma(x, \xi)$ is unbounded, but $\sigma(x, D)$ is a Hilbert – Schmidt operator. For example $\sigma(x, \xi) = a(x)\chi(|x|^\alpha \cdot \xi)$, where $a(x) \in C^\infty(\mathbb{R}^n)$, $\chi \in C_0^\infty(\mathbb{R}^n), \chi \geqslant 0$ and for $|\xi| \leqslant 1, \chi(\xi) = 1$, for $|\xi| \geqslant 2$, $\chi(\xi) = 0$, if $\int |a(x)|^2 |x|^{-n\alpha} dx < \infty$, then the corresponding operator is a Hilbert – Schmidt operator and

$$\|\sigma(x, D)\|_{\mathrm{HS}}^2 = \int |a(x)|^2 |x|^{-\alpha} dx \int \chi(\xi)^2 d\xi.$$

PROBLEM 3. *Find some sufficient conditions for general symbols such that the corresponding operators of these symbols are bounded or compact in $L^2(\mathbb{R}^n)$.*

In this problem the general symbols should include some symbols which do not satisfy the conditions in the Calderón–Vaillancourt theorem.

REFERENCES

1. Fefferman C., *The multiplier problem for the ball*, Ann. Math. **94** (1971), 330–336.
2. Peng L., *Hankel operators on the Paley–Wiener space in disk*, Proc. Cent. Math. Anal. Austral. Natl. Univ. **16** (1988), 173–180.
3. Rochberg R., *Toeplitz and Hankel operators on the Paley–Wiener space*, Integral Equations Oper. Theory **10** (1987), 186–235.
4. Taylor M., *Pseudodifferential Operators*, Princeton University Press, Princeton, N. J., 1981.

DEPARTMENT OF MATHEMATICS
PEKING UNIVERSITY
BEIJING 100871
P. R. CHINA

DEPARTMENT OF MATHEMATICS
PEKING UNIVERSITY
BEIJING 100871
P. R. CHINA

7.7
old

ITERATES OF TOEPLITZ OPERATORS WITH UNIMODULAR SYMBOLS

V. V. PELLER

Each invertible Toeplitz operator T_f on H^2 can be represented as $T_f = T_u T_h$ where h is an outer function with modulus $|f|$ and $u = f/h$ is a unimodular function. The operator T_h, being invertible analytic Toeplitz operator, has simple spectral behaviour. Therefore the Toeplitz operators with unimodular symbols play an especial role (see [1]). I would like to propose the following questions concerning these operators.

Suppose X is one of the function classes $H^\infty + C$, $\mathrm{QC} \stackrel{\text{def}}{=} H^\infty + C \cap \overline{H^\infty + C}$, C, C^k, C^∞.

QUESTION 1. *Let u be a unimodular function in X such that $\ker T_u = \{0\}$. Is it true that there exists f in H^2 with*

$$\inf_{n>0} \|T_u^n f\| > 0?$$

If the answer is positive, it is reasonable to ask whether the following stronger conclusion can be done.

QUESTION 2. *Is it true that under the hypotheses of Question 1*

$$\inf_{n>0} \|T_u^n f\| > 0$$

for every non-zero f in H^2?

It follows from Clark's results [2] that the answer to Question 2 is positive for rational functions u.

In view of T. Wolff's factorization theorem [3] (asserting that each unimodular function u in $H^\infty + C$ can be represented as $u = v\theta$ with $v \in \mathrm{QC}$ and θ inner) it seems plausible that if Question 1 (or 2) has a positive answer for $X = \mathrm{QC}$ then so is for $X = H^\infty + C$. Note that for general unimodular functions u with $\ker T_u = \{0\}$ it may happen that $\lim_{n>0} \|T_u^n f\| = 0$ for any f in H^2. For example, if E is a measurable subset of $\mathbb{T}$, $0 < \operatorname{meas} E < 1$ and $u = 1$ on E and -1 on $\mathbb{T} \setminus E$ then it follows from M. Rosenblum's results [4] that T_u is a selfadjoint operator with absolutely continuous spectrum on $[-1, 1]$.

An affirmative answer to Question 1 for $X = \mathrm{QC}$ would imply the existence of a non-trivial invariant subspaces for Toeplitz operators with unimodular symbols in QC (see some results on invariant subspaces of Toeplitz operators in [5]). Indeed, either one of the kernels $\ker T_u$, $\ker T_u^*$ is non-trivial or T_u and $T_{\bar u}$ satisfy the hypothesis at Question 1 and so both subspaces

$$\mathcal{L}_1 = \{ f : \lim_n \|T_u^n f\| = 0 \}, \qquad \mathcal{L}_2 = \{ f : \lim_n \|T_u^n f\| = 0 \}^\perp$$

are invariant under T_u and $\mathcal{L}_1 \neq H^2$, $\mathcal{L}_2 \neq \{0\}$. Therefore either one of these subspaces is non-trivial or $\mathcal{L}_1 = \{0\}$, $\mathcal{L}_2 = H^2$, i. e. T_u would be a C_{11}-contraction (see [6]). But each C_{11}-contraction has a non-trivial invariant subspace [6].

REFERENCES

1. Sarason D., *Function Theory on the Unit Circle*, Notes for Lectures at Virginia Polytechnic Inst. and State Univ., Blacksburg, Va., 1978.
2. Clark D. N., *On a similarity theory for rational Toeplitz operators*, J. Reine Angew. Math. **320** (1980), 6–31.
3. Wolff T., *Two algebras of bounded functions*, Duke Math. J. **49** (1982), 321–328.
4. Rosenblum M., *The absolute continuity of Toeplitz's matrices*, Pacif. J. Math. **10** (1960), 987–996.
5. Peller V. V., *Invariant subspaces for Toeplitz operators*, Zapiski Nauch. Semin. LOMI **126** (1983), 170–179 (Russian); English transl. in J. Soviet Math. **27** (1984), 2533–2539.
6. Sz.-Nagy B., Foiaş C., *Harmonic Analysis of Operators on Hilbert Space*, North Holland, Amsterdam, 1970.

STEKLOV MATHEMATICAL INSTITUTE
ST. PETERSBURG BRANCH
FONTANKA 27
ST. PETERSBURG, 191011
RUSSIA

AND

DEPARTMENT OF MATHEMATICS
KANSAS STATE UNIVERSITY
MANHATTAN
KANSAS, 66506
USA

EDITORS NOTE

The results of Clark [2] have been improved in [7]; paper [8] could be of interest for the Question raised.

REFERENCES

7. Yakubovich D. V., *Riemann surface models of Toeplitz operators*, Toeplitz Operators and Spectral Function Theory, OT: Adv. Appl., vol. 42, Birkhäuser Verlag, Basel–Boston–Berlin, 1989.
8. Atzmon A. Power regular operator (to appear).

7.8
v.old

LOCALIZATION OF TOEPLITZ OPERATORS

R. G. DOUGLAS

Let H^2 and H^∞ denote the Hardy subspaces of $L^2(\mathbb{T})$ and $L^\infty(\mathbb{T})$, respectively, consisting of the functions with zero negative Fourier coefficients and let P be the orthogonal projection from $L^2(\mathbb{T})$ onto H^2. For φ in $L^2(\mathbb{T})$ the Toeplitz operator with symbol φ is defined on H^2 by $T_\varphi f = P(\varphi f)$. Much of the interest in Toeplitz operators has been directed towards their spectral characteristics either singly or in terms of the algebras of operators which they generate. In particular, one seeks conceptual determinations of why an operator is or is not invertible and more generally Fredholm. One fact which one seeks to explain is the result due to Widom [1] that the spectrum $\sigma(T_\varphi)$ of an arbitrary Toeplitz operator is a connected subset of $\mathbb{C}$ and even [2] the essential spectrum $\sigma_e(T_\varphi)$ is connected. The latter result implies the former in view of Coburn's Lemma.

An important tool introduced in [2], [3] is the algebraic notion of localization. Let $\mathcal{I}$ denote the closed algebra generated by all Toeplitz operators and QC be the subalgebra

$$(H^\infty + C) \cap \overline{(H^\infty + C)}$$

of L^∞, where C denotes the algebra of continuous functions on $\mathbb{T}$. Each ξ in the maximal ideal space M_{QC} of QC determines a closed subset X_ξ of M_{L^∞} and one can show that the closed ideal ϑ_ξ in $\mathcal{I}$ generated by

$$\{\, T_\varphi : \hat{\varphi}|X_\xi \equiv 0 \,\}$$

is proper and that the local Toeplitz operator $T_\varphi + \vartheta_\xi$ in $\mathcal{I}_\xi = \mathcal{I}/\vartheta_\xi$ depends only on $\hat{\varphi}|X_\xi$. Moreover, since $\bigcap_{\xi \in M_{\mathrm{QC}}} \vartheta_\xi$ equals the ideal $\mathcal{K}$ of compact operators on H^2, properties which are true modulo $\mathcal{K}$ can be established "locally". For example, T_φ is Fredholm if and only if $T_\varphi + \vartheta_\xi$ is invertible for each ξ in M_{QC}. These localization results are established [4] by identifying QC as the center of $\mathcal{I}/\mathcal{K}$. One unanswered problem concerning local Toeplitz operators is:

CONJECTURE 1. *The spectrum of a local Toeplitz operator is connected.*

In [4] it was shown that many of the results known for Toeplitz operators have analogues valid for local Toeplitz operators. Unfortunately a proof of the connectedness would seem to require more refined knowledge of the behavior of H^∞ functions on M_{H^∞} than available and the result would imply the connectedness of $\sigma_e(T_\varphi)$.

A more refined localization has been obtained by Axler replacing X_ξ by the subsets of M_{L^∞} of maximal antisymmetry for $H^\infty + C$ using the fact that the local algebras $\mathcal{I}_\xi$ have non-trivial centers and iterating this transfinitely.

There is evidence to believe that the ultimate localization should be to the closed support X_η in M_{L^∞} for the representing measure μ_η for a point η in M_{H^∞}. In particular, one would like to show that if $H^2(\mu_\eta)$ denotes the closure in $L^2(\mu_\eta)$ of the functions

$\hat{\varphi}|X_\eta$ for φ in H^∞, P_η the orthogonal projection from $L^2(\mu_\eta)$ onto $H^2(\mu_\eta)$, then the map

$$T_\varphi \to T_{\hat{\varphi}|X_\eta}$$

extends to the corresponding algebras, where the local Toeplitz operator is defined by

$$T_{\hat{\varphi}|X_\eta} f = P_\eta(\hat{\varphi}|X_\eta)f$$

for f in $H^2(\mu_\eta)$. If η is a point in M_{L^∞}, then $H^2(\mu_\eta) = \mathbb{C}$ and it is a special case of the result [2] that $\mathcal{I}$ modulo its commutator ideal is isometrically isomorphic to L^∞, that the map extends to a character in this case. A generalized spectral inclusion theorem also provides evidence for the existence of this mapping in all cases.

One approach to establishing the existence of this map is to try to exhibit the state on $\mathcal{I}$ which this "representation" would determine. One property that such a state would have is that it would be multiplicative on the Toeplitz operators with symbols in H^∞. Call such states *analytically multiplicative.* Two problems connected with such states seem interesting.

CONJECTURE 2 (GENERALIZED GLEASON-WHITNEY). *If σ_1 and σ_2 are analytically multiplicative states on $\mathcal{I}$ which agree on H^∞ and such that the kernels of the two representations defined by σ_1 and σ_2 are equal, then the representations are equivalent.*

CONJECTURE 3 (GENERALIZED CORONA). *In the collection of analytically multiplicative states the ones which correspond to points of $\mathbb{D}$ are dense.*

One consequence of a localization to X_η when η is an analytic disk $\mathbb{D}_\eta$, would be the following. It is possible for φ in L^∞ that its harmonic extension $\hat{\varphi}|\mathbb{D}_\eta$ agrees with the harmonic extension of a function continuous on $\mathbb{T}$. (Note that this is not the same as saying that $\hat{\varphi}$ is continuous on the boundary of $\mathbb{D}_\eta$ as a subset of M_{H^∞} which is of course always the case.) In that case the invertibility of the local Toeplitz operator would depend on a "winding number" which should yield a subtle necessary condition for T_φ to be Fredholm. Ultimately it may be that there are enough analytic disks in M_{H^∞} on which the harmonic extension $\hat{\varphi}$ is "nice" to determine whether or not T_φ is Fredholm but that would require knowing a lot more about M_{H^∞} than we do now.

REFERENCES

1. Widom H., *On the spectrum of a Toeplitz operator*, Pacif. J. Math. **14** (1964), 365–375.
2. Douglas R. G., *Banach Algebra Techniques in Operator Theory*, Academic Press, New York, 1972.
3. Douglas R. G., *Banach Algebra Techniques in the Theory of Toeplitz Operators*, Conference Board of the Math. Sciences, Regional Conf. Series in Math. 15, Am. Math. Soc., Providence, R. I., 1973.
4. Douglas R. G., *Local Toeplitz operators*, Proc. London Math. Soc. **36** (1978), 243–272.

STATE UNIVERSITY OF NEW YORK
DEPARTMENT OF MATHEMATICS
STONY BROOK, NY 11794, USA

EDITORS NOTE

For more information see [5] p.p. 190–191.

REFERENCE

5. Böttcher A., Silbermann B., *Analysis of Toeplitz operators*, Akademie Verlag, Berlin, 1989.

7.9

PRODUCTS OF TOEPLITZ OPERATORS

Donald Sarason

Problem. *Characterize the pairs of outer functions g, h in H^2 of the unit disk such that the operator $T_g T_{\bar h}$ is bounded on H^2.*

Comments.

1. The problem arose in [6], which contains a class of examples for which the product $T_g T_{\bar h}$ is bounded even though at least one of the factors is not.

2. In case h is unbounded it is probably more convenient if, rather than thinking of $T_{\bar h}$ as an operator in H^2 with a restricted domain, one thinks of it as an operator whose domain is all of H^2 but whose range goes beyond H^2. For that one lets $T_{\bar h} f$, for f in H^2, be the Cauchy integral of $\bar h f$, that is, the holomorphic function in the unit disk defined by

$$(T_{\bar h} f)(z) = \frac{1}{2\pi} \int_{\partial \mathbb{D}} \frac{\overline{h(e^{i\theta})} f(e^{i\theta})}{1 - e^{-i\theta} z} \, d\theta.$$

The meaning of $T_g T_{\bar h} f$ is then clear. If h is bounded, the preceding definition of $T_{\bar h}$ agrees with the usual one.

3. If $T_g T_{\bar h}$ is bounded, then gh is bounded. In fact, for w a point of the unit disk, let k_w be the normalized kernel function in H^2 for the evaluation functional at w:

$$k_w(z) = (1 - |w|^2)^{1/2} (1 - \bar w z)^{-1}.$$

Then $T_{\bar h} k_w = \overline{h(w)} k_w$ and $T_{\bar g} k_w = \overline{g(w)} k_w$, so that

$$\langle T_g T_{\bar h} k_w, k_w \rangle = \langle T_{\bar h} k_w, T_{\bar g} k_w \rangle = g(w) \overline{h(w)} \|k_w\|_2^2.$$

Thus $\|gh\|_\infty \leqslant \|T_g T_{\bar h}\|$.

4. Because of the identity $H_{\bar g}{}^* H_{\bar h} = T_{g\bar h} - T_g T_{\bar h}$, the problem of determining when $T_g T_{\bar h}$ is bounded reduces to the problem of determining when $H_{\bar g}{}^* H_{\bar h}$ is bounded under the assumption that gh is bounded. The latter problem, with the boundedness assumption on gh discarded, is thus more general than the originally stated problem. (It is genuinely more general, as one sees by taking g to be an unbounded function for which $H_{\bar g}$ is bounded, and taking $h = g$.)

The papers [1], [3], [5], [7], [8] study the problem of determining when the product of a Hankel operator with the adjoint of another Hankel operator is compact. Characterizations are obtained, but they are not directly expressed in terms of the structure of the symbols of the operators involved. Progress on the present boundedness question, especially the version for Hankel operators, should provide additional insights into the compactness problem.

5. A solution of the original problem is known for the special case where $h = 1/g$. In fact, if g and $1/g$ are both in H^2, then $T_g T_{1/\bar g}$ is bounded if and only if the operator $T_{\bar g/g}$ is invertible (in which case $T_g T_{1/\bar g}$ is its inverse). Moreover, $T_{\bar g/g}$ is invertible if

and only if the function $|g|^2$ satisfies B. Muckenhoupt's condition (A_2). (References are [2] and [4].) The latter condition can be expressed as follows. For w a point of the unit disk, let P_w denote the corresponding Poisson kernel, and let $P_w(x)$ denote the Poisson integral at w of the L^1 function x. The (A_2) condition on $|g|^2$ is then the condition

$$\sup_{|w|<1} P_w(|g|^2)P_w(|g|^{-2}) < \infty.$$

6. An argument of S. R. Treil' (communicated orally) shows that a natural generalization of the preceding condition is at least necessary for the boundedness of $T_gT_{\bar{h}}$. Suppose $T_gT_{\bar{h}}$ is bounded, and for w a point of the unit disk, let b_w denote the corresponding Blaschke factor. The operator $T_{b_w}T_{\overline{b_w}}$ is then the orthogonal projection onto the orthogonal complement of the normalized kernel function k_w:

$$1 - T_{b_w}T_{\bar{b}_w} = k_w \otimes k_w.$$

Since $T_{b_w}T_gT_{\bar{h}}T_{\overline{b_w}} = T_gT_{b_w}T_{\overline{b_w}}T_{\bar{h}}$, we can conclude that

$$\|T_g(k_w \otimes k_w)T_{\bar{h}}\| \leqslant 2\|T_gT_{\bar{h}}\|.$$

But $T_g(k_w \otimes k_w)T_{\bar{h}} = (gk_w) \otimes (hk_w)$, and

$$\|(gk_w) \otimes (hk_w)\|^2 = \|gk_w\|_2^2\|hk_w\|_2^2 = P_w(|g|^2)P_w(|h|^2).$$

It follows that

$$\sup_{|w|<1} P_w(|g|^2)P_w(|h|^2) < \infty.$$

It is tempting to conjecture that the last condition is also sufficient for the boundedness of $T_gT_{\bar{h}}$.

7. A question related to the original problem is whether, when $T_gT_{\bar{h}}$ is bounded, it must belong to the Toeplitz algebra (the C*-algebra generated by the Toeplitz operators).

8. Another problem that sounds interesting is the modification of the original problem one obtains by replacing the Hardy space of the unit disk by the Bergman space.

References

1. Axler S., Chang S.-Y. A., Sarason D., *Products of Toeplitz operators*, Integral Equations Oper. Theory **1** (1978), 285–309.
2. Garnett J. B., *Bounded Analytic Functions* (1981), Academic Press, New York.
3. Jones P. W., *Estimates for the corona problem*, J. Funct. Anal. **39** (1980), 162–181.
4. Sarason D., *Function Theory on the Unit Circle*, Notes for lectures at Virginia Polytechnic Inst. and State Univ., Blacksburg, Va., 1978.
5. Sarason D., *The Shilov and Bishop decompositions of $H^\infty + C$*, Conference on Harmonic Analysis in Honor of Antoni Zygmund, vol. II, Wadsworth, Belmont, Calif., 1983, pp. 461–474.
6. Sarason D., *Exposed points in H^1*, II, Oper. Theory, Adv. Appl. **48** (1990), 333–347.
7. Volberg A. L., *Two remarks concerning the theorem of S. Axler, S.-Y. A. Chang and D. Sarason*, J. Oper. Theory **8** (1982), 209–218.
8. Wolff T. H., *Some theorems on vanishing mean oscillation*, Dissertation, University of California, Berkeley, Calif., 1979.

Department of Mathematics
University of California
Berkeley, CA 94720
USA

7.10
old

TOEPLITZ OPERATORS ON THE BERGMAN SPACE

C. SUNDBERG

Let A^2 denote the Bergman space of analytic functions in $L^2(\mathbb{D})$, and let P be the orthogonal projection of $L^2(\mathbb{D})$ onto A^2. For $\varphi \in L^\infty(\mathbb{D})$ we define the Toeplitz operator with symbol φ by $T_\varphi = P(\varphi f)$. In general the behaviour of these operators may be quite different from that of the Toeplitz operators on the Hardy space H^2. However it is shown in [1] that Toeplitz operators on A^2 with *harmonic* symbols behave quite similarly to those on H^2, and one can prove analogues for this class of many results about Toeplitz operators on H^2.

An important result about Toeplitz operators on H^2 is Widom's Theorem, which states that the spectrum of such an operator is connected ([2]). This suggests our problem.

CONJECTURE. *A Toeplitz operator on A^2 with harmonic symbol has a connected spectrum.*

In support of this conjecture we mention the following cases for harmonic φ in which the spectra can be explicitly computed.

1) If φ is analytic then $\sigma(T_\varphi) = \operatorname{clos} \varphi(\mathbb{D})$.
2) If φ is real-valued then $\sigma(T_\varphi) = [\inf \varphi, \sup \varphi]$.
3) If φ has piecewise continuous boundary values then $\sigma(T_\varphi)$ consists of the path formed from the boundary values of φ by joining the one-sided limits at discontinuities by straight line segments, together with certain components of the complement of this path.

For proofs of these see [1].

In connection with our conjecture it should be mentioned that there are easy examples of Toeplitz operators on A^2 with disconnected spectra—e.g. $\sigma(T_{1-|z|^2})$ is disconnected since $T_{1-|z|^2}$ is positive and compact. The proof of connectedness in the case of H^2 breaks down almost immediately in the A^2 case. I would expect a solution to the present problem to shed light on Toeplitz operators in general, and perhaps to lead to a different proof and a better understanding of Widom's Theorem.

REFERENCES

1. McDonald G., Sundberg C., *Toeplitz operators on the disc*, Indiana Univ. Math. J. **28** (1979), 595–611.
2. Widom H., *On the spectrum of Toeplitz operators*, Pacific J. Math. **14** (1964), 365–375.

UNIVERSITY OF TENNESSEE
DEPARTMENT OF MATHEMATICS
KNOXVILLE, TN 37916
USA

7.11
old

VECTORIAL TOEPLITZ OPERATORS ON HARDY SPACES

N. Ya. Krupnik, I. E. Verbitskiĭ

Let $\mathcal{M}$ be a separable Hilbert space ($\dim \mathcal{M}$ may be finite), $\mathcal{B}$ be the algebra of all bounded linear operators on $\mathcal{M}$ and $L^p(\mathcal{M})$ be the Banach space of weakly measurable $\mathcal{M}$-valued functions $f : T \to \mathcal{M}$ with the norm

$$\|f\| = \left\{ \int_{\mathbb{T}} |f(t)|^p_{\mathcal{M}} |\, dt| \right\}^{1/p}.$$

We denote by $H^p(\mathcal{M})$ the Hardy space of functions in $L^p(\mathcal{M})$ with zero negatively indexed Fourier coefficients and by P the Riesz projection onto $H^p(\mathcal{M})$ $(1 < p < \infty)$.

Let $L^\infty(\mathcal{B})$ be the space of all essentially bounded $\mathcal{D}$-valued functions and $H^\infty(\mathcal{B})$ be the Hardy space corresponding to $L^\infty(\mathcal{B})$. If $A \in L^\infty(\mathcal{B})$ then the operator $T_A = PA|H^p(\mathcal{M})$ is called a *vectorial Toeplitz operator.*

The following criterion of the invertibility of T_A on $H^2(\mathcal{M})$ has been obtained by Rabindranathan [1].

Theorem 1. *Let $A \in L^\infty(\mathcal{B})$. Then T_A is invertible on $H^2(\mathcal{M})$ if and only if $A = B_1^* U B_2$ where*

(1) *$B_1^{\pm 1}, B_2^{\pm 1} \in H^\infty(\mathcal{B})$;*
(2) *U is a unitary-valued function in $L^\infty(\mathcal{B})$;*
(3) *there exists an operator-valued function B with $B^{\pm 1} \in H^\infty(\mathcal{B})$ such that*

$$\|U - B\|_{L^\infty(\mathcal{B})} < 1. \tag{1}$$

See also [2-3] ($\dim \mathcal{M} = 1$) and [4] ($\dim \mathcal{M} < \infty$).

A sufficient condition for the invertibility of T_A on $H^p(\mathcal{M})$ has been given by the authors [5] (the case $\dim \mathcal{M} = 1$ has been considered earlier by Simonenko [6]).

Theorem 2. *Let $p \in (1, \infty)$, $A = B_1^* U B_2$ and suppose that all conditions of Theorem 1 for U, B_1, B_2 hold except for (1) which has to be replaced by*

$$\|U - B\|_{L^\infty(\mathcal{B})} < \sin\left(\frac{\pi}{\max(p, p')}\right).$$

Then T_A is invertible on $H^p(\mathcal{M})$ (and on $H^{p'}(\mathcal{M})$).

PROBLEM 1. *Are the conditions of Theorem 2 necessary for the invertibility of T_A on both $H^p(\mathcal{M})$ and $H^{p'}(\mathcal{M})$?*

It is shown in [7] that the answer is affirmative if $\dim \mathcal{M} = 1$.

Let us note that the class of operator-valued functions in Theorem 2 admits an equivalent description [8]: $A = B_1^* G B_2$, $G \in L^\infty(\mathcal{B})$, $B_1^{\pm 1}, B_2^{\pm 1} \in H^\infty(\mathcal{B})$ and the numerical range $w(G(t))$ lies in a fixed angle with the vertex at the origin and with the size less than $\pi / \max(p, p')$ a. e. on $\mathbb{T}$.

It is well-known that the problem of invertibility of T_A in $H^p(\mathcal{M})$ can be reduced by means of factorization [2-4,7] to the problem of boundedness of P in weighted L^p-spaces. In the case $\dim \mathcal{M} = 1$ a criterion for boundedness of P was given in [9].

PROBLEM 2. *Let $A \in H^p(\mathcal{M})$ and $A^{-1} \in H^{p'}(\mathcal{M})$ $(1 < p < \infty)$. What are the conditions for APA^{-1} to be bounded on $L^p(\mathcal{M})$?*

We do not know the solution of Problems 1 and 2 even in the case of matrix-valued Toeplitz operators ($\dim \mathcal{M} < \infty$).

REFERENCES

1. Rabindranathan M., *On the inversion of Toeplitz operators*, J. Math. Mech. **19** (1969/70), 195–206.
2. Helson H., Szegö G., *A problem in prediction theory*, Ann. Mat. Pure Appl. **51** (1960), 107–138.
3. Devinatz A., *Toeplitz operators on H^2 space*, Trans. Am. Math. Soc. **112** (1964), 304–317.
4. Pousson H. R., *Systems of Toeplitz operators on H^2*, Trans. Am. Math. Soc. **133** (1968), 527–536.
5. Verbitskiĭ I. È., Krupnik N. Ya., *Exact constants in theorems about boundedness of singular operators in weighted spaces*, Linear operators, Shtiintsa, Kishinev, 1980, pp. 21–35. (Russian)
6. Simonenko I. B., *Riemann's boundary problem for a pair of functions with measurable coefficients and its application to singular integrals in weighted spaces*, Izv. Akad. Nauk SSSR, Ser. Mat. **28** (1964), 277–306. (Russian)
7. Krupnik N. Ya., *Some consequences of the theorem of Hunt-Muckenhoupt-Wheeden*, Operators in Banach spaces, Shtiintsa, Kishinev, 1978, pp. 64–70. (Russian)
8. Spitkovskiĭ I. M., *On factorization of matrix functions whose Hausdorff set lies inside an angle*, Soobshch. Akad. Nauk Gruz. SSR **86** (1977), 561–564. (Russian)
9. Hunt R., Muckenhoupt B., Wheeden R., *Weighted norm inequalities for conjugate function and Hilbert transform*, Trans. Am. Math. Soc. **176** (1973), 227–251.

DEPARTMENT OF MATHEMATICS
BAR ILAN UNIVERSITY
RAMAT GAN
ISRAEL

DEPARTMENT OF MATHEMATICS
WAYNE STATE UNIVERSITY
DETROIT, MI 48202
USA

7.12
old

A FACTORIZATION PROBLEM FOR ALMOST PERIODIC MATRIX-FUNCTIONS AND FREDHOLM THEORY OF TOEPLITZ OPERATORS WITH SEMI-ALMOST PERIODIC MATRIX SYMBOLS

YU. I. KARLOVICH, I. M. SPITKOVSKIĬ

1. We consider $(n \times n)$-matrices G defined on $\mathbb{R}$ with elements in the usual algebra AP of almost periodic functions and Toeplitz operators $T_G = P_- G|\operatorname{im} P_-$ generated by these matrices. Here P_- is the Riesz projection onto the Hardy class H^p_- in the lower half-plane, $1 < p < \infty$.

It is well known (this fact holds for an arbitrary $G \in L_\infty$) that the condition $G^{-1} \in L_\infty$ is necessary for T_G to be semi-Fredholm. In the case $n = 1$ the converse is true. Moreover, T_G is left-invertible if the almost periodic (a. p.) index ν of the function G is non-positive and T_G is right-invertible if $\nu \geqslant 0$. There exists a certain parallel between Fredholm Toeplitz operators and the factorization problem of their symbols, in accordance with which the formula

$$G(t) = G_+(t)\Lambda(t)G^*_-(t) \tag{1}$$

is valid in the case $n = 1$, $G^{\pm 1} \in AP$. Here $\Lambda(t) = e^{i\nu t}$, the functions $(z+i)^{-1}G^{\pm 1}_{\pm}$ belong to Hardy class H^r in the upper half-plane and the operator $G_+P_-G_+^{-1}$ is bounded in all spaces L^r, $1 < r < \infty$.

Formula (1) with $G \in AP$ possessing the above properties (with the natural change of $e^{i\nu t}$ by $\operatorname{diag}[e^{i\nu_1 t}, \dots, e^{i\nu_n t}]$, $\nu_j \in \mathbb{R}$) will be called a P-factorization of G. It is easy to check that the partial a. p. indices $\nu_1, \dots, \nu_n$ are uniquely defined by G provided G admits P-factorization; and it is not difficult to describe the freedom of the choice of $G_\pm$. However in the case $n > 1$ not each matrix G invertible in AP admits P-factorizations. In this connection the following problems appear.

PROBLEM 1. *Obtain a criterion (or at least more or less general sufficient conditions) of existence of P-factorization.*

PROBLEM 2. *Find out whether the existence of P-factorization of G is a necessary condition for T_G to be (semi-)Fredholm. If not, then it is possible to change the definition of P-factorization in such a way as to get an equivalent of the semi-Fredholm property for T_G.*

Note that if G admits a P-factorization then T_G is left- (right-) Fredholm if and only if $\nu_j \leqslant 0$ ($\geqslant 0$), $j = 1, \dots, n$. Consequently an affirmative answer to Problem 2 would mean that "Fredholm character" of T_G is the same in all spaces H^r_- and its Fredholmness implies invertibility. We do not know whether these weaker statements are true if $n > 1$.

2. The class SAP (of semi-almost-periodic functions) is a natural extension of AP. This class has been introduced by D. Sarason [1] and may be defined, for example, as

$$\{g = (0,5+u)f + (0,5-u)h + g_0 : f, h \in AP; u, g_0 \in C(\mathbb{R});$$
$$\lim_{t\to\pm\infty} g_0(t) = 0, \lim_{t\to\pm\infty} u(t) = \pm 0,5\}.$$

The a. p. components f, h (of g) are uniquely determined by g_0.

A criterion for T_G ($G \in$ SAP, $n = 1$) to be semi-Fredholm in the space H^2_- was obtained in [1] and was generalized in [2] to the case of an arbitrary H^p_-, $p \in (1,\infty)$. The case $n > 1$ is considered in [3,4], where the Fredholmness and semi-Fredholmness criteria have been established. These results, however, were obtained under the a priori assumption of existence of P_0-factorization of the a. p. components F, H of G. The latter means that the factors $F_+^{\pm}$, $(F_-^*)^{\pm 1}$ in the P-factorization $F = F_+ \Lambda F_-^*$ belong to the class AP$^+$ of those matrices from AP whose Fourier exponents are all non-negative; the same holds for H. The following problems arise in connection with the question of removing these a priori assumptions.

PROBLEM 3. *Let G be an $n \times n$ matrix in* SAP, $n > 1$, *and let F and H be its a. p. components. Is it true that semi-Fredholmness of T_G implies semi-Fredholmness of T_F and T_H?*

PROBLEM 4. *Is the set of matrices admitting a P_0-factorization dense in the set of all matrices admitting a P-factorization? What should the situation be like if we restricted ourselves to matrices with a fixed (non-zero) set of partial a. p. indices?*

A positive answer to Problems 3, 4 would allow to extend the criterion for T_G to be (semi-)Fredholm [3,4] to the case of arbitrary matrices $G \in$ SAP.

3. Let us consider a triangular matrix $B \in$ AP of the second order. Under some additional assumptions (e. g. absolute convergence of Fourier series of its elements) the P-factorization property of B is reduced to the corresponding question about

$$B_0(t) = \begin{bmatrix} e^{i\nu t} & 0 \\ r_0(t) & e^{-i\nu t} \end{bmatrix} \tag{2}$$

where $\nu > 0$ and the spectrum Ω of $r_0 \in AP$ is contained in $(-\nu, \nu)$. Assuming that card$(\Omega) < \infty$ define a. p. polynomials r_j ($j = 1, 2, \dots$) by the recurrence formula

$$r_{j-2} = q_j r_{j-1} + r_j \quad (r_{-1} = e^{i\nu t}). \tag{3}$$

It is supposed that the sequence μ_j of leading exponents of r_j strictly decreases and $q_j \in$ AP$^+$. Analogously to the case of the usual factorization of continuous triangular matrices [5] there exists an algorithm (say, A) for P-factorization of B_0 which is connected with the continuous fraction expansion of r_{-1}/r_0 or equivalently with the relations (3). The algorithm A unlike in the continuous case does not necessarily lead to the aim. A sufficient condition to make an application possible is $\mu_K + \mu_{K-1} \leqslant 0$ for some $K \in \mathbb{Z}_+$. In this case the factors $B_\pm$ are a. p. polynomials.

PROBLEM 5. *Give conditions for the convergence of the algorithm A to obtain a P_0- (or P-) factorization of matrix* (2). *These conditions have to be formulated in terms of entries of* (2).

The algorithm A can be applied to obtain a P_0-factorization of matrix (2) if, for example, $\Omega \subset (-\nu, 0]$ or the distances between the points of Ω are multiples of a fixed quantity (in particular, with card$(\Omega) \leqslant 2$). Already in the case card $\Omega = 3$, i. e.

$$r_0(t) = ae^{i\alpha t} + be^{i\beta t} + ce^{i\gamma t} \quad (abc \neq 0, \quad -\nu < \alpha < \beta < \gamma < \nu)$$

there exist situations when algorithm A fails. One of such situations is: $\beta = 0$, $\nu = \gamma - \alpha$ and δ $(= -\alpha/\gamma)$ is irrational.

In this case we found another algorithm based on the successive application of the transformation $B \to A_B B C_B^*$ $(A_B^{\pm 1}, C_B^{\pm 1} \in AP^+)$ preserving the structure of B_0 and on the factorization of elements close to the unit matrix. With the help of this algorithm it was established that under the restriction $|c^\delta a| \neq |b^{1+\delta}|$ a P_0-factorization exists with $\nu_1 = \nu_2 = 0$, but $B_\pm$ are no more a. p. polynomials. In the case $|c^\delta a| = |b^{1+\delta}|$ a P-factorization of (2) does not exist. Thus, even in the case card $\Omega = 3$ the following problem is non-trivial.

PROBLEM 6. *Describe the cases when matrix* (2) *admits a P- (or P_0-) factorization, calculate its a. p. partial indices and construct, if possible, corresponding factorizations.*

The interest for matrices of the form (2) is motivated by the fact that they naturally arise in connection with convolution equations on a finite interval with kernels R for which $x^\alpha(\mathcal{F}R)(x)$ has a. p. asymptotics at infinity for some $\alpha \in \mathbb{C}$ [3].

REFERENCES

1. Sarason D., *Toeplitz operators with semi-almost periodic symbols*, Duke Math. J. **44** (1977), 357–364.
2. Saginashvili A. I., *Singular integral equations with coefficients having discontinuities of semi-almost periodic type*, Theory of analytic functions and harmonic analysis, Tr. Tbilis. Mat. Inst. Razmadze **66** (1980), 84–95. (Russian)
3. Karlovich Yu. I., Spitkovskiĭ I. M., *On Noetherness of some singular integral operators with matrix coefficients of class* SAP *and systems of convolution equations on a finite interval associated with them*, Dokl. Akad. Nauk SSSR **269** (1983), 531–535 (Russian); English transl. in Sov. Math. Dokl. **27** (1983), 358–263.
4. Karlovich Yu. I., Spitkovskiĭ I. M., *On Noetherness, n- and d-normality of singular integral operators having discontinuities of semi-almost periodic type*, School of theory of operators in functional spaces, abstracts, Minsk, 1982, pp. 81–82. (Russian)
5. Chebotarev G. N., *Partial indices of the Riemann boundary problem with a triangular matrix of second order*, Uspekhi Mat. Nauk **11** (1956), no. 3, 199–202. (Russian)

HYDROPHYSICAL INSTITUTE OF THE SEA
DIVISION OF THE ECONOMY AND ECOLOGY
OF THE WORLD OCEAN
PROLETARSKIĬ BUL'VAR 29
270044 ODESSA, UKRAINE

DEPARTMENT OF MATHEMATICS
COLLEGE OF WILLIAM AND MARY
WILLIAMSBURG, VA 23187
USA

EDITORS' NOTE

See also the book by G. S. Litvinchuk and I. M. Spitkovskii *Factorization of measurable matrix functions*, OT: Adv. and Appl., vol. 25, Birkhäuser Verlag, Basel–Boston, 1987, or Mathem. Research, vol. 37, Academie-Verlag, Berlin, 1987.

7.13
v.old

FACTORIZATION OF OPERATORS ON $L^2(a,b)$

L. A. SAKHNOVICH

1. A bounded operator S_- (S_+) on $L^2(a,b)$, $-\infty \leqslant a < b \leqslant \infty$, is called *lower triangular* (*upper triangular*) if for every ξ ($a \leqslant \xi \leqslant b$)

$$S_-^* P_\xi = P_\xi S_-^* P_\xi, \qquad S_+ P_\xi = P_\xi S_+ P_\xi,$$

where $P_\xi f \stackrel{\text{def}}{=} \chi_{[a,\xi]} f$.

A bounded operator S on $L^2(a,b)$ is said to *admit left factorization* if $S = S_- S_+$ where S_- and S_+ are lower and upper triangular bounded operators, with bounded inverses.

I. C. Gohberg and M. G. Kreĭn [1] have studied the problem of factorization under the assumption

$$S - I \in \mathfrak{S}_\infty. \tag{1}$$

The operators S_+, S_- have been assumed to be of the form

$$S_+ = I + X_+, \qquad S_- = I + X_-; \qquad X_+, X_- \in \mathfrak{S}_\infty.$$

($\mathfrak{S}_\infty$ is the ideal of compact operators.)

The factorization method has played an essential role in a number of problems of spectral theory. Giving up condition (1) and considering more general triangular operators would essentially widen the scope of applications of this method.

EXAMPLE. Consider [2] the operator

$$S_\beta f = f(x) + \frac{i\beta}{\pi}\text{v.p.}\int_0^\omega \frac{f(t)}{x-t}\,dt, \qquad -1 < \beta < 1.$$

The operator S_β ($\beta \neq 0$) clearly does not satisfy (1). Nevertheless S_β admits a factorization $S_\beta = W_\alpha W_\alpha^*$ with $\alpha = \frac{1}{\pi}\arctan\beta$ and the lower triangular operator W_α defined by the formula:

$$W_\alpha f = \frac{1}{\sqrt{\operatorname{ch}\alpha\pi}}\,\frac{x^{-i\alpha}}{\Gamma(i\alpha+1)}\,\frac{d}{dx}\int_0^x f(t)(x-t)^{i\alpha}\,dt.$$

The following condition is necessary for an operator S to admit left factorization:

(*) $\qquad S_\xi = P_\xi S P_\xi$ *is invertible in* $L^2(a,\xi)$ *for any* ξ ($a \leqslant \xi \leqslant b$).

PROBLEM 1. *For what classes of operators is condition* $(*)$ *sufficient for the existence of the left factorization?*

In the general case $(*)$ is not sufficient. Indeed, the operator S defined by

$$Sf = f(x) + \frac{\tan \pi\beta}{\pi} \text{v.p.} \int_0^\omega \frac{f(t)}{x-t}\, dt, \quad 0 < \beta < 1 \tag{2}$$

satisfies $(*)$ but does not admit factorization, [2]. Note that $(*)$ follows from $(**)$ defined below:

(**) *The operator S is bounded positive and has a bounded inverse.*

An important particular case of Problem 1 is the following

PROBLEM 2. *Does* $(**)$ *imply the existence of factorization?*

If $S - I \in \mathfrak{S}_\omega$ the answer is positive [1] ($\mathfrak{S}_\omega$ is the Matsaev ideal).

2. It is interesting to study Problems 1–2 for operators of convolution type

$$Sf = \frac{d}{dx} \int_0^\omega f(t) s(x-t)\, dt, \quad 0 \leqslant x \leqslant \omega. \tag{3}$$

Let $M(x) = s(x)$, $N(x) = -s(-x)$, $0 \leqslant x \leqslant \omega$. Then [3]

$$(A_0 S - S A_0^*) f = i \int_0^\omega f(t)\big[M(x) + N(t)\big]\, dt, \tag{4}$$

where $A_0 f = i \int_0^x f(t)dt$. If $(*)$ holds, the following matrix functions of second order make sense:

$$W(\xi, z) = I - i \begin{bmatrix} (S_\xi^{-1} P_\xi (A_0 - zI)^{-1} M, 1) & (S_\xi^{-1} P_\xi (A_0 - zI)^{-1} M, \bar{N}) \\ (S_\xi^{-1} P_\xi (A_0 - zI)^{-1} 1, 1) & (S_\xi^{-1} P_\xi (A_0 - zI)^{-1} 1, \bar{N}) \end{bmatrix}$$

$$B(\xi) = \begin{bmatrix} (S_\xi^{-1} P_\xi M, 1) & (S_\xi^{-1} P_\xi M, \bar{N}) \\ (S_\xi^{-1} P_\xi 1, 1) & (S_\xi^{-1} P_\xi 1, \bar{N}) \end{bmatrix}$$

THEOREM 1. *Suppose that the operator S in (3) admits the left factorization. Then the matrix functions $\xi \mapsto W(\xi, z)$ and $\xi \mapsto B(\xi)$ are absolutely continuous and*

$$\frac{dW}{d\xi} = \frac{i}{z} W(\xi, z) H(\xi), \quad H(\xi) = B'(\xi), \tag{5}$$

where the elements $h_{ij}(\xi)$ of the matrix $H(\xi)$ satisfy

$$h_{ij}(\xi) = R_i(\xi)\overline{\Pi_j(\xi)} = 1. \tag{6}$$

and

$$R_1(\xi)\overline{\Pi_1(\xi)} + R_2(\xi)\overline{\Pi_2(\xi)} = 1. \tag{7}$$

The functions R_i, Π_i can be expressed in terms of S_-, S_+:

$$\Pi_1(x) = S_+^{*-1}1,\ \Pi_2(x) = S_+^{*-1}\overline{N},\ R_1(x) = S_-^{-1}M,\ R_2(x) = S_-^{-1}1. \tag{8}$$

Every operator S satisfying (3) and admitting factorization defines (via (6)–(8)) a system of differential equations (5). A procedure of this type in the inverse spectral problem has been developed by M. G. Kreĭn [4] provided $S \geqslant 0$ and $I - S \in \mathfrak{S}_2$. Besides, Theorem 1 means that the *matrix transfer function* [5] $W(\omega, z)$ admits the multiplicative representation

$$W(\omega, z) = \int_0^{\omega}\!\!\!\!\!\!\!\curvearrowleft\ e^{\frac{i}{z}dB(x)}. \tag{9}$$

If S is positive (4) implies that $W(\omega, z)$ is the characteristic matrix-function of the operator $S^{-1/2}A_0S^{1/2}$. Then formulae (5), (9) are known [6, 7]. The equality

$$\operatorname{rank} B'(x) = 1, \quad 0 < x < \omega, \tag{10}$$

is new even in this case.

An immediate consequence of Theorem 1 is the necessity of the following condition for the operator in (3) to admit the factorization:

$$(***) \quad \begin{cases} \textit{The operator } S \textit{ satisfies } (*), \\ \textit{the matrix-function } B(\xi) \textit{is absolutely continuous and } (10) \textit{ holds.} \end{cases}$$

Note that all requirements of (***) but (10) are satisfied in example (2).

PROBLEM 3. *Does (***) imply the existence of factorization?*

THEOREM 2. *If the operator S satisfies both (**) and (***), then it admits the factorization.*

REFERENCES

1. Gohberg I. C. and Kreĭn M. G., *The Theory of Volterra Operators in Hilbert Space and its Applications*, Nauka, Moscow, 1967 (Russian); English translation: Translations of Math. Monographs, Am. Math. Soc., Providence, Rhode Island, 1970.
2. Sakhnovich L. A., *Factorizations of operators in $L^2(a, b)$*, Funkts. Anal. i Prilozh. **13** (1979), no. 3, 40–45 (Russian); English transl. in Funct. Anal. Appl. **13** (1979), 187–192.
3. Sakhnovich L. A., *On an integral equation with kernel depending on the difference of the arguments*, Mat. Issled. **8** (1973), 138–146. (Russian)
4. Kreĭn M. G., *Continuous analogues of propositions about polynomials orthogonal on the unit circle*, Dokl. Akad. Nauk SSSR **106** (1955), 637–640. (Russian)
5. Sakhnovich L. A., *On the factorization of an operator valued transfer operator function*, Dokl. Akad. Nauk SSSR **226** (1976), 781–784 (Russian); English transl. in Sov. Math. Dokl. **17** (1976), 203–207.
6. Livshits M. S., *Operators, Oscillations, Waves. Open Systems*, Nauka, Moscow, 1966 (Russian); English translation: Translations of Math. Monographs, Am. Math. Soc., Providence, Rhode Island, 1973.
7. Potapov V. P., *On the multiplicative structure of J-non-expanding matrix functions*, Tr. Mosk. Mat. O-va **4** (1955), 125–136 (Russian); English translation in Translations of Math. Monographs 15, Am. Math. Soc., Providence, Rhode Island, 1970, 131–243.

PR. DOBROVOLSKOGO 154, 199
ODESSA 270111
UKRAINE

Author's Commentary

Let H be an infinite dimensional Hilbert space. We shall introduce a maximal naturally ordered chain of orthoprojectors $\mathcal{P} = \{P\}$. Let us denote by $(\operatorname{alg}\mathcal{P})$ the set of bounded operators A satisfying the condition

$$AP = PAP. \tag{11}$$

The condition (11) means that the subspaces PH are invariant with respect to A when $P \in \mathcal{P}$. The pair $\{P^+, P^-\}$ is called a chain break if there does not exist any projector in $\mathcal{P}$ located between P^- and P^+. A chain without any breaks is called continuous. We shall say that A is in $(\operatorname{alg}\mathcal{P})^*$ if A^* is in $(\operatorname{alg}\mathcal{P})$. A bounded operator is said to admit a factorization if

$$A = A_+A_-, \tag{12}$$

where

$$A_+, A_+^{-1} \in \operatorname{alg}\mathcal{P},\ A_-, A_-^{-1} \in (\operatorname{alg}\mathcal{P})^*. \tag{13}$$

The following assertion was proved by D. R. Larsson [8] for operators of the form

$$0 < mE \leqslant A \leqslant ME. \tag{14}$$

Theorem. *If the chain $\mathcal{P}$ is continuous, then there is an unfactorizable operator of the form* (4).

This gives *a negative answer* to the question raised in PROBLEM 2. We remark that the requirement $A_+^{-1} \in (\operatorname{alg}\mathcal{P})$, $A_-^{-1} \in (\operatorname{alg}\mathcal{P})^*$, which is contained in [9], is missing in 7.13.

To the unsolved problems 1 and 3 in 7.13 we now add the following.

Problem. *The construction of some concrete examples of unfactorizable operators of the form* (14).

References

8. Larsson D. R., *Nest algebras and similarity transformations*, Ann. Math., II. Ser. **121** (1985), 409–427.
9. Sakhnovich L. A., *Factorization of operators in $L^2(a, b)$*, Funkts. Anal. i Prilozh. **13** (1979), no. 3, 40–45 (Russian); English transl. in Funct. Anal. Appl. **13** (1979), 187–192.

7.14
old

TOEPLITZ OPERATORS IN SEVERAL VARIABLES

L. A. COBURN

For $\mathbb{C}$ the complex numbers, let Ω be a bounded domain in $\mathbb{C}^m$ with closure $\bar{\Omega}$ and with $\partial\Omega$ the Shilov boundary of the uniformly closed algebra $A(\bar{\Omega})$ generated by all polynomials in the complex variables $z = (z_1, z_2, \ldots, z_m)$ on $\bar{\Omega}$. In general, $\partial\Omega$ is a closed subset of the topological boundary of Ω. When Ω is one of the classical domains of Cartan or in other cases of interest, $\partial\Omega$ is a compact manifold with a "natural" volume element $d\mu$ and the space $L^2(\partial\Omega)$ of μ-square integrable complex valued functions is the setting for our analysis.

The closure of $A(\bar{\Omega})$ in $L^2(\partial\Omega)$ is denoted by $H^2(\partial\Omega)$ and this (Hardy) space, together with the (unique) orthogonal projection operator P from $L^2(\partial\Omega)$ onto $H^2(\partial\Omega)$, is a basic object in complex analysis on Ω. For φ essentially bounded on $\partial\Omega(\mu)$, the *Toeplitz operator* T_φ is defined for all f in $H^2(\partial\Omega)$ by $T_\varphi f = P(\varphi f)$. The C*-algebra generated by all T_φ with φ continuous is denoted by $\tau(\partial\Omega)$.

Even for $\Omega = \mathbb{D} \times \mathbb{D} \times \cdots \times \mathbb{D}$ (m times) where $\mathbb{D}$ is the open unit disc in $\mathbb{C}$, many interesting questions about $\tau(\partial\Omega)$ remain open after more than a decade of study. Note that for $\Omega = \mathbb{D} \times \mathbb{D} \times \cdots \times \mathbb{D}$ (m times), $\partial\Omega = \mathbb{T}^m$, the m-torus. The structure of $\tau(\mathbb{T}^m)$ is well-understood for $m = 1, 2$ [1, 2, 3].

In particular, necessary and sufficient conditions for A in $\tau(\mathbb{T}^2)$ to be Fredholm of index r are known [2]. It follows from the analysis of [2] that every Fredholm operator of index r in $\tau(\mathbb{T}^2)$ can be joined by an arc of such operators to

$$\left[T_{z_2}^u (I - T_{z_1} T_{z_1}^*) + T_{z_1} T_{z_1}^*\right] \left[T_{z_1}^{-u-r} (I - T_{z_2} T_{z_2}^*) + T_{z_2} T_{z_2}^*\right].$$

Here, u is an integer and $T_{z_j}^{-|s|} = T_{z_j}^{*|s|}$.

PROBLEM 1. *Classify the arc-components of Fredholm operators in* $\tau(\mathbb{T}^3)$.

This question reduces to:

PROBLEM 2. *Classify the arc-components of invertible elements in* $\tau(\mathbb{T}^3)$.

REFERENCES

1. Coburn L. A., *The C*-algebra generated by an isometry* I, II, Bull. Amer. Math. Soc. **73** (1967), 722–726; Trans. Amer. Math. Soc. **137** (1969), 211–217.
2. Coburn L. A., Douglas R. G., Singer I. M., *An index theorem for Wiener-Hopf operators on the discrete quarter-plane*, J. Diff. Geom. **6** (1972), 587–593.
3. Douglas R. G., Howe R., *On the C*-algebra of Toeplitz operators on the quarter-plane*, Trans. Am. Math. Soc. **158** (1971), 203–217.

DEPARTMENT OF MATHEMATICS
STATE UNIVERSITY OF NEW YORK
AT BUFFALO
BUFFALO, NY 14214
USA

7.15

TOEPLITZ OPERATORS WITH COMPACT RESOLVENT IN BARGMANN SPACES

JAN JANAS

Let μ be the Gaussian measure in $\mathbb{C}^n$ defined by the density $\exp\left(-|z|^2\right)$ with respect to the Lebesgue measure on $\mathbb{C}^n = \mathbb{R}^{2n}$. The Bargmann space $H^2(\mu)$ is given by all entire functions on $\mathbb{C}^n$ which are square integrable with respect to μ. Let P be the orthogonal projection of $L^2(\mu)$ onto $H^2(\mu)$. For a measurable function φ on $\mathbb{C}^n$ the Toeplitz operator T_φ is given by

$$T_\varphi f = P(\varphi \cdot f), \quad f \in H^2(\mu).$$

In recent years Toeplitz operators on the Bargmann space attracted a new interest of mathematicians, see [1], [2], [3], [4], [5], [6]. In general T_φ may be bounded or unbounded depending on a choice of φ. In [5] we considered the question of compactness of the resolvent $R(\lambda, \bar{T}_\varphi)$ of $\bar{T}_\varphi$, where $\bar{T}_\varphi$ denotes the closure of T_φ defined on the domain $D(T_\varphi)$ given by $\{ f : \varphi f \in L^2(\mu) \}$.

The following partial results were obtained in [5].

THEOREM 1. *Suppose that φ is bounded from below,*

$$|\varphi(z)| \leqslant C \exp\left(\frac{|z|^2}{2} - A|z|\right)$$

for any $A > 0$. If, moreover, $\bar{T}_\varphi^* = \bar{T}_\varphi$, and $\lim |\varphi(z)| = \infty$ as $|z| \to \infty$, then $R(\lambda, \bar{T}_\varphi)$ is compact.

THEOREM 2. *If φ defines T_φ such that*

$$(*) \qquad \operatorname{Re} \int \varphi |f|^2 d\mu \geqslant c \int \sum_{k=1}^{n} \left| \frac{\partial f}{\partial z_k} \right|^2 d\mu \quad \text{for } f \in D(T_\varphi),$$

then $R(\lambda, \bar{T}_\varphi)$ is compact.

Remark. As a matter of fact, it is enough to assume that $(*)$ holds for $f \in D$, where D is a core of $\bar{T}_\varphi$.

Since Toeplitz operators in $H^2(\mu)$ are unitarily equivalent to some pseudo-differential operators in $L^2(\mathbb{R}^n)$ [3], it seems to be of a considerable interest to study the problem of compactness of $R(\lambda, \bar{T}_\varphi)$ for different classes of symbols. We have found some specific examples of symbols, which suggest the following

CONJECTURE. *Let φ be a polynomial in z and $\bar{z}$ of the form $\varphi = \psi_N + \varphi_1$, where ψ_N is a homogeneous polynomial of degree N and φ_1 is a polynomial of degree not bigger than $N-1$ in z and $\bar{z}$. Is it true that the compactness of $R(\lambda, \bar{T}_{\psi_N})$ implies the compactness of $R(\lambda, \bar{T}_\varphi)$?*

A related result in this direction was given in [6].

References

1. Berger C. A., Coburn L. A., *Toeplitz operators and quantum mechanics*, J. Funct. Anal. **68** (1986), 273–299.
2. Berger C. A., Coburn L. A., *Toeplitz operators on the Segal–Bargmann space*, Trans. Am. Math. Soc. **301** (1987), 813–829.
3. Guillemin V., *Toeplitz operators in n-dimensions*, Integral Equations Oper. Theory **7** (1984), 145–205.
4. Janas J., *Toeplitz and Hankel operators in Bargmann space*, Glasg. Math. J. **30** (1988), 315–323.
5. Janas J., *Unbounded Toeplitz operators in the Bargmann-Segal space*, Studia Math. (to appear).
6. Shapiro H. S., *An algebraic theorem of E. Fisher, and the holomorphic Goursat problem*, Bull. London Math. Soc. **21** (1989), 513–537.

Instytut Matematyczny PAN
Ul. Solskiego 30
31-027, Krakow
Polska

7.16
old

SOME PROBLEMS CONNECTED WITH THE SZEGÖ LIMIT THEOREMS

M. G. KREĬN, I. M. SPITKOVSKIĬ

1. Every sequence of $r \times r$ matrices $\{C_j\}_{j\in\mathbb{Z}}$ determines a sequence of matrix valued Toeplitz matrices $T_n = \{C_{j-k}\}_{j,k=0}^{n}$. If $\Delta_n \stackrel{\text{def}}{=} \det T_n \neq 0$ for sufficiently large values of n, then the question about the limiting behavior of Δ_{n+1}/Δ_n arises. The analogous question arises about Δ_n/μ^{n+1} provided the non-zero limit $\mu \stackrel{\text{def}}{=} \lim_n \Delta_{n+1}/\Delta_n$ exists.

Both questions were studied for the first time by G. Szegö. He dealt with the case $r = 1$ and supposed that $\{C_j\}_{j\in\mathbb{Z}}$ is a sequence of Fourier coefficients of a positive summable function. See [1] for the precise formulations, for the history of the problem and for its natural generalization

$$C_j = \widehat{M}(j), \tag{1}$$

M being a finite non-negative Borel measure on $\mathbb{T}$. By the Riesz-Herglotz theorem the class of sequences satisfying (1) is the class of positive definite sequences.

We consider here the case when $r \geqslant 1$ and $\{C_j\}$ is an α-sectorial sequence for some $\alpha \in [0, \pi/2)$. The latter means that every T_n is α-sectorial, i. e. its numerical range (Hausdorff set) lies in the angle $\{\, z : |\operatorname{Im} z| \leqslant \tan\alpha \cdot \operatorname{Re} z \,\}$. It is clear that $\{C_j\}$ is an α-sectorial sequence if and only if there exists a measure M satisfying (1) and taking values in the set of α-sectorial $r\times r$ matrices on all arcs of $\mathbb{T}$. The real part M_R of this measure M permits us to construct the Hilbert space $\mathcal{H}_r = L^2(M_R)$ consisting of r-tuples of functions and equipped with the sesquilinear form $A(f,g) = \int_{\mathbb{T}} f(\zeta)dM(\zeta)g^*(\zeta)$.

Employing the factorization theorems in [2,3] we proved in [4] the existence of the limit μ in the case $r \geqslant 1$, $\alpha \geqslant 0$ and obtained the following formula:

$$\mu = \exp\left[\int_{\mathbb{T}} \ln\det G(\zeta)\, dm(\zeta)\right] \tag{2}$$

where $G = dM/dm$ (see [4] for details and information about earlier results by A. Devinatz and B. Gyires). Formula (2) is valid in the case $\ln\det G \notin L^1$ too; this is the only case when $\mu = 0$.

We propose the following as an UNSOLVED PROBLEM: *find an extension of the Szegö second limit theorem to the case of α-sectorial sequences.*

We CONJECTURE that the limit $\lambda \stackrel{\text{def}}{=} \lim\limits_{n\to\infty} \Delta_n/\mu^{n+1}$ (finite or not) exists for every α-sectorial sequence satisfying the regularity condition

$$\ln\det G \in L^1(m). \tag{3}$$

Editor's Note. M. G. Kreĭn regretfully passed away on 17.10.1989.

We are somewhat encouraged in this conjecture by Theorem 2 in Devinatz [5] related to the case provided $r = 1$, M is an absolutely continuous measure and G satisfies some additional restrictions (including the requirement $G \in L^\infty$).

In the case $\alpha = 0$, $r > 1$, we proved in [4] the existence of λ and obtained a formula for its calculation using some geometrical considerations. Before formulating the corresponding result let us remind that under condition (3) there exist two canonical factorizations of the matrix G: the left one $G = G_\ell G_\ell^*$ and the right one $G = G_r^* G_r$ (G_ℓ and G_r are outer matrix functions of class H^2). Let us denote the Toeplitz operator with the (unitary valued) symbol $F = G_\ell^* G_r^{-1}$ by T_F.

THEOREM 1. *Let $\{C_j\}$ be a positive definite sequence of $r \times r$ matrices and let M be the measure connected with it by formula (1), $G = dM/dm$. Then the limit λ ($\leqslant \infty$) of the sequence Δ_n/μ^{n+1} exists under condition (3). This limit is finite if and only if M is absolutely continuous and $\sum_1^\infty k\|\widehat{F}(-k)\|^2 < \infty$. If these conditions are fulfilled then $\lambda = (\det T_F^* T_F)^{-1}$.*

We do not know whether there is any kind of general result in the case $\alpha > 0$, $r > 1$. It is well understood now that the existence of λ (and formulae for its evaluation) may be proved under some additional restrictions with the help of results obtained in another direction (the rejection of positive-definiteness with simultaneous amplification of the restrictions on the smoothness of G) that we do not touch upon here, see [6] and references in it.

2. Considering an α-sectorial matrix measure M concentrated on the line, it is possible to introduce a continuous analogue of the space $\mathcal{H}_r$ and to establish the following result.

THEOREM 2. *The following two statements are equivalent:*

(1) $\int_{-\infty}^{+\infty} \ln|\det G(\lambda)| \dfrac{d\lambda}{1+\lambda^2} > -\infty$, *where* $G = dM/d\lambda$;

(2) *the subspace $\mathcal{N}$ of constant r-tuples has zero intersection with the subspace $\mathcal{L}_s = \bigvee\{ e^{it\lambda}\mathcal{N} : t \geqslant s \}$ for all (at least for one) $s > 0$.*

If these conditions are fulfilled and $\alpha = 0$ then the square of the distance from $\xi \in \mathcal{N}$ to $\mathcal{L}_s$ in the $\mathcal{H}$-metric equals to $\xi \mathcal{D}_s \xi^*$, where $\mathcal{D}_s$ is a "distance matrix", which is calculated by the formula

$$\mathcal{D}_s = \int_0^s \gamma(t)\gamma(t)^*\,dt. \tag{4}$$

Here γ is the inverse Fourier transform of the matrix-function G_+ in the canonical left factorization of G ($G = G_+ G_+^*$ a. e. on $\mathbb{R}$, G_+ is outer and belongs to the Hardy class H^2 in the upper half-plane). In the case $r = 1$ Theorem 2 was proved already in [7]; the discrete analogue of (4) was established in [4] in the general case $\alpha \in [0, \pi/2)$, $r \geqslant 1$. We propose a natural

PROBLEM. *Generalize the second part (concerning formula (4)) of Theorem 2 to the case of distances in the skew A-metric ($\alpha > 0$).*

In this case obscure points already appear after the first attempts to interpret the right hand side of a formula of type (4). It is a fact that the inverse Fourier transform of the factors $G_\pm$ in canonical factorizations of α-sectorial matrix functions [2, 3] are in general not elements of L^2.

The problem to find continuous generalizations for the second Szegö limit theorem admits different formulations and even in the definite case corresponding investigations form an *unordered set* (see [8] and the papers cited there). There are still more unsolved questions in the case $\alpha > 0$ but we shall not go into this matter here.

References

1. Golinskiĭ B. L., Ibragimov I. A., *On the Szegö limit theorem*, Izv. Akad. Nauk SSSR **35** (1972), no. 2, 408–427. (Russian)
2. Kreĭn M. G., Spitkovskiĭ I. M., *On factorizations of α-sectorial matrix functions on the unit circle*, Mat. Issled. **47** (1978), 41–63. (Russian)
3. Kreĭn M. G., Spitkovskiĭ I. M., *Some generalizations of the first Szegö limit theorem*, Anal. Math.. **9** (1983), 23–41. (Russian)
4. Kreĭn M. G., Spitkovskiĭ I. M., *On factorizations of matrix functions in the unit circle*, Dokl. Akad. Nauk SSSR **234** (1977), 287–290 (Russian); English translation in Soviet Math. Dokl. **18** (1977), 287–299.
5. Devinatz A., *The strong Szegö limit theorem*, Illinois J. Math. **11** (1967), 160–175.
6. Basor E., Helton J. W., *A new proof of the Szegö limit theorem and new results for Toeplitz operators with discontinuous symbol*, J. Oper. Theory **3** (1980), 23–39.
7. Kreĭn M. G., *On an extrapolation problem by A. N. Kolmogorov*, Doklady Akad. Nauk SSSR **46** (1945), no. 8, 306–309. (Russian)
8. Mikaelyan L. V., *Continuous matrix analogues of G. Szegö's theorem on Toeplitz determinants*, Izv. Akad. Nauk Arm. SSR **17** (1982), 239–263. (Russian)

Department of Mathematics
College of William and Mary
Williamsburg, VA 23187
USA

Editors' note

See also [9] and [10].

References

9. Böttcher A., Silbermann B., *Analysis of Toeplitz operators*, Akademie-Verlag, Berlin, 1989.
10. Gohberg I., Kaashoek M. A., *Asymptotic formulas of Szegö–Kac–Achiezer type.*, Asymptotic Analysis **5** (1992), no. 3, 187–220.

7.17
old

THE DIOPHANTINE MOMENT PROBLEM, ORTHOGONAL POLYNOMIALS AND SOME MODELS OF STATISTICAL PHYSICS

V. S. VLADIMIROV, I. V. VOLOVICH

1. In [1], [2] it was shown that in investigations of the Ising model in the presence of a magnetic field the following one-parameter Diophantine trigonometrical moment problem (DTMP) appears.

PROBLEM. *Describe all non-negative measures $d\sigma(\theta,\xi)$ on the circle $\mathbb{T} = \{\zeta \in \mathbb{C} : \zeta = e^{i\theta}, \theta \in [-\pi,\pi]\}$, even in θ, depending on a parameter ξ, $0 \leqslant \xi \leqslant 1$, and such that*

$$M_0(\xi) \stackrel{\text{def}}{=} \frac{1}{\pi}\int_0^\pi d\sigma(\theta,\xi) \equiv 1$$

and the moments

$$M_k(\xi) \stackrel{\text{def}}{=} \frac{1}{\pi}\int_0^\pi \cos k\theta \, d\sigma(\theta,\xi), \quad k = 1,2,\ldots$$

are polynomials (in ξ) of degree $\varkappa k$ with integer coefficients; the parity of $M_k(\xi)$ coincides with the parity of $\varkappa k$. Here $\varkappa$ is an integer $\geqslant 2$ ($\varkappa$ is the number of the nearest neighbours in the lattice).

It is known that the description of such measures can be reduced to the description of the corresponding generating functions

$$I(z,\xi) = 1 + 2\sum_{k=1}^\infty M_k(\xi) z^k, \quad |z| < 1.$$

EXAMPLES: 1. $M_k(\xi) = \xi^{\varkappa k}$, $I(z,\xi) = \dfrac{1+z\xi^\varkappa}{1-z\xi^\varkappa}$.

2. $M_k(\xi) = T_k(\xi^\varkappa)$, where T_k are the Chebyshëv polynomials, $I(z,\xi) = \dfrac{1-z^2}{1-2z\xi^2+z^2}$.

3. $M_k(\xi) = T_{k\varkappa}(\xi)$, $I(z,\xi) = \dfrac{1-z^2}{1-2zT_\varkappa(\xi)+z^2}$.

4. $M_k(\xi) = \frac{1}{2}\left[P_k(1-2\xi^2) - P_{k-1}(1-2\xi^2)\right]$, where P_k are the Legendre polynomials, $I(z,\xi) = \dfrac{1-z}{\sqrt{(1-z)^2+4z\xi^2}}$.

We note that in examples 1–3 the generating function is rational, whereas in example 4 it is algebraic (this case corresponds to the one-dimensional Ising model).

QUESTION. *Has the generating function corresponding to a DTMP to be algebraic?*

Fixing a rational value of the parameter ξ, $\xi = \frac{p}{r}$, p, r integers, $0 \leqslant p \leqslant r$, $r \geqslant 1$, we see that our DTMP implies the following "quasi-DTMP":

$$M_k \stackrel{\text{def}}{=} \frac{1}{\pi} \int_0^\pi \cos k\theta \, d\sigma(\theta) = \frac{c_k}{r^{\varkappa k}}, \quad c_k \text{ being integers.}$$

In particular for $\xi = 0$ or 1 ($r = 1$) we obtain the following moment

PROBLEM. *Describe non-negative even measures whose trigonometrical moments are integers.*

This problem is solved by a known theorem due to Helson [3]:

$$d\sigma(\theta) = \sum_{s=0}^{N-1} a_s \delta\left(\theta - \frac{2\pi s}{N}\right) + \sum_{s=0}^{M-1} b_s \cos s\theta \, d\theta$$

under some additional conditions on a_s, b_s [2].

2. It was shown in [4] that the theory of Toeplitz forms and orthogonal polynomials is closely connected with some problems of statistical physics and in particular with the Gauss model on the halfaxis. In this connection some mathematical problems appear whose solution would be useful for the further investigation of such models.

1. Let $f(\theta)$ be an even non-negative summable function of $\mathbb{T}$ satisfying the Szegö condition

$$\int_0^\pi \ln f(\theta) \, d\theta > -\infty.$$

We define the function

$$\pi(z) = \exp\left[-\frac{1}{2\pi} \int_0^\pi \frac{1 - z^2}{1 - 2z\cos\theta + z^2} \ln f(\theta) \, d\theta\right] = \sum_{k=0}^\infty g_k z^k, \quad |z| < 1.$$

PROBLEM. *Find necessary and sufficient conditions for $g_k \to 0$ as $k \to \infty$ (physically the last condition means the absence of a long order parameter).*

It was shown in [4] that the condition $\frac{1}{\sqrt{f}} \in L^1$ implies $g_k \to 0$ as $k \to \infty$ and

$$\sum_{k=0}^\infty \frac{|g_k|}{1+k} < \infty.$$

2. Let $d\sigma(\theta)$ be an even non-negative measure on $\mathbb{T}$ and

$$J_{|k-j|} = \frac{1}{\pi} \int_0^\pi \cos(k-j)\theta \, d\sigma(\theta); \quad k, j = 0, 1, \ldots$$

is the corresponding Toeplitz matrix. We denote by $\left(J_{kj}^{(N)}\right)^{-1}$ the inverse matrix of $J_{|k-j|}^{(N)} = \left(J_{|k-j|}\right)_{k,j=0}^N$.

PROBLEM. *Find an asymptotics for*

$$\sum_{j,k=0}^{N} \left(J_{j,k}^{(N)}\right)^{-1} \quad \text{as } N \to \infty.$$

This problem appears in the study of the free energy in the Gaussian model on the halfaxis with an external field (see [4]). For example, when $d\sigma(\theta) = d\theta + 2\pi a\delta(\theta)$, $a > 0$ this expression tends to $1/a$.

The numbers $\left(J_{0k}^{(N)}\right)^{-1}$, $k = 0, 1, \ldots, N$, are proportional to the coefficients of the orthogonal polynomials $\varphi_N(z)$ (see [4], [5]). This leads to the PROBLEM of a more detailed investigation of the asymptotics of $\varphi_N(e^{i\theta})$ as $N \to \infty$ in the presence of a non-zero singular part of the measure $d\sigma$. As we know only the case of an absolutely continuous measure was considered in detail (see, e.g. [6], [7]).

3. The multidimensional Gaussian model.

PROBLEM. *Calculate the free energy and correlation functions under less restrictive conditions than in* [4], [8].

REFERENCES

1. Barnsley, Bessis D., Moussa P., *The Diophantine moment problem and the analytic structure in the activity of the ferromagnetic Ising model*, J. Math. Phys. **20** (1979), 535–552.
2. Vladimirov V. S., Volovich I. V., *The Ising model with a magnetic field and a Diofantine moment problem*, Teor. Mat. Fiz. **53** (1982), no. 1, 3–15 (Russian); English transl. in Theor. Math. Phys. **50** (1982), 177–185.
3. Helson H., *Note on harmonic functions*, Proc. Am. Math. Soc. **4** (1953), 686–691.
4. Vladimirov V. S., Volovich I. V., *On a model in statistical physics*, Teor. Mat. Fiz. **54** (1983), no. 1, 8–22 (Russian); English transl. in Theor. Math. Phys. **54** (1983), 1–11.
5. Vladimirov V. S., Volovich I. V., *The Wiener-Hopf equation, the Riemann-Hilbert problem and orthogonal polynomials*, Dokl. Akad. Nauk SSSR **266** (1982), 788–792 (Russian); English transl. in Soviet Math. Doklady **26** (1982), 415–419.
6. Szegö G., *Orthogonal Polynomials*, A. M. S. Coll. Publ. 23, revised ed., Am. Math. Soc., New York, 1959.
7. Golinskiĭ B. L., *Asymptotic representation of orthogonal polynomials*, Uspekhi Mat. Nauk **35** (1980), no. 2, 145–196. (Russian)
8. Linnik I. Yu., *A multidimensional analogue of Szegö's theorem*, Izv. Akad. Nauk SSSR, Ser. Mat. **39** (1975), 1393–1403 (Russian); English transl. in Math. USSR, Izv. **9** (1975), 1323–1332.

MIAN
VAVILOVA 42
MOSCOW 117966
RUSSIA

FRUNZENSKAYA NAB. 38/1, 206
MOSCOW
RUSSIA

COMMENTARY BY A. L. SAKHNOVICH

The problem of the greatest value of $\sigma(\theta+0) - \sigma(\theta-0)$ for the spectral functions $\sigma(\theta)$ for the S-node was solved in [9]. On the set of matrix functions $\sigma(\theta)$ allowing the representation

$$S = \int_{-\infty}^{\infty} (E - \theta A)^{-1} \Phi \, d\sigma(\theta) \, \Phi^*(E - \theta A^*)^{-1} S^{-1}$$

of the operator S, the greatest value of $\sigma(\theta + 0) - \sigma(\theta - 0)$ under the conditions in [9] is equal to $\left[\Phi^*(E - \theta A^*)^{-1} S^{-1}(E - \theta A)^{-1} \Phi\right]^{-1}$. Hence in the case of the block Toeplitz matrices $J^{(N)} = \{J_{j,k}\}_{j,k=0}^{N}$ in the representation

$$J_{j,k} = J_{k-j} = \frac{1}{2\pi} \int_{-\pi}^{-\pi} \exp\left[i(k-j)\theta\right] d\sigma(\theta)$$

it follows ([9], [10]) that

$$\text{(1)} \qquad \lim_{N\to\infty} \left[\sum_{j,k=0}^{N} \left(J^{(N)}\right)_{j,k}^{-1} \exp\left[i(k-j)\theta\right] \right]^{-1} = \frac{1}{2\pi} \left[\sigma(\theta + 0) - \sigma(\theta - 0)\right],$$

where $\left(J^{(N)}\right)_{j,k}^{-1}$ are the elements of $\left(J^{(N)}\right)^{-1}$. Formula (1) gives us the solution to the problem in part II.2:

$$\lim_{N\to\infty} \sum_{j,k=0}^{N} \left(J^{(N)}\right)_{j,k}^{-1} = 2\pi \left[\sigma(+0) - \sigma(-0)\right]^{-1}.$$

Related questions were also discussed in the papers of D. Z. Arov and V. M. Adamyan in [11].

References

9. Sakhnovich A. L., *On a class of extremal problems*, Izv. Akad. Nauk SSSR, Ser. Mat. **51** (1987), 436–443 (Russian); English transl. in Math. USSR, Izv. **30** (1988), 411–418.
10. Sakhnovich A. L., *On Toeplitz block matrices and connected properties of the Gauss model on a halfaxis*, Teor. Mat. Fiz. **63** (1985), no. 1, 154–160 (Russian); English transl. in Theor. Math. Phys. **63** (1985), 27–33.
11. Helson H., Sz.-Nagy B., Vasilescu F. H. (ed.), *Linear Operators in Function Spaces*, Birkhäuser, Basel, 1990.

pr. Dobrovolskogo 154, 199
Odessa, 270111
Ukraine

7.18

AXLER-CHANG-SARASON-VOLBERG THEOREMS FOR HARMONIC APPROXIMATION AND STABLE CONVERGENCE

A. BÖTTCHER, B. SILBERMANN

The Toeplitz operator T_a on the Hardy space H^2 of the unit circle $\mathbb{T}$ induced by a function a in L^∞ on $\mathbb{T}$ is the operator acting by the rule $T_a f = P(af)$ for $f \in H^2$, where $P\colon L^2 \to H^2$ denotes the orthogonal projection. The famous Axler-Chang-Sarason-Volberg theorem [1], [5], [4] says that if a and b are in L^∞, then the quasicommutator $T_{ab} - T_a T_b$ is compact if and only if

$$\operatorname{alg}(H^\infty, \bar{a}) \cap \operatorname{alg}(H^\infty, b) \subset C + H^\infty;$$

here $H^\infty := \{\varphi \in L^\infty : \varphi = P\varphi\}$, C stands for the continuous functions on $\mathbb{T}$, $C + H^\infty$ is the familiar Douglas algebra, $\operatorname{alg}(H^\infty, \psi)$ refers to the smallest closed subalgebra of L^∞ containing H^∞ and ψ, and $\bar{a}$ is defined by $\bar{a}(t) = \overline{a(t)}$ for $t \in \mathbb{T}$.

Given a function $a \in L^\infty$, define the r-th Abel-Poisson mean $h_r a \quad (0 < r < 1)$ and the n-th Fejér-Cesàro mean $\sigma_n a \quad (n = 0, 1, 2, \dots)$ by

$$(h_r a)(e^{i\Theta}) = \sum_{k=-\infty}^{\infty} r^{|k|} a_k e^{ik\Theta},$$

$$(\sigma_n a)(e^{i\Theta}) = \sum_{k=-n}^{n} \left(1 - \frac{|k|}{n+1}\right) a_k e^{ik\Theta}$$

where $\{a_k\}_{k=-\infty}^{\infty}$ is the Fourier coefficient sequence of a.

PROBLEM 1. *Let $a, b \in L^\infty$. Find necessary and sufficient conditions for the relations*

$$\lim_{r\to 1} \|h_r(ab) - (h_r a)(h_r b)\|_{L^\infty} = 0 \tag{1}$$

respectively

$$\lim_{n\to\infty} \|\sigma_n(ab) - (\sigma_n a)(\sigma_n b)\|_{L^\infty} = 0 \tag{2}$$

to hold.

PROBLEM 2. *Given $a, b \in L^\infty$, establish a criterion for the existence of a compact operator K on H^2 such that*

$$\lim_{r\to 1} \|T_{h_r(ab)} - T_{h_r a}T_{h_r b} - K\|_{L(H^2)} = 0 \tag{3}$$

respectively

$$\lim_{n\to\infty} \|T_{\sigma_n(ab)} - T_{\sigma_n a}T_{\sigma_n b} - K\|_{L(H^2)} = 0. \tag{4}$$

Problem 1 is concerned with asymptotic multiplicativity of approximate identities and thus with a question that plays an important role throughout the whole theory of Toeplitz operators (see e. g. [2], [4]). Problem 2 is of interest in connection with the application of Banach algebra techniques and local principles to the investigation of stable convergence of "smoothened" Toeplitz operators and hence with formulas that express the index of T_a in terms of the indices of $T_{h_r a}$ and $T_{\sigma_n a}$ as $r \to 1$ and $n \to \infty$, respectively (see [2]).

We remark that (1) to (4) are true if one of the two functions a and b belongs to $\mathrm{QC} := (C + H^\infty) \cap (C + \overline{H^\infty})$. Moreover, one can show that if a and b are any functions in L^∞ such that for each maximal antisymmetric set S for QC either $a|S$ or $b|S$ is constant, then (1) to (4) hold (see [2]). Finally, (1) is also known to be valid if both a and b are in $C + H^\infty$, whereas V. P. Havin and H. Wolf [3] have only recently proved that (2) is in general not in force for arbitrary $a, b \in C + H^\infty$.

CONJECTURE. *Let $a, b \in L^\infty$. For each of the relations (1) to (4) to be true it is necessary and sufficient that*

$$\mathrm{alg}(H^\infty, \bar{a}) \cap \mathrm{alg}(H^\infty, b) \subset C + H^\infty$$

and

$$\mathrm{alg}(H^\infty, a) \cap \mathrm{alg}(H^\infty, \bar{b}) \subset C + H^\infty.$$

REFERENCES

1. Axler S., Chang S. Y. A., Sarason D., *Products of Toeplitz operators*, Integral Equations Oper. Theory **1** (1978), 285–309.
2. Böttcher A., Silbermann B., *Analysis of Toeplitz Operators*, Akademie-Verlag, 1989 and Springer-Verlag, 1990.
3. Havin V. P., Wolf H., *The Poisson kernel is the only approximate identity that is asymptotically multiplicative on H^∞*, Zap. Nauchn. Semin. Leningrad. Otdel. Mat. Inst. Steklova **170** (1989), 82–89. (Russian)
4. Nikol'skiĭ N. K., *Treatise on the Shift Operator*, Grundlehren 273, Springer-Verlag, Berlin–Heidelberg–New York–Tokyo, 1986.
5. Volberg A. L., *Two remarks concerning the theorem of S. Axler, S.-Y. A. Chang and D. Sarason*, J. Oper. Theory **7** (1982), 209–218.

TECHNISCHE UNIVERSITÄT CHEMNITZ
SEKTION MATHEMATIK
PSF 964, CHEMNITZ, D 0-9010,
DEUTSCHLAND

TECHNISCHE UNIVERSITÄT CHEMNITZ
SEKTION MATHEMATIK
PSF 964, CHEMNITZ, D 0-9010,
DEUTSCHLAND

7.19
old

STARKE ELLIPTIZITÄT SINGULÄRER INTEGRALOPERATOREN UND SPLINE-APPROXIMATION*

S. PRÖSSDORF

Sei Γ ein Kurvensystem in $\mathbb{C}$, das aus endlich vielen einfachen geschlossenen oder offenen Ljapunowkurven besteht, die keine gemeinsamen Punkte haben. Des weiteren seien $t_1, \ldots, t_m \in \Gamma$ paarweise verschiedene Punkte, $-1 < \alpha_k < 1$ $(k = 1, \ldots, m)$ und $\varrho(t) \stackrel{\text{def}}{=} \prod_{k=1}^m |t - t_k|^{\alpha_k}$. Mit $L^2(\Gamma, \varrho)$ bezeichnen wir den Hilbertraum aller auf Γ meßbaren Funktionen f mit $\varrho^{1/2} f \in L^2(\Gamma)$. Wir betrachten die singulären Integraloperatoren der Gestalt

$$A_\Gamma \stackrel{\text{def}}{=} aI + bS_\Gamma, \quad (S_\Gamma f)(t) \stackrel{\text{def}}{=} \frac{1}{\pi i} \int_\Gamma \frac{f(\tau)}{\tau - t} d\tau \quad (t \in \Gamma)$$

mit stückweise stetigen Koeffizienten $a, b \in \mathrm{PC}(\Gamma)$. Bekanntlich gilt $A_\Gamma \in \mathcal{L}(L^2(\Gamma, \varrho))$. Mit $\mathcal{A}(\mathrm{PC})$ bezeichnen wir die kleinste abgeschlossene Teilalgebra von $\mathcal{L}(L^2(\Gamma, \varrho))$, die alle Operatoren A_Γ sowie das Ideal $\mathfrak{S}_\infty$ der kompakten Operatoren in $L^2(\Gamma, \varrho)$ enthält, und mit $\operatorname{Sym} A$ das Symbol eines Operators $A \in \mathcal{A}(\mathrm{PC})$ (vgl. [5] oder [9]).

Für stetige Koeffizienten $a, b \in C(\Gamma)$ gilt

$$\operatorname{Sym} A_\Gamma(t, z) = a(t) + b(t) \operatorname{Sym} S_\Gamma(t, z) \quad ((t, z) \in \Lambda),$$

und $\operatorname{Sym} S_\Gamma$ ist eine stetige Funktion auf einer gewissen Raumkurve $\Lambda = \Lambda(\Gamma)$ [6]. Im Falle des Intervalls $\Gamma = [a, b]$ ist Λ der Rand des Rechtecks $[a, b] \times [-1, 1]$. Wenn Γ nur aus geschlossenen Kurven besteht und $\varrho \equiv 1$ ist, dann gilt

$$\operatorname{Sym} A_\Gamma(t, z) = a(t) + b(t)z, \quad t \in \Gamma, z = \pm 1.$$

Die Abbildung Sym ist ein isometrischer Isomorphismus der symmetrischen Algebra $\mathcal{A}(C)/\mathfrak{S}_\infty$ auf $C(\Lambda)$ mit $\operatorname{Sym} A^* = \overline{\operatorname{Sym} A}$ $(A \in \mathcal{A}(C))$ [5], [6].

Der Operator $A \in \mathcal{L}(L^2(\Gamma, \varrho))$ heißt *stark elliptisch*, wenn gilt $\operatorname{Re} A \stackrel{\text{def}}{=} \frac{1}{2}(A + A^*) = D + T$, wobei D positiv definit und $T \in \mathfrak{S}_\infty$ ist. Wir nennen A *θ-stark elliptisch* (auch *lokal stark elliptisch*), wenn eine Funktion $\theta \in C(\Gamma)$, $\theta(t) \neq 0$ $(\forall t \in \Gamma)$, existiert derart, daß θA stark elliptisch ist. Für Operatoren der Algebra $\mathcal{A}(C)$ gelten folgende Kriterien:

1°. $A \in \mathcal{A}(C)$ ist genau dann stark elliptisch, wenn $\operatorname{Re} \operatorname{Sym} A > 0$.

2°. $A_\Gamma = aI + bS_\Gamma$ $(a, b \in C(\Gamma))$ ist genau dann θ-stark elliptisch, wenn

$$a(t) + b(t)\lambda \neq 0, \quad \forall t \in \Gamma, \quad \forall \lambda \in K_\varrho(t), \tag{1}$$

**Chapter Editor's Note.* This is a new edition of the old problem with some additions supplied by the author.

wobei $K_\varrho(t)$ die konvexe Hülle der Menge $\{\operatorname{Sym} S_\Gamma(t,z)\}_{(t,z)\in\Lambda}$ bei festem $t \in \Gamma$ bezeichnet. Im Falle $\varrho \equiv 1$ ist $K_1(t) = [-1,1]$.

Die Hinlänglichkeit der Bedingung $\operatorname{Re}\operatorname{Sym} A > 0$ folgt leicht aus den obengenannten Eigenschaften der Abbildung $\operatorname{Sym} A$ (vgl. [8]); ihre Notwendigkeit ergibt sich aus der Hinlänglichkeit und der Eigenschaft, daß der Operator $A \in \mathcal{A}(C)$ genau dann ein Fredholmoperator ist, wenn $\operatorname{Sym} A \neq 0$ [5], [9]. Die Notwendigkeit von (1) ist eine direkte Folgerung der Eigenschaft 1°; ihre Hinlänglichkeit kann man mit Hilfe einer Einheitszerlegung der Kurve Γ beweisen [3]. Wegen $\sigma_{\text{ess}}(S_\Gamma) = \{\operatorname{Sym} S_\Gamma(t,z)\}_{(t,z)\in\Lambda}$ (vgl. [6]) zieht die Bedingung

$$a(t) + b(t)\lambda \neq 0, \quad \forall t \in \Gamma, \quad \forall \lambda \in \operatorname{conv}\sigma_{\text{ess}}(S_\Gamma), \tag{2}$$

die Bedingung (1) nach sich; für konstante Koeffizienten a, b sind beide Bedingungen (1) und (2) äquivalent.

Die starke Elliptizität ist eine notwendige und hinreichende Bedingung dafür, daß für den invertierbaren Operator A die Reduktionsmethode bezüglich einer beliebigen Orthonormalbasis konvergiert [4]. Wenn $\Gamma = \mathbb{T}$ der Einheitskreis ist, so konvergieren für den singulären Operator $A_\mathbb{T}$ in $L^2(\mathbb{T})$ gewisse Projektionsmethoden mit Spline-Basisfunktionen genau dann, wenn $A_\mathbb{T}$ θ-stark elliptisch ist [11], [12], [4*].

Wir betrachten auf $\mathbb{T}$ die stückweise linearen Splines

$$\varphi_k^{(n)}(t) \stackrel{\text{def}}{=} \begin{cases} (t - t_{k-1})(t_k - t_{k-1})^{-1} & \text{für } t \in (t_{k-1}, t_k), \\ (t_{k+1} - t)(t_{k+1} - t_k)^{-1} & \text{für } t \in (t_k, t_{k+1}), \\ 0 & \text{sonst,} \end{cases}$$

wobei $t_k = t_k^{(n)} = e^{2\pi ik/n}$ $(k = 0, \ldots, n-1)$. Mit P_n bezeichnen wir den Orthoprojektor in $L^2(\mathbb{T})$ auf die lineare Hülle $H^n \stackrel{\text{def}}{=} \operatorname{span}\{\varphi_0^{(n)}, \ldots, \varphi_{n-1}^{(n)}\}$ und mit Q_n den Interpolationsprojektor, der jeder beschränkten Funktion f den Polygonzug

$$(Q_n f)(t) \stackrel{\text{def}}{=} \sum_{k=0}^{n-1} f(t_k)\varphi_k^{(n)}(t)$$

zuordnet. Wenn die Operatoren $A_n \stackrel{\text{def}}{=} Q_n A_\mathbb{T} P_n$ im Unterraum $H^n \subset L^2(\mathbb{T})$ (für alle $n > n_0$) invertierbar sind und $\sup \|A_n^{-1}\| < \infty$ ist, dann schreiben wir $A_\mathbb{T} \in \Pi\{P_n, Q_n\}$; in diesem Falle gilt $A_n^{-1} Q_n f \stackrel{L^2}{\to} A_\mathbb{T}^{-1} f$, $n \to \infty$, für alle f mit $Q_n f \to f$, insbesondere für alle Riemann-integrierbaren Funktionen [11]. Das soeben beschriebene Projektionsverfahren heißt *Kollokationsmethode* (auch Polygonmethode). Analoge Bedeutung hat $\Pi\{P_n, P_n\}$ (*Reduktionsmethode*). Es gilt folgender

SATZ ([11], [12]). *Sei $A_\mathbb{T} = aI + bS_\mathbb{T}$ ($A_\mathbb{T} = aI + S_\mathbb{T} bI$) mit $a, b \in \mathrm{PC}(\Gamma)$. Dann gilt $A_\mathbb{T} \in \Pi\{P_n, Q_n\}$ genau dann, wenn folgende Bedingung erfüllt ist:*

(3)

$$\Big[c(t+0)d(t-0)\mu + c(t-0)d(t+0)(1-\mu)\Big]\nu + d(t+0)d(t-0)(1-\nu) \neq 0,$$
$$\forall t \in \Gamma, \ \forall \mu, \nu \in [0,1]$$

wobei $c \stackrel{\text{def}}{=} a + b$, $d \stackrel{\text{def}}{=} a - b$.

Wir bemerken, daß aus (3) die Invertierbarkeit von $A_\mathbb{T}$ in $L^2(\mathbb{T})$ folgt [6], [9]. Im Falle $a, b \in C(\mathbb{T})$ bedeutet (3) gerade die θ-starke Elliptizität von $A_\mathbb{T}$.

HYPOTHESE 1. *Bedingung (3) ist äquivalent der θ-starken Elliptizität des Operators $A_{\mathbb{T}}$ im Raum $L^2(\mathbb{T})$ $(\theta \in \mathrm{PC}(\mathbb{T}))$.*

Aus der Gültigkeit dieser Hypothese und (3) folgt insbesondere $A_{\mathbb{T}} \in \Pi\{P_n, P_n\}$. Die Schwierigkeiten beim Überprüfen der Hypothese 1 bestehen darin, daß $\mathrm{Sym}\, A_{\mathbb{T}}$ eine Matrixfunktion und $\mathcal{A}(\mathrm{PC})$ eine nichtsymmetrische Algebra ist. In [2*] (siehe auch [4*]) is die Hypothese 1 für den Fall geschlossener Ljapunowkurven Γ unter Benutzung des Symbolkalküls für die Algebra $\mathcal{A}(\mathrm{PC})$ und einer Einheitszerlegung von Γ bewiesen worden. Außerdem wurden in [2*] notwendige und hinreichende Bedingungen für die θ-starke Elliptizität von A_Γ im Falle stückweise stetiger Matrixkoeffizienten a un b angegeben, die allerdings i. a. schwer nachprüfbar sind; für stetige Matrixfunktionen a, b sind sie äquivalent der Bedingung

$$\det[a(t) + \lambda b(t)] \neq 0, \ \forall t \in \Gamma, \ \exists \lambda \in [-1, 1].$$

Im Falle von Kurven Γ mit Ecken ist Hypothese 1 noch offen. In diesem Fall ist die Hinlänglichkeit der Bedingung (3) im vorstehenden Satz von M. Costabel und E. Stephan (mündliche Mitteilung) bewiesen worden; Ihre Notwendigkeit gilt dann allerdings nicht mehr [3*].

Von großem theoretischen und praktischen Interesse sind Bedingungen, die die Konvergenz entsprechender Kollokationsmethoden mit gewichteten Splines auf offenen Kurven garantieren (s.z.B. [2], [7]). Sei der Einfachheit halber $\Gamma = [0, 1]$, $t_j^{(n)} = j/n$ $(j = 0, 1, \dots, n)$, $\varphi_j^{(n)}$ die entsprechenden stückweise linearen Splines und $\psi_j^{(n)} = \varrho^{-1/2}\varphi_j^{(n)}$. Mit P_n^1 bezeichnen wir den Orthoprojektor in $L^2(\Gamma, \varrho)$ auf $\mathrm{span}\{\psi_0^{(n)}, \dots, \psi_n^{(n)}\}$ $(\subset L^2(\Gamma, \varrho))$ und mit Q_n^1 den entsprechenden Interpolationsprojektor.

HYPOTHESE 2. *Sei $A_\Gamma = aI + bS_\Gamma$ $(a, b \in C(\Gamma))$ ein θ-stark elliptischer Operator. Dann gilt $A_\Gamma \in \Pi\{P_n^1, Q_n^1\}$.*

Im Falle $\varrho \equiv 1$ ergibt sich die Richtigkeit der Hypothese 2 aus dem obengenannten Satz durch Abbildung von Γ auf eine Hälfte von $\mathbb{T}$ und anschliessende Fortsetzung der Koeffizienten auf ganz $\mathbb{T}$ (vgl. [9], Seite 86). Partikuläre Lösungen zur Hypothese 2 findet man z. B. in [1*], [3*], [4*].

HYPOTHESE 3. *Hypothese 2 gilt für beliebige polynomiale Splines $\varphi_j^{(n)}$ ungeraden Grades.*

Für geschlossene Kurven Γ wurde die Konvergenz der Kollokations- und Reduktionsmethoden mit Splines beliebigen Grades in nichtgewichteten Sobolewräumen in [1], [10], [13], [4*], [5*] untersucht. Im Falle nicht geschlossener und nicht glatter Kurven Γ siehe [3*], [4*] und die dort zitierte Literatur.

LITERATUR

1. Arnold D. N., Wendland W. L., *On the asymptotic convergence of collocation methods*, Math. Comput. **41** (1983), 349–381.
2. Dang D. Q., Norrie D. H., *A finite element method for the solution of singular integral equations*, Comput. Math. Appl. **4** (1978), 219–224.
3. Elschner J., Prössdorf S., *Über die starke Finite element methods for singular integral equations on an interval*, Engineering Analysis **1** (1984), 83–87.

4. Gohberg I. C., Fel'dman I. A., *Convolution Equations and Projection Method for their Solutions*, Nauka, Moscow, 1971 (Russian); English translation: Amer. Math. Soc., Providence, R. I., 1974
5. Gohberg I. C., Krupnik N. Ya., *Singular integral operators with piecewise continuous coefficients and their symbols*, Izv. Akad. Nauk SSSR **35** (1971), 940–961 (Russian); English translation in Soviet Math., Izv. **5** (1971), 955–979.
6. Gohberg I. C., Krupnik N. Ya., *Introduction to the Theory of one Dimensional Singular Integral Operators*, Shtiintsa, Kishinev, 1973 (Russian); German translation: Birkhäuser, Basel, 1979
7. Ien E., Srivastav R. P., *Cubic splines and approximate solution of singular integral equations*, Math. Comput. **37** (1981), 417–423.
8. Kohn I. I., Nirenberg L. I., *An algebra of pseudodifferential operators*, Commun. Pure and Appl. Math. **18** (1965), 269–305.
9. Michlin S. G., Prössdorf S., *Singuläre Integraloperatoren*, Akademie-Verlag, Berlin, 1980; English translation: Springer-Verlag, Berlin et al, 1986
10. Prössdorf S., *Zur Splinekollokation für lineare Operatoren in Sobolewräumen*, Recent trends in mathematics, Teubner-Texte zur Math. 50, Teubner, Leipzig, 1983, pp. 251–262.
11. Prössdorf S., Schmidt G., *A finite element collocation method for singular integral equations*, Math. Nachr. **100** (1981), 33–60.
12. Prössdorf S., Rathsfeld A., *A spline collocation method for singular integral equations with piecewise continuous coefficients*, Integral Equations Oper. Theory **7** (1984), 536–560.
13. Schmidt G., *On spline collocation for singular integral equations*, Math Nachr. **111** (1982), 177–196.

Supplement zur Literatur

1*. Elschner J., *On spline approximation for singular integral equations on an interval*, Math. Nachr. **139** (1988), 309–319.
2*. Prössdorf S., Rathsfeld A., *On strongly elliptic singular integral operators with piecewise continuous coefficients*, Integral Equations Oper. Theory **8** (1985), 825–841.
3*. Prössdorf S., Rathsfeld A., *Mellin techniques in the numerical analysis for onedimensional singular integral operators*, Report R-MATH-06/88, Karl-Weierstraß-Inst. Math. Akad. Wiss. DDR, Berlin, 1988.
4*. Prössdorf S., Silbermann B., *Numerical Analysis for Integral and Related Operator Equations*, Akademie-Verlag, Berlin, 1991 and Birkhäuser, Basel - Boston - Berlin, 1991.
5*. Saranen J., Wendland W. L., *On the asymptotic convergence of collocation methods with spline functions of even degree*, Math. Comput. **45** (1985), 91–108.

Institut für Angewandte Analysis und Stochastik
Hausvogteiplatz 5–7
O-1086 Berlin
Germany

7.20
old

HOW TO CALCULATE THE DEFECT NUMBERS OF THE GENERALIZED RIEMANN BOUNDARY VALUE PROBLEM

YU. D. LATUSHKIN, G. S. LITVINCHUK

The question concerns the problem of finding functions $\varphi, \psi \in H^p$ satisfying the boundary condition

$$\varphi(\alpha(t)) = a(t)\overline{\psi(t)} + b(t)\psi(t) + h(t), \quad t \in \mathbb{T}. \tag{1}$$

Here $a, b \in L^\infty$, $h \in L^p$, $1 < p < \infty$, α is a non-singular orientation preserving diffeomorphism of $\mathbb{T}$ (*the shift*) with the derivative in $\mathrm{Lip}\mu, 0 < \mu < 1$. The case of orientation-changing shift α comes to this by an evident replacement of $\alpha(t)$ by $\overline{\alpha(t)}$; a by $\bar{b}$; b by $\bar{a}$.

The investigation of (1) and of its generalizations is connected with a number of questions of elasticity theory ([1], Ch. 7), the rigidity problem for piecewise regular surfaces [2], etc., and has already a rather long history, starting with A. I. Markushevich's work of 1946 (see [3] and a detailed bibliography contained therein).

Fredholmness conditions and the index of the operator corresponding to the problem (1) are known and do not depend upon *the shift*. If $a^{-1}b$ is sufficiently small, then under certain additional conditions on a (e.g. $\|a^{-1}b\| < \sin\dfrac{\pi}{\max(p,p')}$, $a \in \mathcal{M}$, $\mathcal{M}$ being the class (introduced in [5]) of multipliers not affecting the factorizability) one of the defect numbers of (1) is equal to zero, and therefore defect numbers don't depend upon *the shift*. I. H. Sabitov's example (see [3], p. 272) shows that this is not the case in general.

PROBLEM. *Calculate the defect numbers of the problem* (1). *Find the conditions on the coefficients a, b, under which the defect numbers do not depend upon* **the shift**.

The defect numbers ℓ and ℓ' of the problem (1) without *shift* ($\alpha(t) = t$, $t \in \mathbb{T}$) are connected [3,4] by the formulas $\ell = \max(\varkappa_1, 0) + \max(\varkappa_2, 0) - 1$; $\ell' = \varkappa_1 + \varkappa_2 - \ell - 2$ with the partial indices $\varkappa_1, \varkappa_2$ of the matrix

$$\bar{a}^{-1}\begin{pmatrix} |b|^2 - |a|^2 & b \\ \bar{b} & 1 \end{pmatrix}$$

(here the defect numbers are calculated over $\mathbb{R}$).

The problem to calculate partial indices of matrix-valued functions even of this special kind, however, is far from its final solution. Under the assumptions

$$p = 2, \ a \in \mathcal{M}; \quad \mathrm{dist}(\omega, H^\infty + C) < 1 \tag{2}$$

(where $\omega = a_+^{-1}b$, a_+ being an outer function with $|a_+| = |a|$ a. e.), as it is shown in [6], $\varkappa_1$ and $\varkappa_2$ are expressed in terms of the multiplicity of the lowest s-number of the Hankel operator $Q\omega P$ (P is the Riesz projection of L^p onto H^p, $Q = I - P$). Using this fact and the results of V. M. Adamyan, D. Z. Arov and M. G. Kreĭn [7,8,9] the defect numbers of problem (1) are expressed in [6] in terms of approximation characteristics of its coefficients. The elimination of restrictions (2), and the generalization of the above-mentioned results to the weighted spaces seems to be of interest.

In the case $\alpha(t) \not\equiv t$ the calculation of defect numbers of (1) may reduce to the problem of calculation of the dimension of the kernel of the block of operator

$$\begin{pmatrix} QW_\alpha^{-1}aQ & QW_\alpha^{-1}bP \\ PW_\alpha^{-1}\bar{b}Q & PW_\alpha^{-1}\bar{a}P \end{pmatrix},$$

composed by *shifted* Hankel and Toeplitz operators (here $W_\alpha f = f \circ \alpha$ is the so-called *shift* (translation) operator). It is interesting to remark, that these operators also appear while investigating the so-called one sided boundary value problems, studied in [10]. Thus, the investigation of the problem $\varphi(\alpha) = a\bar{\varphi} + h$, $\varphi \in H^p$, $h \in L^p$ with an involutory orientation-changing *shift* α under the usual conditions [3] $a \cdot \overline{a(\alpha)} = 1$; $h + a\overline{h(\alpha)} = 0$ can reduce to the study of the operator $QW_\alpha aQ$: both have the same defect numbers, their images are closed simultaneously and so on. Thus, the new information about *shifted* Toeplitz and Hankel operators may be employed in the study of the boundary value problems with the shift.

In conclusion it should be remarked, that by analytical continuation of $\bar{\psi}$ into the domain $\{ z \in \mathbb{C} : |z| > 1 \}$ and by a conformal mapping, problem (1) can be reduced to the problem of finding a pair of functions in Smirnov classes $E_\pm^p$, satisfying a *non-shifted* boundary condition on a certain contour. We note, by the way, that the related question of the change of partial indices of matrix valued functions under a conformal mapping (i. e., practically, the question of calculating defect numbers of vectorial *shifted* Riemann boundary value problem) put by B. Bojarski [11], has so far received no satisfactory solution.

The authors are grateful to I. M. Spitkovskiĭ for useful discussions.

References

1. Vekua N. P., *Systems of Singular Integral Equations and Some Boundary Problems*, 2nd revised ed., Nauka, Moscow, 1970 (Russian); English translation (of 1st ed.): Noordhoff, Groningen, 1967
2. Vekua N. P., *Generalized Analytic Functions*, Fizmatgiz., Moscow, 1959 (Russian); English translation: Pergamon Press, London - Paris - Frankfurt, Addison Wesley, Reading, Mass., 1962
3. Litvinchuk G. S., *Boundary Problems and Singular Integral Equations with Shift*, Nauka, Moscow, 1977. (Russian)
4. Spitkovskiĭ I. M., *On the theory of generalized Riemann boundary value problem in L^p classes*, Ukr. Mat. Zh. **31** (1979), no. 1, 63–73 (Russian); English translation in Ukr. Math. J. **31** (1979), 47–57.
5. Spitkovskiĭ I. M., *On multipliers not affecting factorizability*, Dokl. Akad. Nauk SSSR **231** (1976), 1300–1303 (Russian); English translation in Soviet Math. Doklady **17** (1976), 1733–1739.
6. Litvinchuk G. S., Spitkovskiĭ I. M., *Sharp estimates for the defect numbers of a generalized Riemann boundary value problem, factorization of Hermitian matrix-valued functions and some problems of approximation by meromorphic functions*, Mat. Sb., Nov. Ser. **117** (1982), 196–214 (Russian); English translation in Math. USSR, Sb. **45** (1983), 205–224.

7. Adamyan V. M., Arov D. Z., Kreĭn M. G., *Infinite Hankel matrices and generalizations of the problems Carathéodory-Fejér and F. Riesz*, Funkts. Anal. i Prilozh. **2** (1968), no. 1, 1–19 (Russian); English translation in Funct. Anal. Appl. **2** (1968), 1–20.
8. Adamyan V. M., Arov D. Z., Kreĭn M. G., *Infinite Hankel matrices and generalizations of the problems of Carathéodory-Fejér and I. Schur*, Funkts. Anal. i Prilozh. **2** (1968), no. 4, 1–17 (Russian); English translation in Funct. Anal. Appl. **2** (1968), 269–281.
9. Adamyan V. M., Arov D. Z., Kreĭn M. G., *Analytic properties of the Schmidt pair of a Hankel operator and the generalized Schur-Takagi problem*, Mat. Sb. **86** (1971), 33–73 (Russian); English translation in Math. USSR Sbornik **15** (1971), 31–73.
10. Zverovich È. I., Litvinchuk G. S., *Onesided boundary problems in the theory of analytic functions*, Izv. Akad. Nauk SSSR, Ser. Mat. **28** (1964), 1003–1036. (Russian)
11. Boyarskiĭ B. V. (= Bojarski, B.), *Analysis of the solvability of boundary problems in function theory*, Studies in contemporary problems in the theory of functions of one complex variable, Fizmatgiz, Moscow, 1961, pp. 57–79. (Russian)

Department of Mathematics
University of Missouri
Columbia, MO 65211
USA

Instituto Superior Técnico
Departamento de Matemática
Avenida Rovisco Pais
1096 Lishoa Codex
Portugal

7.21
old

POINCARÉ – BERTRAND OPERATORS IN BANACH ALGEBRAS

M. A. SEMËNOV-TYAN-SHANSKIĬ

Let A be an associative algebra over $\mathbb{C}$. A linear operator $R \in \operatorname{End} A$ is said to satisfy the Poincaré–Bertrand identity if for all $X, Y \in A$

$$R(X \cdot RY + RX \cdot Y) = RX \cdot RY + XY. \tag{1}$$

THEOREM. *Suppose R satisfies* (1). *Then*

(i) *the formula*

$$X \times_R Y = RX \cdot Y + X \cdot RY$$

defines an associative product in A (we denote the corresponding algebra by A_R)*;*

(ii) *The mappings $R \pm 1$ are homomorphisms from A_R into A. Let $A_\pm = \operatorname{Im}(R \pm 1)$, $N_\pm = \operatorname{Ker}(R \pm 1)$. Then $A_\pm \subset A$ is a subalgebra and $N_\pm \subset A_\pm$ is a two-sided ideal. Also, $A = A_+ + A_-$, $N_+ \cap N_- = 0$;*

(iii) *The mapping of the quotient algebras*

$$\theta\colon A_+/N_+ \to A_-/N_-$$

given by $\theta\colon (R+1)X \to (R-1)X$ is an algebra isomorphism;

(iv) *each $X \in A$ can be uniquely decomposed as*

$$X = X_+ - X_-, \; X_\pm \in A_\pm, \; \theta(\bar{X}_+) = \bar{X}_-$$

(we denote by $\bar{X}_\pm$ the residue class of $X_\pm$ modulo $N_\pm$).

EXAMPLE. Let $A = W$ be the Wiener algebra. Define $R \in \operatorname{End} W$ by

$$RX = \begin{cases} X, & \text{if } X \text{ is analytic in } \mathcal{D}, \\ -X, & \text{if } X \text{ is antianalytic in } \mathcal{D}. \end{cases}$$

Then R satisfies (1).

PROBLEM. *For $A = W \otimes \operatorname{Mat}(n, \mathbb{C})$, or $A = L^\infty \otimes \operatorname{Mat}(n, \mathbb{C})$ describe all linear operators satisfying* (1).

Notes. 1. This problem arises as a byproduct from the studies of completely integrable systems. This connection is fully explained in [1]. For partial results in the classification problem, cf. [1], [2]. In most papers the problem is considered in a Lie algebraic setting. In that case equation (1) is replaced by

$$R([X, RY] + [RX, Y]) = [RX, RY] + [X, Y]$$

which is commonly known as the (classical) Yang-Baxter equation.

2. Given a solution of (1) the operator $X \mapsto X \times_R Y$ can be regarded as an analogue of the Toeplitz operator with the symbol Y. It seems interesting to study the corresponding operator calculus in detail.

References

1. Semënov-Tyan-Shanskiĭ M. A., *What is a classical r-matrix?*, Funkts. Anal. i Prilozh. **17** (1983), no. 4, 17–33 (Russian); English transl. in Funct. Anal. Appl. **17** (1983), 259–272.
2. Belavin A. A., Drinfel'd V. G., *On the solutions of the classical Yang-Baxter equation for simple Lie algebras*, Funkts. Anal. i Prilozh. **16** (1982), no. 3, 1–29 (Russian); English transl. in Funct. Anal. Appl. **16** (1982), 159–180.

Steklov Mathematical Institute
St. Petersburg branch
Fontanka 27
St. Petersburg, 191011
Russia

and

Physique-Mathématiques
Université be Bourgogne
B. P. 138
21004 Dijon
France

S.7.22
old

HANKEL OPERATORS ON BERGMAN SPACES

SHELDON AXLER

Let dA be the usual area measure on the open disk $\mathbb{D}$. The Bergman space $L^2_a(\mathbb{D})$ is the subspace of $L^2(\mathbb{D}, dA)$ consisting of functions in $L^2(\mathbb{D}, dA)$ which are analytic on $\mathbb{D}$. Let P denote the orthogonal projection of $L^2(\mathbb{D}, dA)$ onto $L^2_a(\mathbb{D})$. For $f \in L^\infty(\mathbb{D}, dA)$, we define the Toeplitz operator

$$T_f \colon L^2_a(\mathbb{D}) \to L^2_a(\mathbb{D})$$

and the Hankel operator

$$H_f \colon L^2_a(\mathbb{D}) \to L^2(\mathbb{D}, dA) \ominus L^2_a(\mathbb{D})$$

by $T_f = P(fh)$ and $H_f = (I - P)(fh)$.

PROBLEM 1. *For which functions $f \in L^\infty(\mathbb{D}, dA)$ is the Hankel operator compact?*

If we are dealing with Hankel operators on the circle $\mathbb{T}$ rather than the disk $\mathbb{D}$, the answer would be that the symbol must be in the space $H^\infty + C(\mathbb{T})$. On the disk $\mathbb{D}$, it is easy to see that if $f \in H^\infty + C(\bar{\mathbb{D}})$, then H_f is compact. However, it is not hard to construct an open set $S \subsetneqq \mathbb{D}$ with $\bar{S} \cap \mathbb{T} \neq \emptyset$ such that if f is the characteristic function of S, then H_f is compact. Thus the subset of $L^\infty(\mathbb{D}, dA)$ which gives compact Hankel operators is much bigger (in a non-trivial way) than $H^\infty + C(\bar{\mathbb{D}})$, and it is possible that there is no nice answer to the question as asked above.

A more natural question arises by considering only symbols which are complex conjugates of analytic functions:

PROBLEM 2. *For which functions $f \in H^\infty$ is $H_{\bar{f}}$ compact?*

It is believable that this question has a nice answer. A good candidate is that $\bar{f}$ must be in $H^\infty + C(\mathbb{D})$.

The importance of this question stems from the identity

$$T_f^* T_f - T_f T_f^* = H_{\bar{f}}^* H_{\bar{f}},$$

valid for all $f \in H^\infty$. Thus we are asking which Toeplitz operators on the disk with analytic symbols have compact self-commutator.

Readers familiar with a paper of Coifman, Rochberg, and Weiss [1] might think that paper answers the above question. Theorem VIII of [1] seems to determine precisely which conjugates of analytic functions give rise to compact Hankel operators. However, the Hankel operators used in [1] are (unitarily equivalent to) multiplication followed by projection onto $E = \{\bar{f} : f \in L^2_a(\mathbb{D}) \text{ and } f(0) = 0\}$. Since $L^2(\mathbb{D}, dA) \ominus L^2_a(\mathbb{D})$ is far bigger than E, the Hankel operators of [1] are not the same as the Hankel operators defined here.

The Hankel operators as defined in [1] are more natural when dealing with singular integral theory, but the close connection with Toeplitz operators is lost. To determine which analytic Toeplitz operators are essentially normal, the Hankel operators as defined here are the natural objects to study.

Reference

1. Coifman R. R., Rochberg R., Weiss G., *Factorization theorems for Hardy spaces in several variables*, Ann. Math. **103** (1976), 611–635.

Department of Mathematics
Michigan State University
East Lansing, MI 48824
USA

Author's Commentary

In the earlier collection, I posed the problem of characterizing the functions $f \in L^\infty(\mathbb{D}, dA)$ such that the Hankel operator H_f is compact (Problem 1). Thinking that this problem was too hard and that it would not have a nice solution, I also asked an easier question: For which $f \in H^\infty$ is $H_{\bar{f}}$ compact (Problem 2)? (Here H^∞ denotes, as usual, the set of bounded analytic functions on $\mathbb{D}$.) I guessed that for $f \in H^\infty$, the Hankel operator $H_{\bar{f}}$ is compact if and only if the corresponding Hankel operator on the Hardy space is compact.

My guess was wrong. For $f \in H^\infty$, the Hardy space Hankel operator corresponding to $\bar{f}$ is compact if and only if f is in VMOA, the space of analytic functions with vanishing mean oscillation (with respect to arc length measure on the unit circle). Eventually I understood (see [3], Theorem 2) that the "area version" of VMOA is the little Bloch space, which is the space of analytic functions f on $\mathbb{D}$ such that $(1-|z|^2)|f'(z)| \to 0$ as $|z| \to 1$. Then I was able ([3], Theorem 7) to *answer the second question* mentioned above: *For $f \in H^\infty$, the Hankel operator $H_{\bar{f}}$ is compact if and only if f is in the little Bloch space.*

Having found an agreeable answer to my second question, I was still convinced that the *first problem* had no nice solution. Thus I was pleasantly surprised when Karel Stroethoff [4] came up with a pretty characterization of the compact Hankel operators. Stroethoff proved that for $f \in L^\infty(\mathbb{D}, dA)$, the Hankel operator H_f is compact if and only if $\|f \circ \varphi_w - P(f \circ \varphi_w)\|_2 \to 0$ as $|w| \to 1$; here φ_w (with $w \in \mathbb{D}$) denotes the Möbius map $\varphi_w(z) = (w-z)/(1-\bar{w}z)$ and $\|\cdot\|_2$ denotes the norm in $L^2(\mathbb{D}, dA)$.

Extensions and generalizations of the original question have been investigated by several authors. For example, Kehe Zhu ([5], Proposition 6 and Theorem 13) showed that for $f \in L^\infty(\mathbb{D}, dA)$, the Hankel operators H_f and $H_{\bar{f}}$ are both compact if and only if $\|f \circ \varphi_w - \tilde{f}(w)\|_2 \to 0$ as $|w| \to 1$; here $\tilde{f}$ is the Berezin symbol defined on $\mathbb{D}$ by

$$\tilde{f}(w) = \frac{1}{\pi} \int_{\mathbb{D}} f(z) |\varphi'_w(z)|^2 \, dA(z).$$

In another direction, Jonathan Arazy, Stephen Fisher, and Jaak Peetre ([2], Theorem 6.1) showed that for $f \in H^\infty$ and $1 < p < \infty$, the Hankel operator $H_{\bar f}$ is in the Schatten p-class if and only if

$$\int_{\mathbb{D}} |f'(z)|^p (1-|z|)^{p-2}\, dA(z) < \infty.$$

For $1 < p < \infty$, the Hardy space Hankel operators in the Schatten p-class are also characterized by the finiteness of the above integral. However, Arazy, Fisher, and Peetre ([2], Corollary 5.4) proved that for $f \in H^\infty$, the Hankel operator $H_{\bar f}$ is a trace class operator if f is constant. Thus, for $p = 1$ (trace class) as well as $p = \infty$ (compact operators), the Bergman space and Hardy space Hankel operators behave differently.

CHAPTER EDITOR'S NOTE

The Hankel operators considered in [1] are nowadays referred as *small* Hankel operators, while the ones considered by the problem author are called *big* Hankel operators. The phenomenon that there are no non-trivial "Ha-plitz" type operators (of a certain kind) in the Schatten–von Neumann class $\mathfrak{S}_p$ beyond a certain limit p_0 (first encountered in paper by Jansson and Wolff [6] is referred to as the *cut-off*. Recently, compact (big) Hankel operators with *unbounded* symbols have been completely characterized by Luecking [7].

EDITORS' NOTE

Recent paper [8] contains another criterion for boundedness-compactness of big Hankel operators as well as an anologue of the Adamyan–Arov–Kreĭn theorem on the essential norm.

REFERENCES

2. Arazy J., Fisher S., Peetre J., *Hankel operators on weighted Bergman spaces*, Am. J. Math. **110** (1988), 989–1054.
3. Axler S., *The Bergman space, the Bloch space, and commutators of multiplication operators*, Duke Math. J. **53** (1986), 315–332.
4. Stroethoff K., *Compact Hankel operators on the Bergman space*, Ill. J. Math. **34** (1990), 159–174.
5. Zhu K., *VMO, ESV, and Toeplitz operators on the Bergman space*, Trans. Amer. Math. Soc. **302** (1987), 617–646.
6. Janson S., Wolff T., *Schatten classes and commutators of singular integrals*, Ark. Mat. **20** (1982), 301–310.
7. Luecking D., *A characterization of certain classes of Hankel operators on the Bergman space of the unit disk*, preprint, 1990.
8. Lin P., Rochberg R., *The essential norm of Hankel operators on the Bergman space*, Integral Equat. Oper. Theory (to appear).

S.7.23
old

THE BANACH ALGEBRA APPROACH TO THE REDUCTION METHOD FOR TOEPLITZ OPERATORS

B. SILBERMANN

Let H^2 denote the Hardy space on $L^2 = L^2(\mathbb{T})$, consisting of the functions f with $\hat{f}(n) = 0$, $n < 0$, and let P denote the orthogonal projection from L^2 onto H^2. For $a \in L^\infty = L^\infty(\mathbb{T})$ the Toeplitz operator with symbol a is defined on H^2 by $T(a)\varphi = P(a\varphi)$.

Let $\mathcal{B}(H^2)$ be the Banach algebra of linear and bounded operators on H^2. Given a closed subalgebra B of L^∞ denote by $\operatorname{alg} T(B)$ the smallest closed subalgebra of $\mathcal{B}(H^2)$ containing all operators $T(a)$ with $a \in B$. Furthermore, let $Q(B)$ denote the so-called *quasi-commutator ideal* of $\operatorname{alg} T(B)$, i. e. the smallest closed twosided ideal in $\operatorname{alg} T(B)$ containing all operators of the form $T(ab) - T(a)T(b)$ $(a, b \in B)$. It is a rather surprising fact that this ideal plays an important role not only in the Fredholm theory of Toeplitz operators, but also in the theory of the reduction method for operators $A \in \operatorname{alg} T(L^\infty)$ (with respect to the projections P_n defined by $P_n \sum_{k=0}^{\infty} \hat{f}(k)\zeta^k = \sum_{k=0}^{n} \hat{f}(k)\zeta^k$). For $A \in \mathcal{B}(H^2)$ write $A \in \Pi\{P_n\}$ if the projection method is applicable to A (see [3] for a precise definition). Finally, put $Q_n = I - P_n$ and denote by $G\mathcal{A}$ the group of invertible elements in a Banach algebra $\mathcal{A}$ with identity.

For $A \in \mathcal{B}(H^2)$, the following statements are easily seen to be equivalent:

(i) $A \in \Pi\{P_n\}$;

(ii) $A^* \in \Pi\{P_n\}$;

(iii) $A \in G\mathcal{B}(H^2)$, $P_nAP_n + Q_n \in G\mathcal{B}(H^2)$ $(n \geqslant n_0)$, and $\sup_{n \geqslant n_0} \|(P_nAP_n + Q_n)^{-1}\| < \infty$;

(iv) $A \in G\mathcal{B}(H^2)$, $Q_nA^{-1}Q_n + P_n \in G\mathcal{B}(H^2)$ $(n \geqslant n_0)$ and $\sup_{n \geqslant n_0} \|(Q_nA^{-1}Q_n + P_n)^{-1}\| < \infty$;

(v) $A \in G\mathcal{B}(H^2)$, $V_{-n}A^{-1}V_n \in G\mathcal{B}(H^2)$ $(n \geqslant n_0)$ and $\sup_{n \geqslant n_0} \|(V_{-n}A^{-1}V_n)^{-1}\| < \infty$,

where $V_n = T(\zeta^n)$, $V_{-n} = T(\zeta^{-n})$ $(n \in \mathbb{N})$. There is an important estimate closely related to (v) (see [1]†):

$$\Big\|\sum_{i=1}^{k}\sum_{j=1}^{l} T(a_{ij})\Big\| \geqslant \Big\|T\Big(\sum_{i=1}^{k}\prod_{j=1}^{l} a_{ij}\Big)\Big\|, \tag{1}$$

†See also Nikol'skii N. K., *Treatise on the shift operators*, Springer-Verlag, 1986. – Ed.

which holds for every finite collection of functions $a_{ij} \in L^\infty$. Now, given a closed subalgebra B of L^∞ it follows from (1) that

$$S\Big(\sum_{i=1}^{k}\prod_{j=1}^{l} T(a_{ij})\Big) \stackrel{\text{def}}{=} T\Big(\sum_{i=1}^{k}\prod_{j=1}^{l} a_{ij}\Big) \quad (a_{ij} \in B)$$

defines a bounded projection S on $\operatorname{alg} T(B)$. One can show that

$$\operatorname{im} S = \{T(a) : a \in B\}, \quad \ker S = Q(B), \quad S(A) = s\text{-}\lim_{n\to\infty} V_{-n} A V_n.$$

If $A \in \operatorname{alg} T(L^\infty) \cap G\mathcal{B}(H^2)$, then $A \in G \operatorname{alg} T(L^\infty)$, since $\operatorname{alg} T(L^\infty)$ is a C*-algebra. Thus $S(A^{-1})$ makes sense and belongs to $\operatorname{alg} T(L^\infty)$. Moreover, if $A \in \Pi\{P_n\}$ then (ii) and (v) imply the invertibility of $S(A^{-1})$.

CONJECTURE 1. *Let $A \in \operatorname{alg} T(L^\infty)$. Then $A \in \Pi\{P_n\}$ if and only if A and $S(A^{-1})$ are in $G\mathcal{B}(H^2)$.*

The following special cases are of particular interest:

(a) $A = T(a)$;

(b) $A = \sum_{i=1}^{k} \prod_{j=1}^{l} T^{\varepsilon_{ij}}(a_{ij})$, where $\varepsilon_{ij} \in \{-1, 1\}$ and, of course, for $\varepsilon_{ij} = -1$ the invertibility of $T(a_{ij})$ is a part of the hypothesis.

For $a_{ij} \in \mathrm{PC}$ (PC is the algebra of piecewise continuous functions on $\mathbb{T}$ with only finitely many jumps) the case (b) is of importance in connection with the asymptotic behavior of Toeplitz determinants generated by singular functions (cf. [1]). In the case (a) Conjecture 1 is confirmed for $a \in C + H^\infty$ or $a \in \operatorname{clos}_{L^\infty} \mathrm{PC}$ (see [3, 6]), and in the case (b) for $A = (T^\varepsilon(a))^n$ ($\varepsilon = \pm 1$, $n \in \mathbb{N}$, $a \in \operatorname{clos}_{L^\infty} \mathrm{PC}$) (see [4, 7]).

One possible way to attack these problems concerned with the reduction method is to formulate them in the language of Banach algebras and then to use localization techniques (cf. [6]). Define $W_n : H^2 \to H^2$ by

$$W_n \sum_{k=0}^{\infty} \hat{f}(k)\zeta^k = \sum_{k=0}^{n} \hat{f}(k)\zeta^{n-k}$$

and denote by $\mathcal{A}$ the collection of all sequences $\{A_n\}_{n=0}^{\infty}$, $A_n : \operatorname{im} P_n \to \operatorname{im} P_n$ having the following property: there exist two operators $A, \tilde{A} \in \mathcal{B}(H^2)$ such that

$$A_n P_n \to A, \quad A_n^* P_n \to A^*, \quad \tilde{A}_n P_n \stackrel{\text{def}}{=} W_n A_n W_n \to \tilde{A}, \quad \tilde{A}_n^* P_n \to \tilde{A}^*$$

(strong convergence). By defining $\{A_n\} + \{B_n\} = \{A_n + B_n\}$, $\{A_n\}\{B_n\} = \{A_n B_n\}$, $|\{A_n\}| = \sup_n |A_n P_n|$, the set $\mathcal{A}$ becomes a Banach algebra with identity. If $A \in \mathfrak{S}_\infty$ ($\mathfrak{S}_\infty$ is the ideal of compact operators on H^2) or even if $A \in \operatorname{alg} T(L^\infty)$, then $\{P_n A P_n\} \in \mathcal{A}$ (see [5]). Notice that this is obvious for $A = T(a)$, $a \in L^\infty$, since $W_n T(a) W_n = P_n T(\tilde{a}) P_n$, where $\tilde{a}(\zeta) = a(1/\zeta)$. It can be proved that the set

$$J = \{\, \{A_n\} : A_n = P_n T P_n + W_n \tilde{T} W_n + C_n,\ T, \tilde{T} \in \mathfrak{S}_\infty,\ |C_n P_n| \to 0 \,\}$$

actually forms a two-sided closed ideal in $\mathcal{A}$, and that the problem of the applicability of the reduction method to $A \in \operatorname{alg} T(L^\infty)$ admits the following reformulation (see [6]):

$$A \in \Pi\{P_n\} \iff A, \tilde{A} \in G\mathcal{B}(H^2)$$
$$\text{and the coset of } \mathcal{A}/J \text{ containing } \{P_n A P_n\} \text{ is invertible in } \mathcal{A}/J.$$

Note that now localization techniques can very advantageously be applied to study invertibility in the algebra $\mathcal{A}/J$.

There is a construction which is perhaps of interest in connection with the case (a). Denote by $\operatorname{alg}_{\{P_n\}} T(B)$ the smallest closed subalgebra of $\mathcal{A}$ containing all elements of the form $\{P_n T(a) P_n\}$, where $a \in B$. If $C(\mathbb{T}) \subset B$ then $\mathfrak{S}_\infty \subset \operatorname{alg} T(B)$ and $J \subset \operatorname{alg}_{\{P_n\}} T(B)$ (cf. [2]). Assign to each sequence $\{A_n\} \in \operatorname{alg}_{\{P_n\}} T(B)$ the coset $[A] \in \operatorname{alg} T(B)/\mathfrak{S}_\infty$ containing $A \overset{\text{def}}{=} s\text{-}\lim A_n$. In this way a continuous homomorphism φ: $\operatorname{alg}_{\{P_n\}} T(B) \to \operatorname{alg} T(B)/\mathfrak{S}_\infty$ is produced, and one has $\ker \varphi \supset J$.

Conjecture 2. *If $B = L^\infty$ then $\ker \varphi = J$.*

A confirmation of this conjecture would imply that

$$\operatorname{alg}_{\{P_n\}} T(B)/J \cong \operatorname{alg} T(B)/\mathfrak{S}_\infty, \tag{2}$$

which, on its hand, would verify Conjecture 1 for $A = T(a)$, $a \in L^\infty$. It is already of interest to find sufficient conditions for the validity of (2) in the case $B \neq L^\infty$. Note that (2) was proved for $B = C + H^\infty$ or $B = \operatorname{clos}_{L^\infty} \mathrm{PC}$ in [2].

References

1. Böttcher A., Silbermann B., *Invertibility and Asymptotics of Toeplitz Matrice*, Mathematische Forschung 17, Akademie-Verlag, Berlin, 1983.
2. Böttcher A., Silbermann B., *The finite section method for Toeplitz operators on the quarter-plane with piecewise continuous symbols*, Math. Nachr. **110** (1983), 279–291.
3. Gohberg I. C., Fel'dman I. A., *Convolution Equations and Projection Methods for their Solution*, Nauka, Moscow, 1971 (Russian); English translation: Amer. Math. Soc., Providence, R. I., 1974
4. Roch S., Silbermann B., *Das Reduktionsverfahrens für Potenzen von Toeplitzoperatoren mit unstetigem Symbol*, Wiss. Z. Tech. Hochsch. Karl-Marx-Stadt **24** (1982), 289–294.
5. Roch S., Silbermann B., *Toeplitz-like operators, quasicommutator ideals, numerical analysis, Part I*, Math. Nachr. **120** (1985), 141–173; *Part II*, Math. Nachr. **134** (1987), 381–391.
6. Silbermann B., *Lokale Theorie des Reduktionsverfahrens für Toeplitzoperatoren*, Math. Nachr. **104** (1981), 137–146.
7. Verbitskiĭ I. È., *On the reduction method for steps of Toeplitz matrices*, Mat. Issl. **47** (1978), 3–11. (Russian)

Technische Universität Chemnitz
Sektion Mathematik
PSF 964, Chemnitz, D 0-9010,
Deutschland

Author's Commentary

We shall use the same notations as above.

First of all, a result of Treil' [12] (see Böttcher/Silbermann [8]) states that there exist functions $a \in L^\infty$ such that $T(a) \in G\mathcal{B}(H^2)$ but $T(a) \notin \Pi\{P_n\}$. Moreover among these functions there are such which are discontinuous only at one point of $\mathbb{T}$.

Therefore, *Conjectures 1 and 2 are not true.* Nevertheless there are some positive results which I have to mention.

1. Conjecture 1 is true for Toeplitz operators $T(a)$ in case a is locally normal over the algebra QC of all quasicontinuous functions (even in the block case). This was proved by Silbermann [11]. Normal locality of a function a means the following. For a commutative Banach algebra A with identity e let $M(A)$ be the space of maximal ideals. It is well-known that every $\alpha \in M(A)$ can be identified with a multiplicative functional. For $\beta \in M(\mathrm{QC})$, put

$$M_\beta(L^\infty) = \{\alpha \in M(L^\infty) : \alpha|\mathrm{QC} = \beta\}$$

The compact set $M_\beta(L^\infty)$ is referred to as the fiber over β. A function $a \in L^\infty$ is called normal over M_β, if the convex hull $\operatorname{con} a(M_\beta)$ is a line segment (i.e. a set of the form $[z, w] = \{(1-\lambda)z + \lambda w : \lambda \in [0,1]\}$). A function $a \in L^\infty$ is said to be locally normal over QC if a is normal over M_β for each $\beta \in M(\mathrm{QC})$.

2. At the present time, the problem whether $A \in \Pi\{P_n\}$ for $A \in \operatorname{alg} T(\mathrm{PC})$ is completely solved (see Böttcher/Silbermann [8] and the comments made there). It is worth noticing that new conditions occur. More precisely, to each pair $(\tau, A), \tau \in \mathbb{T}$, one has to assign a certain curve Γ_τ in the complex plane. Then we have

$A \in \Pi\{P_n\}$ if and only if

(i) A and $S(A^{-1})$ are in $G\mathcal{B}(H^2)$,

(ii) The curve Γ_τ does not meet the origin and has the winding number 0 for each $\tau \in \mathbb{T}$.

The block case is more difficult. S. Roch [10] pointed out that the conditions (ii) must be replaced by conditions which require that some operators $A_\tau, \tau \in \mathbb{T}$, are invertible. Note that for $A \in \operatorname{alg} T(\mathrm{PQC})$, where PQC refers to the algebra of piecewise quasicontinuous functions, one knows conditions which ensure that $A \in \Pi\{P_n\}$. The question arises: are these conditions necessary? See Chapter 7 in Böttcher/Silbermann [8] for further details. In connection with the second conjecture it should be noted that generally $\ker S \neq J$ (compare the result of Treil'). However, there are also some positive results. So one has

$$\operatorname{alg}_{\{P_n\}} T(\mathcal{B})/J \cong \operatorname{alg} T(\mathcal{B})/\mathfrak{S}_\infty$$

for a variety of algebras $\mathcal{B}$ (Silbermann [11]). For instance this result is true for $\mathcal{B} =$ PQC. Serious difficulties occur if one replaces H^2 by H^p $(1 < p < \infty)$. For example, the structure of $\operatorname{alg}_{\{P_n\}} T(\mathrm{PQC})$ is unknown for general p. Only recently Böttcher and Spitkovsky [9] succeeded in studying the simpler algebra $\operatorname{alg} T(\mathrm{PQC})$ for $p \neq 2$.

References

8. Böttcher A., Silbermann B., *Analysis of Toeplitz Operators*, Springer-Verlag, 1990.
9. Böttcher A., Spitkovsky I. M. (= Spitkovskiĭ, I. M.), *Toeplitz operators with PQC symbols on weighted Hardy spaces*, J. Funct. Anal. **97** (1991), 194–214.
10. Roch S., *Lokale Theorie des Reduktionsverfahrens für singuläre Integraloperatoren mit Carlemanschen Verschiebungen*, Diss. A, Techn. Univ., Karl-Marx-Stadt, 1988.
11. Silbermann B., *Local objects in the theory of Toeplitz operators*, Integral Equations Oper. Theory **9** (1986), 706–738.
12. Treil' S. R., *Invertibility of Toeplitz operators does not imply applicability of the finite section method*, Dokl. Akad. Nauk SSSR **292** (1987), 563–567 (Russian); English transl. in Sov. Math. Doklady **35** (1987), 103–107.

Chapter Editor's note

Apparently, Conjecture 2 should be changed as follows:

Conjecture 2′. *If $B = L^\infty$ then $J = \ker\varphi \cap \ker\psi$, where*

$$\psi : \operatorname{alg}_{\{P_n\}} T(B) \to \operatorname{alg} T(B)/\mathfrak{S}_\infty, \quad \psi\{A_n\} \stackrel{\text{def}}{=} s\text{-}\lim_n W_n A_n W_n.$$

Chapter 8

CLOSE TO NORMAL OPERATORS

Edited by

John B. Conway
Department of Mathematics
University of Tennessee
Knoxville, TN 37996–1300
USA

INTRODUCTION

For any operator T let $D = D(T) = T^*T - TT^*$, the self-commutator of T. An operator T is normal if this self-commutator is the zero operator. There are many ways of thinking of an operator as being close to a normal one, except for the most obvious way. If $B(H)$, the set of all operators on a Hilbert space H, is furnished with the norm topology, then the set of normal operators is closed. Provide $B(H)$ with the weak operator topology and the set of normal operators is dense [2]. If, however, $B(H)$ is furnished with the strong operator topology, the closure of the set of normal operators is the class of subnormal operators ([1], page 36). The problems proposed by Paul McGuire, Robert Olin, C. R. Putnam, and Daoxing Xia all deal with subnormal operators.

Another way to say that an operator is close to normal is to require that the self-commutator D be small in some sense of the word. The usual requirement is that D lie in some ideal of compact operators. One of the problems proposed by Mihai Putinar deals with unitary equivalence of operators for which the self-commutator has rank one. The contribution of Daoxing Xia also includes problems concerned with subnormal operators for which the self-commutator belongs to the trace class.

Another choice for close to normal is to require that the self-commutator be a positive operator rather than 0. In this case, T is said to be hyponormal. If $D(T) \leqslant 0$, then T^* is hyponormal, so that no additional class is required. Every subnormal operator is hyponormal. In the other paper by Putinar he asks whether it is possible to construct a certain type of hyponormal weighted shift that is not subnormal. (The existence of such a weighted shift is known).

Hyponormal operator theory is a rich and extensive theory with a large collection of problems. The lack of problems on hyponormal operators in this chapter is in no way a reflection of the state of the field.

The theory of normal operators is well developed. In fact one can say with justification that it is complete. Unitary invariants that are readily calculated are known for normal operators and there is essentially no question about normal operators that cannot be easily solved. One reason for this is the rich functional calculus associated with every normal operator. It is possible to define $\varphi(T)$ for every bounded Borel function φ on the spectrum of a normal operator T. If T is any operator for which this functional calculus can be defined, then T is similar to a normal operator. It is, nevertheless, appropriate to investigate operators T for which a reasonable functional calculus can be defined. For any operator T the Riesz-Dunford functional calculus defines $\varphi(T)$ for functions φ that are analytic in a neighborhood of the spectrum of T. If K is a compact set that contains the spectrum of T, it is natural to consider the case that $\varphi(T)$ can be defined for any function φ that is the uniform limit on K of functions that are analytic in a neighborhood of K. In this case, K is said to be a spectral set for T (or an M-spectral set if the resulting functional calculus is not contractive). Historically this was not the motivation of von Neumann when he introduced the concept [3]. However it does lead to another concept of being close to a normal operator. If the spectrum of T is itself a spectral set for T, then T is said to be a von Neumann operator. The contribution by this editor proposes some problems concerned with spectral sets and von Neumann operators.

References

1. Conway J. B., *Theory of Subnormal Operators*, Amer. Math. Soc., Providence, 1991.
2. Conway J. B., Szucs J., *The weak sequential closure of certain sets of extreme points in a von Neumann algebra*, Indiana Univ. Math. J. **22** (1973), 763–768.
3. von Neumann J., *Eine Spektraltheorie für allgemeine Operatoren eines unitären Raumes*, Math. Nachr. **4** (1951), 258–281.

8.1

SPECTRAL PICTURES OF IRREDUCIBLE SUBNORMAL OPERATORS

PAUL J. MCGUIRE

In what follows all operators are bounded linear operators on a complex separable Hilbert space. By the spectral picture of an operator T is meant the pair of compact sets $\{\sigma(T), \sigma_e(T)\}$ consisting of the spectrum and essential spectrum of T together with the values of the Fredholm index on the various components of $\sigma(T) \setminus \sigma_e(T)$. The operator S on $\mathcal{H}$ is subnormal if it has a normal extension to a Hilbert space $\mathcal{K}$ containing $\mathcal{H}$. An operator is irreducible if it has no nontrivial reducing subspace, and pure if there is no reducing subspace on which it is normal. If S is a pure subnormal operator, then the index values are negative.

A natural question is which spectral pictures are possible for a pure or irreducible subnormal operator. In Clancey and Putnam [1] and Hastings [2], this question is completely resolved for the pure case. In Olin and Thomson [4] it is shown that K is the spectrum of an irreducible subnormal operator if and only if $R(K)$ has exactly one nontrivial Gleason part G such that K is the closure of G. This result is extended in McGuire [3] to show that K, K_e are the spectrum, essential spectrum respectively of an irreducible subnormal operator if and only if the Olin–Thomson condition holds and $\partial K \subset K_e \subset K$. Additionally it is shown that index values less than or equal to -2 can always be prescribed on a given component. In many instances the prescribing of the index value -1 presents no special difficulties. However, Problem 1 below indicates that a general solution to the -1 index case is far from complete.

PROBLEM 1. *Let C_1 and C_2 be two simple Jordan curves with C_1 contained in the inside of C_2. Let G_1, G_2 denote the bounded components of the complement of $C_1 \cup C_2$ and let ω_1, ω_2 denote harmonic measure for G_1, G_2 respectively. If ω_1 and ω_2 are singular measures, does there exist an irreducible subnormal operator with essential spectrum $C_1 \cup C_2$ and index values -1 on $G_1 \cup G_2$?*

If the measures ω_1, ω_2 above are not singular, then an as yet unpublished result of McGuire and Thomson is that such an irreducible subnormal operator does exist. Problem 2 is a broader question suggested by the known results and to which the answer to Problem 1 is a first step.

PROBLEM 2. *Let G_1 and G_2 be components of $K \setminus K_e$, ω_1, ω_2 harmonic measures for G_1, G_2 respectively, and assume ω_1 is singular to ω_2. If the boundary of G_1 has positive ω_2 measure and the boundary of G_2 has positive ω_1 measure, does this preclude the existence of an irreducible subnormal operator with index values -1 on $G_1 \cup G_2$?*

As the irreducible subnormal operators with index values -1 are the building blocks for all subnormal operators, Problem 3 is of fundamental interest.

PROBLEM 3. *Under what conditions can the index value -1 appear in the spectral picture of an irreducible subnormal operator?*

References

1. Clancey K. F., Putnam C. R., *The local spectral behavior of completely subnormal operators*, Trans. Amer. Math. Soc. **163** (1972), 239–244.
2. Hastings W. W., *The approximate point spectrum of a subnormal operator*, J. Operator Theory **5** (1981), 119–126.
3. McGuire P. J., *On the spectral picture of an irreducible subnormal operator*, Proc. Amer. Math. Soc. **104** (1988), no. 3, 801–808.
4. Olin R. F., Thomson J. E., *Irreducible operators whose spectra are spectral sets*, Pacific J. Math. **91** (1980), 431–434.

Bucknell University
Department of Mathematics
Lewisburg, PA 17837
U.S.A.

8.2

MULTIPLICITY THEORY FOR SUBNORMAL OPERATORS

ROBERT F. OLIN[1]

Using the framework and notation of [3], we can summarize (acknowledging that we are making some serious omissions) J. Thomson's results in [7] as saying that the adjoint of every cyclic irreducible subnormal operator S_μ, belongs to the Cowen, Douglas class $B_1(\Omega)$. In fact, Jim proved that

$$S_\mu^* \in B_1(abpe\,(\mu)).$$

It is natural to ask then the following:

PROBLEM 1. *Given a region Ω, what pure subnormal operators, S, have the property that*

$$S^* \in B_n(\Omega)?$$

We note that the case $n = 1$ and $\Omega = D = \{\,|z| < 1\,\}$ is extremely interesting. In [4] a *non-cyclic* cellular-indecomposable subnormal operator, S, is constructed such that $S^* \in B_1(D)$.

The structure of a cyclic subnormal operator has progressed immensely during the last 20 years. One of the important tools for this case, is J. Bram's characterization (model) of this operator, S_μ, as multiplication by z on a $P^2(\mu)$ space. For example, equipped with Thomson's result, we have the solution to the unitary equivalence problem for cyclic subnormal operators. That is, two irreducible cyclic subnormal operators S_μ and S_ν acting on $P^2(\mu)$ and $P^2(\nu)$, respectively, are unitarily equivalent if and only if

$$\partial\bar{\partial}\log\|k_\mu(\lambda)\| = \partial\bar{\partial}\log\|k_\nu(\lambda)\|.$$

Here k_μ and k_ν are the reproducing kernels for the appropriate P^2 space.

The structure for non-cyclic subnormal operators is far from being understood. The biggest stumbling block is a model for the general case. The following open question from [2], shows how little we still know. Recall that if N is a normal operator, then $\mathcal{S}_p(N)$ denotes the collection of pure subnormal operators that have N as their minimal normal extension.

PROBLEM 2. *When is $\mathcal{S}_p(N) \neq \emptyset$?*

To stimulate interest in finding a model for these non-cyclic critters, we will make the following guess: (See [1] for explanation of notation).

Guess. Every subnormal operator, S, is unitarily equivalent to an operator in the commutant of a rationally cyclic operator. That is, there exist a compactly supported measure μ in the plane, a compact set K, and a function

$$\varphi \in R^2(K,\mu) \cap L^\infty(\mu),$$

[1]This work was supported by a grant from the National Science foundation.

such that S is unitarily equivalent to multiplication by φ on the space $R^2(K,\mu)$.

Note, the guess is true for normal operators. The guess is rather dubious for function algebra reasons. [It probably has to be modified to cover other types of function algebras rather than just the one $R(K)$]. However, the characterization of those operators satisfying the guess would be a valuable step in the theory.

We conclude by asking a question about cyclic subnormal operators. [There is still a lot of work to do with them]. Earlier we stated a criterion that two of these operators are unitarily equivalent. What about a criterion for quasisimilarity?

PROBLEM 3. *Two cyclic subnormal operators are quasisimilar if their essential spectra are equal and their lattices (of closed invariant subspaces) are lattice isomorphic.*

The converse is true. The conclusion of the problem is true if one of the operators is the unilateral shift. In fact, a stronger result holds in this case [6].

REFERENCES

1. Conway J. B., *Theory of Subnormal Operators*, Math. Surveys and Monographs, vol.36, 1991.
2. Conway J. B., Olin R. F., *A functional calculus for subnormal operators*, II, Mem. Amer. Math. Soc. **184** (1977).
3. Cowen M. J., Douglas R. G., *Complex geometry and operator theory*, Acta Math. **141** (1978), 187–261.
4. Olin R. F., Thomson J. E., *Cellular-indecomposable subnormal operators*, Integral Equations and Operator Theory **7** (1984), 392–430.
5. Olin R. F., Thomson J. E., *Cellular-indecomposable operators* II, Integral Equations and Operator Theory **9** (1986), 600–609.
6. Olin R. F., Thomson J. E., *Cellular-indecomposable operators* III, preprint.
7. Thomson J. E., *Approximation in the mean by polynomials*, Ann. Math. **133** (1991), 477–507.

DEPARTMENT OF MATHEMATICS
VIRGINIA TECH
BLACKSBURG, VA 24061-0123
U.S.A.

8.3

REAL PARTS OF SUBNORMAL OPERATORS AND THEIR DUALS

C. R. PUTNAM

Let S be a pure subnormal operator on a separable Hilbert space H with minimal normal extension N on a Hilbert space $K \supset H$. (For properties of subnormal operators see [2, 3]). Since H is invariant under N then $H^{\perp} = K \ominus H$ is invariant under N^*, and $T = N^*|H$, the dual of $S = N|H$, is also a pure subnormal operator with spectrum $\sigma(T) = \{\bar{z} : z \text{ in } \sigma(S)\}$. Further, S is the dual of T; see [1, 2]. Since both S and T are pure subnormal (hence also hyponormal) operators then $\mathrm{Re}(S)$ and $\mathrm{Re}(T)$ are absolutely continuous ([4], pp. 42–43) with corresponding absolutely continuous supports α_S and α_T.

PROBLEM. *Find conditions on S, more specifically, on its selfcommutator $S^*S - SS^* = D\,(\geqslant 0)$, which assure that*

$$\alpha_S = \alpha_T. \tag{1}$$

It is possible that (1) holds in general. It is known that if S is essentially normal, that is, if D is compact, then the von Neumann algebras $W^*(S)$ and $W^*(T)$, generated by S and T respectively, are $*$-isomorphic; [1], p. 209. This suggests that (1) may hold, at least in this case. A somewhat more promising condition for ensuring (1) may lie in the stronger hypothesis of (2) below.

It was shown in [6] that if S is a pure subnormal operator satisfying

$$D^{\frac{1}{2}} \text{ is of trace class,} \tag{2}$$

then $Re(N)$ on K has an absolutely continuous part, $(Re(N))_a$, and one has the unitary equivalence relation

$$(\mathrm{Re}(N))_a \cong \mathrm{Re}(S) \oplus Re(T) \text{ on } K = H \oplus H^{\perp}. \tag{3}$$

In case S is an analytic Toeplitz operator with nonconstant symbol, then $\mathrm{Re}(S)$ and $\mathrm{Re}(T)$ of (3) are even unitarily equivalent ([7]) and, in particular, (1) holds.

If $n(t)$, $n_S(t)$ and $n_T(t)$ denote the respective spectral multiplicity functions of $(\mathrm{Re}(N))_a$, $\mathrm{Re}(S)$ and $\mathrm{Re}(T)$, then, under the hypothesis (2), one has relation (3) and hence also

$$n(t) = n_S(t) + n_T(t). \tag{4}$$

(This relation was used in [8] to determine $n_T(t)$ from the known quantities $n(t)$ and $n_S(t)$ for certain subnormal weighted shifts). Since $\mathrm{tr}(D) = \mathrm{tr}(D_T)$, where $T^*T - TT^* = D_T$, it follows from [5] (even if (2) is not assumed) that

$$\pi\ \mathrm{tr}(D) \leqslant \int_{\alpha_S} n_S(t)F(t)\,dt \ \text{ and } \pi\ \mathrm{tr}(D) \leqslant \int_{\alpha_T} n_T(t)F(t)\,dt,$$

where $F(t)$ denotes the common linear measure of the cross sections $\sigma(S)\cap\{z:\mathrm{Re}(z)=t\}$ and $\sigma(T)\cap\{z:\mathrm{Re}(z)=t\}$. Hence, if (2) is also assumed, then by (4),

$$2\pi\ \mathrm{tr}(D) \leqslant \int_{\alpha_S\cup\alpha_T} n(t)F(t)\,dt. \tag{5}$$

If (1) holds (at least under the restriction (2)), then (4) would be more useful for the calculation of $n_T(t)$, and the estimate (5) would also be sharpened.

References

1. Conway J. B., *The dual of a subnormal operator*, Jour. Operator Theory **5** (1981), 195–211.
2. Conway J. B., *Subnormal operators*, Research Notes in Math., 51; Pitman Adv. Pub. Program, Boston, 1981.
3. Halmos P. R., *A Hilbert space problem book*, 2nd edition, Graduate Texts in Math., vol. 19, Springer-Verlag, 1982.
4. Putnam C. R., *Commutation properties of Hilbert space operators and related topics*, Ergebnisse der Mat., vol. 36, Springer-Verlag, New York, 1967.
5. Putnam C. R., *Trace norm inequalities for the measure of hyponormal spectra*, Ind. Univ. Math. Jour. **21** (1972), 775–779.
6. Putnam C. R., *Real parts of normal extensions of subnormal operators*, Ill. Jour. Math. **31** (1987), 240–247.
7. Putnam C. R., *Analytic Toeplitz operators with self-commutators having trace class square roots*, Jour. Operator Theory **18** (1987), 133–138.
8. Putnam C. R., *Subnormal weighted shifts and spectra*, Jour. Operator Theory (to appear).

Department of Mathematics
Purdue University
West Lafayette, Indiana 47907
U.S.A.

8.4

COMPLETE UNITARY INVARIANT FOR SOME SUBNORMAL OPERATORS

DAOXING XIA

Let $\mathcal{H}$ be a complex separable Hilbert space, $\mathcal{L}(\mathcal{H})$ be the algebra of linear bounded operators on $\mathcal{H}$; let $\mathcal{L}^1(\mathcal{H})$ be the trace ideal of $\mathcal{L}(\mathcal{H})$. Suppose $\mathcal{F} \subset \mathcal{L}(\mathcal{H})$; a set of objects determined by T for every T in $\mathcal{F}$ is said to be a complete unitary invariant for the operator T in $\mathcal{F}$ and is denoted by $CU(T)$, if for every pair of operators S and T in $\mathcal{F}$

$$CU(T) = CU(S)$$

is a necessary and sufficient condition for the existence of a unitary operator U satisfying $S = UTU^{-1}$.

Let $\{a_1, \cdots, a_k\}$ be k–distinct points in the unit disk. Let $\mathcal{G}_0(a_1, \cdots, a_k)$ be the family of all pure subnormal operators S with minimal normal extension N satisfying the conditions that (i) $[S^*, S]^{\frac{1}{2}} \in \mathcal{L}^1(\mathcal{H})$, (ii) $\mathrm{sp}(S)$ is the unit disk and (iii) $\mathrm{sp}(N) = \{ z : |z| = 1 \text{ or } z = a_1, \cdots, a_k \}$.

Example. Let $\mathcal{H}$ be the Hilbert space of all analytic functions in the Hardy space H^2 endowed with the inner product

$$(f,g) = \frac{1}{2\pi} \int_0^{2\pi} f(e^{i\theta})\overline{g(e^{i\theta})}\, d\theta + \sum_{j=1}^{k} \alpha_j f(a_j)\overline{g(a_j)}$$

where $\alpha_1, \cdots, \alpha_k$ are fixed positive numbers. Define $(Sf)(z) = zf(z)$. Then $S \in \mathcal{G}_0(a_1, \cdots, a_k)$.

LEMMA 1 [6]. *For $S \in \mathcal{G}_0(a_1, \cdots, a_k)$, there is a real symmetric $k \times k$ matrix $(\gamma_{mn}(S))$ such that*

$$\mathrm{tr}\big[S^*, (S - \lambda I)^{-1}\big]\big[S^*, (S - \mu I)^{-1}\big] = \frac{i(S)}{\lambda^2 \mu^2} + \sum_{m,n=1}^{k} \frac{\gamma_{mn}(S)}{\lambda\mu(\lambda - a_m)(\mu - a_n)}$$

for $|\lambda| > 1$, $|\mu| > 1$, where $i(S)$ =index $(S^* - \overline{w}I)$ for $w \in \mathrm{sp}(S) \setminus \mathrm{sp}(N)$.

LEMMA 2 [6]. *The set $CU(S) = \{i(S), (\gamma_{mn}(S))\}$ for the operator $S \in \mathcal{G}_0(a_1, \cdots, a_k)$.*

A Jordan curve γ is said to satisfy the condition (CBI) if the univalent analytic mapping function $\phi(\cdot)$ from the interior domain of γ onto the unit disk satisfies the condition that $\phi'(\cdot)$ is bounded. For example, if the Jordan curve γ is smooth and the angle of inclination $\theta(s)$ of the tangent to γ as a function of the arc length s satisfies the Lipschitz condition, then γ satisfies the condition (CBI).

Let $\mathcal{G}$ be the family of all pure subnormal operator S on a Hilbert $\mathcal{H}$ with minimal normal extension N satisfying the conditions that (i) $[S^*, S]^{1/2} \in \mathcal{L}^1(\mathcal{H})$, (ii) the spectrum $\mathrm{sp}(S)$ is the closure of a simple connected domain bounded by a Jordan curve γ satisfying the condition (CBI) and (iii) $\mathrm{sp}(N) = \gamma \cup A$, where A is a finite set in $sp(S) \setminus \gamma$.

THEOREM [7]. *The function*

$$\mathrm{tr}\Big([(\overline{\lambda_0}I - S^*)^{-1}, (\mu_0 I - S)^{-1}]\,[(\overline{\lambda_1}I - S^*)^{-1}, (\mu_1 I - S)^{-1}]\Big)$$

for sufficiently large $|\lambda_j|$, $|\mu_j|$, $j = 0, 1$, *is a complete unitary invariant for* $S \in \mathcal{G}$.

The proofs of the Theorem and Lemmas 1, 2 are mostly based on the analytic model for the subnormal operators (cf. [4]), and some also need the cyclic cohomology (cf. [3], [5]).

PROBLEM 1. *Does the Theorem still hold if in the definition of the* $\mathcal{G}$ *the condition* $[S^*, S]^{1/2} \in \mathcal{L}^1(\mathcal{H})$ *is changed to* $[S^*, S] \in \mathcal{L}^1(\mathcal{H})$.

In [2], Abrahamse and Douglas gave a complete unitary invariant for the subnormal operator with multiply connected spectrum. Connecting [2] and the Theorem in this note, one may ask the following.

PROBLEM 2. *Does the Theorem still hold if, in the definition of the family* $\mathcal{G}$, *the condition of simple connectness of* $\mathrm{sp}(S)$ *is changed to the multiple connectness.*

REFERENCES

1. Conway J. B., *Subnormal Operators*, Pitman ABP, Boston, London, Melbourne, 1981.
2. Abrahamse M. B., Douglas R. G., *A class of subnormal operators related to multiply-connected domains*, Advances in Math. **19** (1976), 106–148.
3. Connes A., *Non-commutative differential geometry*, Inst. Hautes Etu. des Sci. Publ. Math. **62** (1985), 257–360.
4. Xia D., *The analytic model of a subnormal operator*, Integral Equations and Operator Theory **10** (1987), 255–289.
5. Xia D., *Trace formulas for almost commuting operators, cyclic cohomology and subnormal operators*, Integral Equations and Operator Theory **14** (1991), 277–298.
6. Xia D., *Complete unitary invariant for some subnormal operator*, Integral Equations and Operator Theory **15** (1992), 154–166.
7. Xia D.,, *A complete unitary invariant for some subnormal operator with simply connected spectrum*, preprint.

DEPARTMENT OF MATHEMATICS
VANDERBILT UNIVERSITY
NASHVILLE, TN 37235
U.S.A.

8.5

ANALYTICALLY HYPONORMAL WEIGHTED SHIFTS

Mihai Putinar

Let $T \in L(H)$ be a bounded operator acting on the complex Hilbert space H. The operator T is called *analytic hyponormal* if $[f(T)^*, f(T)] \geqslant 0$ for any analytic function f defined in a neighborhood of the spectrum of T. It is obvious that any subnormal operator is analytically hyponormal, but the converse assertion has circulated as an open problem for a while. Recently it was proved that there are analytically hyponormal operators which are not subnormal, see [3].

The proof in [3] is not constructive, so a natural question would be to find a concrete example of such an operator. Moreover, it is also known by qualitative arguments that there exists a unilateral weighted shift S acting on $\ell^2(\mathbb{N})$ which is analytically hyponormal but not subnormal, see [4].

According to the known subnormality criteria for weighted shifts, [5], the problem turns out to be very concrete. We may assume that $\|S\| = 1$ and that the weights $\{w_n\}_{n\in\mathbb{N}}$ of S are positive. One defines a $\mathbb{R}$-linear functional λ_S on the space or real polynomials $\mathbb{R}[t]$ by $\lambda_S(t^n) = w_n$, $n \in \mathbb{N}$.

Problem. *Find a functional $\lambda_S : \mathbb{R}[t] \to \mathbb{R}$ which is non-negative on elements of the form:*

$$\sum_{i\geqslant 0} t^i \Big| b_i + \sum_{j\geqslant 0} a_{i+j} c_j t^j \Big|^2 + \sum_{i\geqslant 1} t^i \Big| \sum_{j\geqslant 0} a_j c_{i+j} t^j \Big|^2 + (1-t) \sum_{j\geqslant 0} |d_j|^2 \, t^j$$

where (a_i), (b_i), (c_i), (d_i) are finitely supported sequences of complex numbers, but which is negative on a polynomial $p \in \mathbb{R}[t]$ which satisfies $p(x) \geqslant 0$ for $x \in [0,1]$.

The form of the previous conditions is explained by Agler's dictionary which identifies contractions T with a cyclic vector ξ, with $\mathbb{C}$-linear functional $\lambda_T \colon \mathbb{C}[z,\bar{z}] \to \mathbb{C}$ which are non-negative on the convex cone generated by $|p|^2$ and $(1-|z|^2)|q|^2$, $p, q \in \mathbb{C}[z]$. Quite specifically, $\lambda_T(P) = \langle P(T^*, T)\xi, \xi\rangle$, where $P \in \mathbb{C}[z,\bar{z}]$, for an ordered non-commutative functional calculus in T and T^*, see [1]. In that dictionary the analytic hyponormality or subnormality conditions correspond to positivity properties of the respective functional. Moreover, for weighted shifts S the reduction of the functional λ_S to rotational invariant polynomials (in $t = |z|^2$) makes possible the above formulation of the open problem, see [3] and [4] for details.

References

1. Agler J., *Hypercontractions and subnormality*, J. Operator Theory **13** (1985), 203–217.
2. Curto R., Fialkow L., *Recursively generated weighted shifts and the subnormal completion problem*, preprint (1990).
3. Curto R., Putinar M., *Existence of non–subnormal polynomially hyponormal operators*, Bull. Amer. Math. Soc. **25:2** (1991), 373–378.

4. McCullough S., Paulsen V., *A note on joint hyponormality*, Proc. Amer. Math. Soc. **107** (1989), 187–195.
5. Shields A., *Weighted shift operators and analytic function theory*, Math. Surveys **13** (1974), 49–128.

INSTITUTE OF MATHEMATICS
P. O. BOX 1–764
BUCHAREST RO 70–700
ROMANIA

AND

DEPARTMENT OF MATHEMATICS
UNIVERSITY OF KANSAS
LAWRENCE, KANSAS 66045
U.S.A.

8.6

ALGEBRAIC OPERATORS WITH RANK-ONE SELF-COMMUTATORS

MIHAI PUTINAR

The class of irreducible Hilbert space operators T with rank-one self-commutator $[T^*, T] = \xi \otimes \xi$ is well understood. These operators are classified up to unitary equivalence by their principal function g_T, whose characteristic properties are: $g_T \in L^\infty(\mathbb{C})$, $\operatorname{supp} g_T = \sigma(T)$ and $0 \leqslant g_T \leqslant 1$, see [5]. Several functional models for T are known, in terms of certain kernels involving g_T as the only free parameter, see [1], [3], [8].

An irreducible operator as above which satisfies in addition $p(T^*)\xi = 0$ for a polynomial $p \in \mathbb{C}[z]$ has special properties. For instance its spectrum $\sigma(T)$ is a compact subset of $\mathbb{C}$ with real algebraic boundary and its principal function g_T is the characteristic function of $\sigma(T)$, [6]. In particular this shows that $T - \lambda$ is Fredholm of index -1 for each $\lambda \in \operatorname{int} \sigma(T)$. By exploiting the before mentioned functional models one remarks that the unitary equivalence class of T depends on at most $\deg(p)^2 + 1$ complex parameters.

PROBLEM. *Classify up to unitary equivalence the irreducible operators T which satisfy $[T^*, T] = \xi \otimes \xi$ and $p(T^*)\xi = 0$ for a $p \in \mathbb{C}[z]$.*

According to the previous remarks the classification of this class of operators reduces to that of their spectra. It is worth mentioning that for $\deg(p) = 1$, a well known theorem of Morrel [4] asserts that these spectra are precisely the disks and the corresponding operators are $aS + b$, where S is the unilateral shift of multiplicity one, $a > 0$ and $b \in \mathbb{C}$. A similar classification has been recently carried out independently by Olin-Thompson-Trent [2] and D. Xia [8] for pure subnormal operators with finite rank self-commutators.

An answer to the above open problem would clarify certain extremal cases of the L-problem of moments in two-dimensions, [6].

REFERENCES

1. Clancey K., *Hilbert space operators with one-dimensional self-commutator*, J. Operator Theory **13** (1985), 265–289.
2. Olin R., Thomson J., Trent T., *Subnormal operators with finite rank self-commutator*, Trans. Amer. Math. Soc. (to appear).
3. Martin M., Putinar M., *Lectures on hyponormal operators*, Birkhäuser Verlag, Basel et al., 1989.
4. Morrel B., *A decomposition for some operators*, Indiana Univ. Math. J. **23** (1973), 497–511.
5. Pincus J. D., *Commutators and systems of singular integral equations I*, Acta. Math. **121** (1968), 219–249.
6. Putinar M., *The L-problem of moments in two dimensions*, J. Funct. Analysis **94** (1990), 288–307.
7. Xia D., *On the kernels associated with a class of hyponormal operators*, Integral Eq. Operator Theory **6** (1983), 134–157.
8. Xia D., *The analytic model of a subnormal operator*, Integral Eq. Operator Theory **10** (1987), 258–289.

INSTITUTE OF MATHEMATICS
P. O. BOX 1–764
BUCHAREST RO 70–700, ROMANIA

AND

DEPARTMENT OF MATHEMATICS
UNIVERSITY OF KANSAS
LAWRENCE, KANSAS 66045, USA

8.7

ON THE FUNDAMENTAL PROBLEM FOR SPECTRAL SETS

JOHN B. CONWAY

A bounded operator T on a Hilbert space $\mathcal{H}$ has a compact subset K of the complex numbers $\mathbb{C}$ as a spectral set if K contains the spectrum of T and $\|r(T)\| \leqslant \sup\{|r(z)| : z \in K\} \equiv \|r\|_K$ for every rational function r with poles off K. Let $\mathrm{Rat}(K)$ be the collection of rational functions with poles off K. If K is the closed unit disk, $\mathrm{cl}\,\mathbb{D} \equiv \{z : |z| \leqslant 1\}$, it is a well known result of von Neumann that every contraction on $\mathcal{H}$ has K as a spectral set.

What are some further examples of spectral sets?

Let $\mathcal{H}$ and $\mathcal{K}$ be Hilbert spaces with $\mathcal{H} \subseteq \mathcal{K}$. If $T \in \mathcal{B}(\mathcal{H})$, the bounded operators on $\mathcal{H}$, and $A \in \mathcal{B}(\mathcal{K})$, then A is an $R(K)$-dilation of T if $\sigma(T) \subseteq K$, $\sigma(A) \subseteq K$, and $r(T) = Pr(A)|\mathcal{H}$ for every function r in $\mathrm{Rat}(K)$, where P is the projection of $\mathcal{K}$ onto $\mathcal{H}$. This implies that $\mathcal{H}$ is an $R(K)$-semi-invariant subspace for A. That is, $r(PA|\mathcal{H}) = Pr(A)|\mathcal{H}$ for every r in $\mathrm{Rat}(K)$. This in turn implies that $\mathcal{H} = \mathcal{M} \cap \mathcal{N}^{\perp}$ where $\mathcal{M}$ and $\mathcal{N}$ are subspaces of $\mathcal{H}$ that are invariant for $\{r(A) : r \in \mathrm{Rat}(K)\}$ and $\mathcal{M} \supseteq \mathcal{N}$. Note that if K is a spectral set for A and A is an $R(K)$-dilation of T, then K is a spectral set for T. This gives a way for finding examples of operators with K as a spectral set.

Moreover, the Sz.-Nagy Dilation Theorem [8] states that if T is any contraction on $\mathcal{H}$, there is a unitary $R(\mathrm{cl}\,\mathbb{D})$-dilation of T. This has led to the phrasing of the fundamental problem for spectral sets. If $\mathcal{SS}(K)$ is the collection of all operators with K as a spectral set, is it true that each operator in $\mathcal{SS}(K)$ has a normal $R(K)$-dilation with spectrum contained in the boundary of K? The Sz.-Nagy Dilation Theorem says that this is precisely the case if K is the closed unit disk. The only other general theorem of this type is due to Agler [2] who showed that the question has an affirmative answer if K is an annulus. For a survey of the theory of spectral sets, see [6].

There are two things that strike me about this problem. First is that it is basically an attempt to know whether all the operators in $\mathcal{SS}(K)$ arise from one way of generating examples of operators with K as a spectral set. The second is that the problem may not be properly phrased. What happens if K has no interior? (I think this objection will be answered later in this paper. On the other hand why worry about sets without interior when the only results known are for disks and annuli?) In this note alternate ways of generating examples are presented and it will be shown that for finitely connected sets they give the same collection of examples obtained by taking operators with normal $R(K)$-dilations.

Let $\mathcal{SS}_{\partial n}(K)$ be the collection of all operators T that have a normal $R(K)$-dilation whose spectrum is contained in ∂K. From what was said before we have that $\mathcal{SS}_{\partial n}(K) \subseteq \mathcal{SS}(K)$ and the fundamental problem is to decide if $\mathcal{SS}(K) = \mathcal{SS}_{\partial n}(K)$.

Another possibility is to consider the class $\mathcal{SS}_n(K)$ of all operator T with normal $R(K)$-dilation N whose spectrum is only required to be included in K. Clearly

$$\mathcal{SS}_{\partial n}(K) \subseteq \mathcal{SS}_n(K) \subseteq \mathcal{SS}(K).$$

Here is another source of examples. Let S_1 and S_2 be subnormal operators on $\mathcal{H}_1$ and $\mathcal{H}_2$ whose spectrum is contained in K and K^*, respectively, and put $T = S_1 \oplus S_2^*$. It is easy to see that $T \in \mathcal{SS}(K)$. In fact, $T \in \mathcal{SS}_n(K)$. To see this let N_1 and N_2 be the minimal normal extensions and put $A = N_1 \oplus N_2^*$. It is trivial that $\mathcal{H}_1 \oplus \mathcal{H}_2$ is $R(K)$-semi-invariant for A and A is a normal $R(K)$-dilation of T. Similarly the compression of T to any $R(K)$-semi-invariant subspace is also in $\mathcal{SS}_n(K)$. Let $\mathcal{SS}_s(K)$ be the collection of all operators that have an $R(K)$-dilation of the form $S_1 \oplus S_2^*$ for some subnormal operators S_1 and S_2 whose spectra lie in K and K^*, respectively. What we just proved is that $\mathcal{SS}_s(K) \subset \mathcal{SS}_n(K)$. The other inclusion is also easy to see for a trivial reason: every normal operator is subnormal. Thus

$$\mathcal{SS}_s(K) = \mathcal{SS}_n(K).$$

The preceding discourse points out an elementary fact. If you try to generate a set $\mathcal{E}$ of examples of operators in $\mathcal{SS}(K)$ and have hope that $\mathcal{E} = \mathcal{SS}(K)$, then you must have that the compression of any operator T in $\mathcal{E}$ to an $R(K)$-semi-invariant subspace also belongs to the class $\mathcal{E}$.

The next method of generating examples only works for special sets K. Let Ω be a bounded open subset of $\mathbb{C}$, $K = \operatorname{cl}\Omega$, and assume that $\Omega = \operatorname{int} K$. Let $\tau\colon \mathbb{D} \to \Omega$ be the universal analytic covering map. Since Ω is bounded, $\tau \in H^\infty$. Thus if A is a completely non-unitary contraction on a Hilbert space $\mathcal{H}$, $T = \tau(A)$ is a well defined operator. It is easy to see that $\sigma(T) \subseteq K$ and if $r \in \operatorname{Rat}(K)$, then $\|r(T)\| = \|r \circ \tau(A)\| \leqslant \|r \circ \tau\|_{\mathbb{D}} = \|r\|_K$ so that K is a spectral set for T. It is also easy to see that $T \in \mathcal{SS}_{\partial n}(K)$. In fact, let U be an unitary power dilation of A operating on the Hilbert space $\mathcal{K}$ and put $N = \tau(U)$. It follows that N is a normal operator, $\sigma(N) \subset \partial K = \partial\Omega$, and N is an $R(K)$-dilation of T.

Suppose f is an analytic function on $\mathbb{D}$ such that $f(\mathbb{D}) \subseteq \Omega$. The same reasoning as above shows that $f(A) \in \mathcal{SS}_{\partial n}(K)$. However a standard result from function theory states that there is an analytic function $g : \mathbb{D} \to \mathbb{D}$ such that $f = \tau \circ g$, so that $f(A) = \tau(B)$, where B is the completely non-unitary contraction $g(A)$. So confining our attention to the universal analytic covering map does not diminish the class of examples.

On the other hand, if $\tau\colon \mathbb{D} \to \Omega$ is the universal analytic covering map and A is a completely non-unitary contraction, then the compression of $\tau(A)$ to an $R(K)$-semi-invariant subspace does not necessarily have a representation $\tau(B)$ for some other completely non-unitary contraction B. Also if T is the Bergman operator for Ω, then K is a spectral set for T but there is no completely non-unitary contraction A such that $T = \tau(A)$.

In light of these comments we extend the class of operators under consideration by defining the class $\mathcal{SS}_\ell(K)$ to consist of all those operators $T = T_1 \oplus T_2$, where T_1 has an $R(K)$-dilation of the form $\tau(A)$ for some completely non-unitary contraction A and T_2 is normal with spectrum contained in ∂K. We have shown that

$$\mathcal{SS}_\ell(K) \subseteq \mathcal{SS}_{\partial n}(K).$$

THEOREM. *If K is a finitely connected set with ∂K consisting of a finite number of smooth Jordan curves and $\Omega = \operatorname{int} K$ connected, then*

$$\mathcal{SS}_{\partial n}(K) = \mathcal{SS}_n(K) = \mathcal{SS}_s(K) = \mathcal{SS}_\ell(K).$$

Proof. We already have $\mathcal{SS}_\ell(K) \subseteq \mathcal{SS}_{\partial n}(K) \subseteq \mathcal{SS}_n(K)$ and $\mathcal{SS}_s(K) = \mathcal{SS}_n(K)$. The proof is completed by showing $\mathcal{SS}_n(K) \subseteq \mathcal{SS}_{\partial n}(K)$ and $\mathcal{SS}_{\partial n}(K) \subseteq \mathcal{SS}_\ell(K)$.

To begin assume that $T \in \mathcal{B}(\mathcal{H})$ and N is a normal $R(K)$-dilation of T acting on $\mathcal{K}$. Let $N = \int z\, dE(z)$ be the spectral decomposition of N. For vectors h and k in $\mathcal{K}$, denote by $E_{h,k}$ the scalar-valued measure defined by $E_{h,k}(\Delta) = \langle E(\Delta) h, k\rangle$ for all Borel sets Δ. Let $F_{h,k}$ be the sweep of $E_{h,k}$ to ∂K. So $F_{h,k}$ is the unique measure on ∂K such that if $u \in C(\partial K)$,

$$\int_{\partial K} u\, dF_{h,k} = \int_K \hat{u}\, dE_{h,k} = \langle \hat{u}(N) h, k\rangle,$$

where $\hat{u}$ denotes the solution of the Dirichlet problem for K with boundary values u. For a fixed Borel set Δ contained in ∂K, it is easy to verify that $[h,k] \equiv F_{h,k}(\Delta)$ is a sesquilinear form on $\mathcal{K}$ and $\|[h,k]\| \leqslant \|h\|\,\|k\|$ for all vectors h and k. Thus there is an operator $F(\Delta)$ on $\mathcal{K}$ such that $\|F(\Delta)\| \leqslant 1$ and $\langle F(\Delta) h, k\rangle = F_{h,k}(\Delta)$ for all h and k. Because $E_{h,h} \geqslant 0$, $F_{h,h}$ is a positive measure and so $F(\Delta)$ is a positive operator. It follows that $\Delta \to F(\Delta)$ is a positive operator-valued measure on ∂K with $F(\partial K) = 1$. By a result of Naimark (see [7]) there is a Hilbert space $\mathcal{L}$ containing $\mathcal{K}$ and a spectral measure G on ∂K with values in $\mathcal{B}(\mathcal{L})$ such that for every Borel subset Δ of ∂K, $F(\Delta) = QG(\Delta)|\mathcal{K}$, where Q is the projection of $\mathcal{L}$ onto $\mathcal{K}$. Let $M = \int z dG(z)$. So M is a normal operator with $\sigma(M) \subseteq \partial K$. If $r \in \operatorname{Rat}(K)$, then $r = \hat{r}$ and so for x and y in $\mathcal{H}$,

$$\begin{aligned}
\langle r(T) x, y\rangle &= \langle r(N) x, y\rangle \\
&= \int r\, dE_{x,y} \\
&= \int r\, dF_{x,y} \\
&= \int r\, dG_{x,y} \\
&= \langle r(M) x, y\rangle.
\end{aligned}$$

Thus M is a normal $R(K)$-dilation of T and $\sigma(M) \subseteq \partial K$. This shows that $\mathcal{SS}_n(K) \subseteq \mathcal{SS}_{\partial n}(K)$.

Now let T be an operator on $\mathcal{H}$ that belongs to $\mathcal{SS}_{\partial n}(K)$. It suffices to assume that T has no normal summand. Let N be a normal $R(K)$-dilation acting on $\mathcal{K}$ with $\sigma(N) \subseteq \partial K$. Let m be arc length measure on ∂K. Since T is assumed not to have a normal summand, it follows that N can be taken so that its scalar valued spectral measure is absolutely continuous with respect to m.

Let $\mathcal{M}$ and $\mathcal{N}$ be subspaces of $\mathcal{K}$ that are invariant for $r(N)$ for all rational functions in $\operatorname{Rat}(K)$ such that $\mathcal{H} = \mathcal{M} \cap \mathcal{N}^\perp$. Thus $S \equiv N|\mathcal{M}$ is a subnormal operator with normal extension N. Thus the spectrum of the minimal normal extension of S is contained in ∂K. By results of [1], we may take $\mathcal{M}$ to be a subspace of vector-valued H^2 space, $H^2_{\mathcal{L}}$, such that $\mathcal{M}$ is invariant under multiplication by $r \circ \tau$ for all r in $\mathrm{Rat}(K)$ and $S = M_\tau | \mathcal{M}$. With this convention, $(r \circ \tau)\mathcal{N} \subseteq \mathcal{N}$. Let $A = M_z \left| H^2_{\mathcal{L}} \right.$. So A is a completely non-unitary contraction (in fact a pure shift) and $\tau(A) = M_\tau \left| H^2_{\mathcal{L}} \right.$. Since S is an $R(K)$-dilation of T, it follows that $\tau(A)$ is an $R(K)$-dilation of T. □

Now for a rephrasing of the fundamental problem of spectral sets.

PROBLEM 1. *If K is a compact subset of $\mathbb{C}$, is $\mathcal{SS}(K) = \mathcal{SS}_n(K)$?*

By the preceding theorem this is equivalent to the original problem for nice sets, but it also covers those compact sets that have no interior. If K is such that $R(K) = C(K)$, then every operator in $\mathcal{SS}(K)$ is normal and so Problem 1 has a positive answer in this case.

PROBLEM 2. *Is there a Swiss cheese K with $R(K) \neq C(K)$ for which Problem 1 has a positive answer?*

Let Q be the set of non-peak points for $R(K)$ and let $\mathcal{B}_Q$ be the band of measures on K generated by the representing measures for points in Q (see [3]). If N is a normal operator with $\sigma(N) \subseteq K$ and if μ is a scalar valued spectral measure for N and if μ does not belong to $\mathcal{B}_Q$, then there is a reducing subspace $\mathcal{R}$ for N such that every subspace $\mathcal{M}$ of $\mathcal{R}$ that is invariant for $r(N)$ for all r in $\mathrm{Rat}(K)$ reduces $r(N)$. So if it were the case that N were a normal $R(K)$-dilation of some operator T on $\mathcal{H}$, then either $N \left| \mathcal{M}^\perp \right.$ is also an $R(K)$-dilation of T or there is a subspace $\mathcal{L}$ of $\mathcal{H}$ that reduces $r(T)$ for all r in $\mathrm{Rat}(K)$ and such that $T | \mathcal{L}$ is normal. In particular, in the latter of these cases T is not pure. So if T is pure and $T \in \mathcal{SS}_n(K)$, a normal $R(K)$-dilation can be found with scalar valued spectral measure in $\mathcal{B}_Q$. In the case that K is as in the theorem where a normal dilation is sought with spectrum in ∂K, this is reflected by the fact that for a pure operator this dilation with spectrum in ∂K has its spectral measure absolutely continuous with respect to arc length measure.

PROBLEM 3. *If T is a von Neumann operator with $\sigma(T) = K$, is $T \in \mathcal{SS}_n(K)$?*

It is possible that the functional calculus developed for von Neumann operators in [4] may be of help in Problem 2.

PROBLEM 4. *If T is compact and $T \in \mathcal{SS}(K)$, is $T \in \mathcal{SS}_n(K)$? What about matrices?*

In [5] this question is answered in the affirmative for 2×2 matrices.

The next problems will use the notation and terminology of [4].

Suppose T is a pure von Neumann operator and $K = \sigma(T)$. Also assume that T has a normal $R(K)$-dilation N acting on $\mathcal{K}$. If μ is a scalar valued spectral measure for N, then it follows that μ belongs to the band of measures $\mathcal{B}_K$. We therefore have the following commutative diagram of dual algebra homomorphisms,

$$\begin{array}{ccc} H^\infty(\mathcal{B}_K) & \xrightarrow{\kappa} & R^\infty(K;T) \\ \pi \downarrow & & \uparrow \eta \\ H^\infty(\mu) & \xrightarrow{\kappa_N} & R^\infty(K;N) \end{array}$$

Here π is the natural projection, κ is the functional calculus defined in [4], κ_N is the usual functional calculus for normal operators, and η is the map defined by $\eta(A) = P_{\mathcal{H}} A | \mathcal{H}$.

PROBLEM 5. *Is η an isometry?*

PROBLEM 6. *If N is a minimal normal $R(K)$-dilation for T and $f \in H^\infty(\mathcal{B}_K)$, is $f(N)$ a minimal $R(K)$-dilation for $f(T) = \kappa(f)$?*

REFERENCES

1. Abrahamse M. B., Douglas R. G., *A class of subnormal operators related to multiply connected domains*, Adv. Math. **19** (1976), 106–148.
2. Agler J., *Rational dilation on an annulus*, Ann. Math. **121** (1985), 537–563.
3. Conway J. B., *Theory of Subnormal Operators*, Amer. Math. Soc., Providence, 1991.
4. Conway J. B., Dudziak J. J., *Von Neumann operators are reflexive*, J. Reine Angew. Math. **408** (1990), 34–56.
5. Paulsen V. I., *K-spectral values for some finite matrices*, J. Operator Theory **18** (1987), 249–263.
6. Paulsen V. I., *Toward a theory of K-spectral sets*, Surveys of some Recent Results in Operator Theory, vol. 1 (J.B. Conway and B.B. Morrel, eds.), Research Notes in Mathematics, Longman, London, 1988, pp. 221–240.
7. Steinespring W. F., *Positive functions on C*-algebras*, Proc. Amer. Math. Soc. **6** (1955), 211–216.
8. Sz-Nagy B., *Sur les contractions de l'espace de Hilbert*, Acta. Sci. Math. **15** (1953), 87–92.

DEPARTMENT OF MATHEMATICS
UNIVERSITY OF TENNESSEE
KNOXVILLE, TN 37996-1300
U.S.A.

8.8
old

ALMOST-NORMAL OPERATORS MODULO $\mathfrak{S}_p$

D. VOICULESCU

0. **Notations.** H = separable complex Hilbert space of infinite dimension; $(\mathcal{L}(H), \|\cdot\|)$ = (bounded operators on H, uniform norm); $(\mathfrak{S}_p, |\cdot|_p)$ = (Schatten – von Neumann p-class, p-norm); $\mathcal{R}_1^+ = \{T \in \mathcal{L}(H) : T \text{ finite rank}, 0 \leqslant T \leqslant I\}$, $\mathcal{P} = \{P \in \mathcal{R}_1^+ : P = P^2\}$. $\mathrm{AN}(H)$ = (almost normal operators on H) = $\{T \in \mathcal{L}(H) : [T^*, T] \in \mathfrak{S}_1\}$. For $T \in \mathrm{AN}(H)$ we shall denote by P_T its Helton–Howe measure and by G_T its Pincus G-function (see [6], [10], [5]) so that $P_T = (2\pi)^{-1} G_T \, d\lambda$ where $d\lambda$ is Lebesgue measure on $\mathbb{R}^2$.

I. **Basic Analogy.** It is known that for $T \in \mathrm{AN}(H)$ we have: index $(T - zI) = G_T(z)$ for $z \in \mathbb{C}$ such that $T - zI$ is Fredholm. In [13] we noticed several instances which suggest that this relation is part of a far reaching analogy, in which the G-function plays the same role for $\mathfrak{S}_2$-perturbations of almost normal operators as the index for compact perturbations of essentially normal operators.

II. **Invariance of P_T under $\mathfrak{S}_2$-perturbations.** This should correspond to the invariance of the index under compact perturbations. In [13] we proved that if $T, S \in \mathrm{AN}(H)$, $T - S \in \mathfrak{S}_2$ and T or S has finite multicyclicity then $P_T = P_S$. Our proof in [13] depended on the use of the quasidiagonality relative to $\mathfrak{S}_2$. In fact one needs less. Consider

$$k_2(T) = \liminf_{A \in \mathcal{R}_1^+} \big|[A, T]\big|_2$$

where the lim inf is with respect to the natural order on $\mathcal{R}_1^+$ (see [12]).

PROPOSITION. *Let $T, S \in \mathrm{AN}(H)$ and suppose $k_2(T) = 0$ and $T - S \in \mathfrak{S}_2$. Then we have $P_S = P_T$.*

Proof. As in [13] (Prop. 3) the proof reduces to showing that $\mathrm{Tr}[T^*, T] = \mathrm{Tr}[S^*, S]$. Since $k_2(T) = 0$ there are $A_n \in \mathcal{R}_1^+$, $A_n \uparrow I$ such that $\big|[A_n, T]\big|_2 \to 0$ as $n \to \infty$ and the same holds also for T replaced by S. Denoting $X = T - S$ we have:

$$\begin{aligned}
&\big|\mathrm{Tr}\big([T, T^*] - [S, S^*]\big)\big| = \\
&\quad \big|\mathrm{Tr}\big([X, T^*] + [X^*, S]\big)\big| = \\
&\quad \lim_{n \to \infty} \Big|\mathrm{Tr}\big(A_n([X, T^*] + [X^*, S])\big)\Big| \leqslant \\
&\quad \limsup_{n \to \infty} \Big|\mathrm{Tr}\big([A_n X, T^*] + [A_n X^*, S]\big)\Big| + \\
&\quad \limsup_{n \to \infty} \Big(\big|[T^*, A_n]\big|_2 \big|X\big|_2 + \big|[S, A_n]\big|_2 \big|X^*\big|_2\Big) = 0.
\end{aligned}$$

We proved in [13] that $k_2(T) = 0$ for $T \in \mathrm{AN}(H)$ with finite multiplicity.

CONJECTURE 1. $T \in \mathrm{AN}(H) \Longrightarrow k_2(T) = 0$.

III. **Quasitriangularity.** A refinement of Halmos' notion of quasitriangularity [7] was considered in [11]. The corresponding generalization of Apostol's modulus of quasitriangularity is

$$q_p(T) = \liminf_{P \in \mathcal{P}} \left|(I - P)TP\right|_p$$

where the lim inf is with respect to the natural order on $\mathcal{P}$. We proved in [13] that for $T \in \mathrm{AN}(H)$, $q_2(T) = 0 \Longrightarrow P_T \leqslant 0$. We conjecture an analogue of the Apostol–Foiaş–Voiculescu theorem on quasitriangular operators [1].

CONJECTURE 2. *For $T \in \mathrm{AN}(H)$, we have*

$$\left(q_2(T)\right)^2 = 2 \iint_{\mathbb{R}^2} dP_T^+$$

where P_T^+ is the positive part of P_T.

This would imply in particular that $P_T \leqslant 0 \Longrightarrow q_2(T) = 0$.

Some results for subnormal and cosubnormal operators supporting the conjecture have been obtained in [13].

IV. **Analogue of the BDF theorem.** The following conjecture concerns an analogue of the Brown–Douglas–Fillmore theorem [3] on essentially normal operators.

CONJECTURE 3. *Let $T_1, T_2 \in \mathrm{AN}(H)$ be such that $P_{T_1} = P_{T_2}$. Then there is a normal operator $N \in \mathcal{L}(H)$ and a unitary $U \in \mathcal{L}(H \oplus H)$ such that*

$$U(T_1 \oplus N)U^* - T_2 \oplus N \in \mathfrak{S}_2.$$

This conjecture implies the following

CONJECTURE 4. *If $T \in \mathrm{AN}(H)$ then there is $S \in \mathrm{AN}(H)$ such that $T \oplus S =$ normal + Hilbert–Schmidt.*

Note that this last statement corresponds to an important part in the proof of the BDF theorem, the existence of inverses in Ext or equivalently the completely positive lifting part in the " Ext is a group" theorem (see [2]). Even for almost normal weighted shifts, there is only a quite restricted class for which Conjecture 4 has been established [8].

Note also that Conjecture 4 implies Conjecture 1 and that by analogy with the proof of the Choi–Effros completely positive lifting theorem one should expect the vanishing of k_2 to be an essential ingredient in establishing Conjecture 4.

REFERENCES

1. Apostol C., Foiaş C., Voiculescu D., *Some results on non-quasitriangular operators* VI, Rev. Roum. Math. Pures Appl. **18** (1973), 1473–1494.
2. Arveson W. B., *A note on essentially normal operators*, Proc. Royal Irish Acad. **74** (1974), 143–146.
3. Brown L. G., Douglas R. G., Fillmore P. A., *Unitary equivalence modulo the compact operators and extensions of C^*-algebras*, Lect. Notes in Math., vol. 345, 1973, pp. 58–128.

4. Carey R. W., Pincus J. D., *Commutators, symbols and determining functions*, J. Funct. Anal. **19** (1975), 50–80.
5. Clancey K., *Seminormal operators*, Lect. Notes in Math., vol. 742, 1979.
6. Helton J. W., Howe R., *Integral operators, commutator traces, index and homology*, Lect. Notes in Math., vol. 345, 1973, pp. 141-209.
7. Halmos P. R., *Quasitriangular operators*, Acta Sci. Math. (Szeged) **29** (1968), 283–293.
8. Pasnicu C., *Weighted shifts as direct summands* mod $\mathfrak{S}_2$ *of normal operators*, INCREST preprint, 1982.
9. Pearcy C., *Some recent developments in operator theory*, CBMS, Regional Conference Series in Mathematics no. 36, Amer. Math. Soc., Providence, 1978.
10. Pincus J. D., *Commutators and systems of integral equations*, I, Acta Math. **121** (1968), 219–249.
11. Voiculescu D., *Some extensions of quasitriangularity*, Rev. Roum. Math. Pures Appl. **18** (1973), 1303–1320.
12. Voiculescu D., *Some results on norm-ideal perturbations of Hilbert space operators*, J. Operator Theory **2** (1979), 3–37.
13. Voiculescu D., *Remarks on Hilbert–Schmidt perturbations of almost-normal operators*, Topics in Modern Operator Theory, Birkhäuser, 1981.

Department of Mathematics
University of California at Berkeley
Berkeley, CA 94720
USA

8.9
old

HYPONORMAL OPERATORS AND SPECTRAL ABSOLUTE CONTINUITY

C. R. PUTNAM

In the sequel only bounded operators on an infinite dimensional, separable Hilbert space H will be considered. An operator T on H is said to be hyponormal if $T^*T-TT^* \geqslant 0$. Such an operator is said to be completely hyponormal if, in addition, T has no normal part, that is, if there is no subspace $\neq \{0\}$ reducing T on which T is normal.

If A is selfadjoint with the spectral family $\{E_t\}$ then the set of vectors x in H for which $\|E_t x\|^2$ is an absolutely continuous function of t is a subspace, $H_a(A)$, reducing A (see, e.g., Kato [1], p.516). If $H_a(A) \neq \{0\}$, then $A|H_a(A)$ is called the absolutely continuous part of A, and if $H_a(A) = H$ then A is said to be absolutely continuous. Similar concepts can be defined for a unitary operator.

If T is completely hyponormal then its real and imaginary parts are absolutely continuous. In addition, if T has a polar factorization

$$(1) \qquad T = U|T|, \quad U \text{ unitary and } |T| = \left(T^*T\right)^{1/2},$$

then U is also absolutely continuous. (See [2], p. 42 and [3], p. 193. Incidentally, such a polar factorization (1) exists, and is unique, if and only if 0 is not in the point spectrum of T^*; see [4], p. 277).

In general, if T is completely hyponormal, then its absolute value $|T| = \left(T^*T\right)^{1/2}$ need not be absolutely continuous or even have an absolutely continuous part. Probably the simplest example is the simple unilateral shift V for which V^*V is the identity. Of course, V does not have a polar factorization (1), but, nevertheless, there are simple examples of completely hyponormal T satisfying (1) for which $|T|$ has no absolutely continuous part. Moreover, it has recently been shown by K. F. Clancey and the author [5] that if P is selfadjoint on H, then there exists a completely hyponormal T satisfying $|T| = P$ and having the polar factorization (1) if and only if (i) $P \geqslant 0$ and $\sigma(P)$ contains at least two points, (ii) 0 is not in the point spectrum of P, and (iii) neither $\max \sigma(P)$ nor $\min \sigma(P)$ is in the point spectrum of P with a finite multiplicity.

Let a nonempty compact set of the complex plane be called radially symmetric if whenever z_1 is in the set then so is the entire circle $|z| = z_1$. All examples known to the author of completely hyponormal operators T for which $|T|$ does not have an absolutely continuous part, and whether or not (1) obtains, seem to have radially symmetric spectra. For instance, if $\sigma(|T|)$ has Lebesgue linear measure zero, then $\sigma(T)$ is surely radially symmetric; see [6], p. 426, also [7]. At the other extreme, if T is completely hyponormal and if there exists some open wedge $W = \{z : z \neq 0 \text{ and } 0 < \arg z < \text{const} < 2\pi\}$ (or rotated set $e^{i\theta}W$) which does not intersect $\sigma(T)$ then $|T|$ is absolutely continuous. In certain other instances also one can show at least that $H_a(|T|) \neq \{0\}$; see [7] and the references cited there. The following conjecture was made in [7]:

CONJECTURE 1. *Let T be completely hyponormal with polar factorization* (1). *Suppose that $\sigma(T)$ is not radially symmetric, so that some circle $|z| = r$ intersect both $\sigma(T)$ and its complement in nonempty sets. Then $H_a(|T|) \neq \{0\}$.*

The following stronger statement was also indicated in [7] and is set forth here, but with somewhat less conviction than the preceding conjecture, as

CONJECTURE 2. *Conjecture 1 remains true without the hypothesis* (1) .

REFERENCES

1. Kato T., *Perturbation theory for linear operators*, Springer-Verlag, New York, 1967.
2. Putnam C. R., *Commutation properties of Hilbert space operators and related topics*, Ergebnisse der Math., 36, Springer-Verlag, New York, 1967.
3. Putnam C. R., *A polar area inequality for hyponormal spectra*, J. Operator Theory **4** (1980), 191–200.
4. Putnam C. R., *Absolute continuity of polar factors of hyponormal operators*, Amer. J. Math., Suppl. (1981), 277–283.
5. Clancey K. F., Putnam C. R., *Nonnegative perturbations of selfadjoint operators*, J. Funct. Anal. **50** (1983), no. 3, 306–316.
6. Putnam C. R., *Spectra of polar factors of hyponormal operators*, Trans. Amer. Math. Soc. **188** (1974), 419–428.
7. Putnam C. R., *Absolute values of hyponormal operators with asymmetric spectra*, Mich. Math. Jour. **30** (1983), no. 1, 89–96.

DEPARTMENT OF MATHEMATICS
PURDUE UNIVERSITY
WEST LAFAYETTE, IN 47907
USA

8.10
v.old

OPERATORS, ANALYTIC NEGLIGIBILITY, AND CAPACITIES

C. R. PUTNAM

Let $\sigma(T)$ and $\sigma_p(T)$ denote the spectrum and point spectrum of a bounded operator T on a Hilbert space H. Such an operator is said to be *subnormal* if it has a normal extension on a Hilbert space K, $K \supset H$. For the basic properties of subnormal operators see [1]. A subnormal T on H is said to be *completely subnormal* if there is no nontrivial subspace of H reducing T on which T is normal. If T is completely subnormal then $\sigma_p(T)$ is empty. A necessary and sufficient condition in order that a compact subset of $\mathbb{C}$ be the spectrum of a completely subnormal operator was given in [2].

If X is a compact subset of $\mathbb{C}$, let $R(X)$ denote the functions on X uniformly approximable on X by rational functions with poles off X. A compact subset Q of X is called a *peak set* of $R(X)$ if there exists a function f in $R(X)$ such that $f = 1$ on Q and $|f| < 1$ on $X \setminus Q$; see [3], p. 56. The following result was proved in [4].

THEOREM. *Let T be subnormal on H with the minimal normal extension $N = \int z\, dE_z$ on K, $K \supset H$. Suppose that Q is a non-trivial proper peak set of $R(\sigma(T))$ and that $E(Q) \neq \mathbb{O}$. Then $E(Q)H \neq \{\mathbb{O}\}$ and H, the space $E(Q)H$ reduced T, $T|E(Q)H$ is subnormal with the minimal normal extension $E(Q)N$ on $E(Q)K$, and $\sigma(T|E(Q)H) \subset Q$. Further, if it is also assumed that $R(Q) = C(Q)$ then $T|E(Q)H$ is normal.*

Thus, in dealing with reducing subspaces of subnormal operators T, it is of interest to have conditions assuring that a subset of a compact $X(=\sigma(T))$ be a peak set of $R(X)$.

PROBLEM 1. *Let X be a compact subset of $\mathbb{C}$ and let C be a rectifiable simple closed curve for which $Q = \operatorname{clos}((\text{exterior of } C) \cap X)$ is not empty and $C \cap X$ has Lebesgue arc length measure 0. Does it follow that Q must be a peak set of $R(X)$?*

In case C is of class C^2 (or piecewise C^2), the answer is affirmative and was first demonstrated by Lautzenheiser [5]. A modified version of his proof can be found in [6], pp. 194–195. A crucial step in the argument is an application of a result of Davie and Øksendal [7] which requires that the set $C \cap X$ be analytically negligible. (A compact set E is said to be analytically negligible if every continuous function on $\mathbb{C}$ which is analytic on an open set V can be approximated uniformly on $V \cup E$ by functions continuous on $\mathbb{C}$ and analytic on $V \cup E$; see [3], p. 234). The C^2 hypothesis is then used to ensure the analytic negligibility of $C \cap X$ as a consequence of a result of Vitushkin [8]. It may be noted that the collection of analytically negligible sets has been extended by Vitushkin to include Liapunov curves (see [9], p. 115) and by Davie ([10], section 4) to include "hypo-Liapunov" curves. Thus, for such curves C, the answer to Problem 1 is again yes. The question as to whether a general rectifiable curve, or even one of class C^1, for instance, is necessarily analytically negligible, as well as the corresponding question in Problem 1, apparently remains open however.

As already noted, Problem 1 is related to questions concerning subnormal operators. The problem also arose in connection with a possible generalization of the notion of an "areally disconnected set" as defined in [6] and with a related rational approximation question. Problem 2 below deals with some estimates for the norms of certain operators associated with a bounded operator T on a Hilbert space and with two capacities of the set $\sigma(T)$.

Let $\gamma(E)$ and $\alpha(E)$ denote the analytic capacity and the continuous analytic capacity (or AC capacity) of a set E in $\mathbb{C}$. (For definitions and properties see, for example, [3, 11, 9]. A brief history of both capacities is contained in [9], pp. 142–143, where it is also noted that the concept of continuous analytic capacity was first defined by Dolzhenko [12]). It is known that for any Borel set E,

$$\left(\pi^{-1}\mathrm{meas}_2\, E\right)^{1/2} \leqslant \alpha(E) \leqslant \gamma(E)$$

see [11], pp. 9, 79. The following was proved in [13].

THEOREM. *Let T be a bounded operator on a Hilbert space and suppose that*

$$(T-z)(T-z)^* \geqslant D \geqslant \mathbb{O} \tag{1}$$

holds for some nonnegative operator D and for all z in the unbounded component of the complement of $\sigma(T)$. Then $|D|^{1/2} \leqslant \gamma(\sigma(T))$. If, in addition, (1) holds for all z in $\mathbb{C}$ and if, for instance, $\sigma_p(T)$ is contained in the interior of $\sigma(T)$ (in particular, if $\sigma_p(T)$ is empty), then also $|D|^{1/2} \leqslant \alpha(\sigma(T))$.

PROBLEM 2. *Does condition (1), if valid for all z in $\mathbb{C}$, but without any restriction on $\sigma_p(T)$, always imply that $|D|^{1/2} \leqslant \alpha(\sigma(T))$, or possibly even that $|D| \leqslant \pi^{-1}\mathrm{meas}_2\,\sigma(T)$?*

It may be noted that if T^* is hyponormal, so that $TT^* - T^*T \geqslant \mathbb{O}$ then (1) holds for all z in $\mathbb{C}$ with $D = TT^* - T^*T$ and, moreover, $|D| \leqslant \pi^{-1}\mathrm{meas}_2(\sigma(T))$; see [14].

REFERENCES

1. Halmos P. R., *A Hilbert space problem book*, van Nostrand Co., 1967.
2. Clancey K. F., Putnam C. R., *The local spectral behavior of completely subnormal operators*, Trans. Amer. Math. Soc. **163** (1972), 239–244.
3. Gamelin T. W., *Uniform algebras*, Prentice–Hall, Inc., 1969.
4. Putnam C. R., *Peak sets and subnormal operators*, Ill. Jour. Math. **21** (1977), 388–394.
5. Lautzenheiser R. G., *Spectral sets, reducing subspaces, and function algebras*, Thesis, Indiana Univ., 1973.
6. Putnam C. R., *Rational approximation and Swiss cheeses*, Mich. Math. Jour. **24** (1977), 193–196.
7. Davie A. M., Øksendal B. K., *Rational approximation on the union of sets*, Proc. Amer. Math. Soc. **29** (1971), 581–584.
8. Vitushkin A. G., *Analytic capacity of sets in problems of approximation theory*, Uspekhi Matem. Nauk **22** (1967), no. 6, 141–199.
9. Zalcman L., *Analytic capacity and rational approximation*, Lecture Notes in Math., 50, Springer–Verlag, 1968.
10. Davie A. M., *Analytic capacity and approximation problems*, Trans. Amer. Math. Soc. **171** (1972), 409–444.
11. Garnett J., *Analytic capacity and measure*, Lecture Notes in Math., 297, Springer–Verlag, 1972.

12. Dolzhenko E. P., *On approximation on closed regions and sets of measure zero*, Doklady Akad. Nauk SSSR **143** (1962), no. 4, 771–774.
13. Putnam C. R., *Spectra and measure inequalities*, Trans. Amer. Math. Soc. **231** (1977), 519–529.
14. Putnam C. R., *An inequality for the area of hyponormal spectra*, Math. Zeits. **116** (1970), 323–330.

Department of Mathematics
Purdue University
West Lafayette, IN 47907
USA

Editors' note

Two papers of Valskii (Valskii R. E., Doklady Akad. Nauk SSSR, **173** no. 1 (1967), 12–14; Sibirskii Matem. Zhurn., **8** no. 6 (1967), 1222–1235) contain some results concerning the themes discussed in this section.

8.11
old

PERTURBATION OF CONTINUOUS SPECTRUM AND NORMAL OPERATORS

N. Makarov, N. Nikolski

Let T be an invertible bounded operator on Hilbert space $\mathcal{H}$. The continuous spectrum $\sigma_c(T)$ of T is defined as $\sigma(T)\setminus\sigma_o(T)$, where $\sigma_o(T)$ stands for the set of all isolated points of the spectrum $\sigma(T)$ whose spectral subspaces are finite-dimensional. If the origin lies in the unbounded component of $\mathbb{C}\setminus\sigma(T)$ then $0 \notin \sigma_c(T+K)$ for any compact operator K. On the other hand, if $\sigma(T)$ separates 0 and ∞, then for any symmetrically-normed ideal $\mathfrak{S}$, $\mathfrak{S} \neq \mathfrak{S}_1$ ($\mathfrak{S}_p$ denotes throughout the Schatten–von Neumann class, $0 < p \leqslant \infty$) there exists $K \in \mathfrak{S}$ such that $0 \in \sigma_c(T+K)$ [1]. The question we are interested in concerns the stability of the continuous spectrum under "small" perturbations (e.g., finite rank, nuclear, etc.) For rank perturbations, the problem can be solved easily in terms of the lattice Lat of invariant subspaces of the operator.

THEOREM 1. *Suppose 0 does not belong to the unbounded component of $\sigma(T)$ and $0 \notin \sigma(T)$. Then a rank one operator K with $0 \in \sigma_c(T+K)$ exists if and only if* $\operatorname{Lat} T^{-1} \not\subset \operatorname{Lat} T$.

See [2], for the proof. Given an operator T denote by $\mathcal{R}(T)$ the weekly closed algebra of operators on $\mathcal{H}$ generated by T and the identity I. Suppose N is a normal operator on $\mathcal{H}$. Then $\operatorname{Lat} N^{-1} \subset \operatorname{Lat} N$ iff $N \in \mathcal{R}(N^{-1})$, see [3]. Therefore Theorem 1 together with a theorem of Sarason [4] imply the following criterion for the stability of the continuous spectrum under finite rank one perturbations. Denote by G_M the Sarason hull of the spectral measure of a normal operator M, defined in [4].

THEOREM 2. *Let N be the invertible normal operator. The following are equivalent:*

1) $0 \notin \sigma_c(N+K)$ *for every* K, $\operatorname{rank}(K) = 1$;
2) $0 \notin G_{N^{-1}}$.

In particular, the continuous spectrum of a unitary operator is stable iff the Lebesgue measure on $\mathbb{T}$ is absolutely continuous with respect to its spectral measure. It is shown in [5] that the continuous spectrum of a unitary operator is stable under perturbations of rank one iff it is stable under nuclear perturbations. This result can be extended to normal operators with essential spectra on smooth curves. At the same time it does not hold for arbitrary normal operators [1].

QUESTION 1. *Given an invertible operator are the following equivalent?*

1) $0 \notin \sigma_c(T+K)$ *for every K of rank one.*
2) $0 \notin \sigma_c(T+K)$ *for every K of finite rank.*

QUESTION 2. *Are 1) and 2) equivalent for an arbitrary normal operator $N = T$? Is it true that they are equivalent to*

3) $0 \notin \sigma_c(N+K)$ *for every* $K \in \bigcup_{p<1} \mathfrak{S}_p$?

Note that for $T = U + K$, where U is unitary and $K \in \mathfrak{S}_1$, the answer to Question 1 is affirmative [2].

QUESTION 3. *Is either of the following implications* $0 \notin \sigma_c(T+K)$, $\forall K$, $\operatorname{rank}(K) < +\infty \Longleftrightarrow T \in \mathcal{R}(T^{-1})$ *true?*

The inclusion $T \in \mathcal{R}(T^{-1})$ being equivalent to the series of inclusions

$$\operatorname{Lat} \underbrace{T^{-1} \oplus \cdots \oplus T^{-1}}_{n} \subset \operatorname{Lat} \underbrace{T \oplus \cdots \oplus T}_{n},$$

$n = 1, 2, \ldots$ (see [3]), it is natural to ask

QUESTION 4. *Are the following statements equivalent for any integer n?*

1) $0 \notin \sigma_c(T+K)$, $\forall K$, $\operatorname{rank}(K) \leqslant n$.

2) $\operatorname{Lat} \underbrace{T^{-1} \oplus \cdots \oplus T^{-1}}_{n} \subset \operatorname{Lat} \underbrace{T \oplus \cdots \oplus T}_{n}$.

By the way, we do not know the answer even to the following question.

QUESTION 5. *Is it true that*

$$T \in \mathcal{R}(T^{-1}) \quad \text{iff} \quad \operatorname{Lat} T^{-1} \subset \operatorname{Lat} T?$$

Many interesting problems arise when considering special perturbations of normal operators. Recall that the problem of stability of continuous spectra in the case of normal operators is reduced to the calculation of Sarason hulls.

QUESTION 6. *Let N and $N + K$ be normal operators. Is it true that*

$$\operatorname{rank} K < +\infty \Longrightarrow G_N = G_{N+K}?$$

$$K \in \mathfrak{S}_p,\ p < 1 \Longrightarrow G_N = G_{N+K}?$$

For what normal operators N

$$K \in \mathfrak{S}_1 \Longrightarrow G_N = G_{N+K}? \tag{1}$$

It was noted in [1] that there are normal operators not satisfying (1).

If U and $U + K$ are unitary then (1) holds with $U = N$. This is so because the absolutely continuous parts of U and $U + K$ are unitarily equivalent and the Sarason hull of a unitary operator depends on its absolutely continuous part only. Therefore Question 6 may be considered as a question of "scattering theory of normal operators".

Consider a narrower class of perturbations, namely we assume henceforth that N commutes with $N + K$. Then the symmetric difference $G_N \,\Delta\, G_{N+K}$ consists of the points of the point spectrum of N or of $N + K$. This reduces the question to the investigation of metric properties of the harmonic measure. L. Carleson has proved in [6] that the harmonic measure of any simply connected domain is absolutely continuous with respect to Hausdorff measure Λ_β, where $\beta > 1/2$ is an absolute constant. Using this result, it can be proved that $G_N = G_{N+K}$ if $K \in \mathfrak{S}_\beta$ and N commutes with $N + K$.

QUESTION 7. *Let N and $N+K$ be commuting normal operators. Is it true that*

$$K \in \mathfrak{S}_p,\ \ p<1 \Longrightarrow G_N = G_{N+K}\,?$$

REFERENCES

1. Makarov N. G., *Determining subsets, support of a harmonic measure and perturbations of the spectrum of operators on a Hilbert space*, Dokl. Akad. Nauk SSSR **274** (1984), no. 5, 1033–1037; English transl. in Soviet Math. Dokl. **29** (1984), 103–106.
2. Makarov N. G., Vasjunin V. I., *A model for noncontractions and stability of the continuous spectrum*, Complex Analysis and Spectral Theory, Lecture notes in Math., vol. 864, 1981, pp. 365–412.
3. Sarason D., *Invariant subspaces and unstarred operator algebras*, Pacific J. Math. **17** (1966), 511–517.
4. Sarason D., *Weak-star density of polynomials*, J. Reine und Angew. Math. **252** (1972), 1–15.
5. Nikol'skii N. K., *The spectrum perturbations of unitary operators*, Matem. Zametki (1969), no. 5, 341–349.
6. Carleson L., *On the distortion of sets on a Jordan curve under conformal mapping*, Duke J. Math. **40** (1973), 547–559.

STEKLOV MATHEMATICAL INSTITUTE
ST. PETERSBURG BRANCH
FONTANKA 27
ST. PETERSBURG, 191011
RUSSIA

AND

CALIF. INSTIT. OF TECHNOLOGY
DEPARTMENT OF MATHEATICS
PASADENA
CA 91125
USA

STEKLOV MATHEMATICAL INSTITUTE
ST. PETERSBURG BRANCH
FONTANKA 27
ST. PETERSBURG, 191011
RUSSIA

AND

UNIVERSITÉ BORDEAUX-I
UFR MATHÉMATIQUES
351, COURS DE LA LIBÉRATION
33405 TALENCE CEDEX
FRANCE

COMMENTARY BY N. G. MAKAROV

Among other results the implication (1) of Question 6 is proved in [7] for the normal operators whose spectrum is a subset of a rectifiable curve. The affirmative answer is given to Question 7 as well. Other questions seem to remain open.

REFERENCE

7. Makarov N. G., *Perturbations of normal operators and stability of continuous spectrum*, Izvestiya AN SSSR, ser. matem. **50** (1986), no. 6, 1178–1203 (Russian); English transl. in Math. USSR – Izvestiya **29** (1987), no. 3, 535–558.

Chapter 9

FUNCTIONAL MODELS

Edited by

N. K. Nikolskii
Université Bordeaux-I
UFR de Mathématiques
351, cours de la Libération
33405 Talence CEDEX
France
and
Steklov Mathematical Institute
St. Petersburg branch
Fontanka 27
St. Petersburg, 191011
Russia

V. I. Vasyunin
Steklov Mathematical Institute
St. Petersburg branch
Fontanka 27
St. Petersburg, 191011
Russia

PREFACE

In this chapter four problems from the previous edition of the Problem Book are collected, and two problems are new. Two of the old problems are completely solved. The new problems being connected with some properties of the lattices of invariant subspaces could be placed in Chapter 5 "General operator theory" as well. However many results concerning this subject were obtained in framework of Szőkefalvi-Nagy–Foiaş functional model. By this reason these problems appear in this chapter.

9.1
v.old

OPERATORS AND APPROXIMATION

N. K. NIKOLSKII

1. What is a "Blaschke product"? As long as we are concerned with scalar-valued analytic functions in the unit disc $\mathbb{D}$, the answer is well-known: this is a function B satisfying one of the following equivalent statements.

(i) B can be represented as a product of elementary factors b_λ:

$$B = \prod_{\lambda \in \mathbb{D}} b_\lambda^{k(\lambda)}, \qquad b_\lambda = \frac{|\lambda|}{\lambda} \cdot \frac{\lambda - z}{1 - \overline{\lambda} z},$$

here k is a function from $\mathbb{D}$ to nonnegative integers with $\sum_\lambda k(\lambda)(1 - |\lambda|) < +\infty$.

(ii) B is inner (in Beurling's sense) and the part of the shift operator $\mathbf{z}^*$ on the invariant subspace K, $K = K_B \stackrel{\text{def}}{=} H^2 \ominus \mathbf{B}H^2$, has a complete (in K) family of root subspaces. Here H^2 stands for the standard Hardy class and $\mathbf{z}$, $\mathbf{B}$ are operators of multiplication by z and B respectively.

(iii) The same for the operator $\mathbf{T}_B \stackrel{\text{def}}{=} P_K \mathbf{z}|K$ adjoint of $\mathbf{z}^*|K$ (P_K stands for the orthogonal projection onto K).

(iv) B is inner and

$$\lim_{r \uparrow 1} \int_{\mathbb{T}} \log |B(r\xi)| \, dm(\xi) = 0.$$

The spectral interpretation of (i) and (iv) is of importance for studying operators in terms of their characteristic functions and the problems discussed in this section are essentially those of a "correct" choice of the notion of Blaschke product in the general case, when operator-valued inner functions are considered (the equality $|B(\xi)| = 1$ a.e. is replaced in this case by the requirement that $B(\xi)$ be a unitary operator on an auxiliary coefficient space E; H^2 is replaced by $H^2(E)$, and so on). Statements (i)–(iv), appropriately modified, are still equivalent for operators $\mathbf{T}_B$ having a determinant (i.e. when $I - \mathbf{T}_B^* \mathbf{T}_B$ is nuclear). If this is no longer true, (ii) AND (OR) (iii) seem to be the most natural definitions of a Blaschke product.

QUESTION 1. *Is it true in case of an arbitrary operator valued inner function that one of Conditions* (ii), (iii) *implies the other one?*

The definition under consideration should presume a metric criterion for a characteristic function B to belong to the class of "Blaschke products" (that is a criterion for $\mathbf{T}_B$ AND (OR) $\mathbf{T}_B^*$ to be complete).

QUESTION 2. *Do the following conditions give such criteria:*

$$\lim_{r \to 1} \int_{\mathbb{T}} \log \|B(r\xi)\| \, dm(\xi) = 0 \quad \text{or} \quad \lim_{r \to 1} \int_{\mathbb{T}} \log B(r\xi)^* B(r\xi) \, dm(\xi) = \mathbb{O}?$$

If we restrict ourselves to the case when $\mathbf{T}_B^* = \mathbf{z}^*|K$ has a simple spectrum, the problem of the "proper" definition of Blaschke product reduces to the following one.

QUESTION 3. *How to construct a "Blaschke product" B describing a subspaces*

$$\text{(v)} \qquad \operatorname{span}\left(\frac{\Delta_\lambda E}{1-\bar{\lambda}z} : \lambda \in \mathbb{D}\right)$$

in its canonical form $K_B = H^2 \ominus \mathbf{B}H^2$

Here $\{\Delta_\lambda : \lambda \in \mathbb{D}\}$ is a family of orthogonal projections on E. The well-known results by V. Potapov, Yu. Ginzburg, V. Brodskii, I. Gohberg and M. Krein give such a construction for the case when $1-(\mathbf{z}^*|K)^*(\mathbf{z}^*|K)$ belongs to the Matsaev ideal of compact operators, see [1] for references.

It seems important to know when the space (v) coincides with $H^2(E)$, i.e.

QUESTION 4. *For which families $\{\Delta_\lambda : \lambda \in \mathbb{D}\}$ the conditions $f \in H^2(E)$ and $\Delta_\lambda f(\lambda) = \mathbb{O}$, $\lambda \in \mathbb{D}$, imply $f \equiv 0$?*

If $\Delta_\lambda = \mathbb{O}$ or I, Question 4 clearly reduces to the scalar uniqueness theorem $\sum_{\Delta_\lambda \neq 0}(1-|\lambda|) = \infty$. The last condition remains necessary in general case. Perhaps the answer to Question 4 is the following: $\sum(1-|\lambda|)\|\Delta_\lambda e\|^2 = \infty$ for each $e \in E$. As to the subspaces (v) it is worth to mention that in the case $\dim E = 1$ they can be described in terms of meromorphic pseudocontinuation of functions in (v) (M. M. Djrbashyan, G. C. Tumarkin, R. Douglas, H. Shapiro, A. Shields and others, see [2] for references). Possibly the same language fits for $\dim E > 1$.

2. Weak generators of the algebra $\operatorname{Alg}(\mathbf{T}_\Theta)$**.** In this section Θ is a scalar inner function and $\operatorname{Alg}(*)$ is the weakly closed algebra of operators generated by the operators $*$ and I. It is known (D. Sarason) that $A \in \operatorname{Alg}(\mathbf{T}_\Theta)$ iff $A = \varphi(\mathbf{T}_\Theta)$ for some φ in H^∞. (The operator $\varphi(\mathbf{T}_\Theta)$ acts in K_Θ by the rule $\varphi(\mathbf{T}_\Theta)f = P_K\varphi f$, $f \in K_\Theta$.) The description of weak generators φ of $\operatorname{Alg}(\mathbf{z}) = H^\infty$ is also known (D. Sarason) and can be expressed in a geometrical language, in terms of properties of the (necessarily univalent) image $\varphi(\mathbb{D})$. Since the algebras $\operatorname{Alg}(\mathbf{T}_\Theta)$ and $H^\infty/\Theta H^\infty$ are isometrically isomorphic, it is plausible that the Sarason theorem could admit "projecting":

QUESTION 5. *How to describe weak generators φ of the algebra $\operatorname{Alg}(\mathbf{T}_\Theta) \cong H^\infty/\Theta H^\infty$ in term of the quotient class $\varphi + \Theta H^\infty$? Is it true that $\varphi + \Theta H^\infty$ is a generator if $P_K 1$ is a cyclic vector of $\varphi(\mathbf{T}_\Theta)$? Is it true that $\operatorname{Alg}(\varphi(\mathbf{T}_\Theta)) = \operatorname{Alg}(\mathbf{T}_\Theta)$ if and only if $\varphi + \Theta H^\infty$ contains a generator of the algebra H^∞?*

QUESTION 6. *Which operators $\varphi(\mathbf{T}_\Theta)$ have simple spectrum? (In other words, for which φ there exists f in K_Θ with $\operatorname{span}(P_K\varphi^n f : n \geqslant 0) = K_\Theta$?)*

If φ is a generator of H^∞ then the cyclic vectors f from Question 6 do exist and can be easily described. In the partial case $\Theta = \exp a\dfrac{z+1}{z-1}$ Question 6 can be reduced (for some functions φ at least) to the question whether $\varphi(\mathbf{T}_\Theta)$ is unicellular (or to the same question about the operator $x \mapsto \int_0^t x(s)K(t-s)\,ds$ on $L^2(0,a)$, G. E. Kisilevskii, [3]). Related to this matter are a paper of J. Ginsberg and D. Newman [4] and the problem 11.17 of this Collection. Other references, historical comments and more discussion can be found in [1], [5].

References

1. Nikol'skii N. K., *Invariant subspaces in operator theory and function theory*, Itogi Hauki i Tekhniki: Mat. Anal. **12** (1974), 199–412, VINITI, Moscow (Russian); English transl. in J. Soviet Math. **5** (1976), no. 2.
2. Nikolskii N. K., *Treatise on the shift operator*, Springer-Verlag, 1986.
3. Kisilevskii G., *Invariant subspaces of an integration (convolution) operator*, Funktsional. Anal. i Prilozhen. **8** (1974), no. 4, 85–86 (Russian); English transl. in Funct. Anal. Appl..
4. Ginsberg J., Newman D., *Generators of certain radical algebras*, J. Approx. Theory **3** (1970), no. 3, 229–235.
5. Frankfurt R., Rovnyak J., *Recent results and unsolved problems on finite convolution operators*, Linear spaces and approximation, (Proc. Conf., Math. Res. Inst. Oberwolfach, 1977) Lect. Notes in Biomath., vol. 21, Springer-Verlag, 1978, pp. 133–150.

Steklov Mathematical Institute
St. Petersburg branch
Fontanka 27
St. Petersburg, 191011
Russia

and

Université Bordeaux-I
UFR de Mathématiques
351, cours de la Libération
33405 Talence CEDEX
France

Commentary

B. M. Solomyak has answered the last part of Question 5 in the negative (oral communication). For the sake of convenience we replace here the unit disc by the upper half-plane $\Pi = \{\zeta : \operatorname{Im}\zeta > 0\}$ and consider the corresponding spaces H^2_Π, H^∞_Π. Let $\Theta = e^{iz}$, $\varphi = \left(\dfrac{e^{iz}-1}{iz}\right)^n$, $n > 2$. Clearly $\varphi \in H^\infty_\Pi$ and it is proved in [6] that $\operatorname{Alg}(\varphi(\mathbf{T}_\Theta)) = \operatorname{Alg}(\mathbf{T}_\Theta)$. (This fact is just equivalent to the unicellularity of J^n, $(Jf)(x) = \int_0^x f(t)\,dt$ in $L^2(0,1)$.) On the other hand, for every $g \in H^\infty_\Pi$, $h = \varphi + \Theta g = (i/z)^n + e^{iz}g_1$; $g_1(z + i\eta) \in H^\infty_\Pi$, $\eta > 0$. Since $(i/z)^n$, $n > 2$ is nonunivalent in $\{\operatorname{Im} z > \eta\}$ and e^{iz} tends to zero rapidly as $\operatorname{Im} z \to +\infty$, it can be easily verified that h is also nonunivalent. Thus h cannot be an H^∞_Π-generator. □

Another counterexample for $\Theta = B$, an interpolation Blaschke product, was constructed by N. G. Makarov.

Reference

6. Frankfurt R., Rovnyak J., *Finite convolution operators*, J.Math. Anal. Appl. **49** (1975), 347–374.

9.2
v.old

SIMILARITY PROBLEM AND THE STRUCTURE OF THE SINGULAR SPECTRUM OF NON-DISSIPATIVE OPERATORS

S. N. Naboko

The similarity problems under consideration are to find necessary and sufficient conditions for a given operator on a Hilbert space to be similar to a selfadjoint (or dissipative) operator. For the first problem an answer was found in terms of the integral growth of the resolvents [3] (see also [2]):

PROPOSITION 1. *An operator L is similar to a selfadjoint operator if and only if*

$$\sup_{\varepsilon>0} \varepsilon\left(\int_{\mathbb{R}} \|(L-k+i\varepsilon)^{-1}u\|^2\,dk + \int_{\mathbb{R}} \|(L^*-k+i\varepsilon)^{-1}u\|^2\,dk\right) \leqslant c\|u\|^2, \quad u \in \mathcal{H}.$$

The second problem is not yet solved. Here we discuss an approach based on the notion of the characteristic function of an operator [1]. For a *dissipative* operator L ($\operatorname{Im} L = (2i)^{-1}(L-L^*) \geqslant \mathbb{O}$) there exists a criterion of similarity to a selfadjoint operator (due to B. Sz.-Nagy and C. Foiaş) in terms of its characteristic function S, namely

$$\sup_{\operatorname{Im}\lambda>0} \| S^{-1}(\lambda)\| < +\infty.$$

The main tool in the proof of this result was the Sz.-Nagy–Foiaş functional model which yields a complete spectral description of a dissipative operator*. For a non-dissipative operator L an analogous condition on its characteristic function Θ

$$\sup_{\operatorname{Im}\lambda>0} \|\Theta(\lambda)\| < +\infty, \qquad \sup_{\operatorname{Im}\lambda>0} \|\Theta^{-1}(\lambda)\| < +\infty$$

is sufficient for L to be similar to a selfadjoint operator (L. A. Sakhnovich), but not necessary. It is possible to give counterexamples on finite-dimensional spaces which show that operators whose characteristic function does not satisfy the above condition can be similar to selfadjoint operators (cf. [4], where related problems of similarity to a unitary operator and to a contraction are discussed).

To be more precise, let us consider the characteristic function

$$S \stackrel{\text{def}}{=} I + 2i(|V|)^{1/2}(A - i|V| - \lambda)^{-1}(|V|)^{1/2}, \qquad S(\lambda): E \to E, \quad \operatorname{Im}\lambda > 0,$$

of an auxiliary dissipative operator $A + i|V|$, where $A = \operatorname{Re} L$, $V = \operatorname{Im} L$, $E = \operatorname{clos}\operatorname{Range} V$ and let $V = J \cdot |V|$, $J = \operatorname{sign} V$, be the polar decomposition of V. The

*Nevertheless this result can be obtained without using the functional model [3, 5].

latter operator for the sake of simplicity is assumed to be bounded. The characteristic function Θ of L and the function S are connected by a triangular factorization

$$\Theta(\lambda) \stackrel{\text{def}}{=} I + iJ(|V|)^{1/2}(L^* - \lambda)^{-1}(|V|)^{1/2} = (\chi_- + \chi_+ S(\lambda))(\chi_+ + \chi_- S(\lambda))^{-1}$$

where $\chi_\pm = (I \pm J)/2$, [6]. Note that $\|\chi_\mp S(\lambda)\chi_\pm\| \leqslant 1$ for $\operatorname{Im}\lambda > 0$. Under the additional condition

$$\sup_\lambda \max\{\|\chi_- S(\lambda)\chi_+\|, \|\chi_+ S(\lambda)\chi_-\|\} < 1 \tag{1}$$

the above mentioned condition of bounded invertibility of Θ is necessary for L to be similar to a selfadjoint operator and the condition $\sup_{\operatorname{Im}\lambda>0}\|\Theta(\lambda)\| < +\infty$ is necessary and sufficient for the similarity to a dissipative operator [7]. In this case corresponding selfadjoint and dissipative operators can be constructed explicitly in terms of the Sz.-Nagy–Foiaş model for $A + i|V|$.

In general case (beyond (1)) serious obstacles appear. The reason is that it is difficult *to obtain a complete description of the spectral subspaces of L corresponding to the singular real spectrum.* The solution of the LATTER PROBLEM would be of independent interest.

Let us dwell upon this question. The operator L is supposed to act on the model space K which can be defined as follows. Let $\mathcal{H}$ be the Hilbert space of pairs $(\tilde{g}, g)$ of E-valued functions on $\mathbb{R}$ square summable with respect to the matrix weight $\begin{pmatrix} I & S^* \\ S & I \end{pmatrix}$, $S(k) \stackrel{\text{def}}{=} S(k + i0)$ being the boundary values (in the strong topology) of the analytic operator-valued function S. Then

$$K = \mathcal{H} \ominus (\mathcal{D}_- \oplus \mathcal{D}_+), \quad \mathcal{D}_+ \stackrel{\text{def}}{=} (H^2(E), \mathbb{O}), \quad \mathcal{D}_- \stackrel{\text{def}}{=} (\mathbb{O}, H^2_-(E)),$$

where $H^2_\pm(E)$ are the Hardy classes of E-valued functions in the upper and lower half-planes.

The absolutely continuous subspace N_e in the model representation of L has the following form [6]

$$N_e = N_e(L) = \operatorname{clos} P_K\left(\mathcal{H} \ominus \left(\operatorname{clos}(\chi_- L^2(E), \chi_+ L^2(E))\right)\right), \tag{2}$$

where P_K is the orthogonal projection from $\mathcal{H}$ onto K, $L^2(E) = L^2(E, \mathbb{R})$. The singular subspace N_i is defined by $N_i = K \ominus N_e^*$, where $N_e^* = N_e(L^*)$. It is natural to distinguish in N_i two subspaces N_i^+ and N_i^-, the first one corresponds to the point spectrum in the upper half-plane and a part of the real singular continuous spectrum, the second one corresponds to the point spectrum in the lower half-plane and another (in general) part of the real singular continuous spectrum:

$$\begin{aligned} N_i^+(L) &\stackrel{\text{def}}{=} N_i^+ \stackrel{\text{def}}{=} \operatorname{clos} P_k\left(H^2_-(E) \ominus (\chi_- + S^*\chi_+)H^2_-(E), \mathbb{O}\right), \\ N_i^-(L) &\stackrel{\text{def}}{=} N_i^- \stackrel{\text{def}}{=} \operatorname{clos} P_k\left(\mathbb{O}, H^2_+(E) \ominus (\chi_+ + S\chi_-)H^2_+(E)\right). \end{aligned} \tag{3}$$

The subspace $N_i^+ (N_i^-)$ is analogous to the subspace corresponding to the singular spectrum of a dissipative (adjoint of a dissipative) operator. Nevertheless if (1) fails, N_i

does not necessarily coincide with $\operatorname{clos}\{N_i^+ + N_i^-\}$. In particular the eigenvectors and root vectors of the real point spectrum do not belong to $\operatorname{clos}\{N_i^+ + N_i^-\}$.

Therefore in general it is necessary to introduce a "complementary spectral component" N_i^0 ($N_i^0 \subset N_i$) which would permit us to take into account the real spectrum of L. Put

$$N_i^0 \stackrel{\text{def}}{=} N_i \ominus \operatorname{clos}\{N_i^{+*} + N_i^{-*}\}$$

where $N_i^{\pm *} \stackrel{\text{def}}{=} N_i^{\pm}(L^*)$.

PROBLEM 1. *When $N_i = \operatorname{clos}\{N_i^+ + N_i^- + N_i^0\}$? To estimate angles between N_i^0 and $N_i^{\pm}$, N_e in terms of the characteristic function Θ. To give an explicit description of N_i^0 (similar to that of $N_i^{\pm}, N_e$), for example as the closure of the projection onto K of some linear manifold in $\mathcal{H}$ described in terms of characteristic function.*

PROBLEM 2. *Find the factor of Θ corresponding to N_i^0. Investigate its further factorization. Describe properties of this factor and connection of its roots with the spectrum of $L|N_i^0$. How to separate the spectral singularities of $L|N_e$ from the spectrum of $L|N_i^0$?*

PROBLEM 3. *Make clear the spectral structure of N_i^0, i.e., in terms of the model space construct spectral projection onto intervals of the real singular spectrum and onto the root-space corresponding to the real point spectrum.*

PROBLEM 4. *Let L be similar to a dissipative operator. Then does N_i equal to $\operatorname{clos}\{N_i^+ + N_i^- + N_i^0\}$? Does N_i^0 coincide with the subspace of K corresponding to the singular continuous spectrum plus point spectrum of the selfadjoint part of this dissipative operator?*

PROBLEM 5. *Consider concrete examples (Friedrichs model with rank one perturbation, Schrödinger operator on $\mathbb{R}$ with a powerlike decreasing potential) and describe N_i^0 for such operators.*

Besides the real discrete spectrum the space N_i^0 apparently can contain one more spectral component. The elements of N_i^0 no longer have the "smoothness" properties as those of $N_i^{\pm}$ (namely for a dense set of vectors $u \in N_i^{\pm}$ we have $(|V|)^{1/2}(L-\lambda I)^{-1}u \in H^2_{\mp}(E)$). Perhaps, the structure of N_i^0 is similar to that of the singular continuous component of a selfadjoint operator (namely, one should take into account the "non-orthogonality" caused by the non-selfadjointness of L). It is also important for the similarity problem to know what is the factor of the characteristic function Θ corresponding to N_i^0. Note also that all difficulties of the problem appear already in the case when the imaginary part of V is of finite rank ($\dim E \geqslant 2$).

Let us present here one more assertion closely related to the problem discussed above and especially to the spectral decomposition of L.

PROPOSITION 2. *An operator L on a Hilbert space $\mathcal{H}$ is similar to a dissipative operator L_{diss} if and only if there exists an operator M such that*

$$\|u\|^2 \asymp \lim_{\varepsilon \to 0} \varepsilon \int \left\|(L-k+i\varepsilon)^{-1}u\right\|^2 dk + \left\|M(L-\lambda)^{-1}u\right\|^2_{H^2_-(\mathcal{H})}, \quad u \in \mathcal{H}.$$

Moreover, if such L_{diss} exists then this M can be chosen satisfying the additional inequality $\operatorname{rank} M \leqslant \operatorname{rank} \operatorname{Im} L_{\mathrm{diss}}$.

REFERENCES

1. Sz.-Nagy B., Foiaş C., *Harmonic Analysis of Operators on Hilbert Space*, North Holland–Akadémiai Kiadó, Amsterdam–Budapest, 1970.
2. Sz.-Nagy B., *On uniformly bounded linear transformations in Hilbert space*, Acta Sci. Math. **11** (1947), 152–157.
3. Naboko S. N., *Conditions for similarity to unitary and self-adjoint operators*, Funktsional. Anal. i Prilozhen. **18** (1984), no. 1, 16–27 (Russian); English transl. in Funct. Anal. Appl. **18** (1984), no. 1, 13–22.
4. Davis Ch., Foiaş C., *Operators with bounded characteristic functions and their J-unitary dilation*, Acta Sci. Math. **32** (1971), no. 1–2, 127–139.
5. van Castern J., *A problem of Sz.-Nagy*, Acta Sci. Math. **42** (1980), no. 1–2, 189–194.
6. Naboko S. N., *Absolutely continuous spectrum of a nondissipative operator and a functional model. II*, Zapiski nauchn. semin. LOMI **73** (1977), 118–135 (Russian); English transl. in J. Soviet Math. **34** (1986), no. 6, 2090–2101.
7. Naboko S. N., *Singular spectrum of a non-self-adjoint operator*, Zapiski nauchn. semin. LOMI **113** (1981), 149–177 (Russian); English transl. in J. Soviet Math. **22** (1983), no. 6, 1793–1813.

UL. ORDZHONIKIDZE 37–1, 5
ST. PETERSBURG 196234
RUSSIA

COMMENTARY BY S. NABOKO AND V. VESELOV

At present in the solution of this Problem a considerable progress was achieved. However completely the problem has not been solved and many questions remain without answers.

Problem 1. In [8] the following sufficient condition of completeness (i.e. of the equality $\overline{N_i + N_e} = K$) was obtained in terms of the characteristic function Θ of an operator L:

$$(*)\qquad \operatorname{v.sup}_{k\in\mathbb{R}}\{\|\Theta^*(k)J\Theta(k)\| + \|\Theta^{*-1}(k)J\Theta^{-1}(k)\|\} < \infty.$$

This condition provides also $\sin(N_e, N_i) > 0$. Note that under condition $(*)$ the subspace N_i^0 can be nontrivial. For Friedrichs model operator with a one-dimensional perturbation in [8] the "evident" estimations of angles between subspaces N_e and N_i was established directly in terms of the perturbation.

In [9] making use the corona theorem estimates of the angles between invariant subspaces were established in terms of the corresponding factors of Θ (in particular between $N_i^{\pm}$, N_e and N_i^0). In [9] also a rough suffucient conditions of completeness of invariant subspaces was obtained.

Problem 2. The factor $\hat\Theta$ of the characteristic function corresponding to the invariant subspace N_i^0 can be characterized as follows [10]:

1) $\hat\Theta(k)$ is J-unitary (i.e., $J - \hat\Theta^*(k)J\hat\Theta(k) = J - \hat\Theta(k)J\hat\Theta^*(k) = 0$) for a.e. $k \in \mathbb{R}$

2) $\hat\Theta(k)$ is an outer operator-valued function in the half-planes $\mathbb{C}_\pm$ simultaneously.

However a detailed factorisation of the factor $\hat\Theta$ which has the spectral sense so far in the proper form was not accomplished. The last circumstance connects with significant difficulties of spectral analysis of operators which have $N_i^0 = K$. However in the recent years the considerable progress was achieved in the investigation of these operators (for the case rank $\operatorname{Im} L = 2$). The spectral singularities σ_0 of the operator $L|N_e$ were investigated in [8]. In particular, it was shown that the condition $\operatorname{dist}(\sigma_0, \sigma(L|N_i)) > 0$ is sufficient to separate the subspaces N_e and N_i.

Problem 3. As to the real point spectrum of the operator L that we can refer to [11] where a detailed investigation of the eigenvectors corresponding to the real spectrum of L was carrying out. In [11] also the necessary and sufficient condition are given for the set of eigenvectors to be uniformly minimal or to be a Riesz basis. The triangular model of operators [12] and Potapov's multiplicative representation of the characteristic function are used. This approach allowed us to understand in details (cf. [13]) the connections between the triangular and functional models of an operator.

Problem 4. For the operators with a trace class imaginary part in [10] the positive answers were given to both questions of problem 4.

Problem 5. The concrete examples of operators with nontrivial subspaces N_i^0 were considered in papers [9,10]. It is necessary to note that the detailed spectral analysis of such operators is far from to be complete.

References

8. Veselov V.F., *On separatibility of absolutely continuous and singular subspaces of nondissipative operator*, Vestnik LGU, ser.1 (1988), no. 2(8), 11–17 (Russian); English transl. in Vestnik Leningrad Univ. Math. **21** (1988), no. 2, 13–20.
9. Veselov V. F., *Separatibility of invariant subspaces of a nondissipative operator*, Vestnik LGU, Mat. Mekh. Astronom. (1988), no. 4(22), 19–24 (Russian); English transl. in Vestnik Leningrad Univ. Math. **21** (1988), no. 4, 19–24.
10. Veselov V. F., Naboko S. N., *The determinant of the characteristic function and the singular spectrum of a nonselfadjoint operator*, Mat. sbornik **129(171)** (1986), no. 1, 20–39 (Russian); English transl. in Math. USSR – Sbornik **57** (1987), no. 1, 24–41.
11. Veselov V. F., *Bases of the eigensubspaces of a nondissipative operator and the characteristic function*, Funktsional. Anal. i Prilozhen. **22** (1988), no. 4, 76–77 (Russian); English transl. in Funct. Anal. Appl. **22** (1989), no. 4, 320–322.
12. Livsic M. S., *On the spectral decomposition of linear nonselfadjoint operators*, Mat. sbornik **34(76)** (1954), no. 1, 144–199. (Russian)
13. Veselov V. F., *Relationship between triangular and functional models of a non-self-adjoint operator*, Funktsional. Anal. i Prilozhen. **21** (1987), no. 4, 66–68 (Russian); English transl. in Funct. Anal. Appl. **21** (1988), no. 4, 312–314.

9.3

TWO PROBLEMS ABOUT COMMUTANTS

V. V. KAPUSTIN

For an operator T on a Hilbert space H we consider the commutant $\{T\}'$ and the bicommutant $\{T\}''$.

1. The first problem is about continuity properties of lifting of the commutant[1]. Let T be a contraction. Then T has a unitary dilation U acting on a space $\mathcal{H} \supset H$ [1]. The space $\mathcal{H}$ can be represented as $\mathcal{H} = G_* \oplus H \oplus G$, where $UG \subset G$, $U^*G_* \subset G_*$, and $T = P_H U|H$ (P_H denotes the orthogonal projection onto H).

It is well known [1] that for any $A \in \{T\}'$ there exists a $B \in \{U\}'$ for which

$$(1) \qquad BG \subset G, \qquad B^*G_* \subset G_*$$

and $A = P_H B|H$. Such a B is not unique. Let $\mathcal{B}$ denote the set of $B \in \{U\}'$ satisfying (1), and let $\mathcal{Z} = \{B \in \mathcal{B} : P_H B|H = 0\}$. Thus we have a natural map $\mathcal{F}$ from $\{T\}'$ onto the factor algebra $\mathcal{B}/\mathcal{Z}$ which sends an operator $A \in \{T\}'$ to the class of its liftings. An operator B such that $A = P_H B|H$ can be chosen so that $\|A\| = \|B\|$. This means that $\mathcal{F}$ is isometric, and in particular, it is continuous with respect to the norm topology.

PROBLEM 1. *Is $\mathcal{F}$ continuous with respect to the weak operator topology (WOT)? More precisely, is it continuous if $\{T\}'$ is endowed with the WOT and $\mathcal{B}/\mathcal{Z}$ with the factor topology generated by the WOT in $\mathcal{B}$?*

2. We say that T is an *almost isometric* (AI) operator (on a Hilbert space) if $I - T^*T$ is of trace class or, equivalently, T is a trace class perturbation of an isometry. An operator T is said to have *the bicommutant property* if $\{T\}'' = \operatorname{Alg} T$, where $\operatorname{Alg} T$ denotes the minimal WOT-closed algebra containing I and T. For AI contractions a criterion for an operator to have the bicommutant property can be found in [2] and [3]. If T is not a contraction, then our knowledge of algebras related with T is less than satisfactory. A solution of the following problem would probably help to understand the structure of commutants of AI operators.

PROBLEM 2. *Characterize AI operators for which $\{T\}'' = \operatorname{Alg} T$.*

It is natural to generalize the conditions for AI contractions to have the bicommutant property as follows.

CONJECTURE A. *Let T be an AI operator. Then $\{T\}'' = \operatorname{Alg} T$ if and only if at least one of the following two conditions holds:*

a) *The spectrum $\sigma(T)$ covers the whole unit disc.*

b) *There exists a subset e of the unit circle such that the Lebesgue measure of e is positive, $\Theta(z)$ is a J-unitary operator for every $z \in e$, and $E(e) = 0$, where E is the spectral measure of the unitary part of T (Θ denotes the characteristic function of T, see, e.g., [4]).*

[1] This problem was stated jointly with A. V. Lipin

It is evident that the resolvents $R_z = (T - zI)^{-1}$, $z \notin \sigma(T)$, belong to $\{T\}''$ for any operator T. Once again, let T be an AI operator. Suppose that $\sigma(T)$ does not cover the unit disc, and let $|z| < 1$, $z \notin \sigma(T)$. In [4] it is proved that $R_z \in \operatorname{Alg} T$ if and only if condition b) of Conjecture A holds. The statement of Conjecture A would follow from

CONJECTURE B. *Let T be an AI operator. Then the WOT-closure of the linear span of the set $\{ R_z : z \notin \sigma(T) \}$ coincides with $\{T\}''$.*

REFERENCES

1. Szőkefalvi-Nagy B., Foiaş C., *Harmonic Analysis of Operators on Hilbert Space*, North Holland–Akadémiai Kiadó, Amsterdam–Budapest, 1970.
2. Wu P. Y., *Bi-invariant subspaces of weak contractions*, J. Operator Theory **1** (1979), 261–272.
3. Takahashi K., *Contractions with the bicommutant property*, Proc. Amer. Math. Soc. **93** (1985), 91–95.
4. Makarov N. G., Vasyunin V. I., *A model for non-contractions and stability of the continuous spectrum*, Lect. Notes in Math. **864** (1981), 365–412.

STEKLOV MATHEMATICAL INSTITUTE
ST. PETERSBURG BRANCH
FONTANKA 27
ST. PETERSBURG, 191011
RUSSIA

QUASI-SIMILARITY INVARIANCE OF REFLEXIVITY

LÁSZLÓ KÉRCHY

The notion of reflexivity was introduced into operator theory by D. Sarason in [16]. An operator T acting on a complex Hilbert space H is said to be reflexive if the invariant subspaces of T determine the weakly closed algebra $W(T)$ generated by T and the identity I. That means that T has plenty of invariant subspaces. To be more precise, let us introduce the following notation. The invariant subspace lattice of T is denoted by $\operatorname{Lat} T$ and $\operatorname{Alg}\operatorname{Lat} T$ stands for the set of all (bounded, linear) operators S on H having the property $\operatorname{Lat} S \supset \operatorname{Lat} T$. The operator T is *reflexive* if $\operatorname{Alg}\operatorname{Lat} T = W(T)$.

In [16] Sarason proved that normal operators and analytic Toeplitz operators are reflexive. Many operators have been investigated from the point of reflexivity since then. For some recent results and further references see [6].

Generalizing the results of [4], in [1] H. Bercovici, C. Foiaş and B. Sz.-Nagy showed, among other things, that reflexivity is quasi-similarity invariant in the class of C_0 contractions. We recall that an absolutely continuous contraction T is of class C_0 if $u(T) = 0$ for a non-zero function $u \in H^\infty$. An operator A is a quasi-affine transform of an operator B, in notation: $A \prec B$, if there exists a one-to-one (bounded) linear transformation X with dense range such that $XA = BX$; and A and B are *quasi-similar*, in notation: $A \sim B$, if $A \prec B$ and $B \prec A$ hold simultaneously. (For details see [17].) The problem of characterizing reflexive C_0 operators has been finally settled recently by V. V. Kapustin [9].

C_0 contractions belong to the larger class C_{00}; that means that the sequences $\{T^n\}_{n=1}^\infty$ and $\{T^{*n}\}_{n=1}^\infty$ converge to zero in the strong operator topology for every C_0 contraction T. It is natural to ask what happens in the opposite case when the sequences $\{\|T^n h\|\}_{n=1}^\infty$ and $\{\|T^{*n} h\|\}_{n=1}^\infty$ do not converge to zero, for every non-zero vector h in H. This class of contractions was introduced and called C_{11} in [17]. It is known that a contraction T is of class C_{11} if and only if T is quasi-similar to a unitary operator U_T, which is uniquely determined (up to unitary equivalence). The unitary operator U_T can be represented as the $*$-residual part $R_{*,T}$ of the unitary dilation of T or as the unitary asymptote $T^{(a)}$ of T (see [17] and [12]). The latter operator $T^{(a)}$ can be associated with any contraction T in the following way. Let us introduce a new semi-inner product on H by $[x, y] \stackrel{\text{def}}{=} \lim_{n\to\infty}\langle T^n x, T^n y\rangle$. After factorization and completion we obtain a Hilbert space $H_{+,T}$ and an isometry $T_+^{(a)}$ on $H_{+,T}$ such that $X_T T = T_+^{(a)} X_T$, where X_T is the natural embedding of H into $H_{+,T}$. The minimal unitary extension of $T_+^{(a)}$ is called the *unitary asymptote* of T, and is denoted by $T^{(a)}$.

If T is of class C_{11} then X_T is a quasi-affinity and $T \sim T^{(a)}$. *Does T inherit the reflexivity property from $T^{(a)}$?*

PROBLEM A. *Is every C_{11} contraction T reflexive?*

An affirmative answer was obtained by P. Y. Wu in the case when the defect index $d_T = \operatorname{rank}(I - T^*T)$ is finite [20]. Dropping the assumption on the defect index, the reflexivity of T was proved by the author in the case when the unitary asymptote $T^{(a)}$

is not reductive, i.e. if $T^{(a)}$ has a non-reducing invariant subspace [11]. This result followed from the following

THEOREM 1. *If T is an absolutely continuous contraction of class C_{11} such that $T^{(a)}$ is not reductive, then* $\operatorname{Alg}\operatorname{Lat} T = H^\infty(T)$ $\left(\stackrel{\text{def}}{=} \{u(T) : u \in H^\infty\}\right)$.

This theorem it proven in [11] by establishing that there exists a subspace $N \in \operatorname{Lat} T$ such that the restriction $T^{(a)}|(X_T N)^-$ is the simple unilateral shift S (which is a special analytic Toeplitz operator). Since $T|N \prec S$, it follows by a result of H. Bercovici and K. Takahashi [3] that $\operatorname{Alg}\operatorname{Lat}(T|N) = H^\infty(T|N)$, and the statement of the theorem can be derived from this relation. Theorem 1 was generalized by K. Takahashi [19] dropping the assumption $T \in C_{11}$ but keeping the non-reductivity of $T^{(a)}$.

In view of Theorem 1 our problem can be reduced to the following

PROBLEM B. *Is every C_{11} contraction T with reductive unitary asymptote $T^{(a)}$ reflexive?*

A reasonable generalization of reductivity for C_{11} contractions is the following. A C_{11} contraction T is *quasi-reductive* if $\operatorname{Lat} T = \operatorname{Lat}_1 T$, where $\operatorname{Lat}_1 T$ consists of the quasi-reducing subspaces of T, where a subspace $N \in \operatorname{Lat} T$ is called *quasi-reducing* if the restriction $T|N$ is of class C_{11}. Clearly, for unitary operators reductivity and quasi-reductivity are exactly the same properties.

It is not difficult to show (see [10]) that the C_{11} contraction T is quasi-reductive if $T^{(a)}$ is reductive and the characteristic function of T has a scalar multiple. What is more, it can be proved in that case that the orthogonal sum $T^{(n)}$ of n copies of T is also quasi-reductive, for every $n \in \mathbb{N}$. Since quasi-reducing subspaces are invariant for the bicommutant $\{T\}''$ of T (see [10]) it follows that $\operatorname{Lat} T^{(n)} = \operatorname{Lat}\{T^{(n)}\}''$ holds, for every n. Thus we conclude that $W(T) = \{T^{(n)}\}''$ (see [15]). It was proved by K. Takahashi [18] that every operator T which is quasi-similar to a normal operator N has a reflexive bicommutant $\{T\}''$. Therefore we obtain the following

THEOREM 2. *If T is a C_{11} contraction with a reductive unitary asymptote $T^{(a)}$ and the characteristic function of T has a scalar multiple, then T is reflexive and $W(T) = \{T\}''$.*

This result raises the following

PROBLEM C. *Is the quasi-reductivity of a C_{11} contraction T inherited by the orthogonal multiples $T^{(n)} (n \in \mathbb{N})$?*

A positive answer to this problem would imply the reflexivity of quasi-reductive C_{11} contractions. From the proof of Theorem 1 it can be seen that if T is quasi-reductive then $T^{(a)}$ is reductive. However, as the examples given in [2] and [10] show, there are non-quasi-reductive C_{11} contractions with reductive unitary asymptotes. Hence an affirmative answer to Problem C does not settle Problem B. This latter problem should be attacked first in the cyclic case.

The unitary asymptote $T^{(a)}$ can be defined for power bounded operators T as well, using a generalized Banach limit, and $T \sim T^{(a)}$ if the power bounded operator T of class C_{11} (see [17], [12]). Reflexivity results obtained for contractions would have immediate extensions to this more general setting if we had affirmative answer to the following

PROBLEM D. *Is every power bounded operator T of class C_{11} similar to a contraction?*

Of course, there are known power bounded operators which are not similar to contractions, but the examples given in [5], [7] and [14] do not belong to the class C_{11}. A special case of Problem D is the following

PROBLEM E. *Let T be a power bounded operator of class C_{11} such that the unitary asymptote $T^{(a)}$ is a singular unitary operator. Is T similar to $T^{(a)}$?*

If the answer to Problem A is affirmative, one can pose the following

PROBLEM F. *Is every operator T, which is quasi-similar to a normal operator N, reflexive?*

Moreover, since even subnormal operators are know to be reflexive [13], it can be asked

PROBLEM G. *Is every operator T, quasi-similar to a subnormal operator M, reflexive?*

Finally, we remark that reflexivity is not a quasi-similarity invariant in general. An example of a reflexive operator which is quasi-similar to a non-reflexive operator can be given using orthogonal sums of finite dimensional Jordan blocks (for such a construction see [8]).

REFERENCES

1. Bercovici H., Foiaş C., Sz.-Nagy B., *Reflexive and hyper-reflexive operators of class C_0*, Acta Sci. Math. (Szeged) **43** (1981), 5–13.
2. Bercovici H., Kérchy L., *Quasisimilarity and properties of the commutant of C_{11} contractions*, Acta Sci. Math. (Szeged) **45** (1983), 67–74.
3. Bercovici H., Takahashi K., *On the reflexivity of contractions on Hilbert space*, J. London Math. Soc. (2) **32** (1985), 149–156.
4. Deddens J. A., Fillmore P. A., *Reflexive linear transformations*, Linear Algebra and Appl. **10** (1975), 89–93.
5. Foguel S. R., *A counterexample to a problem of Sz.-Nagy*, Proc. Amer. Math. Soc. **15** (1964), 788–790.
6. Hadwin D. W., Nordgren E. A., *Reflexivity and direct sums*, Acta Sci. Math. (Szeged) **55** (1991), no. 1–2, 181–197.
7. Halmos P. R., *On Foguel's answer to Nagy's question*, Proc. Amer. Math. Soc. **15** (1964), 791–793.
8. Hoover T. B., *Quasi-similarity of operators*, Illinois J. Math. **16** (1972), 678–686.
9. Kapustin V. V., *Reflexivity of operators: general methods and a criterion for almost isometric contractions*, Algebra i Analiz **4** (1992), no. 2, 141–160 (Russian); English transl. in St. Petersburg Math. J. **4** (1993), no. 2.
10. Kérchy L., *Contractions being weakly similar to unitarie*, Operator Theory: Advances and Applications, vol. 17, Birkhäuser Verlag, Basel, 1986, pp. 187–200.
11. Kérchy L., *Invariant subspaces of C_1-contractions with non-reductive unitary extensions*, Bull. London Math. Soc. **19** (1987), 161–166.
12. Kérchy L., *Isometric asymptotes of power bounded operators*, Indiana Univ. Math. J. **38** (1989), 173–188.
13. Olin R., Thomson J., *Algebras of subnormal operators*, J. Funct. Anal. **37** (1980), 271–301.
14. Peller V. V., *Estimates of functions of power bounded operators on Hilbert space*, J. Operator Theory **7** (1982), 341–372.
15. Radjavi H., Rosenthal P., *Invariant Subspaces*, Springer Verlag, New York, 1973.
16. Sarason D., *Invariant subspaces and unstarred operator algebras*, Pacific J. Math. **17** (1966), 511–517.
17. Sz.-Nagy B., Foiaş C., *Harmonic Analysis of Operators on Hilbert Space*, North Holland – Akadémiai Kiadó, Amsterdam – Budapest, 1970.
18. Takahashi K., *Double commutant of operators quasi-similar to normal operators*, Proc. Amer. Math. Soc. **92** (1984), 404–406.

19. Takahashi K., *The reflexivity of contractions with nonreductive $*$-residual parts*, Michigan Math. J. **34** (1987), 153–159.
20. Wu P. Y., *C_1 contractions are reflexive*, Proc. Amer. Math. Soc. **77** (1979), 68–72.

Bolyai Institute, University of Szeged
Aradi Vértanúk tere 1, 6720 Szeged, Hungary

Commentary by W. R. Wogen

L. Kérchy has posed the question of whether a Hilbert space operator T must be reflexive provided T is quasisimilar to a normal operator. A counterexample can be constructed by a slight modification of the main example in the paper [25] of Larson and Wogen. (See also [22], [23], and [24].) In that paper, a sequence $\{T_n\}$ of rank one idempotents in $\mathcal{B}(\mathcal{H})$ is constructed with several special properties. First of all, $\{T_n\}$ is algebraically orthogonal; that is, $T_n T_m = 0$ if $m \neq n$. In addition, the closed span of the union of the ranges of the idempotents $\{T_n\}$ is $\mathcal{H}$, and the same is true of $\{T_n^*\}$. It follows from a result of Apostol [21] that each operator $T = \sum \lambda_n T_n$ (assume here, e.g., that the partial sums converge in norm) is quasisimilar to a diagonal normal operator with eigenvalues $\{\lambda_n\}$.

In [25], $\|T_n\| \approx 2^n$, so taking $\lambda_n = 4^{-n}$, we get a compact operator $T = \sum 4^{-n} T_n$. It is shown [25, Lemma 3.3] that T is reflexive, but that I is not in the weakly closed span of $\{T_n\}$ [25, Lemma 3.6]. However, $\forall x \in \mathcal{H}$, x is in the closed span of $\{T_n x\}$ [25, Lemma 3.2]. Now consider $T \oplus 0 \in \mathcal{B}(\mathcal{H} \oplus \mathbb{C})$. The above properties of T yield that $I \oplus 0 \notin W(T \oplus 0)$, but $I \oplus 0 \in \operatorname{Alg}\operatorname{Lat}(T \oplus 0)$. Thus $T \oplus 0$ is not reflexive, [25, Theorem 3.7]. Note that $T \oplus 0$ is quasisimilar to a self-adjoint operator.

It is now straightforward to produce a counterexample to Kérchy's question. Choose a sequence $\{\lambda_n\}$ so that $|\lambda_n| = 1$ and $|\lambda_n - 1| = 4^{-n}$, $\forall n$. If $S_0 = \left(\sum(\lambda_n - 1)T_n\right) \oplus 0 \in \mathcal{B}(\mathcal{H} \oplus \mathbb{C})$, then the proof of [25, Theorem 3.7] shows that S_0 is not reflexive but is quasisimilar to a diagonal operator with spectrum $\{\lambda_n - 1\} \bigcup \{0\}$. Let $S = (S_0 + I) \oplus I$. Then S is quasisimilar to the (unitary) diagonal operator with spectrum $\{\lambda_n\} \bigcup \{1\}$.

Note that S is not a contraction, so there remains the possibility that every contraction which is quasisimilar to a unitary must be reflexive.

The paper [25] provided the first example of a reflexive operator T whose direct sum with 0 is not reflexive. The methods were then used to settle in the negative several other questions which had been open.

References

21. Apostol C., *Operators quasisimilar to a normal operator*, Proc. Amer. Math. Soc. **53** (1975), 104–106.
22. Argyros S., Lambrou M., Longstaff W., *Atomic Boolean subspace lattices and applications to the theory of bases*, Mem. Amer. Math. Soc. (1991).
23. Azoff E., Sheheda H., *Algebras generated by mutually orthogonal idempotent operators*, J. Operator Theory (to appear).
24. Katavolos A., Lambrou M., Papadakis M., *On some algebras diagonalized by M-bases of ℓ^2*, preprint.
25. Larson D. R., Wogen W. R., *Reflexivity properties of $T \oplus 0$*, J. Functional Anal. **92** (1990), 448–467.

Department of Mathematics
University of North Carolina
Chapel Hill, NC 27599-3250, USA

S.9.5
v.old

SPECTRAL DECOMPOSITIONS AND THE CARLESON CONDITION

N. K. Nikol'skii, B. S. Pavlov, V. I. Vasyunin

Completely nonunitary contractions can be included into the framework of the Szőkefalvi-Nagy–Foiaş model [1]. Especially simple is the case when $\sigma(T)$ does not cover the unit disc clos$\mathbb{D}$ and $d = d_* < \infty$, where $d = \dim(I - T^*T)H$, $d_* = \dim(I - TT^*)H$ are the defect numbers of T. If $s\text{-lim}(T^*)^n = \mathbb{O}$ and the above conditions are fulfilled then T is unitarily equivalent to its model, i.e. to the operator

$$P\mathbf{z}|K, \qquad K = H^2(E) \ominus \Theta H^2(E),$$

where E is an auxiliary Hilbert space with $\dim E = d$, Θ is a bounded analytic $(E \to E)$-operator valued function in $\mathbb{D}$ whose boundary values are unitary almost everywhere on the unit circle $\mathbb{T}$, $H^2(E)$ is the Hardy space of E-valued functions, $\mathbf{z}$ is the multiplication operator $f \mapsto zf$, P is the orthogonal projection onto K. The function Θ is called the characteristic function of T. It is connected very closely with the resolvent of T, e.g. $|R(\lambda, T)| \asymp (1 - |\lambda|)^{-1}|\Theta(\lambda)^{-1}|$, $\lambda \in \mathbb{D}$. An operator T can be investigated in details in terms of its characteristic function Θ. Namely, it is possible to find its spectrum, point spectrum $\sigma_p(T)$, eigenvectors and root vectors, to calculate the angles between maximal spectral subspaces etc. (cf. [1–4]). In particular an operator T is complete (i.e., the linear hull of its eigenvectors and root vectors is dense in K) iff $\det\Theta$ is a Blaschke product.

A more detailed spectral analysis should include, however, not only a description of spectral subspaces but also methods of recovering T from its restrictions to the spectral subspaces. The strongest method of recovering yields the unconditionally convergent spectral decomposition generated by a given decomposition of the spectrum. For a complete operator T the question is whether its root subspaces $\{ K_\lambda : \lambda \in \sigma_p(T) \}$ form an unconditional basis. In the case of a simple point spectrum necessary and sufficient conditions of such "spectrality" (i.e. in the case under consideration for the operator to be similar to a normal one) were found in [2], [3]. These conditions are as follows: the *vectorial Carleson condition*

$$\inf\{ |\widetilde{\Delta}_\lambda \Theta_\lambda(\lambda)^{-1} \Delta_\lambda|_E^{-1} : \lambda \in \sigma_p(T) \} > 0 \tag{1}$$

holds and the following *imbedding theorems* are valid:

$$\begin{aligned} &\sum_{\lambda \in \sigma_p(T)} (1 - |\lambda|) \|\Delta_\lambda f(\lambda)\|_E^2 < \infty, \\ &\sum_{\lambda \in \sigma_p(T)} (1 - |\lambda|) \|\widetilde{\Delta}_\lambda f(\lambda)\|_E^2 < \infty, \end{aligned} \qquad \forall f \in H^2(E). \tag{2}$$

Here $\widetilde{\Delta}_\lambda$ is the orthoprojection from E onto the subspace $\ker\Theta(\lambda)$ and Δ_λ is the orthoprojection from E onto $\ker\Theta(\lambda)^*$; $\Theta = \Theta_\lambda\big[b_\lambda\widetilde{\Delta}_\lambda + (I-\widetilde{\Delta}_\lambda)\big]$ is the factorization of Θ corresponding to the eigenspace $K_\lambda = \ker(T-\lambda I)$, $b_\lambda \stackrel{\text{def}}{=} \frac{|\lambda|}{\lambda}(\lambda - z)(1-\overline{\lambda}z)^{-1}$. From a geometrical point of view condition (1) means nothing else as the so called uniform minimality of the family $\{\, K_\lambda : \lambda \in \sigma_p \,\}$. Moreover,

$$|\widetilde{\Delta}_\lambda\Theta_\lambda(\lambda)^{-1}\Delta_\lambda|_E^{-1} = \sin\big(\widehat{K_\lambda, K^\lambda}\big),$$

where

$$K^\lambda = \operatorname{span}\big(K_\mu : \mu\in\sigma_p(T)\setminus\{\lambda\}\big),$$

cf. [2]. In the case $d = d_* = 1$ L. Carleson proved that (1) implies (2) (cf. [4]), in the case $d = d_* = \infty$ this is no longer true ([3]).

PROBLEM 1. *Prove or disprove the implication* (1) $\Longrightarrow$ (2) *in the case* $1 < d = d_* < \infty$.

The case $d = d_* = 1$ seems to be an exceptional one, because of an arbitrary family of subspaces the property to be uniformly minimal is very far (in the general case) from the property to form an unconditional basis. However for $d = d_* = 1$ these conditions coincide not only for eigenspaces but for root subspaces as well and, moreover, for arbitrary families of spectral subspaces of a contraction T [5]. The proofs of this equivalence we are aware of (cf. [5], [4]) represent some kinds of analytical tricks and depend on the evaluation of the angles between pairs of "complementary" spectral subspaces K_ϑ and $K^\vartheta \stackrel{\text{def}}{=} K_{\vartheta'}$ corresponding to a divisor ϑ of Θ . Here ϑ and ϑ' are left divisors of Θ corresponding to a given pair of subspaces; if $d = d_* = 1$ then $\vartheta' = \frac{\Theta}{\vartheta}$. A divisor ϑ is called spectral divisor if K_ϑ is a spectral subspace. So the main part of the above mentioned trick consists in the following implication [5]; let $\dim E = 1$ and let $\{\vartheta\}$ be an arbitrary family of spectral divisors of Θ, then the condition

$$\inf_{\vartheta}\ \inf_{\substack{e\in E\\ \|e\|=1}}\ \inf_{\xi\in\mathbb{D}}\{\, \|\vartheta(\xi)e\|_E + \|\vartheta'(\xi)e\|_E \,\} > 0 \tag{3}$$

implies the following one

$$\inf_{\sigma}\ \inf_{\substack{e\in E\\ \|e\|=1}}\ \inf_{\xi\in\mathbb{D}}\{\, \|\vartheta_\sigma(\xi)e\|_E + \|\vartheta'_\sigma(\xi)e\|_E \,\} > 0 \tag{4}$$

where ϑ_σ is the inner function corresponding to the subspace $\operatorname{span}\{\, K_\sigma : \vartheta\in\sigma \,\}$, σ being an arbitrary subset of $\{\vartheta\}$. The proof uses a lower estimate of $\|\Theta(\xi)e\|_E$ depending on $\|\vartheta(\xi)e\|_E$ and $\|\vartheta'(\xi)e\|_E$ only. However such an estimate is impossible for $\dim E > 1$ (L. E. Isaev, private communication).

PROBLEM 2. *Let* $1 < d = d_* < \infty$. *Prove or disprove the implication* (3) $\Longrightarrow$ (4) *for an arbitrary family* $\{\vartheta\}$ *of spectral divisors of* Θ.

References

1. Szőkefalvi-Nagy B., Foiaş C., *Harmonic Analysis of Operators on Hilbert Space*, North Holland–Akadémiai Kiadó, Amsterdam–Budapest, 1970.
2. Nikolskii N. K., Pavlov B. S. Eigenvector bases of completely nonunitary contractions and the characteristic function, Izv. AN SSSR, ser. mat. **34** (1970), no. 1, 90–133; English transl. in Math. of the USSR - Izvestija.
3. Nikolskii N. K., Pavlov B. S., *Decompositions with respect to the eigenvectors of nonunitary operators and the characteristic function*, Zapiski nauchn. sem. LOMI **11** (1968), 150–203. (Russian)
4. Nikolskii N. K., *Treatise on the shift operator*, Springer-Verlag, 1986.
5. Vasyunin V. I., *Unconditionally convergent spectral decomposition and interpolation problems*, Trudy Matem. Inst. Steklov **130** (1978), 5–49; English transl. in Proc. Steklov Inst. Math. (1979), no. 4, 1-54.

Steklov Mathematical Institute
St. Petersburg branch
Fontanka 27
St. Petersburg, 191011
Russia

and

Université Bordeaux-I
UFR de Mathématiques
351, cours de la Libération
33405 Talence CEDEX
France

Dept. of Physics
St. Petersburg State University
Staryi Peterhof, St. Petersburg, 198904
Russia

Steklov Mathematical Institute
St. Petersburg branch
Fontanka 27
St. Petersburg, 191011
Russia

Commenary

The problem is completely solved by R. Treil. In [6], [7] the implication (1) $\Longrightarrow$ (2) is proved for the case when the set of unit vectors $\left\{\frac{\Delta_\lambda e}{\|\Delta_\lambda e\|} : \lambda \in \sigma_p(T), e \in E\right\}$ is relatively compact in E. In particular this is true if $\dim E < \infty$, as in Problem 1.

In [8] the implication (3) $\Longrightarrow$ (4) is proved also in a bit more general situation, namely for arbitrary, not obligatory equal, finite defect indices.

References

6. Treil S. R., *Space compact system of eigenvectors forms a Riesz basis if it is uniformly minimal*, Dokl. Akad. Nauk SSSR **288** (1986), no. 2, 308–312 (Russian); English transl. in Soviet Math. Doklady **33** (1986), no. 3, 675–679.
7. Treil S. R., *Geometric methods in spectral theory of vector-valued functions: some recent results*, Operator Theory: Advances and Appl., vol. 42, 1989, pp. 209–280.
8. Treil S. R., *Hankel operators, imbeddings theorems and bases of co-invariant subspaces of the multiple shift operator.*, Algebra i analiz **1** (1989), no. 6, 200–234 (Russian); English transl. in Leningrad Math. J. **1** (1990), no. 6.

S.9.6
old

ON EXISTENCE OF INVARIANT SUBSPACES OF C_{10}-CONTRACTIONS

R. TEODORESCU, V. I. VASYUNIN

Let T be a completely nonunitary C_{10}-contraction* with the characteristic function $\Theta \in H^\infty(E, E_*)$. So T^* can be supposed acting on the space

$$K_\Theta = H^2(E_*) \ominus \mathbf{\Theta} H^2(E)$$

as follows

$$T^* f = \frac{f - f(0)}{z}, \qquad f \in K_\Theta.$$

Recall that $T \in C_{10}$ iff Θ is inner and $*$-outer.

CONJECTURE. *For every inner $*$-outer function Θ there exists a nonzero noncyclic vector for T^* (defined by (1)) of the form $P_+ h_*$ where $h_* \in \operatorname{Ker} \mathbf{\Theta}^*$.*

REFERENCES

1. Sz.-Nagy B., Foiaş C., *Harmonic Analysis of Operators on Hilbert Space*, North Holland–Akadémiai Kiadó, Amsterdam–Budapest, 1970.
2. Nikol'skii N. K., *Treatise on the Shift Operator*, Springer-Verlag, 1986.

UNIVERSITY LAVAL
DEPARTMENT OF MAHTEMATICS
QUEBEC, G1K 7P4
CANADA

STEKLOV MATHEMATICAL INSTITUTE
ST. PETERSBURG BRANCH
FONTANKA 27
ST. PETERSBURG, 191011
RUSSIA

COMMENTARY

In [3] L. Kérchy has shown that the conjecture is true for the case when so called minimal isometric extension contains a uniteral shift. In [4] K. Takahashi has proved that such contractions are even reflexive. However in general situation the answer is negative. Namely, in [5] L. Kérchy has constructed a family of analytic Toeplitz operators whose restrictions to any cyclic invariant subspaces are C_{10}-operators without the conjectured property.

REFERENCES

3. Kérchy L., *Invariant subspaces of C_1.-contractions with non-reductive unitary extension*, Bull. London Math. Soc. **19** (1987), 161–166.
4. Takahashi K., *The reflexivity of contractions with non-reductive $*$-residual parts*, Michigan Math. J. **34** (1987), 153–159.
5. Kérchy L., *On a conjecture of Teodorescu and Vasyunin*, Operator Theory: Adv. and Appl., vol. 28, Birkhäuser Verlag, Basel, 1988, pp. 169–172.

*All used terminology can be found in [1] or [2].

Chapter 10

SINGULAR INTEGRALS, BMO, H^p

Edited by

E. M. Dyn′kin
Department of Mathematics
Technion
32000 Haifa
Israel

S. V. Kisliakov
St. Petersburg branch of
Steklov Mathematical Institute
Fontanka 27
St. Petersburg, 191011
Russia

INTRODUCTION

This chapter splits naturally into two parts of approximately equal length. The first part has been composed of problems concerning singular integrals proper. The second one consists of related questions of the theory of analytic functions of one complex variable. An important role is played here, e.g., by Blaschke products, but traditionally these questions are attributed to the field under consideration. Hardy spaces H^p appear as "real variable Hardy classes" in the first part and as classical "complex" classes in the second one.

The problem of L^p-boundedness of Cauchy integral, opening the corresponding chapter in the preceding edition, has lost its exceptional position in the theory since that time. Nevertheless, questions on Cauchy integrals can be found among "new" problems as well (see, e.g., 10.3, 10.4), so that this subject still remains topical.

10.1
v.old

SOME PROBLEMS CONCERNING CLASSES OF DOMAINS DETERMINED BY PROPERTIES OF CAUCHY TYPE INTEGRALS

G. C. TUMARKIN

Investigation of boundary properties of analytic functions representable by Cauchy–Stieltjes type integral in a given planar domain G (i.e. functions of the form $z \mapsto \int_{\partial G} (\zeta - z)^{-1}\, d\mu(\zeta)$; if $d\mu = \omega\, d\zeta$ we denote this function by $\mathcal{K}^\omega$), as well as some other problems of function theory (approximation by polynomials and rational fractions, boundary value problems, etc.) have led to introduction of some classes of domains. These classes are defined by conditions that the boundary singular integral $S_\Gamma \omega$ ($\Gamma = \partial\Omega$) should exist and belong to a given class of functions on Γ or (which is in many cases equivalent) that analytic functions representable by Cauchy type integrals should belong to a given class of analytic functions in G. See [1] for a good survey on solutions of boundary value problems. An important role is played by the class of curves (denoted in [1] by R_p) for which the singular integral operator is continuous on $L^p(\Gamma)$: $\Gamma \in R_p$ if and only if

$$\forall \omega \in L^p(\Gamma) \quad \|S_\Gamma(\omega)\| \leqslant C_p \|\omega\|_p. \tag{1}$$

This means that M. Riesz theorem (well-known for the circle) holds for Γ. Some sufficient conditions for (1) were given by B. V. Khvedelidze, A. G. Džvarsheishvily, G. A. Khuskivadze and others. I. I. Danilyuk and V. Yu. Shelepov (a detailed exposition can be found in the monograph [2]) have shown that (1) is true for all $p > 1$ for simple rectifiable Jordan curves Γ with bounded rotation and without cusps. Some general properties of the class R_p were described by V. P. Khavin, V. A. Paatashvily, V. M. Kokilashvily and others. It was shown, e.g., that (1) is equivalent to the following condition:

$$\forall \omega \in L^p(\Gamma) \quad f \stackrel{\text{def}}{=} \mathcal{K}^\omega \in E_p(G),$$

$E_p(G)$ being the well-known V. I. Smirnov class (cf. e.g. [3]) of functions f analytic in G and such that integrals of $|f|^p$ over some system of closed curves $\{\gamma_i\}$ (with $\gamma_i \subset G$, $\gamma_i \to \Gamma$) are bounded. That is, R_p can be characterized by the property

$$\forall \omega \in L^p(\Gamma) \quad (f \circ \varphi) \sqrt[p]{\varphi'} \in H_p \; (f = \mathcal{K}^\omega), \tag{2}$$

where φ is a conformal mapping of $\mathbb{D}$ onto G.

Another class of domains (denoted by K) has been introduced and investigated earlier by the author (cf. [4] and references to other author's papers therein). We quote a definition of K that is closely connected with definition (2) of R_p: $G \in K$ if for any function f in G, analytic and representable as a Cauchy type integral, the function $(f \circ \varphi)\varphi'$

(φ being a conformal mapping of $\mathbb{D}$ onto G) is also representable as a Cauchy type integral:*

$$f = \mathcal{K}^\omega,\ \omega \in L^1(\Gamma) \implies (f \circ \varphi)\varphi' = \mathcal{K}^\Omega,\ \Omega \in L^1(\mathbb{T}).$$

Note that by Riesz theorem it is sufficient for $\Gamma \in R_p$ that the function $(f \circ \varphi)\sqrt[p]{\varphi'}$ in (2) be representable in the form $\mathcal{K}^\Omega$ with $\Omega \in L^p(\mathbb{T})$. This allows one to consider K as a counterpart of R_p for $p = 1$ (it is well known that to use (2) directly is impossible for $p = 1$ even for $\Gamma = \mathbb{T}$). It is established in [5] (see also [1]), using Cotlar's approach, that the classes R_p coincide for $p > 1$. Thus the following problem arises naturally.

PROBLEM 1. *Do the classes R_p ($p > 1$) and R_1 coincide? If not, what geometric conditions guarantee $\Gamma \in R_p \cap R_1$?*

Note that for $G \in K$ it becomes easier to transfer many theorems, known for the disc, on approximation by polynomials or by rational fractions in various metrics (cf. references in [4]). For such domains, it is possible to obtain conditions that guarantee convergence of boundary values of Cauchy type integrals [4]. As I have proved, K is a rather wide class containing in particular all domains G bounded by curves with finite rotation (cusps are allowed) [4]. At the same time, it follows from characterizations of K proved by me earlier that K coincides with the class of Faber domains, introduced and used later by Dyn'kin (cf. e.g. [6], [7]) to investigate uniform approximations by polynomials and by Anderson and Ganelius [8] to investigate uniform approximation by rational fractions with fixed poles. This fact seems to have stayed unnoticed by the authors of these papers, because they reprove for the class of Faber domains some facts established earlier by me (the fact that domains with bounded rotation and without cusps belong to this class, conditions on the distribution of poles guaranteeing completeness, etc.). The following question is of interest.

PROBLEM 2. *Suppose that the interior domain G^+ of a curve Γ belongs to $K(= R_1)$. Is it true that the exterior domain G^- also belongs to K?* (Of course, we use here a conformal mapping of G^- onto$\{|w| > 1\}$).

For R_p with $p > 1$ the positive answer to the analogous question is evident. At the same time the similar problem formulated in [9] for the class S of Smirnov domains still remains open. At last it is of interest to study the relationship between the classes S of Smirnov domains and A_0 of Ahlfors domains (bounded by quasicircles [10]), on the one hand, and K and R_p (considered here) on the other. See [9] for more details on S and A_0. It is known that $R_p \subset S$, $K \subset S$ ([4], [11]). At the same time there exist domains with rectifiable boundary in A_0 which do not belong to S (cf. [3], [9]). Simple examples of domains bounded by piecewise differentiable curves with cusp points show that $K \setminus A_0 \neq \emptyset$.

PROBLEM 3. *Find geometric conditions guaranteeing*

$$G \in K \cap R_p \cap A_0.$$

Once these conditions are satisfied, it follows from the papers cited above and [12], [13] that many results known for the unit disc can be generalized.

One of such conditions is that Γ should be of bounded rotation and without cusps.

*In virtue of a well-known V. I. Smirnov theorem, the analog of this property for Cauchy integrals is always true.

References

1. Khvedelidze B. V., *The method of Cauchy type integrals in discontinuous boundary problems of the theory of holomorphic functions of one complex variable*, Modern problems of mathematics **7** (1975), Moscow, 5–162 (Russian); English transl. in J. Soviet Math.
2. Danilyuk I. I., *Non-regular boundary problems on the plane*, Nauka, Moscow, 1975. (Russian)
3. Duren P. L., Shapiro H. S., Shields A. L., *Singular measures and domains not of Smirnov type*, Duke Math. J. **33** (1966), no. 2, 247–254.
4. Tumarkin G. Ts., *Boundary properties of analytic functions representable by a Cauchy type integral*, Mat. Sb. **84(126)** (1971), no. 3, 425–439 (Russian); English transl. in Math. USSR-Sb. **13** (1971), no. 3, 419–434.
5. Paatashvili V. A., *On singular integrals of Cauchy*, Soobshch. Akad. Nauk Gruz. SSR **53** (1969), no. 3, 529–532. (Russian)
6. Dyn′kin E. M., *On the uniform approximation by polynomials in the complex domain*, Zap. Nauchn. Sem. Leningrad. Otdel. Mat. Inst. Steklov (LOMI) **56** (1975), 164–165 (Russian); English transl. in J. Soviet Math. **9** (1978), 269–271.
7. ______, *On the uniform approximation of functions on Jordan domains*, Sibirsk. Mat. Zh. **18** (1977), no. 4, 775–786 (Russian); English transl. in Sib. Math. J. **18** (1978), 548–557.
8. Andersson Jan-Erik, Ganelius Tord, *The degree of approximation by rational function with fixed poles*, Math. Z. **153** (1977), no. 2, 161–166.
9. Tumarkin G. Ts., *Boundary properties of conformal mappings of certain classes of domains*, In: Some questions of modern function theory, Novosibirsk, 1976, pp. 149–160. (Russian)
10. Ahlfors L., *Lectures on quasiconformal mappings*, Van Nostrand Reinhold, 1966.
11. Havin V. P., *Boundary properties of Cauchy type integrals and harmonic conjugate functions in domains with rectifiable boundaries*, Mat. Sb. **68(110)** (1965), 499–517 (Russian); English transl. in Amer. Math. Soc. Transl. **74** (1968), no. 2, 40–60.
12. Belyi V. I., Miklyukov V. M., *Some properties of conformal and quasiconformal mappings, and direct theorems of the constructive function theory*, Izv. Akad. Nauk SSSR Ser. Mat. (1974), no. 6, 1343–1361 (Russian); English transl. in Math. USSR Izv. **8** (1974).
13. Belyi V. I., *Conformal mappings and approximation of analytic functions in domains with quasiconformal boundary*, Mat. Sb. **102** (1977), no. 3, 331–361 (Russian); English transl. in Math. USSR-Sb. **31** (1977), no. 3, 289–317.

ul. Borisa Galuskina 17, 363
Moscow 129301
Russia

Commentary

A complete geometric description of the class R_p, $1 < p < \infty$, has been obtained by Guy David. Cf. [14].

Reference

14. David G., *Opérateurs intégraux singuliérs sur certaines courbes du plan complexe*, Ann. Sci. E.N.S. **17** (1984), no. 1, 157–189.

10.2
old

BILINEAR SINGULAR INTEGRALS AND MAXIMAL FUNCTIONS

PETER W. JONES

While the boundedness of Cauchy integrals on curves is now fairly well understood [1], there remain some difficult one dimensional problems in this area. One such example is the operator

$$T_1(f,g)(x) = \text{p.v.} \int_{-\infty}^{\infty} f(x+t)g(x-t)\,\frac{dt}{t}.$$

Is T_1 a bounded operator from $L^2 \times L^2$ to L^1? A. P. Calderón first considered these operators during the 1960's, when he noticed (unpublished) that the boundedness of T_1 implies the boundedness of the first commutator (with kernel $\dfrac{A(x)-A(y)}{(x-y)^2}$, $A' \in L^\infty$) as an operator from L^2 to L^2. In order to make sense out of T_1, it seems that one must first study the related maximal operator

$$T_2(f,g)(x) = \sup_{h>0} \frac{1}{2h} \int_{-h}^{h} f(x+t)g(x-t)\,dt,$$

and see whether T_2 is a bounded operator from $L^2 \times L^2$ to L^1. It is easy to see that T_2 maps to weak L^1.

REFERENCE

1. Coifman R. R., McIntosh A., Meyer Y., *L'intégrale de Cauchy définit un opérateur borné sur L^2 pour les courbes Lipschitziennes*, Ann.Math. **116** (1982), 361–387.

DEPARTMENT OF MATHEMATICS
YALE UNIVERSITY
BOX 2155 YALE STATION
NEW HAVEN, CT 06520
USA

LIMITS OF INTEGRALS OF THE CAUCHY TYPE

J. KRÁL

Let $A \subset \mathbb{R}^2 \equiv \mathbb{C}$ be a Borel set with compact boundary ∂A such that, for any $z \in \partial A$ and $r > 0$, the disk $B(z,r) = \{\zeta \in \mathbb{R}^2; |\zeta - z| < r\}$ meets both A and $G \equiv \mathbb{R}^2 \setminus A$ in a set of positive Lebesgue measure (denoted by λ_2), and put

$$E = \{\zeta \in \mathbb{R}^2; \limsup_{r\to 0+} r^{-2}[B(\zeta,r) \cap A] > 0,\ \limsup_{r\to 0+} r^{-2}[B(\zeta,r) \cap G] > 0\}.$$

C_0^1 is the class of all complex-valued continuously differentiable functions with compact support in $\mathbb{R}^2$, $C^1(\partial A)$ consists of all restrictions to ∂A of functions in C_0^1, $\Re C_0^1$ and $\Re C^1(\partial A)$ are classes of real-valued functions in C_0^1 and $C^1(\partial A)$, respectively. Put $\overline{\partial} = \frac{1}{2}(\partial_1 + i\partial_2)$, where ∂_j stands for the partial derivative with respect to the j-th variable ($j = 1,2$). If $z \in \mathbb{R}^2 \setminus \partial A$ and $f \in C^1(\partial A)$, then choose $\varphi_f \in C_0^1$ vanishing at z such that $\varphi_f = f$ on ∂A and define

$$\mathcal{K} f(z) = \frac{1}{\pi i} \int_G \frac{\overline{\partial}\varphi_f(\zeta)}{(\zeta - z)}\, d\lambda_2(\zeta).$$

$\mathcal{K} f(z)$ does not depend on the choice of the corresponding φ_f and defines a holomorphic function of the variable z on $\mathbb{C} \setminus \partial A$. For $f \in \Re C^1(\partial A)$ put

$$W f(z) = \Im\mathcal{K} f(z), \qquad P f(z) = \Re\mathcal{K} f(z),$$

so that Wf, Pf are harmonic (compare [1] and §2 in [4]). Suppose that $Q \geqslant 0$ is a real-valued bounded lower-semicontinuous function on ∂A which is strictly positive off a fixed point $\eta \in \partial A$ and consider the space $\mathcal{C}(\partial A, Q)$ of all continuous functions f on ∂A subject to

$$|f(\zeta) - f(\eta)| = o(Q(\zeta)) \quad \text{as} \quad \zeta \to \eta, \zeta \in \partial A;$$

the norm $\|f\|_Q$ in $\mathcal{C}(\partial A, Q)$ is defined as the supremum of all the quantities

$$|f(\zeta)| \quad (\zeta \in \partial A), \quad (|f(\zeta) - f(\eta)|)/Q(\zeta), \quad (\zeta \in \partial A \setminus \{\eta\}).$$

If $\operatorname{int} A = A \setminus \partial A \neq \emptyset$ and λ_1 denotes the length, then

$$\int_E Q(\zeta)\, d\lambda_1(\zeta) < \infty \tag{1}$$

is equivalent with any of the following assertions (A_1), (A_2):

(A_1) The operator $f \mapsto \mathcal{K} f$ is continuous from $C^1(\partial A) \cap \mathcal{C}(\partial A, Q)$ to the space of all holomorphic functions on $\operatorname{int} A$ equipped with the topology of uniform convergence on compact subsets of $\operatorname{int} A$.

(A_2) The operator $f \mapsto W f$ is continuous from $\Re C^1(\partial A) \cap \mathcal{C}(\partial A, Q)$ into the space of all harmonic functions on $\operatorname{int} A$ equipped with the topology of uniform convergence on compact subsets of $\operatorname{int} A$.

Assuming (1) one may naturally extend the definition of $\mathcal{K} f$ to any $f \in \mathcal{C}(\partial A, Q)$ and the definitions of $W f$, $P f$ to any $f \in \Re\mathcal{C}(\partial A, Q)$. For investigation of limits of $\mathcal{K} F$, $W f$, $P f$ at η the following geometric quantities appear to be useful (cf. [2], [3]). Denote for $\rho > 0$

$$\mathcal{U}^Q(\rho;\eta) = \sum_{\zeta} Q(\zeta), \qquad \zeta \in E, \qquad |\zeta - \eta| = \rho;$$

then $\rho \mapsto \mathcal{U}^Q(\rho;\eta)$ is a Baire function of the variable ρ so that the definition

$$\mathcal{U}_r^Q(\eta) = \int_0^r \rho^{-1}\mathcal{U}^Q(\rho;\eta)\,d\rho \qquad (\rho > 0)$$

is justified. Consider also, for $\zeta \in \mathbb{C}$, $\gamma \in [0, 2\pi[$ and $r > 0$, the segment $S_r^\gamma(\zeta) = \{\zeta + \rho\exp(i\gamma); 0 < \rho < r\}$ and introduce the sums

$$V_r^Q(\gamma;\zeta) = \sum_{\xi} |\xi - \zeta| Q(\xi), \qquad \xi \in E \cap S_r^\gamma(\zeta),$$

$$v^Q(\gamma;\zeta) = \sum_{\xi} Q(\xi), \qquad \xi \in E \cap S_\infty^\gamma(\zeta);$$

again, $\gamma \mapsto V_r^Q(\gamma;\zeta)$ and $\gamma \mapsto v^Q(\gamma;\zeta)$ are Baire functions which justifies the notation

$$V_r^Q(\zeta) = \int_0^{2\pi} V_r^Q(\gamma;\zeta)\,d\gamma, \qquad v^Q(\zeta) = \int_0^{2\pi} v^Q(\gamma;\zeta)\,d\gamma.$$

Let now $S \subset \operatorname{int} A$ be a connected set whose closure meets ∂A at η only such that the contingent of S at η (in the sense of [5]) reduces to a single half-line H with end-point η such that neither H nor its reflection at η belongs to the contingent of ∂A at η.

CONJECTURE. *The finite* $\lim\limits_{\zeta\to\eta,\,\zeta\in S} W f(\zeta)$ *exists for each* $f \in \Re\mathcal{C}(\partial A, Q)$ *iff*

$$v^Q(\eta) + \sup_{r>0} r^{-1}\mathcal{U}_r^Q(\eta) < \infty.$$

The finite $\lim\limits_{\zeta\to\eta,\,\zeta\in S} P f(\eta)$ *exists for each* $f \in \Re\mathcal{C}(\partial A, Q)$ *iff*

$$\mathcal{U}_\infty^Q(\eta) + \sup_{r>0} r^{-1} V_r^Q(\eta) < \infty.$$

The finite $\lim\limits_{\zeta\to\eta,\,\zeta\in S} \mathcal{K} f(\eta)$ *exists for each* $f \in \mathcal{C}(\partial A, Q)$ *iff*

$$\mathcal{U}_\infty^Q(\eta) + v^Q(\eta) < \infty.$$

References

1. Astala K., *Calderón's problem for Lipschitz classes and the dimension of quasicircles*, Revista Matemática Iberoamericana **4** (1988), 469–486.
2. Dont M., *Non-tangent limits of the double layer potentials*, Časopis pro pěstování matematiky **97** (1972), 231–258.
3. Král J., Lukeš J., *Integrals of the Cauchy type*, Czechoslovak Math. J. **22** (1972), 663–682.
4. Král J., *Integral operators in Potential theory*, Lecture Notes in Math., vol. 823, Springer-Verlag, 1980.
5. Saks S., *Theory of the integral*, Dover Pub., 1964.

Mathematical Institute
Czech Academy of Sciences
Žitná 25
115 67 Prague 1
Czech Republic

10.4

CHORD ARC CURVES AND GENERALIZED NEUMANN–POINCARÉ OPERATOR C_1^Γ

J. G. KRZYŻ

A rectifiable Jordan curve Γ of length $|\Gamma|$ is called a chord-arc (or Lavrentiev) curve if there exists a constant $M \geqslant 1$ such that for any complementary subarcs Γ_1, Γ_2 of Γ with end points z_1, z_2 we have $\min\{|\Gamma_1|, |\Gamma_2|\} \leqslant M|z_1 - z_2|$. Zinsmeister characterized chord-arc curves as quasicircles that are AD-regular (i.e. regular in the sense of Ahlfors–David), cf. [3]. If $h \mapsto C^\Gamma h$ is the Cauchy singular integral operator, i.e.

$$(C^\Gamma h)(\zeta_0) = (\pi i)^{-1}\, p.v. \int_\Gamma \frac{h(\zeta)\, d\zeta}{\zeta - \zeta_0}, \qquad \zeta_0 \in \Gamma,$$

then according to G. David, $C^\Gamma h$ is a bounded linear operator on $L^p(\Gamma) = \{(h\colon \Gamma \to \overline{\mathbb{C}}) : \int_\Gamma |h(\zeta)|^p |d\zeta| < +\infty\}$ if and only if $p > 1$ and Γ is AD-regular, cf. [1]. Consequently, the operator $h \mapsto C_1^\Gamma h = \mathrm{Re}(C^\Gamma h)$ is bounded on $L_0^2(\Gamma) = \{(h\colon \Gamma \to \overline{\mathbb{R}}) : (\int_\Gamma |h(\zeta)|^2 |d\zeta| < +\infty) \wedge (\int_\Gamma f(\zeta)\, |d\zeta| = 0)\}$ for any chord-arc curve Γ. If $\|C_1^\Gamma\|_{L_0^2(\Gamma)} < 1$ for an AD-regular Γ then Γ is chord-arc, cf. [2].

PROBLEM. *Is it true that for any chord-arc curve Γ in the finite plane we have*

$$\|C_1^\Gamma\|_{L_0^2(\Gamma)} < 1\ ?$$

Remark. If the answer is in the positive, some other interesting properties of chord-arc curves would follow.

REFERENCES

1. David G., *Opérateurs intégraux singuliers sur certaines courbes du plan complexe*, Ann. Sci. École Norm. Sup. **17** (1984), 157–189.
2. Krzyż J. G., *Generalized Neumann–Poincaré operator and chord-arc curves*, Ann. Univ. M. Curie-Skłodowska Sect. A **43** (1989), 69–78.
3. Zinsmeister M., *Domaines de Lavrentiev*, Publ. Math. d'Orsay, Paris, 1985.

DEPARTMENT OF MATHEMATICS
MARIA CURIE-SKLODOWSKA UNIVERSITY
PLAC M. CURIE-SKLODOWSKIEJ 1
PL 20031 LUBLIN, POLAND

10.5
old

WEIGHTED NORM INEQUALITIES

Benjamin Muckenhoupt

The problems to be discussed here are of the following type. *Given p satisfying $1 < p < \infty$ and two operators T and S, determine all pairs of nonnegative functions U, V such that*

$$\int_{\mathbb{R}^n} |Tf(x)|^p U(x)\,dx \leqslant C \int_{\mathbb{R}^n} |Sf(x)|^p V(x)\,dx,$$

throughout this paper C denotes a constant independent of f but not necessarily the same at each occurence. There is a question of what constitutes a solution to this sort of problem; it is to be hoped that the conditions are simple and that it is possible to decide easily whether a given pair U, V satisfies the conditions. In some cases, particularly with the restriction $U = V$, this problem has been solved; for a survey of such results and references to some of the literature see [3]. Some of the most interesting unsolved and partially solved problems of this type are as follows.

1. *For $1 < p < \infty$ find all nonnegative pairs U and V such that*

$$\int_0^\infty\int_0^\infty \left|\int_0^x\int_0^y f(t,u)\,du\,dt\right|^p U(x,y)\,dx\,dy \leqslant C \int_0^\infty\int_0^\infty |f(x,y)|^p V(x,y)\,dx\,dy. \tag{1}$$

This two dimensional version of Hardy's inequality appears easy because f can be assumed nonnegative and no cancellation occurs on the left. The solution of the one dimensional case is known; the obvious two dimensional version of the one dimensional characterization is

$$\left[\int_s^\infty\int_t^\infty U(x,y)\,dy\,dx\right]\left[\int_0^s\int_0^t V(x,y)^{-\frac{1}{p-1}}\,dy\,dx\right]^{p-1} \leqslant C \tag{2}$$

for $0 < s,t < \infty$. This condition is necessary for (1) but not sufficient except for $p = 1$. See [7] for a proof that (2) is not sufficient for (1) and for additional conditions under which (2) does imply (1).

2. *For $1 < p < \infty$ find a simple characterization of all nonnegative pairs U, V such that*

$$\int_{-\infty}^{+\infty} [Mf(x)]^p U(x)\,dx \leqslant C \int_{-\infty}^{+\infty} |f(x)|^p V(x)\,dx, \tag{3}$$

where $Mf(x) = \sup_{y \neq x} (y-x)^{-1} \int_x^y |f(t)|\,dt$ *is the Hardy–Littlewood maximal function.* This problem was solved by Sawyer in [5]; his condition is that for every interval I

$$\text{(4)} \qquad \int_I \left[M(\chi_I(x)V(x)^{-\frac{1}{p-1}})\right]^p U(x)\,dx \leqslant C \int_I V(x)^{-\frac{1}{p-1}}\,dx$$

with C independent of I. It seems that there should be a characterization that does not use the operator M. One CONJECTURE is that (3) *holds if and only if for every interval I and every subset E of I with $|E| = |I|/2$ we have*

$$\text{(5)} \qquad \left[\int_I U(x)\,dx\right]\left[\frac{1}{|I|}\int_I V(x)^{-\frac{1}{p-1}}\,dx\right]^p \leqslant C \int_E V(x)^{-\frac{1}{p-1}}\,dx$$

with C independent of E and I. Condition (5) does give the right pairs for some of the usual troublesome functions and is not satisfied by the counter example in [5] to an earlier conjecture.

3. *For $1 < p < \infty$ find a simple characterization of all nonnegative pairs U, V such that*

$$\text{(6)} \qquad \int_{-\infty}^{+\infty} |Hf(x)|^p U(x)\,dx \leqslant C \int_{-\infty}^{+\infty} |f(x)|^p V(x)\,dx,$$

where $Hf(x) = \lim_{\varepsilon \to 0+} \int_{|y|>\varepsilon} f(x-y)/y\,dy$ *is the Hilbert transform.* There is a complicated solution to the periodic version of this by Cotlar and Sadosky in [1]. One CONJECTURE here is that *a pair U, V satisfies* (6) *if and only if U, V satisfy* (3) *and*

$$\text{(7)} \qquad \int_{-\infty}^{+\infty} [Mf(x)]^{p'} V(x)^{-\frac{1}{p-1}}\,dx \leqslant C \int_{-\infty}^{+\infty} |f(x)|^{p'} U(x)^{-\frac{1}{p-1}}\,dx,$$

where $p' = p/(p-1)$.

4. *For $1 \leqslant p < +\infty$ find a simple characterization of all nonnegative pairs U, V for which the weak type inequality*

$$\text{(8)} \qquad \int_{|Hf(x)|>a} U(x)\,dx \leqslant Ca^{-p} \int_{-\infty}^{+\infty} |f(x)|V(x)\,dx$$

is valid for $a > 0$. A CONJECTURED SOLUTION is that (7) *is a necessary and sufficient condition for* (8).

5. *For $1 < p < +\infty$ find a simple characterization of all nonnegative functions U such that*

$$\int_{-\infty}^{+\infty} |Hf(x)|^p U(x)\,dx \leqslant C \int_{-\infty}^{+\infty} [Mf(x)]^p U(x)\,dx. \tag{9}$$

A necessary condition for (8) is the existence of positive constants C and ε such that for all intervals I and subsets E of I

$$\int_E U(x)\,dx \leqslant C\left[\frac{|E|}{|I|}\right]^{\varepsilon} \int_{-\infty}^{+\infty} \frac{|I|^q U(x)\,dx}{|I|^q + |x - x_I|^q}, \tag{10}$$

where x_I denotes the center of I and $q = p$. In [6] it is shown that if (10) holds for some $q > p$, then (9) holds. It is CONJECTURED that (10) *with $q = p$ is also sufficient for* (9).

6. *For $1 < p \leqslant q < \infty$ find all nonnegative pairs U, V for which*

$$\left[\int_{\mathbb{R}^n} |\hat{f}(x)|^q U(x)\,dx\right]^{1/q} \leqslant C \left[\int_{\mathbb{R}^n} |f(x)|^p V(x)\,dx\right]^{1/p} \tag{11}$$

It was shown by Jurkat and Sampson in [2] that if for $r > 0$

$$\left[\int_0^r U^*(x)\,dx\right]^{1/q} \left[\int_0^{1/r} \left(V(x)^{-\frac{1}{p-1}}\right)^* dx\right]^{1/p'} \leqslant C, \tag{12}$$

where * indicates the nonincreasing rearrangement and $p' = p/(p-1)$, then (11) holds. Furthermore, if (11) holds for all rearrangements of U and V, then (12) is true. However, (12) is not a necessary condition for (11) as shown in [4]. This problem is probably difficult since if $p = q = 2$ and $V(x) = |x|^a$, $0 < a < 1$, then the necessary and sufficient condition on U is a capacity condition. Its difficulty is also suggested by the fact that a solution would probably solve the restriction problem for the Fourier transform.

REFERENCES

1. Cotlar M., Sadosky C., *On some L^p versions of the Helson–Szegö theorem*, Conference on Harmonic Analysis in Honor of Antoni Zygmund, Wadsworth, Belmont, California, 1983, pp. 306–317.
2. Jurkat W. B., Sampson G., *On rearrangement and weight inequalities for the Fourier transform*, Indiana Univ. Math. J. **33** (1984), no. 2, 257–270.
3. Muckenhoupt B., *Weighted norm inequalities for classical operators*, Proc. Symp. in Pure Math. **35** (1979), no. 1, 69–83.
4. Muckenhoupt B., *Weighted norm inequalities for the Fourier transform*, Trans. Amer. Math. Soc. **276** (1983), 729–742.
5. Sawyer E., *Two weight norm inequalities for certain maximal and integral operators*, Harmonic Analysis, Lecture Notes in Math., vol. 908, Springer, Berlin, 1982, pp. 102–127.
6. Sawyer E., *Norm inequalities relating singular integrals and the maximal function*, Studia Math. **75** (1983), no. 3, 253–263.
7. Sawyer E., *Weighted Lebesgue and Lorentz norm inequalities for the Hardy operator*, Trans. Amer. Math. Soc. **281** (1984), no. 1, 329–337.

MATH. DEPT.
RUTGERS UNIVERSITY
NEW BRUNSWICK
N.J. 08903, USA

10.6
old

THE NORM OF THE RIESZ PROJECTION

I. È. VERBITSKY, N. YA. KRUPNIK

The operator of the harmonic conjugation S and the Riesz projection P (i.e. the orthogonal projection onto H^2 in $L^2(\mathbb{T})$) are connected by the simple formula $S = 2P - I$. It has been proved in [1] that

$$(1)\quad |S|_{L^p} = \cot\frac{\pi}{2p}\ (p = 2^n),\quad |S|_{L^p} \geqslant \cot\frac{\pi}{2r},\quad |P|_{L^p} \geqslant \sin^{-1}\frac{\pi}{p}\quad (r = \max(p, p')).$$

In [1] it has been also conjectured that the inequalities in (1) can be replaced by equalities. In the case of operator S this conjecture has been proved in [2,3], but for P the question remains open. The following refinement of the main inequality of [2] has been obtained in [4]:

$$(2)\qquad \frac{1}{\cos\pi/2r}\|\operatorname{Im} h\|_{L^p} \leqslant \|h\|_{H^p} \leqslant \frac{1}{\sin\pi/2r}\|\operatorname{Re} h\|_{L^p},$$

where $h \in H^p$, $\operatorname{Im} h(0) = 0$. The right-hand side of (2) gives the norm of the restriction of P onto the space of all real-valued functions in L^p satisfying $\hat{f}(0) = 0$.

The same situation occurs for the weighted L^p spaces

$$L^p(w) = \left\{ f\colon \int_{\mathbb{T}} |f|^p w\, dm < +\infty \right\},\quad w(t) = |t - t_0|^\beta,$$

where $t_0 \in \mathbb{T}$, $-1 < \beta < p - 1$. The formula for the norm of S in $L^p(w)$ has been obtained in [5]. For P it is known only that (see [6])

$$(3)\qquad |P| \geqslant \sin^{-1}\pi/r,\quad r = \max(p, p', p/(1+\beta), p/(p-1-\beta)).$$

CONJECTURE. $|P| = \dfrac{1}{\sin\frac{\pi}{r}}$.

The Conjecture holds for $p = 2$ because in this case the problem can be reduced to the calculation of the norm of the Hilbert matrix $\left\{\frac{1}{j+k+\lambda}\right\}_{j,k\geqslant 0}$. Here is a

Sketch of the proof. Let

$$a(t) = t^{-\beta/2},\ a_+(t) = (1-t)^{\beta/2}\quad (t \in \mathbb{T},\ -1 < \beta < 1)$$

and $T_a = Pa \mid H^2$. It is known [6] that the Toeplitz operator T_a is invertible in H^2 and $T_a^{-1} = a_+ P \bar{a}_+^{-1}$. Consequently

$$|T_a^{-1}| = |\, a_+ P \bar{a}_+^{-1}|_{L^2} = |\, P|_{L^2(\rho)}.$$

The operator T_a is invertible and $|a| = 1$, therefore (see [7]) $\|T_a^{-1}\|^{-2} = 1 - \| PaQ\|^2$. Here $Q = 1 - P$ and $PaQ = (\hat{a}(j+k+1))_{j,k\geqslant 0}$ is a Hankel operator. Let us note that

$$\hat{a}(k+1) = \frac{\exp(\pi i\beta/2)}{\pi}\,\frac{\sin \pi\beta/2}{k+1+\beta/2}.$$

It is known [8] that the norm of the matrix $\left\{\frac{1}{j+k+\lambda}\right\}_{j,k\geqslant 0}$ equals π if $\lambda > 1/2$. We have $\|PaQ\| = |\sin \pi\beta/2|$, $\|P\|_{L^2(\rho)} = \cos^{-1} \pi\beta/2 = \sin^{-1} \pi/r$. $\square$

Let Γ be a simple closed oriented Lyapunov curve; $t_1, \dots, t_n$ be points on Γ, $\|P\|_{\mathrm{ess}}$ ($\|P\|_{\mathrm{ess}}^{(k)}$) be the essential norm of P in the space $L^p(\Gamma, \rho)$ ($L^p(\Gamma, \rho_k)$) on Γ with the weight $\rho(T) = \Pi|t - t_k|^{\beta_k}$ ($\rho_k(t) = |t - t_k|^{\beta_k}$); here $-1 < \beta_k < p - 1$, $1 < p < \infty$. In [6] it was proved that $\|P\|_{\mathrm{ess}} \geqslant \max \gamma(p, \beta_k)$ ($\gamma(p,\beta) \overset{\mathrm{def}}{=} \sin^{-1} \pi/r$, r being defined by (3)). Then in [5] it was proved that $\|P\|_{\mathrm{ess}} = \max_k \| P\|_{\mathrm{ess}}^{(k)}$. If our Conjecture is true then $\|P\|_{\mathrm{ess}} = \max_k \gamma(p, \beta_k)$.

In conclusion we note that in the space L^p on the circle $\mathbb{T}$ (without weight) $\|P\|_{\mathrm{ess}} = \| P\|$ ([3]). But in general the norm $\|P\|$ depends on the weight and on the contour Γ ([3], [5]).

References

1. Gohberg I. Ts., Krupnik N. Ya., *Norm of the Hilbert transformation in the space L^p*, Funktsional. Anal. i Prilozhen. **2** (1968), no. 2, 91–92 (Russian); English transl. in Functional. Anal. Appl. **2** (1968), no. 2, 180–181.
2. Pichorides S. K., *On the best values of the constants in the theorems of M. Riesz, Zygmund and Kolmogorov*, Studia Math. **44** (1972), no. 2, 165–179.
3. Krupnik N. Ya.,Polonskii E. P., *On the norm of an operator of singular integration*, Funktsional. Anal. i Prilozhen. **9** (1975), no. 4, 73–74 (Russian); English transl. in Finctional. Anal. Appl. **9** (1975), no. 4, 337–339.
4. Verbitsky I. È., *Estimate of the norm of a function in a Hardy space in terms of the norms of its real and imaginary parts*, In: Matem. issledovaniya, Shtiintsa, Kishinev, 1980, pp. 16–20. (Russian)
5. Verbitsky I. È., Krupnik N. Ya., *Sharp constants in the theorems of K. I. Babenko and B. V. Khvedelidze on the boundedness of a singular operator*, Soobshch. Akad. Nauk Gruz.SSR **85** (1977), no. 1, 21–24. (Russian)
6. Gohberg I. C., Krupnik N. Ya., *Einführung in die Theorie der eindimensionalen singulären Integraloperatoren*, Birkhäuser Verlag, 1979 (transl. from Russian).
7. Nikol'skii N. K., *Treatise on the Shift Operator*, Springer-Verlag, 1986 (transl. from Russian).
8. Hardy G. H., Littlewood J. E., Pólya G., *Inequalities*, 2nd ed., Cambridge Univ.Press, London and New York, 1952.

Department of Mathematics
Wayne State University
Detroit, MI 48202
USA

Department of Mathematics
Bar Ilan University
Ramat Gan
Israel

10.7
old

IS THIS OPERATOR INVERTIBLE?

STEPHEN SEMMES

Let G denote the group of increasing locally absolutely continuous homeomorphisms h of $\mathbb{R}$ onto itself such that h' lies in the Muckenhoupt class A^∞ of weights. Let V_h denote the operator defined by $V_h(f) = f \circ h$, so that V_h is bounded on BMO($\mathbb{R}$) if and only if $h \in G$ (Jones [3]). Suppose that P is the usual projection of BMO onto BMOA. *For which $h \in G$ is it true that there exists a $c > 0$ such that $\|PV_h(f)\|_{\text{BMO}} \geqslant c\|f\|_{\text{BMO}}$ for all $f \in$BMOA? Is this true for all $h \in G$?*

This question asks about a quantitative version of the notion that a direction-preserving homeomorphism cannot take a function of analytic type to one of antianalytic type. For nice functions and homeomorphisms this can be proved using the argument principle, but there are examples where it fails; see Garnett–O'Farell [2].

We should point out that the natural predual formulation of this problem takes place not on H^1 but on $H^1_1 = \{f\colon f' \in H^1\}$ because V_h is self-adjoint with respect to the pairing $<f,g> = \int \overline{fg'}\,dx$. This also has the advantage of working with analytic functions whose boundary values trace a rectifiable curve.

An equivalent reformulation of the problem is to ask when $H + V_h^{-1} H V_h$ is invertible on BMO, if H denotes the Hilbert transform. This question is related to certain conformal mapping estimates; see the proof of Theorem 2 in [1]. In particular, it is shown there that this operator is invertible if $\|\log h'\|_{\text{BMO}}$ is small enough.

REFERENCES

1. David G., *Courbes corde-arc et espaces de Hardy généralisés*, Ann. Inst. Fourier (Grenoble) **32** (1982), 227–239.
2. Garnett J., O'Farell A., *Sobolev approximation by a sum of subalgebras on the circle*, Pacific J. Math. **65** (1976), 55–63.
3. Jones P., *Homeomorphisms of the line which preserve BMO*, Ark. Mat. **21** (1983), no. 2, 229–231.

DEPARTMENT OF MATHEMATICS
WIESS SCHOOL OF NATURAL SCIENCES
RICE UNIVERSITY
P.O. BOX 1892
HOUSTON, TEXAS 77251
USA

COMMENTARY BY C. BISHOP

The question "Is it true for all $h \in G$" is answered in the negative by Charles Moore in his UCLA thesis [4] and independently by myself in [5]. I construct a biLipshitz homeomorphism h of the line and a nonconstant, bounded, continuous function f in BMOA such that $\overline{f \circ h}$ is in BMOA (i.e., the operator in question has nontrivial kernel).

References

4. Moore C. N., *Some Applications of Cauchy Integrals on Curves*, University of California, Los Angeles, 1986.
5. Bishop C. J., *A counterexample in conformal welding concerning Hausdorff dimension*, Michigan Math. J. **35** (1988), 151–159.

Department of Mathematics
State University of New York
at Stony Brook
Stony Brook, NY11794
USA

10.8
old

AN ESTIMATE OF BMO NORM IN TERMS OF AN OPERATOR NORM

RICHARD ROCHBERG

Let b be a function in $\mathrm{BMO}(\mathbb{R}^n)$ with norm $\|b\|_*$ and let K be a Calderon–Zygmund singular integral operator acting on $L^2(\mathbb{R}^n)$. Define K_b by $K_b(f) = e^{-b}K(e^b f)$. The theory of weighted norm inequalities insures that K_b is bounded on L^2 if $\|b\|_*$ is small. In fact the map of b to K_b is an analytic map of a neighborhood of the origin in BMO into the space of bounded operators (for instance, by the argument on p.611 of [3]). Much less is known in the opposite direction.

QUESTION. *Given b, K; if $|K - K_b|$ is small, must $\|b\|_*$ be small?*

The hypothesis is enough to insure that $\|b\|_*$ is finite but the naive estimates are in terms of $|K| + |K_b|$.

If $n = 1$ and K is the Hilbert transform then the answer is yes. This follows from the careful analysis of the Helson–Szegö theorem given by Cotlar, Sadosky, and Arocena (see, e.g. Corollary (III.d) of [1]).

A similar question can be asked in more general contexts, for instance with the weighted projections of [2]. In that context one would hope to estimate the operator norm of the commutator $[M_b, P]$ (defined by $[M_b, P](f) = bPf - P(bf)$) in terms of the operator norm of $P - P_b$.

REFERENCES

1. Arocena R., *A refinement of the Helson–Szegö theorem and the determination of the extremal measures*, Studia Math. **LXXI** (1981), 203–221.
2. Coifman R., Rochberg R., *Projections in weighted spaces, skew projections, and inversion of Toeplitz operators*, Integral Equations and Operator Theory **5** (1982), 145–159.
3. Coifman R., Rochberg R., Weiss G., *Factorization theorems for Hardy spaces in several variables*, Ann. Math. **103** (1976), 611–635.

DEPARTMENT OF MATHEMATICS
WASHINGTON UNIVERSITY
ST LOUIS, MO 63130
USA

10.9
old

EQUIVALENT NORMS IN H^p

PETER W. JONES

Let H^p denote the real variables Hardy space on $\mathbb{R}^n$. Let K_j be a Fourier multiplier operator whose symbol Θ_j is $C^\infty(\mathbb{R}^n \setminus \{0\})$ and homogeneous of degree zero. *For which families* $\{K_j\}_{j=1}^m$ *is it true that*

$$\|f\|_{H^p} \sim \sum_{j=1}^{m} \|K_j f\|_{L^p}$$

for all $f \in H^p \cap L^2$? This problem was solved for $p = 1$ in [1] and the results were extended in [2] to the case where p is only slightly less than one. A subproblem is to decide whether the above equivalence holds for all $p < 1$ when the family consists of the identity operator and the first order Riesz kernels. See [3] for related results.

REFERENCES

1. Uchiyama A., *A constructive proof of the Fefferman–Stein decomposition of* BMO($\mathbb{R}^n$), Acta Math. **148** (1982), 215–241.
2. Uchiyama A., *The Fefferman–Stein decomposition of smooth functions and its application to* $H^p(\mathbb{R}^n)$, University of Chicago, Ph.D. thesis, 1982.
3. Calderón A. P., Zygmund A., *On higher gradients of harmonic functions*, Studia Math. **24** (1964), 211–226.

DEPARTMENT OF MATHEMATICS
YALE UNIVERSITY
BOX 2155 YALE STATION
NEW HAVEN, CT 06520
USA

10.10

THE HARDY SPACES $H^p(\mathbb{R}^d)$ AND APPROXIMATION IN $L^p(\mathbb{R}^d)$ FOR $p < 1$

A. B. Aleksandrov, P. P. Kargaev

Let u be a harmonic function in the half-space $\mathbb{R}_+^{d+1} \stackrel{\text{def}}{=} \{(t,x) : t > 0, x \in \mathbb{R}^d\}$. By definition the function u belongs to $H^p(\mathbb{R}^d)$ $(0 < p < +\infty)$ if the non-tangential maximal function Mu belongs to $L^p(\mathbb{R}^d)$, $\|u\|_{H^p} \stackrel{\text{def}}{=} \|Mu\|_{L^p}$ (see [1] for details). Denote by u^* the boundary values of u, $u^*(x) \stackrel{\text{def}}{=} \lim_{t\to 0} u(t,x)$. Let us define the operator $T\colon H^p(\mathbb{R}^d) \to (L^p(\mathbb{R}^d))^{d+1}$, $Tu \stackrel{\text{def}}{=} (u^*, (R_1u)^*, \ldots, (R_du)^*)$, where R_j $(1 \leqslant j \leqslant d)$ are the Riesz transforms. It is well known that $\operatorname{Ker} T = \{0\}$ for $p \geqslant \frac{d-1}{d}$ and T is an isomorphic embedding for $p > \frac{d-1}{d}$ (see [2]).

Question 1. *For which numbers p and d is T an isomorphic embedding?*

T. Wolff [3] proved that T is not an isomorphic embedding if p is small enough and $d \geqslant 2$. T. Wolff [3] proved also that T is onto (thus, $\operatorname{Ker} T \neq \{0\}$) for $d = 2$ and for small p. Moreover, [3] contains implicitly the following fact: there exists a number $p(d) \in [0, \frac{d-1}{d}]$ such that T is an onto operator iff $0 < p < p(d)$. The authors proved [4] that $p(d) \geqslant \frac{d-1}{d+1}$.

Problem 1. *Find $p(d)$.*

Conjecture. *$\operatorname{Ker} T = \{0\}$ for $p \geqslant p(d)$ and T is an isomorphic embedding for $p > p(d)$.*

Let $\mathcal{X}(\mathbb{R}^d)$ be the linear span of the family $\left\{\frac{x-a}{|x-a|^{d+1}}\right\}_{a\in\mathbb{R}^d}$ of $\mathbb{R}^d$-valued functions. Denote by $\mathcal{X}^p(\mathbb{R}^d)$ the closure of $\mathcal{X}(\mathbb{R}^d) \cap (L^p(\mathbb{R}^d))^d$ in the space $(L^p(\mathbb{R}^d))^d$.

Theorem 0.2 in [5] (see also [6]) implies the following

Theorem 1. *The following equalities are equivalent:*

1) $T(H^p(\mathbb{R}^d)) = (L^p(\mathbb{R}^d))^{d+1}$;
2) $\mathcal{X}^p(\mathbb{R}^d) = (L^p(\mathbb{R}^d))^d$.

Corollary. *The equality $\mathcal{X}^p(\mathbb{R}^d) = (L^p(\mathbb{R}^d))^d$ holds iff $p < p(d)$.*

If $d = 2$, $\mathcal{X}(\mathbb{R}^2)$ is the $\mathbb{R}$-linear span of the family $\left\{\frac{z-a}{|z-a|^3}\right\}_{a\in\mathbb{C}}$. Let $\mathcal{X}(\mathbb{C})$ denote the $\mathbb{C}$-linear span of this family. Denote by $\mathcal{X}^p(\mathbb{C})$ the closure of $\mathcal{X}^p(\mathbb{C}) \cap L^p(\mathbb{C})$ in the complex space $L^p(\mathbb{C})$. It is clear that $\mathcal{X}^p(\mathbb{C}) = L^p(\mathbb{C})$ for $p < p(2)$. The authors do not know whether $\mathcal{X}^p(\mathbb{C}) \neq L^p(\mathbb{C})$ for all $p \geqslant p(2)$. Results of E. M. Stein and G. Weiss [7] (see also [8]) imply that $\mathcal{X}^p(\mathbb{C}) \neq L^p(\mathbb{C})$ for all $p \geqslant \frac{1}{2}$.

Let $\mathcal{X}_n(\mathbb{C})$ $(n \geqslant 0)$ be the $\mathbb{C}$-linear span of the family $\left\{\frac{(z-a)^n}{|z-a|^{n+2}}\right\}_{a\in\mathbb{C}}$. Denote by $\mathcal{X}_n^p(\mathbb{C})$ the closure of $\mathcal{X}_n(\mathbb{C}) \cap L^p(\mathbb{C})$ in the space $L^p(\mathbb{C})$.

QUESTION 2. *For which numbers p and n does the equality*

$$\mathcal{X}_n^p(\mathbb{C}) = L^p(\mathbb{C}) \tag{1}$$

hold?

It is interesting to note that for even n (1) holds for all $p < 1$. The proof for $n \geqslant 2$ may be obtained from the following equality for $N = \frac{n}{2}$ and $a \in \mathbb{C}$

$$\int_{\mathbb{T}} \frac{(z-\zeta)^{2N}}{|z-\zeta|^{2N+2}} \operatorname{Re}(a\zeta^N)\, dm(\zeta) = \begin{cases} 0, & |z| < 1 \\ \frac{a}{2N} \sum\limits_{k=0}^{N-1} \frac{(-1)^k (2N+k)!\, z^{N-k-1}}{(k+N)!\,(N-k-1)!\,k!\,\bar{z}^{2N+k+1}}, & |z| > 1 \end{cases},$$

where $\mathbb{T}$ is the unit circle in $\mathbb{C}$, m is the normalized Lebesgue measure on $\mathbb{T}$. This equality allows us to prove also a real analogue of (1) (for $\mathbb{R}$-linear spans).

Let Q be a cube in $\mathbb{R}^d$. Set

$$\varphi_Q(x) = \begin{cases} \int_Q \frac{dy}{|x-y|^{d+1}}, & x \notin Q \\ -\int_{\mathbb{R}^d \setminus Q} \frac{dy}{|x-y|^{d+1}}, & x \in Q \end{cases}.$$

Denote by $Y(\mathbb{R}^d)$ the linear span of the family $\{\varphi_Q\}$, where Q runs through the set of all cubes with sides parallel to coordinate axes. Let $Y^p(\mathbb{R}^d)$ be the closure of $Y(\mathbb{R}^d) \cap L^p(\mathbb{R}^d)$ in the space $L^p(\mathbb{R}^d)$. We proved in [4] that

$$Y^p(\mathbb{R}^d) = \{\, u^* : u \in H^p(\mathbb{R}^d),\ (R_j u)^* = 0 \text{ a.e. } (1 \leqslant j \leqslant d)\,\}.$$

This equality implies the following

THEOREM 2. *The following equalities are equivalent:*

1) $T\big(H^p(\mathbb{R}^d)\big) = \big(L^p(\mathbb{R}^d)\big)^{d+1}$;
2) $Y^p(\mathbb{R}^d) = L^p(\mathbb{R}^d)$.

Theorem 2 deals with the approximation by scalar functions, while Theorem 1 deals with the approximation by $\mathbb{R}^d$-valued functions.

Let $X_\alpha(\mathbb{R}^d)$ denote the linear span of the family $\left\{\frac{1}{|x-a|^\alpha}\right\}_{a \in \mathbb{R}^d}$. Denote by $X_\alpha^p(\mathbb{R}^d)$ the closure of $X_\alpha(\mathbb{R}^d) \cap L^p(\mathbb{R}^d)$ in the space $L^p(\mathbb{R}^d)$. It is clear that $X_{d+1}^p(\mathbb{R}^d) \subset Y^p(\mathbb{R}^d)$. Hence, the equality $X_{d+1}^p(\mathbb{R}^d) = L^p(\mathbb{R}^d)$ implies the surjectivity of T. In fact it is easy to prove this implication directly (not using Theorem 2).

QUESTION 3. *Does the surjectivity of T imply $X_{d+1}^p(\mathbb{R}^d) = L^p(\mathbb{R}^d)$?*

QUESTION 4. *For which numbers α, p, d does the equality $X_\alpha^p(\mathbb{R}^d) = L^p(\mathbb{R}^d)$ hold?*

It suffices to consider the case $pd < p\alpha < d$, because $X_\alpha^p(\mathbb{R}^d) = L^p(\mathbb{R}^d)$ for $0 < \alpha \leqslant d$ and $X_\alpha^p(\mathbb{R}^d) = \{0\}$ for $\alpha p \geqslant d$.

THEOREM 3. *Let $pd < p\alpha < d$. Then the following assertions are equivalent:*

1) $X_\alpha^p(\mathbb{R}^d) = L^p(\mathbb{R}^d)$;
2) *there exists a function $f \in X_\alpha(\mathbb{R}^d)$ satisfying the following inequality:*

$$\int_{\mathbb{R}^d} \left(|1 - f(x)|^p - 1\right) dx < 0. \tag{2}$$

Sketch of the proof. 1) $\Longrightarrow$ 2) is trivial. To prove 2) $\Longrightarrow$ 1) it is enough to obtain (2) for some function $g \in X_\alpha^p(\mathbb{R}^d)$. We may construct a discrete measure μ on $\mathbb{R}^d$ with finite support such that $\mu(\mathbb{R}^d) = 0$ and (2) holds for $f * \mu$. By induction we may construct a finite sequence $\{\mu_k\}_{k=1}^N$ of discrete measures on $\mathbb{R}^d$ such that $\mu_k(\mathbb{R}^d) = 0$, μ_k has finite support ($1 \leqslant k \leqslant N$) and (2) holds for $g = f * \mu_1 * \mu_2 * \cdots * \mu_N$. It remains to note that $g \in X_\alpha^p(\mathbb{R}^d)$ if $(\alpha + N)p > d$. $\square$

COROLLARY 1. *If $p < \frac{2d}{\alpha} - 1$, then $X_\alpha^p(\mathbb{R}^d) = L^p(\mathbb{R}^d)$.*

Proof. It suffices to note that $\int_{\mathbb{R}^d} \left(\left|1 - \frac{1}{|x|^\alpha}\right|^p - 1\right) dx < 0$ if $p < \frac{2d}{\alpha} - 1$. $\square$

COROLLARY 2. *The operator T is onto for $p < \frac{d-1}{d+1}$.*

QUESTION 5. *For which numbers α, p, d does the following inequality*

$$\int_{\mathbb{R}^d} \left(\left|1 - \sum_{j=1}^n \lambda_j \frac{1}{|x - a_j|^\alpha}\right|^p - 1\right) dx \geqslant 0 \tag{3}$$

hold for all $n \in \mathbb{N}$, $\lambda_j \in \mathbb{R}$, $a_j \in \mathbb{R}^d$ ($1 \leqslant j \leqslant n$)?

Theorem 3 asserts that (3) is equivalent to $X_\alpha^p(\mathbb{R}^d) \neq L^p(\mathbb{R}^d)$. It is easy to see that there exists a number $p(d, \alpha) \in [0, \frac{d}{\alpha}]$ such that (3) holds iff $p \geqslant p(d, \alpha)$.

PROBLEM 2. *Find $p(d, \alpha)$.*

A positive answer to Question 3 is equivalent to $p(d) = p(d, d+1)$. It should be noted that Question 4, Question 5 and Problem 2 are not obvious (at least for us) even for $d = 1$.*

THEOREM 4 (see [4]). *Suppose that there exists a non-zero function $u \in H^p(\mathbb{R}^d)$ such that $u^* \geqslant 0$ a.e. on $\mathbb{R}^d$ and $(R_j u)^* = 0$ a.e. on $\mathbb{R}^d$ ($1 \leqslant j \leqslant d$). Then $p < \frac{p(d)+1}{2}$.*

The case $d = 1$ is well known (see, for example, [9]). We do not know whether the converse theorem holds for $d \geqslant 2$. In [4] we construct a function u satisfying conditions of Theorem 4 for $p < \frac{d}{d+1}$.

Let T_n denote an analogue of T for the Riesz transforms up to the order n. For example, for $n = 2$ it is natural to consider the operator

$$T_2 \colon H^p(\mathbb{R}^d) \to \left(L^p(\mathbb{R}^d)\right)^{\frac{d(d+3)}{2}},$$

$$T_2 u \stackrel{\text{def}}{=} \left((R_j u)^*, (R_k R_l u)^*\right)_{1 \leqslant j \leqslant d, 1 \leqslant k \leqslant l \leqslant d}.$$

*Recently A. I. Sergeev has proved that $X_\alpha^p(\mathbb{R}) \neq L^p(\mathbb{R})$ for all $\alpha \geqslant 2$, i.e., (3) holds for $d = 1$ and for all $\alpha \geqslant 2$. Consequently $p(1, \alpha) = 0$ for all $\alpha \geqslant 2$ (private communication).

It is well known (see [2]) that $\operatorname{Ker} T_n = \{0\}$ for $p \geqslant \frac{d-1}{d+n-1}$ and T_n is an isomorphic embedding for $p > \frac{d-1}{d+n-1}$.

We do not know the answers to the following questions. *Can T_2 be not an isomorphic embedding? Can T_2 be not one-to-one ? Can T_2 be onto?*

References

1. Fefferman C., Stein E. M., *H^p spaces of several variables*, Acta Math. **129** (1972), 137–193.
2. Stein E. M., Weiss G., *Introduction to Fourier Analysis on Euclidean Spaces*, Princeton University Press, 1971.
3. Wolff T., *Counterexamples with harmonic gradient in $\mathbb{R}^3$*, preprint.
4. Aleksandrov A. B., Kargaev P. P., *Hardy classes of functions harmonic in the half-space*, Algebra i Analiz **5** (1993), no. 2, 1–70. (Russian)
5. Aleksandrov A. B., *Essays on non locally convex Hardy classes*, Lect. Notes Math. **864** (1981), 1–89.
6. Aleksandrov A. B., *Approximation by rational functions and an analog of the M. Riesz theorem on conjugate functions for L^p-spaces with $p \in (0,1)$*, Matem. Sbornik **107(149)** (1978), no. 1(9), 3–19 (Russian); English transl. in Math. USSR – Sbornik **35** (1979), 301–316.
7. Stein E. M., Weiss G., *Generalization of the Cauchy–Riemann equations and representation of the rotation group*, Amer. Math. J. **90** (1968), 163–196.
8. Alekseev R. B., *On a subharmonic property of solutions of generalized Cauchy–Riemann equations*, Vestnik LGU (to appear). (Russian)
9. Garnett J. B., *Bounded Analytic Functions*, Academic Press, 1981.

Steklov Mathematical Institute
St. Petersburg branch
Fontanka 27
St. Petersburg, 191011
Russia

Dept. of Mathematics and Mechanics
St. Petersburg State University
Bibliotechnaya pl. 2
198904, Staryi Peterhof
Russia

10.11

PERMUTATION OF THE HAARSYSTEM

PAUL F. X. MÜLLER

Let us first describe the setting in which we are working. $\mathcal{D}$ denotes the set of all dyadic intervals contained in the unit interval. $\pi\colon \mathcal{D} \to \mathcal{D}$ denotes a permutation of the dyadic intervals. The operator induced by π is determined by the equation

$$T_\pi h_I = h_{\pi(I)}$$

where the h_I are the L_∞ normalized Haar functions. The conjecture to be discussed here concerns "geometric" conditions on π which are necessary **and** sufficient for the boundedness of T_π on dyadic BMO.

For special types of permutations the following is known (see [M] and [S]):

If for every $I \in \mathcal{D}$ we have $|I| = |\pi(I)|$ then $T_\pi\colon BMO \to BMO$ is an isomorphism if and only if there exist $K > 0$ such that for every $\mathcal{B} \subset \mathcal{D}$

$$\frac{|\mathcal{B}^*|}{K} \leqslant |\pi(\mathcal{B})^*| \leqslant K|\mathcal{B}^*|$$

where $\mathcal{B}^*$ denotes the point set which covered by $\mathcal{B}$.

In order to study general permutations we need to consider a scale invariant measure for the size of collections of dyadic intervals $\mathcal{B}$.

The following sufficient condition on general permutations can be obtained:

If there exists $K > 0$ such that for every $\mathcal{B} \subset \mathcal{D}$:

$$\frac{1}{K}\frac{1}{|\pi(\mathcal{B})^*|}\sum_{J\in\mathcal{B}}|\pi(J)| \leqslant \frac{1}{|\mathcal{B}^*|}\sum_{J\in\mathcal{B}}|J| \leqslant \frac{K}{|\pi(\mathcal{B})^*|}\sum_{J\in\mathcal{B}}|\pi(J)|$$

then $T_\pi\colon BMO \to BMO$ is an isomorphism.

Examples show, that the above is not a necessary condition. It is however **related** to a necessary one:

If we let

$$CC(\mathcal{B}) = \sup_{I\in\mathcal{B}} \frac{1}{|I|} \sum_{\substack{J\subseteq I\\ J\in\mathcal{B}}} |J|$$

and if $T_\pi\colon BMO \to BMO$ is an isomorphism then for every $\mathcal{B} \subset \mathcal{D}$

$$\frac{CC(\mathcal{B})}{||T_\pi^{-1}||^2} \leqslant CC\big(\pi(\mathcal{B})\big) \leqslant ||T_\pi||^2 CC(\mathcal{B}).$$

This leads to the following

PROBLEM. *Is it true that $T_\pi\colon BMO \to BMO$ is an isomorphism if and only if there exists $K > 0$ such that for every $\mathcal{B} \subset \mathcal{D}$*

$$\frac{1}{K}CC(\mathcal{B}) \leqslant CC\big(\pi(\mathcal{B})\big) \leqslant K\, CC(\mathcal{B})?$$

References

[M] Müller P. F. X., *Permutations of the Haar system*, in GAFA Seminar (V. Milman & J. Lindenstrauss, eds.), Lect. Notes Math., vol. 1469, Springer-Verlag, 1991, pp. 125–126.

[S] Schipp F., *On equivalence of rearrangements of the Haar system in dyadic Hardy and BMO spaces*, Analytic Mathematica **16** (1990), 135–143.

Institut für Mathematik
J. Kepler Universität
A-4040 Linz
Austria
email: K318290@alijku11.bitnet

10.12
old

A SUBSTITUTE FOR THE WEAK TYPE (1,1) INEQUALITY FOR MULTIPLE RIESZ PROJECTIONS

S. V. Kisliakov

Let $C_A(\mathbb{T}^n)$ denote the polydisc algebra, i.e. the subspace of $C(\mathbb{T}^n)$ consisting of the restrictions to the n-dimensional torus $\mathbb{T}^n$ of functions analytic in the open polydisc $\mathbb{D}^n$ and continuous in $\operatorname{clos}\mathbb{D}^n$. By $H^2(\mathbb{T}^n)$ we denote the closure of $C_A(\mathbb{T}^n)$ in $L^2(\mathbb{T}^n)$ and by $i\colon C_A(\mathbb{T}^n) \to H^2(\mathbb{T}^n)$ the identity operator. The space $H^2(\mathbb{T}^n)^*$ will be identified with $\overline{H^2(\mathbb{T}^n)}$, the bar standing for the complex conjugation (we use throughout the duality established by the pairing $<f,g> = \int_{\mathbb{T}^n} f(\Theta)g(\Theta)\,d\Theta$).

Problem 1. *Does there exist a positive function φ on $(0,1]$ with $\varphi(\varepsilon) \searrow 0$ as $\varepsilon \searrow 0$ such that for each $g \in \overline{H^2(\mathbb{T}^n)}$ with $\|g\|_2 = 1$ the following inequality holds:*

$$\|g\|_1 \leqslant \varphi(\|i^* g\|_{C_A(\mathbb{T}^n)^*})? \tag{1}$$

(throughout $\|\cdot\|_p$ denotes the L^p-norm).

If $n = 1$ the answer is evidently "yes".

Indeed, in this case the Riesz projection $\mathbb{P}_-$ (i.e. the orthogonal projection of $L^2(\mathbb{T}^1)$ onto $\overline{H^2(\mathbb{T}^1)}$) is of weak type (1,1) and so the above function g satisfies the estimate

$$m\{|g| > t\} \leqslant \frac{\text{const}}{t} \|i^* g\|_{C_A(\mathbb{T}^1)^*}.$$

Using this and $\|g\|_2 = 1$ it can be shown by means of a simple calculation that (1) holds with $\varphi(\varepsilon) = \text{const}\,\varepsilon(1 + \log \varepsilon^{-1})$ (and moreover, for all p, $1 < p \leqslant 2$ we have $\|g\|_p \leqslant c(p-1)^{-1}\|i^* g\|^\theta$ with θ given by $p^{-1} = \theta + (1-\theta)/2$).

For $n \geqslant 2$ the orthogonal projection of $L^2(\mathbb{T}^n)$ onto $\overline{H^2(\mathbb{T}^n)}$ (which is nothing else as the n-fold tensor product $\mathbb{P}_- \otimes \cdots \otimes \mathbb{P}_-$) is no longer of weak type (1,1) and the above argument fails. Nevertheless for $n = 2$ Problem 1 also has a positive solution. This was proved by the author [1] with φ a power function. Using the same idea as in [1] but more careful calculations it can be shown that for $n = 2$ and $1 < p < 2$ we have

$$\|g\|_p \leqslant \text{const}\,(p-1)^{-2}\|i^* g\|^\theta$$

(Θ is the same as for $n = 1$) provided $g \in \overline{H^2(\mathbb{T}^2)}$ and $\|g\|_2 = 1$. Consequently, (1) holds for $n = 2$ with $\varphi(\varepsilon) = \text{const}\,\varepsilon(1+\log\varepsilon^{-1})^2$ (to see this, set $p = 1+(\log\|i^* g\|^{-1})$ in the preceding inequality).

Nothing is known for $n > 2$. It seems plausible that for such n Problem 1 should also have a positive solution. Moreover, I think that for $1 < p \leqslant 2$ the inequality $\|g\|_p \leqslant c_n(p-1)^n\|i^* g\|^\theta$ should be true (and so (1) should hold with $\varphi(\varepsilon) = c_n\varepsilon(1+(\log\varepsilon^{-1})^n)$).

Estimate (1) for $n = 2$ was used in [1] to carry over from $C_A(\mathbb{T})$ to $C_A(\mathbb{T}^2)$ some results whose standard proofs for $C_A(\mathbb{T})$ use the weak type (1,1) inequality for $\mathbb{P}_-$. (For example, it was established in [1] that, given a Λ_2-subset E of $(\mathbb{Z}_+)^2$, the operator S, $Sf = \{\hat{f}(n)\}_{n\in E}$ maps $C_A(\mathbb{T}^2)$ onto $\ell^2(E)$. It is still unknown if the same is true with $\mathbb{T}$ and $(\mathbb{Z}_+)^2$ replaced by $\mathbb{T}^n$ and $(\mathbb{Z}_+)^n$, $n \geqslant 2$.) So (1) is really a substitute for the weak type inequality. Profound generalizations of inequality (1) for $n = 2$ with very interesting applications can be found in [2] (some of these applications are quoted in Commentary to Problem S.1 in the 1984 edition).

The proof of (1) for $n = 2$ in [1] is based on the weak type (1,1) inequality for $\mathbb{P}_-$ and a complex variable trick, and essentially the same trick appears in [2]. Investigating the case $n > 2$ one may seek a more complicated trick that also involves analyticity. But to seek a purely real variable proof is probably more promising from different viewpoints. The solution of the following problem might be the first step in this direction.

PROBLEM 2. *Find a real-variable proof of* (1) *for* $n = 2$.

In connection with Problem 2 we formulate another problem which is also rather vague but probably clarifies what is meant in the former. The inequality (1) is clearly equivalent to the following one:

$$\|(\mathbb{P}_- \otimes \cdots \otimes \mathbb{P}_-)h\|_1 \leqslant \varphi(\|h\|_1)$$

provided $h \in L^1(\mathbb{T}^n)$ and $\|(\mathbb{P}_- \otimes \cdots \otimes \mathbb{P}_-)h\|_2 = 1$.

PROBLEM 3. *At least for* $n = 2$, *find and prove a "right" analog of the above inequality involving* n*-fold tensor products of operators of the form* $f \mapsto f * (\mu_i + K_i)$, μ_i *being a measure on a multidimensional torus and* K_i *being a Calderon–Zygmund kernel on the same torus, rather than tensor products of Riesz projections.*

REFERENCES

1. Kisliakov S. V., *Fourier coefficients of boundary values of functions analytic in the disc and in the bidisc*, Trudy Mat. Inst. Steklov **155** (1981), 77–94 (Russian); English transl. in Proc. Steklov Inst. Math. **155** (1983), 75–92.
2. Bourgain J., *Extensions of H^∞-valued functions and bounded bianalytic functions*, preprint, 1982.

STEKLOV MATHEMATICAL INSTITUTE
ST. PETERSBURG BRANCH
FONTANKA 27
ST. PETERSBURG, 191011
RUSSIA

10.13

EXTENSION OF OPERATORS BOUNDED IN THE WEAK L^1 QUASINORM

PETER SJÖGREN

This problem is taken from the author's paper [1]. For simplicity, we consider only the case of $\mathbb{R}$, although the problem makes sense in any locally compact group.

Weak L^1, denoted $L^{1,\infty}$, is the space of measurable functions f on $\mathbb{R}$ verifying

$$\left|\{ x : |f(x)| > \alpha \}\right| \leqslant C/\alpha$$

for all $\alpha > 0$ and some constant C. Here the outer vertical bars denote Lebesgue measure. The smallest admissible value of C is the $L^{1,\infty}$ quasinorm of f, denoted $\| f \|_{1,\infty}$. Notice that L^1 is a subspace of $L^{1,\infty}$ which is not dense.

Now assume that S is a linear operator $L^1 \to L^{1,\infty}$ commuting with translation and bounded for the $L^{1,\infty}$ quasinorm. This boundedness means that

$$\| Sf \|_{1,\infty} \leqslant C \| f \|_{1,\infty}$$

for all $f \in L^1$, where C is a constant.

PROBLEM. *Does any such operator S have a bounded linear extension*

$$S' : L^{1,\infty} \to L^{1,\infty}$$

which also commutes with translations?

As explained in [1], a good deal can be said about S. In fact, S is a convolution operator and given by $Sf = \mu \star f$ for $f \in L^1$, for some finite measure μ. Moreover, μ is discrete, i.e., given by

$$\mu = \sum m_i \delta_{t_i},$$

where the t_i are distinct points of $\mathbb{R}$ and $m = (m_i)$ is a sequence in ℓ^1. In particular, S actually maps L^1 into itself.

When the sequence m is in $\ell \log \ell$, which means that

$$\sum |m_i| \log(2 + |m_i|^{-1}) < \infty,$$

S can trivially be extended. This is because then the convolution $|\mu| \star |f|$ is finite almost everywhere and belongs to $L^{1,\infty}$. Indeed, the addition theorem of weak L^1 allows us to sum the multiples of translates of $|f|$ which make up this convolution. In [1], it is proved that actually $m \in \ell \log \ell$ if the mass positions t_i are linearly independent over $\mathbb{Q}$, and also if all the masses m_i are positive.

If neither of these last conditions is satisfied, it may happen that m is not in $\ell \log \ell$. Ref. [1] contains an example of this, where one can define the extension S' by grouping together certain of the point masses and exploiting the cancellation that will occur within the corresponding parts of the convolution $\mu \star f$, for $f \in L^{1,\infty}$. Notice that for this cancellation to be arithmetically possible, the mass positions have to be linearly dependent over $\mathbb{Q}$.

The problem amounts to finding out whether masses can be grouped together in the general case, in such a way that there will be enough cancellation. Another reasonable approach would be to use more abstract, functional analytic methods to construct the extension S'.

Reference

1. Sjögren P., *Translation-invariant operators on weak* L^1, J. Funct. Anal. **89** (1990), 410-427.

Department of Mathematics
Chalmers University of Technology
and
University of Göteborg
S-412 96 Göteborg
Sweden

10.14
v.old

SOME OPEN PROBLEMS CONCERNING H^∞ AND BMO

JOHN GARNETT

1. *An interpolating Blaschke product* is a Blaschke product having distinct zeros which lie on an H^∞ interpolating sequence. *Is H^∞ the uniformly closed linear span of the interpolating Blaschke products?* See [1], [2]. It is known that the interpolating Blaschke products separate the points of the maximal ideal space (Peter Jones, thesis University of California, Los Angeles 1978).

2. Let φ be a real locally integrable function on $\mathbb{R}$. Assume that for every interval

$$m(\{x \in I : |\varphi(x) - \varphi_I| > \lambda\}) \leqslant Ce^{-\lambda}|I|,$$

where φ_I is the mean value of φ over I, and where C is a constant. *Does it follow that $\varphi = u + Hv$, where $u \in L^\infty$ and $\|v\|_\infty \leqslant \pi/2$?* ($H$ denotes the Hilbert transform). This is the limiting case of the equivalence of the Muckenhoupt (A_2) condition with the condition of Helson and Szegö. See [3] and [4]. This question is due to Peter Jones. A positive solution should have several applications.

3. Let f be a function of bounded mean oscillation on $\mathbb{R}$. *Construct L^∞ functions u and v so that $f = u + Hv$, $\|u\|_\infty + \|v\|_\infty \leqslant C\|f\|_{BMO}$ with C a constant not depending on f.* See [5] and [6].

4. Let $T_1, T_2, \ldots, T_n$ be singular integral operators on $\mathbb{R}^n$. See [7]. *Find necessary and sufficient conditions on $\{T_1, T_2, \ldots, T_n\}$ such that $f \in H^1(\mathbb{R}^n)$ if and only if $|f| + \sum_{j=1}^n |T_j f| \in L^1(\mathbb{R}^n)$.* See [5] and [8].

REFERENCES

1. Marshall D., *Blaschke products generate H^∞*, Bull. Amer. Math. Soc. **82** (1976), 494–496.
2. Marshall D., *Subalgebras of L^∞ containing H^∞*, Acta Math. **137** (1976), 91–98.
3. Hunt R. A., Muckenhoupt B., Wheeden R. L., *Weighted norm inequalities for the conjugate function and Hilbert transform*, Trans. Amer. Math. Soc. **176** (1973), 227-251.
4. Helson H., Szegö G., *A problem in prediction theory*, Ann. Math. Pure Appl. **51** (1960), 107–138.
5. Fefferman C., Stein E. M., *H^p spaces of several variables*, Acta Math. **129** (1972), 137–193.
6. Carleson L., *Two remarks on H^1 and BMO*, Advances in Math. **22** (1976), 269–277.
7. Stein E. M., *Singular integrals and differentiability properties of functions*, Princeton N. J., 1970.
8. Janson S., *Characterization of H^1 by singular integral transforms on martingales and $\mathbb{R}^n$*, Math. Scand. **41** (1977), 140–152.

UNIVERSITY OF CALIFORNIA
LOS ANGELES, CALIFORNIA
90024 U.S.A.

COMMENTARY BY THE AUTHOR

QUESTION 2 has been answered in the negative by T. Wolff [9].

QUESTION 3 has been solved by P. Jones [10]. Other constructive (and more explicit) decompositions were given later in [11], [12] and [13]. One more constructive decomposition of BMO functions can be obtained from a remarkable paper [14]. See also [15], [16].

QUESTION 4 has the following answer found by A. Uchiyama in [12] (he obtained a more general result). Let $T_j f = K_j \star f$, $1 \leqslant j \leqslant m$, M_j be the Fourier transform of K_j. Suppose M_j are homogeneous of degree zero and C^∞ on the unit sphere S^{n-1} of $\mathbb{R}^n$. Then

$$f \in H^1(\mathbb{R}^n) \iff \sum_{j=1}^{m} |T_j f| \in L^1(\mathbb{R}^n)$$

if and only if the matrix

$$\begin{pmatrix} M_1(\xi), & \dots, & M_m(\xi) \\ M_1(-\xi), & \dots, & M_m(-\xi) \end{pmatrix}$$

if of rank 2 everywhere on S^{n-1}. The "only if" part is essentially due to S. Janson [8]. In particular, $f \in H^1(\mathbb{R}^n) \iff |f| + \sum_{j=1}^{m} |T_j f| \in L^1(\mathbb{R}^n)$ iff for any $\xi \in S^{n-1}$ there exists j such that $M_j(\xi) \neq M_j(-\xi)$.

In connection with this result see also Problem 10.9.

NOTE

It is well known that all Blaschke products span a dense subset of H^∞. In [17] it is shown that every Blaschke product whose zeros accumulate at a closed subset E of $\mathbb{T}$ with length 0, is in the uniform algebra qenerated by the interpolating Blaschke products.

REFERENCES

9. Wolff T., *Counterexamples to two variants of the Helson–Szegö theorem*, preprint, vol. 11, Institut Mittag–Leffler, 1983.
10. Jones P., *Carleson measures and the Feffermann–Stein decomposition of $BMO(\mathbb{R})$*, Ann. of Math. **111** (1980), 197–208.
11. Jones P., *L^∞-estimates for the $\overline{\partial}$-problem in a half-plane*, Acta Math. **150** (1983), no. 1-2, 137–152.
12. Uchiyama A., *A constructive proof of the Fefferman–Stein decompositions of $BMO(\mathbb{R}^n)$*, Acta Math. **148** (1982), 215–241.
13. Stray A., *Two applications of the Schur–Nevanlinna algorithm*, Pacif. J. of Math. **91** (1980), no. 1, 223–232.
14. Rubio de Francia J. L., *Factorization and extrapolation by weights*, Bull. Amer. Math. Soc. **7** (1982), no. 2, 393–395.
15. Amar E., *Représentation des fonctions de BMO et solutions de l'équation $\overline{\partial}_b$*, preprint, Univ. Paris XI Orsay, 1978.
16. Coifman R., Jones P. W., Rubio de Francia J. L., *Constructive decomposition of BMO functions and factorization of A_p weights*, Proc. Amer. Math. Soc. **87** (1983), no. 4, 675–680.
17. Marshall D., Stray A., *Interpolating Blaschke products*, Pacif. J. Math. (to appear).

10.15
v.old

TWO CONJECTURES BY ALBERT BAERNSTEIN

ALBERT BAERNSTEIN

In [1] I proved a factorization theorem for zero-free univalent functions in the unit disk $\mathbb{D}$. Let S_0 denote the set of all functions F analytic and 1-1 in $\mathbb{D}$ with $0 \notin F(\mathbb{D})$, $F(0) = 1$.

THEOREM 1. *If $F \in S_0$, then, for each λ, $\lambda \in (0,1)$, there exist functions B and Q analytic in $\mathbb{D}$ such that*

$$F(z)^\lambda = B(z)Q(z), \qquad z \in \mathbb{D},$$

where $B \in H^\infty$, $1/B \in H^\infty$, and $|\arg Q| < \pi$.

The "Koebe function" for the class S_0 is $K(z) = \left(\frac{1+z}{1-z}\right)^2$ which maps $\mathbb{D}$ onto the slit plane $\{ w \in \mathbb{C} : |\arg w| < \pi \}$. This suggests that it might be possible to let $\lambda \to 1$ in Theorem 1.

CONJECTURE 1. *If $F \in S_0$, then there exist functions B and Q analytic in $\mathbb{D}$ such that*

$$F(z) = B(z)Q(z), \qquad z \in \mathbb{D},$$

where $B \in B^\infty$, $1/B \in H^\infty$, and $|\arg Q| < \pi$.

We do not insist that B or Q be univalent, nor that $Q(0) = 1$. However, when the functions are adjusted so that $|Q(0)| = 1$, then $\|B\|_\infty$ and $\|B^{-1}\|_\infty$ should be bounded independently of F.

Using the fact that $Q^{1/2}$ has positive real part, it is easy to show that the power series coefficients $\{a_n\}$ of Q satisfy $|a_n| \leqslant 4n$, $n \geqslant 1$, with equality when $Q(z) = K(z)$. *Littlewood's Conjecture* asserts that this inequality is true for coefficients of functions in S_0. A proof of Conjecture 1 could possibly tell us something new about how to attempt Littlewood's conjecture, and this in turn might lead to fresh ideas about how to prove (the stronger) Bieberbach's conjecture.

Theorem 1 is easily deduced from a decomposition theorem obtained by combining results of Helson and Szegö [2] and Hunt, Muckenhoupt and Wheeden [3]. Suppose $f \in L^1(\mathbb{T})$ and f real valued. Consider the zero-free analytic function F defined by $F(z) = \exp\big(f(z) + i\tilde{f}(z)\big), z \in \mathbb{D}$, where $f(z)$ denotes the harmonic extension of $f(e^{i\Theta})$ and $\tilde{f}$ the conjugate of f. Also, let $S(F)$ denote the set of all functions obtained by "hyperbolically translating" F and then normalizing,

$$S(F) = \left\{ F\left(\frac{z+a}{1+\bar{a}z}\right) F(a)^{-1} : a \in \mathbb{D} \right\}$$

and let H^p denote the usual Hardy space. Part of Theorem 1 of [3] can be phrased in the following way.

THEOREM 2. *For $f \in L^1(\mathbb{T})$ the following are equivalent.*

(1) $f = u_1 + \tilde{u}_2$, *where* $u_1, u_2 \in L^\infty(\mathbb{T})$ *and* $\|u_2\|_\infty < \frac{\pi}{2}$;

(2) $S(F) \cup S(1/F)$ *is a bounded subset of* H^1.

Theorem 1 follows, since $F^{\lambda/2}$ satisfies (2) when $F \in S_0$ and $0 < \lambda < 1$. □

Theorem 2 may be regarded as a sharpened form of the theorem of Fefferman and Stein [4], which asserts that $f = u_1 + \tilde{u}_2$ for **some** pair of bounded functions if and only if f is of bounded mean oscillation.

To obtain Conjecture 1 in the same fashion as Theorem 1, we need a result like Theorem 2 in which the $< \pi/2$ of (1) is replaced by $\leqslant \pi/2$. Consideration of $F(z) = \frac{1+z}{1-z}$ leads to the following guess.

CONJECTURE 2. *For $f \in L^1(\mathbb{T})$ the following are equivalent.*

(1′) $f = u_1 + \tilde{u}_2$, *where* $u_1, u_2 \in L^\infty(\mathbb{T})$ *and* $\|u_2\|_\infty \leqslant \pi/2$.

(2′) $S(F) \cup S(1/F)$ *is bounded subset of weak* H^1.

Statement (2′) means the following: There is a constant C such that for every t, $t \in R_+$, and every G, $G \in S(F) \cup S(1/F)$

$$m\{\Theta : |G(e^{i\Theta})| > t\} \leqslant Ct^{-1}.$$

It is not hard to prove, using subordination, that (1′) implies (2′). If the implication (2′) $\Longrightarrow$ (1′) is true, then so is Conjecture 1.

Condition (2′) can be restated in a number of equivalent ways. We mention one which is closely related to the subharmonic maximal type function used by the author in [5] and elsewhere.

(2″) *There is a constant C such that*

$$\int_E \left[f\left(\frac{e^{i\Theta} + a}{1 + \bar{a}e^{i\Theta}} \right) - f(a) \right] d\Theta \leqslant \int_{-\frac{1}{2}mE}^{\frac{1}{2}mE} \log \left| \frac{1 + e^{i\Theta}}{1 - e^{i\Theta}} \right| d\Theta + C|E|$$

for every measurable set E, $E \subset \mathbb{T}$, and every a, $a \in \mathbb{D}$.

For $F \in S_0$, Theorem 6 of [5] asserts that (2″) holds with $C = 0$.

In both the Fefferman-Stein and Helson-Szegö theorems the splitting $f = u_1 + \tilde{u}_2$ is accomplished via duality and pure existence proofs from functional analysis. It would be of considerable interest if, given f, $f \in BMO$, one could show how to actually **construct** the bounded functions u_1 and u_2. We remark that if $f \in BMO$ then some constant multiple of f satisfies (2″).

I can prove that (2″) $\Longrightarrow$ (1′) provided we assume also that f is **monotone** on $\mathbb{T}$, i.e., there exist $\Theta_1 < \Theta_2 < \Theta_1 + 2\pi$ such that

$$f(e^{i\Theta}) \downarrow \text{ as } \Theta \uparrow \text{ on } (\Theta_1, \Theta_2) \text{ and } f(e^{i\Theta}) \uparrow \text{ as } \Theta \uparrow \text{ on } (\Theta_2, \Theta_1 + 2\pi).$$

By composing with a suitable Möbius transformation, we may assume $\Theta_1 = 0$, $\Theta_2 = \pi$. Then, when $C = 0$, u_2 can be constructed as follows. Let $\Theta \in (0, \pi)$ and $x \in (-1, 1)$ be related by $(1 + x)(1 - x)^{-1} = |1 + e^{i\Theta}||1 - e^{i\Theta}|^{-1}$. Let V be the harmonic function in $\mathbb{D}$ with boundary values $V(e^{i\Theta}) = f(x)$, $0 < \Theta < \pi$, and $V(e^{-i\Theta}) = V(e^{i\Theta})$. Then it turns out that $|\widetilde{V}| \leqslant \pi/2$ and $f - V = o(1)$, so that $u_2 = -\widetilde{V}$ gives us (1′).

It follows that Conjecture 1 is true for functions F, $F \in S_0$, which map $\mathbb{D}$ onto the complement of a "monotone slit".

References

1. Baernstein A. II., *Univalence and bounded mean oscillation*, Mich. Math J. **23** (1976), 217–223.
2. Helson H., Szegö G., *A problem in prediction theory*, Ann. Mat. Pura Appl. **51** (1960), no. 4, 107–138.
3. Hunt R., Muckenhoupt B., Wheeden R., *Weighted norm inequalities for the conjugate function and Hilbert transform*, Trans. Amer. Math. Soc **176** (1973), 227–251.
4. Fefferman C., Stein E. M., *H^p spaces of several variables*, Acta Math. **129** (1972), 137–193.
5. Baernstein A. II., *Integral means, univalent functions and circular symmetrization*, Acta Math. **133** (1974), 139–169.

Washington University
St. Louis, Missouri 63130
USA

Commentary

Conjecture 2 has been disproved by T. Wolff (see ref. [9] after the Commentary to Problem 10.14).

10.16
v.old

BLASCHKE PRODUCTS IN $\mathcal{B}_0$

DONALD SARASON

The class $\mathcal{B}_0$ consists of those functions f that are holomorphic in $\mathbb{D}$ and satisfy $\lim_{|z|\to 1}(1-|z|)|f'(z)| = 0$. It can be described alternatively as the class of functions in $\mathbb{D}$ that are derivatives of holomorphic functions having boundary values in the Zygmund class (the class of uniformly smooth functions) [1, p.263]. It is a subclass of the class $\mathcal{B}$ of Bloch functions (those holomorphic f in $\mathbb{D}$ satisfying $\sup_{|z|<1}(1-|z|)|f'(z)| < \infty$); see, for example, [2]. It contains $VMOA$, the class of holomorphic functions in $\mathbb{D}$ whose boundary values have vanishing mean oscillation [3]. The class $\mathcal{B}_0 \cap H^\infty$ has an interesting interpretation: it consists of those functions in H^∞ that are constant on each Gleason part of H^∞.

It is not too hard to come up with an example to show that the inclusion $VMOA \subset \mathcal{B}_0$ is proper. Indeed, it is known that λ_* contains functions that are not of bounded variation [1, p. 48]. If u is the Poisson integral of such a function and v is its harmonic conjugate, then the derivative of $u + iv$ will be such an example. In connection with a problem in prediction theory mentioned in [4], I was interested in having an example of a *bounded* function in $\mathcal{B}_0$ which is not in $VMOA$, and that seems somewhat more difficult to obtain. Eventually I realized one can produce such an example on the basis of a result of H. S. Shapiro [5] and J.-P. Kahane [6]. They showed, by rather complicated constructions, that there exist positive singular measures on $\mathbb{T}$ whose indefinite integrals are in λ_*. It is easy to check that the singular inner function associated with such a measure is in $\mathcal{B}_0$. That does it, because the only inner functions in $VMOA$ are the finite Blaschke products.

If f is an inner function in $\mathcal{B}_0$ and $|c| < 1$, then $\frac{f-c}{1-\bar{c}f}$ is also an inner function in $\mathcal{B}_0$, and it is a Blaschke product for "most" values of c. Thus, $\mathcal{B}_0$ contains infinite Blaschke products. I should like to propose THE PROBLEM *of characterizing the Blaschke products in $\mathcal{B}_0$ by means of the distribution of their zeros.* One has the feeling that the zeros of a Blaschke product in $\mathcal{B}_0$ must, in some sense, be "spread smoothly" in $\mathbb{D}$. A natural first step in trying to find the correct condition would be to try to give a direct construction of an infinite Blaschke product in $\mathcal{B}_0$. The only information I can offer on the problem is very meagre: A Blaschke product in $\mathcal{B}_0$ cannot have an isolated singularity on $\mathbb{T}$. The proof, unfortunately , is too involved to indicate here. As a TEST QUESTION one might ask *whether a Blaschke product in $\mathcal{B}_0$ can have a singular set which meets some subarc of $\mathbb{T}$ in a nonempty set of measure zero.*

ANOTHER QUESTION, admittedly vague, concerns the abundance of Blaschke products in $\mathcal{B}_0$. For instance, a Blaschke product should be in $\mathcal{B}_0$ if its zeros are evenly spread throughout $\mathbb{D}$. One is led to suspect that, in some sense, a Blaschke product with random zeros will be almost surely in $\mathcal{B}_0$.

References

1. Zygmund A., *Trigonometric Series*, vol. I, Cambridge Univ. Press, Cambridge, 1959.
2. Anderson J. M., Clunie J., Pommerenke Ch., *On Bloch functions and normal functions*, J. Reine Angew. Math. **270** (1974), 12–37.
3. Pommerenke Ch., *On univalent functions, Bloch functions and VMOA*, Math. Ann. **236** (1978), no. 3, 199–208.
4. Sarason D., *Functions of vanishing mean oscillation*, Trans. Amer. Math. Soc **207** (1975), 391–405.
5. Shapiro H. S., *Monotonic singular functions of high smoothness*, Michigan Math. J. **15** (1968), 265–275.
6. Kahane J.-P., *Trois notes sur les ensembles parfaits linéaires*, Enseignement Math. **15** (1969), no. 2, 185–192.

Dept. Math. University of California
Berkeley, California, 94720
USA

Commentary by the Author

The problem is still open. T. H. Wolff has pointed out that the measures constructed by Kahane and Shapiro can be taken with supports of Lebesgue measure 0, so there do exist infinite Blaschke products in $\mathcal{B}_0$ whose singularities form a set of measure 0. (The author was remiss in failing to notice this.) Wolff (unpublished) has shown that the set of singularities on the unit circle of an inner function in $\mathcal{B}_0$ meets each open subarc either in the empty set or in a set of positive logarithmic capacity. He conjectures that "positive logarithmic capacity" can be replaced by "Hausdorff dimension 1".

Further Commentary by the Author

C. J. Bishop [7] has solved the main problem by obtaining a characterization of the functions in the unit ball of H^∞ that belong to $\mathcal{B}_0$. When specialized to Blaschke products, his characterization describes the ones in $\mathcal{B}_0$ in terms of the distribution of their zeros. He also gives a direct construction of an infinite Blaschke product in $\mathcal{B}_0$.

G. J. Hungerford [9] has used one of Bishop's preliminary results to prove that the set of singularities of an infinite Blaschke product in $\mathcal{B}_0$ has Hausdorff dimension 1. (One can show from this, by standard reasoning, that the set of singularities intersects each open subarc of $\mathbb{T}$ either in the empty set or in a set of Hausdorff dimension 1.)

W. G. Cochran [8] has refuted, or at least cast in serious doubt, the suggestion at the end of the original proposal that random Blaschke products should in some sense be almost surely in $\mathcal{B}_0$. His result: Let $(r_n)_1^\infty$ be an increasing sequence in $(0,1)$ such that $\sum(1-r_n)<\infty$. Let$(\theta_n)_1^\infty$ be an independent sequence of random variables each uniformly distributed over $[0,2\pi)$. The Blaschke product with zero sequence $(r_n\exp(i\theta_n))_1^\infty$ is almost surely *not* in $\mathcal{B}_0$.

References

7. Bishop C. J., *Bounded functions in the little Bloch space*, Pacific J. Math. **142** (1990), 209–225.
8. Cochran W. G., *Random Blaschke products*, Trans. Amer. Math. Soc. **322** (1990), 731–755.
9. Hungerford G. J., *Boundaries of smooth sets and singular sets of Blaschke products in the little Bloch class*, Thesis, California Institute of Technology, Pasadena, California, 1988.

EDITORS' NOTE

We supplement the commentary by the following quotation from C. Bishop's letter:

A characterization of the Blaschke products in $\mathcal{B}_0$ is given in [7] as a special case of a characterization of all bounded functions in the little Bloch space. For Blaschke products the answer if as follows: given a sequence $\{z_n\}$ in the unit disk define a measure μ by placing delta mass of size $1-|z_n|$ at z_n. Then the corresponding Blaschke product is in the little Bloch space iff for every $\varepsilon > 0$ there exist $N, \delta > 0$ such that for every Carleson square Q with side length $l(Q) \leqslant \delta$ and every sub-Carleson square Q' with $l(Q') = \frac{1}{2}l(Q)$ we have either

$$\frac{\mu(Q)}{l(Q)} > \varepsilon^{-1}$$

or

$$\left|\frac{\mu(Q)}{l(Q)} - \frac{\mu(Q')}{l(Q')}\right| < \varepsilon \quad \text{and} \quad \int_{(NQ)^c} P_Q(z)\,d\mu(z) < \varepsilon.$$

Here $(NQ)^c$ is the complement of the N times enlargement of Q and $P_Q(z) = (1-|z_0|^2)\times |1-\bar{z}z_0|^{-2}$ is the Poisson kernel with respect to a point z_0 in the top half of Q.

10.17
old

ANALYTIC FUNCTIONS WITH FINITE DIRICHLET INTEGRAL

SUN-YUNG A. CHANG

If f is an analytic function defined on $\mathbb{D}$, let

$$\mathcal{D}(f) = \left(\iint_{\mathbb{D}} |f'(z)|^2 \frac{dxdy}{\pi} \right)^{1/2}$$

be the Dirichlet integral of f. In [1], the following theorem is proved.

THEOREM. *There is a constant $C_0 < \infty$, such that if f is analytic on $\mathbb{D}$, $f(0) = 0$ and $\mathcal{D}(f) \leqslant 1$ then*

$$\int_0^{2\pi} e^{\alpha |f(e^{i\theta})|^2} \, d\theta \leqslant C_0 \qquad \text{for all } \alpha \leqslant 1.$$

It would be interesting to know the size of C_0 and also the extremal functions (if exist) which correspond to the sharp constant C_0. Actually, the above theorem is only a part of results similar to Moser's sharp form of the Trudinger inequality (see [2]). It would actually be interesting to see if there is a general form of extremal functions which correspond to Moser's sharp inequalities.

REFERENCES

1. Chang S.-Y. A., Marshall D., *A sharp inequality concerning the Dirichlet integral*, Am. J. Math. **107** (1985), no. 5, 1015–1034.
2. Moser J., *A sharp form of an inequality by N. Trudinger*, Ind. Univ. Math. J. **20** (1971), 1077–1092.

DEPARTMENT OF MATHEMATICS
UNIVERSITY OF CALIFORNIA
AT LOS ANGELES
LOS ANGELES, CA 90024
USA

10.18
old

SUBALGEBRAS OF $L^\infty(\mathbb{T}^2)$ CONTAINING $H^\infty(\mathbb{T}^2)$

SUN-YUNG A. CHANG

Let $H^\infty(\mathbb{T})$ denote the Hardy space of boundary values of bounded analytic functions defined on $\mathbb{D}$. There has been a systematic study of the subalgebras (called the Douglas algebras) between $L^\infty(\mathbb{T})$ and $H^\infty(\mathbb{T})$ in the past 10 years. (For a survey article, see [1].) In particular, it has been noticed there is a parallel relationship between subalgebras of $L^\infty(\mathbb{T})$ containing $H^\infty(\mathbb{T})$ to subspaces of BMO (functions of bounded mean oscillations) which contain VMO (functions of vanishing mean oscillations). For example, based on the fact that on $\mathbb{T}$, BMO $= L^\infty + H(L^\infty)$, where H denotes the Hilbert transform, one can deduce that each Douglas algebra can be written as $H^\infty(\mathbb{T}) +$ some C^*-algebra. There are some indications that relations of this type may still hold on the bi-disc $\mathbb{D}^2$ (with distinguished boundary $\mathbb{T}^2$). For example, if one views BMO$(\mathbb{T}^2)$ as $L^\infty(\mathbb{T}^2) + H_1(L^\infty) + H_2(L^\infty) + H_1H_2(L^\infty)$, where the H_i, $i = 1,2$ are Hilbert transforms acting on z_i variables independently with $(z_1, z_2) \in \mathbb{D}^2$ and H_1H_2 is the composition of H_1 with H_2, one can ask the question whether each subalgebra of $L^\infty(\mathbb{T}^2)$ containing $H^\infty(\mathbb{T}^2)$ has the structure of $H^\infty(\mathbb{T}^2)+$some other three C^*-algebras. It seems this problem can be studied independently of the maximal ideal structure of $H^\infty(\mathbb{T}^2)$. So far the only case which has been worked out is the subalgebra of $L^\infty(\mathbb{T}^2)$ generated by $H^\infty(\mathbb{T}^2)$ and $C(\mathbb{T}^2)$ (see [2]).

REFERENCES

1. Sarason D., *Algebras between L^∞ and H^∞*, Lect. Notes in Math., Springer-Verlag **512** (1976), 117–129.
2. Chang S.-Y. A., *Structure of some subalgebra of L^∞ of the torus*, Proc. Symposia in Pure Math. **35,** Part 1 (1979), 421-426.

DEPARTMENT OF MATHEMATICS
UNIVERSITY OF CALIFORNIA
AT LOS ANGELES
LOS ANGELES, CA 90024
USA

10.19
old

INNER FUNCTIONS WITH DERIVATIVE IN H^p, $0 < p < 1$

PATRICK AHERN

Let φ be an inner function defined in the unit disc $\mathbb{D}$. For $\alpha \in \mathbb{D}$ let $\varphi_\alpha(z) = (\varphi(z) - \alpha) \,/\, (1 - \bar{\alpha}\varphi(z))$. Let $\{z_n(\alpha)\}_{n=1}^\infty$ denote the zero set of φ_α. From [1], theorem 6.2, we have:

THEOREM. *Suppose that $\varphi(z) = \sum_{n=0}^\infty a_n z^n$ is an inner function and that $1/2 < p < 1$. Then the following are equivalent:*

1. $\varphi' \in H^p$;
2. $\sum_{n=0}^\infty |a_n|^2 n^p < \infty$;
3. $\sum_{n=1}^\infty \bigl(1 - |z_n(\alpha)|\bigr)^{1-p} < \infty$ *for all $\alpha \in \mathbb{D}$ with the exception of a set of capacity zero.*

For $0 < p < 1/2$ the situation is quite different. It is still true that 1 implies 2 and 3. However, as is pointed out in [1], page 342, there is a Blaschke product $\varphi(z) = \sum_{n=0}^\infty a_n z^n$ such that 2 and 3 hold for φ for all p, $0 < p < 1/2$, but φ' is not a function of bounded characteristic.

PROBLEM. *Find a condition on the Taylor coefficients or on the distribution of values of an inner function φ that is equivalent to the condition $\varphi' \in H^p$, $0 < p < 1/2$.*

REFERENCE

1. Ahern P., *The mean modulus and the derivative of an inner function*, Indiana Univ. Math. J. **28** (1979), no. 2, 311–347.

DEPARTMENT OF MATHEMATICS
UNIVERSITY OF WISCONSIN
MADISON, WI 53706
USA

EDITORS' NOTE

I. È. Verbitskii has informed us about his result pertaining to the Problem.

THEOREM. *If $\frac{1}{p} - 1 < s < \frac{1}{p}$ then the following are equivalent:*

1) $\varphi^{(s)} \in H^p$;
2) $\sum |a_n|^2 n^{sp} < \infty$;
3) $\sum \bigl(1 - |z_n(\alpha)|\bigr)^{1-sp} < \infty$ *for all $\alpha \in \mathbb{D}$ with the exception of a set of capacity zero;*
4) $\varphi \in B_p^s$.

Here $\varphi^{(s)}$ denotes the fractional derivative of φ of order s, B_p^s is the Besov class, i.e.

$$B_p^s \stackrel{\text{def}}{=} \left\{ f \text{ analytic in } \mathbb{D} : \iint_{\mathbb{D}} \left|f^{(n)}(z)\right|^p (1-|z|)^{(n-s)p-1}\, dx dy < \infty \right\},$$

n being any integer $> s$, $z = x + iy$.

This theorem is implied by results of [1] when $p \leqslant 2$, $0 < s \leqslant 1$, $sp > \frac{1}{2}$. It is not valid when $s \leqslant \frac{1}{p} - 1$ and no analogous result seems to be known in that case.

Commentary by K. M. Dyakonov

The following theorem (see [2] for the proof) seems to provide some information on the problem.

Theorem. *Suppose Θ is an inner function, $0 < p < 1$. The following are equivalent.*

1°. $\Theta' \in H^p$.

2°. $\displaystyle\int_{\mathbb{T}} (1 - |\Theta(r\zeta)|)^p \, dm(\zeta) = O((1-r)^p)$ as $r \to 1-0$.

3°. $\displaystyle\int_{\mathbb{T}} \left(\sum_{n=0}^{\infty} |R_n\Theta(\zeta)|^2 \right)^p dm(\zeta) < +\infty,$

where R_n stands for the remainder of the Fourier series: $R_n\Theta(\zeta) \stackrel{\text{def}}{=} \sum_{k=n+1}^{\infty} \widehat{\Theta}(k)\zeta^k$.

In fact, the equivalence of 1° and 3° is an immediate consequence of the amusing identity

$$|\Theta'(\zeta)| = \sum_{n=0}^{\infty} |R_n\Theta(\zeta)|^2$$

that holds for any point ζ, $\zeta \in \mathbb{T}$, for which both limits $\lim\limits_{r\to 1-0} \Theta(r\zeta) \stackrel{\text{def}}{=} \Theta(\zeta)$ and $\lim\limits_{r\to 1-0} \Theta'(r\zeta) = \Theta'(\zeta)$ exist, and $|\Theta(\zeta)| = 1$.

Furthermore, in case $1/2 < p < 1$ each of the three conditions above is equivalent to

4°. $\displaystyle\int_{\Gamma_\varepsilon} \frac{|dz|}{(1-|z|)^p} < +\infty,$

where Γ_ε is the so-called Carleson curve associated with Θ and an arbitrary fixed number ε, $0 < \varepsilon < 1$. (See [3], chapter VIII.)

References

2. Dyakonov K. M., *Smooth functions and coinvariant subspaces of the shift operator*, Algebra i Analiz **4** (1992), no. 5, 117–147 (Russian); English transl. in St. Petersburg Math. J. **4** (1993), no. 5.
3. Garnett J. B., *Bounded Analytic Functions*, Academic Press, NY, 1981.

Prospekt Khudozhnikov,
24–1–412
St. Petersburg 194295,
Russia

10.20

HARDY CLASSES AND RIEMANN SURFACES OF PARREAU–WIDOM TYPE

MORISUKE HASUMI

The theory of Hardy classes on the unit disc and its abstract generalization have received considerable attention in recent years. The case of compact bordered surfaces has also been studied in detail. In this article we wish to consider the case of *infinitely connected* Riemann surfaces.

1. Surfaces of Parreau–Widom type.

BASIC QUESTION. *For which class of Riemann surfaces can one get a fruitful extension of the Hardy class theory on the disk?*

The author's original aim was to extend Beurling's invariant subspace theorem. A most promising candidate so far is the class of Riemann surfaces of Parreau–Widom type.

Definition. Let R be a hyperbolic Riemann surfaces, $G(a,z)$ the Green function for R with pole at a point $a \in R$ and $B(a,\alpha)$ the first Betti number of the region $R(a,\alpha) = \{z \in R : G(a,z) > \alpha\}$ with $\alpha > 0$. We say that R is of *Parreau–Widom type* if $\int_0^\infty B(a,\alpha)\,d\alpha < \infty$.

We first sketch some relevant properties shared by such surfaces, R.

(1) *Parreau* [1]. (a) Every positive harmonic function on R has a limit along almost every Green line issuing from any fixed point in R. (b) The Dirichlet problem on Green lines on R for any bounded measurable boundary function has a unique solution, which converges to the boundary data along almost all Green lines.

(2) *Widom* [2]. For a hyperbolic Riemann surface R it is of Parreau–Widom type if and only if the set $H^\infty(R,\xi)$ of all bounded holomorphic sections of any given complex flat unitary line bundle ξ over R has nonzero elements.

(3) *Hasumi* [3]. (a) Every surface of Parreau–Widom type is obtained by deleting a discrete subset from a surface of Parreau–Widom type, R, which is regular in the sense that $\{z \in R : G(a,z) \geqslant \varepsilon\}$ is compact for any $\varepsilon > 0$. (b) Brelot–Choquet's problem (cf. [4]) concerning the relation between Green lines and Martin's boundary has a completely affirmative solution for any surface of Parreau–Widom type. (c) The inverse Cauchy theorem holds for R.

In view of (3)–(a), *we assume* in the following that *R is a regular surface of Parreau–Widom type.* The Parreau–Widom condition stated in the definition is then equivalent to the inequality $\sum\{G(a,w) : w \in Z(a)\} < \infty$, where $Z(a)$ denotes the set of critical points, repeated according to multiplicity, of the function $z \mapsto G(a,z)$. We set $g^{(a)}(z) = \exp(\sum\{G(z,w) : w \in Z(a)\})$. Moreover, let Δ_1 be Martin's minimal boundary of R and $d\chi_a$ the harmonic measure, carried by Δ_1, at the point a. Look at the following

STATEMENT (DCT). *Let h be a meromorphic function on R such that $|h|g^{(a)}$ has a harmonic majorant on R. Then $h(a) = \int_{\Delta_1} \hat{h}(b)\, d\chi_a(b)$, where $\hat{h}$ denotes the fine boundary function for h.* (Note: DCT stands for Direct Cauchy Theorem.)

(4) *Hayashi* [5]. (a) (DCT) is valid if and only if each β-closed ideal of $H^\infty(R)$ is generated by some (multiple-valued) inner function on R. (b) Beurling's invariant subspace theorem admits a direct generalization if and only if (DCT) holds. (c) There exist surfaces of Parreau–Widom type for which (DCT) fails.

2. Problems related to surfaces of Parreau–Widom type.

(A) *Find simple sufficient conditions for a surfaces of Parreau–Widom type to satisfy* (DCT). Further analysis of (DCT) is important, for it is, among others, equivalent to the Beurling invariant subspace theorem. There have been no practical tests.

(B) *Is there any criterion for a surface of Parreau–Widom type to satisfy the Corona Theorem ?* A Parreau–Widom surface with a corona has been constructed by Nakai [6].

(C) (1) *Does $H^\infty(R, \xi)$ for any ξ have only constant common inner factors ?* (2) *Is a generalized F. and M. Riesz theorem true for measures on Wiener's harmonic boundary, which are orthogonal to H^∞ ?* (M. Hayashi)

(D) *Characterize those surfaces R for which $H^\infty(R, \xi)$ for every ξ has an element without zero.* (H. Widom) The condition will hopefully yield a better subclass of Parreau–Widom surfaces.

On the other hand, plane domains of Parreau–Widom type are not very well known.

(E) *Characterize closed subsets E of the Riemann sphere S for which $S \setminus E$ is of Parreau–Widom type.* (Cf. [7; 8].)

Finally we note that interesting observations may be found in work of Pommerenke, Stanton, Pranger and others.

REFERENCES

1. Parreau, M., *Théorème de Fatou et problème de Dirichlet pour les lignes de Green de certaines surfaces de Riemann*, Ann. Acad. Sci. Fenn. Ser. A. I, Math. (1958), no. **250/25**, 8 pp.
2. Widom, H., *$\mathcal{H}_p$ sections of vector bundles over Riemann surfaces*, Ann. of Math. **94** (1971), 304–324.
3. Hasumi, M., *Hardy Classes on Infinitely Connected Riemann surfaces*, Lecture Notes in Mathematics, Vol. **1027**, Springer-Verlag, 1983.
4. Brelot, M., *Topology of R. S. Martin and Green lines*, Lectures on Functions of a Complex Variable, Univ. of Michigan Press, Ann Arbor, 1955, pp. 105–121.
5. Hayashi, M., *Invariant subspaces on Riemann surfaces of Parreau–Widom type*, Trans. Amer. Math. Soc. **279** (1983), 737–757.
6. Nakai, M., *Corona problems for Riemann surfaces of Parreau–Widom type*, Pacific J. Math. **103** (1982), 103–109.
7. Voichick, M., *Extreme points of bounded analytic functions on infinitely connected regions*, Proc. Amer. Math. Soc. **17** (1966), 1366–1369.
8. Neville, C., *Invariant subspaces of Hardy classes on infinitely connected open surfaces*, Memoirs of the Amer. Math. Soc. No. **160**, 1975.

DEPARTMENT OF MATHEMATICS
IBARAKI UNIVERSITY
MITO, IBARAKI 310
JAPAN

10.21
old

INTERPOLATING BLASCHKE PRODUCTS

PETER G. CASAZZA

If $B = \prod_{n=1}^{\infty} \frac{\bar{\alpha}_n}{|\alpha_n|} \frac{\alpha_n - z}{1 - \bar{\alpha}_n z}$ is a Blaschke product, the interpolation constant of B, denoted $\delta(B)$ is $\inf_m \prod_{\substack{n=1\\ n\neq m}} \left|\frac{\alpha_n - \alpha_m}{1 - \bar{\alpha}_n \alpha_m}\right|$. A well known result of L. Carleson asserts that B is an interpolating Blaschke product if and only if $\delta(B) > 0$. It is also well known *that the following open problems are equivalent:*

PROBLEM 1. *Can every inner function be uniformly approximated by interpolating Blaschke products?* i.e. *Given any Blaschke product B and an $\varepsilon > 0$ is there an interpolating Blaschke product B_1 such that $\|B - B_1\| < \varepsilon$?*

PROBLEM 2. *Is there a function $f(\varepsilon)$ so that for any finite Blaschke product B and any $\varepsilon > 0$, there is a (finite) Blaschke product B_1 such that $\|B - B_1\| < \varepsilon$ and $\delta(B_1) \geqslant f(\varepsilon)$?*

These problems are stronger than Problem 1 posed by John Garnett in "Some open problems concerning H^∞ and BMO" in this problem book, 10.14.

If these problems are eventually answered in the negative, then the obvious question is to classify those inner functions which can be so approximated. T. Trant and P. Casazza have observed (and this may already be known) that changing convergence in norm to convergence uniform on compacta produces satisfactory classifications. For example,

PROPOSITION 3. *The following are equivalent for a function $F \in H^\infty$:*

(1) *There is a sequence $\{B_n\}$ of finite Blaschke products which converge to F uniformly on compacta for which $\inf_n \delta(B_n) > 0$,*

(2) *$F = BG$, where B is an interpolating Blaschke product and G is an outer function satisfying*

$$0 < \inf_{z\in\partial D} |G(z)| \leqslant \sup_{z\in\partial D} |G(z)| \leqslant 1.$$

The proof that (2) $\Longrightarrow$ (1) follows by calculating the interpolation constants of the approximating Blaschke products given in the proof of Frostman's Theorem. By using some techniques developed in [1], it is easily shown that (1) $\Longrightarrow$ (2).

I am particularly interested in the form of the function $f(\varepsilon)$ given in Problem 2. A variation of this relates to a problem stated in [1]. If K is a compact subset of the unit circle with Lebesgue measure zero, let A_K denote the ideal in the disc algebra A consisting of the functions which vanish on K. The most general closed ideals in A have the well known form $J_F = \{g \cdot F\colon g \in A_K\}$, where F is an inner function continuous on the complement of K in the closed disc. A sequence $\{z_n\}$ in the open disc is called a Carleson sequence if $M = \sup\left\{ \sum_{n=1}^{\infty} (1 - |z_n|^2)|f(z_n)| : f \in H^1,\ \|f\| \leqslant 1 \right\} < \infty$. In [1], the following problem appeared:

PROBLEM 4. *If $\{z_n\}$ is a Carleson sequence and B the Blaschke product with zeroes $\{z_n\}$ continuous off K, does there exist absolute constants a and A so that*

$$a \log M \leqslant \inf\{ \|Q\| : Q\colon A_K \to J_B \text{ is a projection onto}\} \leqslant A \log M?$$

I have since discovered that the left hand inequality is true (there does exist a universal constant a) but the right hand inequality is false (there does not exist a universal constant A). A new conjecture for the norm of the best projection onto an ideal in A is needed. The calculations involved in computing this seem to be related to those needed for problems (1) and (2) above.

REFERENCE

1. Casazza P. G., Pengra R., Sundberg C., *Complemented ideals in the disk algebra*, Israel J. Math. **37** (1980), no. 1–2, 76–83.

DEPARTMENT OF MATHEMATICS
UNIVERSITY OF MISSOURI–COLUMBIA
COLUMBIA, MISSOURI 65211
USA

10.22

QUASI-PROPER MAPS OF 2-SHEETED COVERINGS OF THE DISC

Frank Forelli

I wish to describe a problem concerning function theory on certain Riemann surfaces. Namely, those surfaces which may be mapped bivalently onto the open unit disc $\mathbb{D}$. However, it will be best to begin without the hypothesis that our surface is a 2-sheeted covering of the disc. So let X simply be an open Riemann surface. Also, let θ be a proper holomorphic map of X to $\mathbb{D}$. (Most Riemann surfaces do not have such maps, but let's suppose ours does.) The term **proper** means that $\{\theta \in E\}$ is bounded in X whenever E is bounded in $\mathbb{D}$. (A set is bounded if it is contained in a compact set.) Here is a version of the classical Schwarz lemma, but for functions on X instead of $\mathbb{D}$.

Let f be holomorphic in X; in symbols, $f \in \mathcal{O}(X)$. Then f and the product θf have the same supremum:

$$\sup_X |f| = \sup_X |\theta f|. \tag{1}$$

Let's drop the hypothesis that θ is proper, and simply suppose that θ is a holomorphic map of X to $\mathbb{D}$. We will say that θ is quasi-proper if (1) holds for every $f \in \mathcal{O}(X)$. Notice that if θ is quasi-proper, then θ is not a constant (since $|\theta| < 1$), and θ must vanish somewhere in X (since otherwise, $|1/\theta| \leqslant 1$). Also, if θ_1 and θ_2, like θ, are holomorphic maps of X to $\mathbb{D}$, and if $\theta = \theta_1\theta_2$, then θ is quasi-proper iff θ_1 and θ_2 are.

When X is the disc, the quasi-proper maps are what one would expect them to be:

Lemma. *Let $X = \mathbb{D}$. Then θ is quasi-proper iff θ is a Blaschke product.*

Suppose now that our Riemann surface X is a 2-sheeted covering of the disc $\mathbb{D}$, and that the number of branch points is infinite. This is to say, suppose there is a proper holomorphic map φ of X to $\mathbb{D}$, that φ vanishes twice (counting multiplicities), and that $d\varphi$ vanishes infinitely often. We obtain additional proper holomorphic maps of X to $\mathbb{D}$ by composing finite Blaschke products with φ, and since the number of branch points is infinite, we obtain all such maps in this way [2]. Our problem is this:

Find (or characterize) the quasi-proper maps of X.

Just as we obtain (all) proper maps by composing finite Blaschke products with φ, so we obtain quasi-proper maps by composing Blaschke products with φ. But we will see that this does not give all quasi-proper maps, provided the traces of branch points satisfy the Blaschke condition:

$$\sum_{\varphi'(x)=0} (1 - |\varphi(x)|) < \infty. \tag{2}$$

(If (2) does not hold, and f in $\mathcal{O}(X)$ is bounded, then $f = g(\varphi)$ with $g \in \mathcal{O}(\mathbb{D})$, so if θ is quasi-proper, then by the lemma, $\theta = b(\varphi)$ with b a Blaschke product.)

We will identify $\mathcal{O}(\mathbb{D})$ with $\{ g(\varphi) \mid g \in \mathcal{O}(\mathbb{D}) \}$; then $\mathcal{O}(X)$ is an overring of $\mathcal{O}(\mathbb{D})$. Let σ be the sheet interchange automorphism:

$$\sigma \in \operatorname{Aut} X, \quad \varphi(\sigma) = \varphi, \quad \sigma(\sigma) = \iota, \quad \sigma \neq \iota.$$

(Iota (ι) is the identity automorphism.) The automorphism σ serves to tell who is in $\mathcal{O}(\mathbb{D})$. The test is this: let $f \in \mathcal{O}(X)$; then $f \in \mathcal{O}(\mathbb{D})$ iff $f(\sigma) = f$.

Suppose now that θ is quasi-proper. Then so is the composition $\theta(\sigma)$, hence the product $\theta \cdot \theta(\sigma)$ is too. But by the test, the product is in $\mathcal{O}(\mathbb{D})$, so by the lemma, it is a Blaschke product. This proves (half of) the following.

CRITERION. *Let $\theta \in \mathcal{O}(X)$ with $|\theta| < 1$. Then θ is quasi-proper iff $\theta \cdot \theta(\sigma) = b(\varphi)$ with b a Blaschke product.*

To see that not every quasi-proper map is in $\mathcal{O}(\mathbb{D})$, let B be the Blaschke product whose zeros (all simple) coincide with the traces of branch points (here we use (2)). Then $\sqrt{B} \in \mathcal{O}(X)$ (this is implicit in Lemma 4 of [2]); in other words, there is $\vartheta \in \mathcal{O}(X)$ such that

$$\vartheta^2 = B(\varphi). \tag{3}$$

By (3), ϑ is quasi-proper (since $|\vartheta| < 1$), but $\vartheta \neq b(\varphi)$ since B is not a square (its zeros are simple). (Or since the zeros of ϑ are simple.)

If $\theta = b(\varphi)$ or $\theta = \vartheta$, then $\theta^2 \in \mathcal{O}(\mathbb{D})$. To see that this is not true of every quasi-proper map, suppose $b(0) \neq 0$, and let b^* be the Blaschke product whose zeros coincide with the squares of those of the Blaschke product b. (For example, if ξ is a zero of order 2 of b and $-\xi$ a zero of order 3, then ξ^2 is a zero of order 5 of b^*.) Alternatively,

$$b^*(w^2) = b(w)b(-w). \tag{4}$$

(If $b(0) = 0$, then (4) may be off by a factor of -1.) Let u (in $\mathcal{O}(\mathbb{D})$) be unitary (or inner). Then by the criterion, plus $\vartheta(\sigma) = -\vartheta$, we have this rule:

> *The composition $b(u\vartheta)$ is quasi-proper if (and only if) the composition $b^*(u^2B)$ is a Blaschke product.*

Choose ξ, $0 < |\xi| < 1$, so that $(\xi^2 - B)/(1 - \bar{\xi}^2 B)$, like B, is a Blaschke product. Then by the rule, $(\xi - \vartheta)/(1 - \bar{\xi}\vartheta)$ is quasi-proper, but its square is not in $\mathcal{O}(\mathbb{D})$.

The criterion characterizes the quasi-proper maps of X, but in an unsatisfying way, so here is a more specific problem.

> *Are there quasi-proper maps other than products (finite or infinite) whose factors come from this list:*
>
> (i) b (or $b(\varphi)$)
>
> (ii) $\vartheta = \sqrt{B}$
>
> (iii) $b(u\vartheta)$ *provided $b^*(u^2B)$ is a Blaschke product.*
>
> *(Here, of course, b is Blaschke and u is unitary.)*

By (the proof of) a theorem of Ahern and Rudin, every quasi-proper map is the product (finite or infinite) of irreducible quasi-proper maps [1]. (A quasi-proper map is irreducible if it is not the product of two quasi-proper maps.) So we may paraphrase the question just asked:

Does the list include every irreducible quasi-proper map?

The second item in the list, namely, ϑ, is irreducible. Presumably, so is $\frac{\xi-\vartheta}{1-\xi\vartheta}$, but I do not have a proof.

Here is the proof that ϑ is irreducible. Suppose $\vartheta = \theta\psi$ with each factor of modulus less than 1. Then θ is quasi-proper, hence so is $\theta(\sigma)$. But θ and $\theta(\sigma)$ have the same zeros, so (since they are quasi-proper) $\theta(\sigma) = \xi\theta$ with $|\xi| = 1$. Then $\xi = \pm 1$ (since $\sigma(\sigma) = \iota$), but $\xi = 1$ is not possible since the zeros of θ are simple. Likewise, $\psi(\sigma) = -\psi$. But then $\vartheta(\sigma) = \vartheta$, and this is not so. $\square$

Two more comments.

(i) If θ is quasi-proper, and vanishes only finitely often, then by the criterion, θ is proper, and so is in the list.

(ii) Let p be a proper holomorphic map of X to itself. Then p commutes with σ:

$$p(\sigma) = \sigma(p). \tag{5}$$

This is in [2]. By (5), the composition $\theta(p)$ is quasi-proper if θ is, and if θ is in the list, so is $\theta(p)$.

References

1. Ahern P., Rudin W., *Factorizations of bounded holomorphic functions*, Duke Math. J. **39** (1972), 767–777.
2. Forelli F., *Two-sheeted coverings of the disc*, Math. Scand. **61** (1987), 17–38.

Department of Mathematics
University of Wisconsin-Madison
Madison, WI 53706
USA

10.23

SMOOTH FUNCTIONS AND INNER FACTORS

K. M. DYAKONOV

Let Λ^α, $0 < \alpha < +\infty$, denote the Lipschitz (Zygmund for $\alpha \in \mathbb{N}$) space of functions on the circle:

$$\Lambda^\alpha = \{ f \in C(\mathbb{T}) : \omega_m(f, \delta) = O(\delta^\alpha) \},$$

where m is any integer for which $m > \alpha$, and $\omega_m(f, \cdot)$ stands for the m-th order continuity modulus of the function f. The endpoint $\alpha = 0$ will also be considered, in which case we set $\Lambda^0 \stackrel{\text{def}}{=} \text{BMO}$. Finally, we define the class Λ^α_A, $\alpha \geqslant 0$, to be the "analytic subspace" of Λ^α: $\Lambda^\alpha_A \stackrel{\text{def}}{=} \Lambda^\alpha \cap H^1$. (More generally, given a space X of functions on the circle, we set $X_A \stackrel{\text{def}}{=} X \cap H^1$; elements of X_A are also treated as functions analytic in $\mathbb{D}$.)

In this note a few questions are posed that arise in connection with the following theorem due to the author [1,2].

THEOREM 1. *Suppose $f \in \Lambda^\alpha_A$ and Θ is an inner function. Let $n \in \mathbb{N}$, $n > \alpha$. The following are equivalent.*

1. $f\overline{\Theta}^n \in \Lambda^\alpha$.
2. $f\Theta^n \in \Lambda^\alpha_A$.
3. $f\Theta^k \in \Lambda^\alpha$ *for all* $k \in \mathbb{Z}$.
4. *For some (every)* ε, $0 < \varepsilon < 1$, *we have*

$$\sup\{ |f(z)|(1 - |z|)^{-\alpha} : z \in \mathbb{D}, |\Theta(z)| < \varepsilon \} < +\infty. \tag{1}$$

The original proof of the theorem [1,2] relied heavily on the $(H^{1/(1+\alpha)}, \Lambda^\alpha_A)$ duality, $0 \leqslant \alpha < +\infty$. Another proof involving $\bar\partial$-techniques has recently been suggested by E. M. Dyn'kin. In what follows we deal with some generalizations of the Λ^α spaces that probably need a new approach.

Now let $\omega(t)$ be a continuous non-decreasing positive function (a "continuity modulus") defined for $0 < t < 1$. Consider the space

$$\text{BMO}_\omega \stackrel{\text{def}}{=} \left\{ f \in L^1(\mathbb{T}) : \sup_{I: I \subset \mathbb{T}} \frac{1}{|I|\omega(|I|)} \int_I |f - f_I|\, dm < +\infty \right\},$$

where I ranges over all open subarcs of $\mathbb{T}$, $|I| \stackrel{\text{def}}{=} m(I)$ and $f_I \stackrel{\text{def}}{=} |I|^{-1} \int_I f\, dm$. (Cf. [7].)

CONJECTURE 1. *Suppose X is any of the spaces BMO_ω or $BMO_\omega \cap L^\infty$. (Perhaps, some hypotheses on ω are needed.) Let $f \in X_A$ and Θ be an inner function. Then the following are equivalent.*

1. $f\overline{\Theta} \in X$.
2. $f\Theta \in X_A$.
3. *For some (every) ε, $0 < \varepsilon < 1$, we have*

$$\sup\left\{\frac{|f(z)|}{\omega(1-|z|)} : z \in \mathbb{D}, |\Theta(z)| < \varepsilon\right\} < +\infty. \tag{2}$$

Let us discuss some special cases when Conjecture 1 is known to be true.

a) Let $\omega(t) \equiv 1$, so that BMO_ω is just BMO, and $\mathrm{BMO}_\omega \cap L^\infty = L^\infty$. For $X = \mathrm{BMO}$ Conjecture 1 follows from Theorem 1 (set $\alpha = 0$, $n = 1$), and the case $X = L^\infty$ is trivial.

b) Let $\omega(t) = t^\alpha$, $0 < \alpha < 1$, so that $\mathrm{BMO}_\omega = \mathrm{BMO}_\omega \cap L^\infty = \Lambda^\alpha$. Once again Conjecture 1 reduces to Theorem 1 (set $n = 1$).

c) Let $\omega(t) = \left(\log \frac{1}{t}\right)^{-1}$, so that $\mathrm{BMO}_\omega \cap L^\infty$ is the space of all BMO multipliers [8]. The validity of Conjecture 1 for the space $X = \mathrm{BMO}_\omega \cap L^\infty$ can easily be deduced from the last fact and from the BMO part of Theorem 1 (see [1] for details).

The next conjecture is a generalization of the preceding one.

CONJECTURE 2. *Let $n \in \mathbb{N}$, $X = \{ f \in C^{n-1}(\mathbb{T}) : f^{(n-1)}$ is absolutely continuous, and $f^{(n)} \in BMO_\omega \}$, or let X be the space defined in a similar way with BMO_ω replaced by $BMO_\omega \cap L^\infty$. Suppose $f \in X_A$ and Θ is an inner function. Then each of the inclusions*

$$f\overline{\Theta}^{n+1} \in X \tag{3}$$

and

$$f\Theta^{n+1} \in X_A \tag{4}$$

is equivalent to the following condition

$$\sup\left\{\frac{|f(z)|}{(1-|z|)^n \omega(1-|z|)} : z \in \mathbb{D}, |\Theta(z)| < \varepsilon\right\} < +\infty \tag{5}$$

for some (every) ε, $0 < \varepsilon < 1$.

Remarks. 1) The case $\omega(t) = t^\alpha$, $0 < \alpha < 1$, is contained in Theorem 1.

2) In the case when $\omega(t) \equiv 1$, $X = \{ f : f^{(n)} \in \mathrm{BMO} \}$ the equivalence of (3) and (5) has recently been established by the author.

Now let

$$\Lambda^\omega = \{ f \in C(\mathbb{T}) : \omega_1(f, \delta) = O(\omega(\delta)) \}.$$

Division and multiplication by inner factors in the space Λ^ω_A were studied by N. A. Shirokov [3,4,5]. In particular, he proved [3] that, given a function f in Λ^ω_A and Θ inner, the inclusion $f\Theta \in \Lambda^\omega_A$ holds if and only if $m(\operatorname{spec}\Theta) = 0$ and

$$|f(\zeta)| = O(\omega(|\Theta'(\zeta)|^{-1})), \qquad \zeta \in \mathbb{T} \setminus \operatorname{spec}\Theta. \tag{6}$$

Here $\operatorname{spec}\Theta$ denotes the boundary spectrum of Θ, i.e. the smallest closed set E, $E \subset \mathbb{T}$, such that Θ is analytic across $\mathbb{T} \setminus E$.

QUESTION. a) *Is Shirokov's condition* (6) *on the pair* (f,Θ) ($f\in\Lambda_A^\omega$, Θ *inner*) *equivalent to* (2) *for an arbitrary continuity modulus* ω?

b) *Is any of these conditions equivalent to the inclusion* $f\overline{\Theta}\in\Lambda^\omega$ *for an arbitrary* ω? (Division without analyticity is not discussed in [3].)

In the event that the answer to part a) is NO, a problem would arise to describe the ω's for which (2) and (6) are equivalent. It would also be nice to find a direct proof of the fact that (1) is equivalent to (6) in case $\omega(t)=t^\alpha$, $0<\alpha<1$. (Indirectly, this follows by a juxtaposition of Theorem 1 and Shirokov's theorem cited above.)

One more problem is connected with the "Bloch type" space

$$B_\omega \stackrel{\text{def}}{=} \left\{ f \text{ analytic in } \mathbb{D} : |f'(z)|(1-|z|) \underset{|z|\to 1}{=} O(\omega(1-|z|)) \right\}$$

PROBLEM 1. a) *Describe the* ω*'s for which the inclusion*

$$f\Theta\in B_\omega \tag{7}$$

is equivalent to (2), *whenever* $f\in B_\omega$ *and* Θ *is inner.*

b) *The same for* $B_\omega\cap H^\infty$ *instead of* B_ω.

Since $B_{t^\alpha}=\Lambda_A^\alpha$, $0<\alpha<1$, Theorem 1 shows that the continuity moduli $\omega(t)=t^\alpha$ possess the desired property. On the other hand, if $\omega(t)$ tends to 0 slowly enough as $t\to+0$, B_ω is known to contain a nontrivial inner function Θ (see [6], exercise 11 to chapter X). Since $B_\omega\cap H^\infty$ is an algebra and $\Theta\in B_\omega$, (7) holds for any f in $B_\omega\cap H^\infty$. Therefore (7) does not imply (2) unless $\omega(t)$ vanishes rapidly enough as $t\to 0$.

In order to pose our last problem, we go back to the Lipschitz–Zygmund spaces Λ_A^α and introduce the following notion.

DEFINITION. An inner function Θ is called α-stable iff the implication

$$f\Theta\in\Lambda_A^\alpha \implies f\Theta^k\in\Lambda_A^\alpha \quad \forall k\in\mathbb{N}$$

holds whenever $f\in\Lambda_A^\alpha$.

It should be noted that every inner function is α-stable for $0<\alpha<1$, and every singular inner function is α-stable for all α, $0<\alpha<+\infty$. These facts are implied by Shirokov's results [4,5] or by Theorem 1. However, for any $\alpha>1$ there exist Blaschke products that are not α-stable. (A more subtle result is contained in [4,5].)

PROBLEM 2. *Let* $\alpha\geqslant 1$. *Describe the* α*-stable Blaschke products in terms of their zeros.*

Here is some information concerning Problem 2. Let $n=[\alpha]+1$. It follows from Theorem 1 (see also [4,5]) that a Blaschke product B is α-stable, provided its zeros are all of multiplicity $\geqslant n$. In [2] a better sufficient condition (a bit too cumbersome to be reproduced here) is obtained. In particular, it is true that B is α-stable if its zero sequence can be partitioned into countably many finite "constellations", containing at least n zeros each, whose noneuclidean diameters are bounded away from 1. It is also proved in [2] that a Blaschke product B is α-stable for $1<\alpha<2$, provided its zero sequence $\{z_j\}$ is separated (in the sense that $\inf\limits_{j\neq k}\rho(z_j,z_k)>0$, where ρ is the pseudohyperbolic distance) and tends to 1 nontangentially.

References

1. Dyakonov K. M., *Division and multiplication by inner functions and embedding theorems for star-invariant subspaces*, Amer. J. Math. (to appear).
2. Dyakonov K. M., *Invariant subspaces of the backward shift operator in Hardy spaces*, Thesis, St. Petersburg State University, St. Petersburg, 1991. (Russian)
3. Shirokov N. A., *Ideals and factorization in algebras of analytic functions smooth up to the boundary*, Tr. Mat. Inst. Steklova **130** (1978), 196–222. (Russian)
4. Shirokov N. A., *Division and multiplication by inner functions in spaces of analytic functions smooth up to the boundary*, Lect. Notes in Math., vol. 864, 1981, pp. 413–439.
5. Shirokov N. A., *Analytic functions smooth up to the boundary*, Lect. Notes in Math., vol. 1312, 1988.
6. Garnett J. B., *Bounded Analytic Functions*, Academic Press, New York, 1981.
7. Janson S., *On functions with conditions on the mean oscillation*, Ark. Mat. **14** (1976), 189–196.
8. Stegenga D. A., *Bounded Toeplitz operators on H^1 and applications of the duality between H^1 and the functions of bounded mean oscillation*, Amer. J. Math. **98** (1976), 573–589.

Prospekt Khudozhnikov,
24–1–412
St. Petersburg 194295,
Russia

S.10.24
v.old

ALGEBRAS CONTAINED WITHIN H^∞

J. M. Anderson

Let $A = \{f : f \text{ analytic in } \mathbb{D},\ f \text{ continuous in } \operatorname{clos}\mathbb{D} = \mathbb{D} \cup \mathbb{T}\}$. Then A is an algebra contained within H^∞, but there are two intermediate algebras that present some interest. First we require some notation.

Let B denote the Banach space of functions f, analytic in $\mathbb{D}$ for which the norm

$$\|f\|_B = |f(0)| + \sup_{|z|<1} (1 - |z|^2)|f'(z)|$$

is finite. This is called the *Bloch* space. We also define

$$B_0 = \{f : f \in B, f'(z) = o(1 - |z|^2)^{-1}, |z| \to 1\}.$$

For a survey of these spaces see [1]. The following facts are easily established:

a) $H^\infty \subset B$,
b) $H^\infty \not\subset B_0$,
c) $H^\infty \cap B_0 \stackrel{\text{def}}{=} X$ is a subalgebra of H^∞.

Similarly we define BMOA (*analytic functions of bounded mean oscillation*) to be the space of those functions f, analytic in $\mathbb{D}$ for which the norm

$$\|f\| = |f(0)| + \sup_{|\zeta|<1} \|f_\zeta\|_2$$

is finite. Here $\|\cdot\|_2$ is the ordinary H^2 norm and

$$f_\zeta(z) = f\left(\frac{z+\zeta}{1+\bar{\zeta}z}\right) - f(\zeta), \qquad z \in \mathbb{D}.$$

Similarly

$$\mathrm{VMOA} = \{f : f \in \mathrm{BMOA}, \|f_\zeta\|_2 = o(1), |\zeta| \to 1\}.$$

The space VMOA consists of those analytic functions in $\mathbb{D}$ whose boundary values on $\mathbb{T}$ have vanishing mean oscillation (see [2], p.591). It is also easy to see that

d) $H^\infty \subset \mathrm{BMOA}$,
e) $H^\infty \not\subset \mathrm{VMOA}$,
f) $H^\infty \cap \mathrm{VMOA} \stackrel{\text{def}}{=} Y$ is a subalgebra of H^∞.

It is not difficult to establish the following relation (see e.g. [3])

$$A \subsetneqq Y \subsetneqq X \subsetneqq H^\infty.$$

The algebra X has already been studied. It was shown by Behrens, unpublished, that X consists precisely of those f, $f \in H^\infty$, whose Gelfand transform $\widehat{f}$ is constant on all the non-trivial Gleason parts of the maximal ideal space of H^∞. It is also known [3] that X does not possess the f-property or K-property in the sense of Havin [4].

It would be nice to have a similar study made of Y. The space Y cannot contain any inner functions [3], other than finite Blaschke products, in contrast to X. But Y does, of course, contain functions having an inner factor—for example the function of [5], p.29 belongs to A.

References

1. Anderson J. M., Clunie J., Pommerenke Ch., *On Bloch functions and normal functions*, J. Reine Angew. Math. **270** (1974), 12–37.
2. Pommerenke Ch., *Schlichte Funktionen und analytische Funktionen von beschränkter mittlerer Oszillation*, Comment. Math. Helv. **52** (1977), 591–602.
3. Anderson J. M., *On division by inner factors*, Comment. Math. Helv. **54** (1979), no. 2, 309–317.
4. Havin V. P., *On the factorization of analytic functions smooth up to the boundary*, Zap. Nauchn. Sem. Leningrad. Otdel. Mat. Inst. Steklov (LOMI) **22** (1971), 202–205. (Russian)
5. Gurarii V. P., *On the factorization of absolutely convergent Taylor series and Fourier integrals*, Zap. Nauchn. Sem. Leningrad. Otdel. Mat. Inst. Steklov (LOMI) **30** (1972), 15–32. (Russian)

Department of Mathematics
University College London
Gower Street
London, WC1E 6BT
England

Editors' Note

The fact that QA has the f-property was shown by P. Gorkin in [6]. See also Lemma 2.3 in [7].

Some new facts on division and multiplications by inner factors can be found in [8], [9], especially p. 511.

Commentary by H. Hedenmalm

We use here the more standard notation $X = COP$ and $Y = QA$. It is well-known that QA consists of those functions f in H^∞ whose Gelfand transform $\hat{f}$ is constant on so-called QC-level sets; this corresponds to the above-mentioned description of COP. In [10], it was shown that QA has the f-property, but fails to have the K-property; see also [11]. Given an inner function u in H^∞, there exists an outer function g in QA such that $gu \in QA$, by a theorem of Thomas Wolff [12]. Thus, every inner function appears as the inner factor of some QA-function.

References

6. Gorkin P., *Prime ideals in closed subalgebras of L^∞*, Michigan Math. J. **33** (1986), 315–323.
7. Gorkin P., Hedenmalm H., Mortini R., *A Beurling–Rudin theorem for H^∞*, Illinois Math. J. **31** (1987), 629–644.
8. Dyakonov K. M., *Division and multiplication by inner functions and embedding theorems for star-invariant subspaces*, Amer. J. Math. (to appear).
9. Mortini R., Gorkin P., *F-ideals in QA_B*, J. London Math. Soc. **37** (1988), 509–519.
10. Hedenmalm H., *On the f- and K-properties of certain function spaces*, Contemporary Mathematics **91** (1989), 89–91.
11. Izuchi K., *VMO has the K-property*, preprint.
12. Wolff T., *Two algebras of bounded functions*, Duke Math. J. **49** (1982), 321–328.

S.10.25
old

ON THE DEFINITION OF $H^p(\mathbb{R}^n)$

A. B. ALEKSANDROV AND V. P. HAVIN

Suppose T is a distribution on $\mathbb{R}^n$, φ a compactly supported C^∞-function on $\mathbb{R}^n$, $\int_{\mathbb{R}^n} \varphi = 1$, $p > 0$. Put

$$\varphi_{\varepsilon,x}(y) = \frac{1}{\varepsilon^n}\varphi\left(\frac{x-y}{\varepsilon}\right) \qquad (x, y \in \mathbb{R}^n,\ \varepsilon > 0),$$

$$T^+_\varphi(x) = \sup\{\,|T(\varphi_{\varepsilon,x})|: \varepsilon > 0\,\}$$

(the *radial* maximal function of T corresponding to the mollifier φ).

QUESTION. *Does the inclusion $T^+_\varphi \in L^p(\mathbb{R}^n)$ imply $T \in H^p(\mathbb{R}^n)$?*

The answer is YES if $p \geqslant 1$ or under the supplementary assumption $T \in S'(\mathbb{R}^n)$ [1] or if we replace T^+_φ by T^*_φ, the *angular* maximal function (because then the inclusion $T \in S'(\mathbb{R}^n)$ is easy to prove). We were unable to answer the question following the patterns of [1].

REFERENCE

1. Fefferman C., Stein E. M., *H^p spaces of several variables*, Acta Math. **129** (1972), 137–193.

STEKLOV MATHEMATICAL INSTITUTE
ST. PETERSBURG BRANCH
FONTANKA 27
ST. PETERSBURG, 191011
RUSSIA

DEPT. OF MATHEMATICS AND MECHANICS
ST. PETERSBURG STATE UNIVERSITY
BIBLIOTECHNAYA PL. 2
STARYI PETERHOF, 198904
RUSSIA

COMMENTARY BY A. UCHIYAMA

This Problem was solved by A. Uchiyama [2]. Namely,

Suppose T is a distribution on $\mathbb{R}^n$, φ a compactly supported C^∞-function on $\mathbb{R}^n$, $\int_{\mathbb{R}^n} \varphi = 1$, $p > 0$. Put

$$\varphi_{\varepsilon,x}(y) = \frac{1}{\varepsilon^n}\varphi\left(\frac{x-y}{\varepsilon}\right) \qquad (x, y \in \mathbb{R}^n,\ \varepsilon > 0),$$
$$T^+_\varphi(x) = \sup\{\,|T(\varphi_{\varepsilon,x})| : \varepsilon > 0\,\}.$$

Then, if $T^+_\varphi \in L^p(\mathbb{R}^n)$, then $T \in H^p(\mathbb{R}^n)$.

REFERENCE

2. A. Uchiyama, *On the radial maximal function of distributions*, Pacific J. of Math. **121** (1986), 467–483.

SUBJECT INDEX

SUBJECT INDEX

SUBJECT INDEX

SUBJECT INDEX

SUBJECT INDEX

SUBJECT INDEX

SUBJECT INDEX

SUBJECT INDEX

AUTHOR INDEX

AUTHOR INDEX

STANDARD NOTATION

Symbols $\mathbb{N}$, $\mathbb{Z}$, $\mathbb{R}$, $\mathbb{C}$ denote respectively the set of positive integers, the set of all integers, the real line, and the complex plane;

$$\mathbb{R}_+ = \{ t \in \mathbb{R} : t \geq 0 \};$$
$$\mathbb{Z}_+ = \mathbb{Z} \cap \mathbb{R}_+;$$
$$\mathbb{T} = \{ z \in \mathbb{C} : |z| = 1 \};$$
$$\mathbb{D} = \{ z \in \mathbb{C} : |z| < 1 \};$$

$\widehat{\mathbb{C}}$ stands for the one-point compactification of $\mathbb{C}$;
m denotes the normed Lebesgue measure on $\mathbb{T}$ ($m(\mathbb{T}) = 1$);
$f|X$ is the restriction of a mapping (function) f to X;
$\operatorname{clos}(\cdot)$ is the closure of the set $(\cdot)$;
$\vee(\cdot)$ is the closed span of the set $(\cdot)$ in a linear topological space;
$|T|$ denotes the norm of the operator T;
$\hat{f}(\cdot)$ denotes the sequence of Fourier coefficients of f;
$\mathcal{F}f$ denotes the Fourier transform of f;
$\mathfrak{S}_p$ is a class of operators A on a Hilbert space satisfying trace $(A^*A)^{p/2} < +\infty$;
H^p is a Hardy class in $\mathbb{D}$, i.e., the space of all holomorphic functions on $\mathbb{D}$ with

$$\|f\|_p \stackrel{\text{def}}{=} \sup_{0<r<1} \left(\int_{\mathbb{T}} |f(r\zeta)|^p \, dm(\zeta) \right)^{1/p} < +\infty, \qquad p > 0 .$$

Lecture Notes in Mathematics

For information about Vols. 1–1394
please contact your bookseller or Springer-Verlag

Vol. 1395: K. Alladi (Ed.), Number Theory, Madras 1987. Proceedings. VII, 234 pages. 1989.

Vol. 1396: L. Accardi, W. von Waldenfels (Eds.), Quantum Probability and Applications IV. Proceedings, 1987. VI, 355 pages. 1989.

Vol. 1397: P.R. Turner (Ed.), Numerical Analysis and Parallel Processing. Seminar, 1987. VI, 264 pages. 1989.

Vol. 1398: A.C. Kim, B.H. Neumann (Eds.), Groups – Korea 1988. Proceedings. V, 189 pages. 1989.

Vol. 1399: W.-P. Barth, H. Lange (Eds.), Arithmetic of Complex Manifolds. Proceedings, 1988. V, 171 pages. 1989.

Vol. 1400: U. Jannsen. Mixed Motives and Algebraic K-Theory. XIII, 246 pages. 1990.

Vol. 1401: J. Steprans, S. Watson (Eds.), Set Theory and its Applications. Proceedings, 1987. V, 227 pages. 1989.

Vol. 1402: C. Carasso, P. Charrier, B. Hanouzet, J.-L. Joly (Eds.), Nonlinear Hyperbolic Problems. Proceedings, 1988. V, 249 pages. 1989.

Vol. 1403: B. Simeone (Ed.), Combinatorial Optimization. Seminar, 1986. V, 314 pages. 1989.

Vol. 1404: M.-P. Malliavin (Ed.), Séminaire d´AIgèbre Paul Dubreil et Marie-Paul Malliavin. Proceedings, 1987–1988. IV, 410 pages. 1989.

Vol. 1405: S. Dolecki (Ed.), Optimization. Proceedings, 1988. V, 223 pages. 1989. Vol. 1406: L. Jacobsen (Ed.), Analytic Theory of Continued Fractions III. Proceedings, 1988. VI, 142 pages. 1989.

Vol. 1407: W. Pohlers, Proof Theory. VI, 213 pages. 1989.

Vol. 1408: W. Lück, Transformation Groups and Algebraic K-Theory. XII, 443 pages. 1989.

Vol. 1409: E. Hairer, Ch. Lubich, M. Roche. The Numerical Solution of Differential-Algebraic Systems by Runge-Kutta Methods. VII, 139 pages. 1989.

Vol. 1410: F.J. Carreras, O. Gil-Medrano, A.M. Naveira (Eds.), Differential Geometry. Proceedings, 1988. V, 308 pages. 1989.

Vol. 1411: B. Jiang (Ed.), Topological Fixed Point Theory and Applications. Proceedings. 1988. VI, 203 pages. 1989.

Vol. 1412: V.V. Kalashnikov, V.M. Zolotarev (Eds.), Stability Problems for Stochastic Models. Proceedings, 1987. X, 380 pages. 1989.

Vol. 1413: S. Wright, Uniqueness of the Injective III_1 Factor. III, 108 pages. 1989.

Vol. 1414: E. Ramirez de Arellano (Ed.), Algebraic Geometry and Complex Analysis. Proceedings, 1987. VI, 180 pages. 1989.

Vol. 1415: M. Langevin, M. Waldschmidt (Eds.), Cinquante Ans de Polynômes. Fifty Years of Polynomials. Proceedings, 1988. IX, 235 pages.1990.

Vol. 1416: C. Albert (Ed.), Géométrie Symplectique et Mécanique. Proceedings, 1988. V, 289 pages. 1990.

Vol. 1417: A.J. Sommese, A. Biancofiore, E.L. Livorni (Eds.), Algebraic Geometry. Proceedings, 1988. V, 320 pages. 1990.

Vol. 1418: M. Mimura (Ed.), Homotopy Theory and Related Topics. Proceedings, 1988. V, 241 pages. 1990.

Vol. 1419: P.S. Bullen, P.Y. Lee, J.L. Mawhin, P. Muldowney, W.F. Pfeffer (Eds.), New Integrals. Proceedings, 1988. V, 202 pages. 1990.

Vol. 1420: M. Galbiati, A. Tognoli (Eds.), Real Analytic Geometry. Proceedings, 1988. IV, 366 pages. 1990.

Vol. 1421: H.A. Biagioni, A Nonlinear Theory of Generalized Functions, XII, 214 pages. 1990.

Vol. 1422: V. Villani (Ed.), Complex Geometry and Analysis. Proceedings, 1988. V, 109 pages. 1990.

Vol. 1423: S.O. Kochman, Stable Homotopy Groups of Spheres: A Computer-Assisted Approach. VIII, 330 pages. 1990.

Vol. 1424: F.E. Burstall, J.H. Rawnsley, Twistor Theory for Riemannian Symmetric Spaces. III, 112 pages. 1990.

Vol. 1425: R.A. Piccinini (Ed.), Groups of Self-Equivalences and Related Topics. Proceedings, 1988. V, 214 pages. 1990.

Vol. 1426: J. Azéma, P.A. Meyer, M. Yor (Eds.), Séminaire de Probabilités XXIV, 1988/89. V, 490 pages. 1990.

Vol. 1427: A. Ancona, D. Geman, N. Ikeda, École d'Eté de Probabilités de Saint Flour XVIII, 1988. Ed.: P.L. Hennequin. VII, 330 pages. 1990.

Vol. 1428: K. Erdmann, Blocks of Tame Representation Type and Related Algebras. XV. 312 pages. 1990.

Vol. 1429: S. Homer, A. Nerode, R.A. Platek, G.E. Sacks, A. Scedrov, Logic and Computer Science. Seminar, 1988. Editor: P. Odifreddi. V, 162 pages. 1990.

Vol. 1430: W. Bruns, A. Simis (Eds.), Commutative Algebra. Proceedings. 1988. V, 160 pages. 1990.

Vol. 1431: J.G. Heywood, K. Masuda, R. Rautmann, V.A. Solonnikov (Eds.), The Navier-Stokes Equations – Theory and Numerical Methods. Proceedings, 1988. VII, 238 pages. 1990.

Vol. 1432: K. Ambos-Spies, G.H. Müller, G.E. Sacks (Eds.), Recursion Theory Week. Proceedings, 1989. VI, 393 pages. 1990.

Vol. 1433: S. Lang, W. Cherry, Topics in Nevanlinna Theory. II, 174 pages.1990.

Vol. 1434: K. Nagasaka, E. Fouvry (Eds.), Analytic Number Theory. Proceedings, 1988. VI, 218 pages. 1990.

Vol. 1435: St. Ruscheweyh, E.B. Saff, L.C. Salinas, R.S. Varga (Eds.), Computational Methods and Function Theory. Proceedings, 1989. VI, 211 pages. 1990.

Vol. 1436: S. Xambó-Descamps (Ed.), Enumerative Geometry. Proceedings, 1987. V, 303 pages. 1990.

Vol. 1437: H. Inassaridze (Ed.), K-theory and Homological Algebra. Seminar, 1987–88. V, 313 pages. 1990.

Vol. 1438: P.G. Lemarié (Ed.) Les Ondelettes en 1989. Seminar. IV, 212 pages. 1990.

Vol. 1439: E. Bujalance, J.J. Etayo, J.M. Gamboa, G. Gromadzki. Automorphism Groups of Compact Bordered Klein Surfaces: A Combinatorial Approach. XIII, 201 pages. 1990.

Vol. 1440: P. Latiolais (Ed.), Topology and Combinatorial Groups Theory. Seminar, 1985–1988. VI, 207 pages. 1990.

Vol. 1441: M. Coornaert, T. Delzant, A. Papadopoulos. Géométrie et théorie des groupes. X, 165 pages. 1990.

Vol. 1442: L. Accardi, M. von Waldenfels (Eds.), Quantum Probability and Applications V. Proceedings, 1988. VI, 413 pages. 1990.

Vol. 1443: K.H. Dovermann, R. Schultz, Equivariant Surgery Theories and Their Periodicity Properties. VI, 227 pages. 1990.

Vol. 1444: H. Korezlioglu, A.S. Ustunel (Eds.), Stochastic Analysis and Related Topics Vl. Proceedings, 1988. V, 268 pages. 1990.

Vol. 1445: F. Schulz, Regularity Theory for Quasilinear Elliptic Systems and – Monge Ampère Equations in Two Dimensions. XV, 123 pages. 1990.

Vol. 1446: Methods of Nonconvex Analysis. Seminar, 1989. Editor: A. Cellina. V, 206 pages. 1990.

Vol. 1447: J.-G. Labesse, J. Schwermer (Eds), Cohomology of Arithmetic Groups and Automorphic Forms. Proceedings, 1989. V, 358 pages. 1990.

Vol. 1448: S.K. Jain, S.R. López-Permouth (Eds.), Non-Commutative Ring Theory. Proceedings, 1989. V, 166 pages. 1990.

Vol. 1449: W. Odyniec, G. Lewicki, Minimal Projections in Banach Spaces. VIII, 168 pages. 1990.

Vol. 1450: H. Fujita, T. Ikebe, S.T. Kuroda (Eds.), Functional-Analytic Methods for Partial Differential Equations. Proceedings, 1989. VII, 252 pages. 1990.

Vol. 1451: L. Alvarez-Gaumé, E. Arbarello, C. De Concini, N.J. Hitchin, Global Geometry and Mathematical Physics. Montecatini Terme 1988. Seminar. Editors: M. Francaviglia, F. Gherardelli. IX, 197 pages. 1990.

Vol. 1452: E. Hlawka, R.F. Tichy (Eds.), Number-Theoretic Analysis. Seminar, 1988–89. V, 220 pages. 1990.

Vol. 1453: Yu.G. Borisovich, Yu.E. Gliklikh (Eds.), Global Analysis – Studies and Applications IV. V, 320 pages. 1990.

Vol. 1454: F. Baldassari, S. Bosch, B. Dwork (Eds.), p-adic Analysis. Proceedings, 1989. V, 382 pages. 1990.

Vol. 1455: J.-P. Françoise, R. Roussarie (Eds.), Bifurcations of Planar Vector Fields. Proceedings, 1989. VI, 396 pages. 1990.

Vol. 1456: L.G. Kovács (Ed.), Groups – Canberra 1989. Proceedings. XII, 198 pages. 1990.

Vol. 1457: O. Axelsson, L.Yu. Kolotilina (Eds.), Preconditioned Conjugate Gradient Methods. Proceedings, 1989. V, 196 pages. 1990.

Vol. 1458: R. Schaaf, Global Solution Branches of Two Point Boundary Value Problems. XIX, 141 pages. 1990.

Vol. 1459: D. Tiba, Optimal Control of Nonsmooth Distributed Parameter Systems. VII, 159 pages. 1990.

Vol. 1460: G. Toscani, V. Boffi, S. Rionero (Eds.), Mathematical Aspects of Fluid Plasma Dynamics. Proceedings, 1988. V, 221 pages. 1991.

Vol. 1461: R. Gorenflo, S. Vessella, Abel Integral Equations. VII, 215 pages. 1991.

Vol. 1462: D. Mond, J. Montaldi (Eds.), Singularity Theory and its Applications. Warwick 1989, Part I. VIII, 405 pages. 1991.

Vol. 1463: R. Roberts, I. Stewart (Eds.), Singularity Theory and its Applications. Warwick 1989, Part II. VIII, 322 pages. 1991.

Vol. 1464: D. L. Burkholder, E. Pardoux, A. Sznitman, Ecole d'Eté de Probabilités de Saint- Flour XIX-1989. Editor: P. L. Hennequin. VI, 256 pages. 1991.

Vol. 1465: G. David, Wavelets and Singular Integrals on Curves and Surfaces. X, 107 pages. 1991.

Vol. 1466: W. Banaszczyk, Additive Subgroups of Topological Vector Spaces. VII, 178 pages. 1991.

Vol. 1467: W. M. Schmidt, Diophantine Approximations and Diophantine Equations. VIII, 217 pages. 1991.

Vol. 1468: J. Noguchi, T. Ohsawa (Eds.), Prospects in Complex Geometry. Proceedings, 1989. VII, 421 pages. 1991.

Vol. 1469: J. Lindenstrauss, V. D. Milman (Eds.), Geometric Aspects of Functional Analysis. Seminar 1989-90. XI, 191 pages. 1991.

Vol. 1470: E. Odell, H. Rosenthal (Eds.), Functional Analysis. Proceedings, 1987-89. VII, 199 pages. 1991.

Vol. 1471: A. A. Panchishkin, Non-Archimedean L-Functions of Siegel and Hilbert Modular Forms. VII, 157 pages. 1991.

Vol. 1472: T. T. Nielsen, Bose Algebras: The Complex and Real Wave Representations. V, 132 pages. 1991.

Vol. 1473: Y. Hino, S. Murakami, T. Naito, Functional Differential Equations with Infinite Delay. X, 317 pages. 1991.

Vol. 1474: S. Jackowski, B. Oliver, K. Pawałowski (Eds.), Algebraic Topology, Poznań 1989. Proceedings. VIII, 397 pages. 1991.

Vol. 1475: S. Busenberg, M. Martelli (Eds.), Delay Differential Equations and Dynamical Systems. Proceedings, 1990. VIII, 249 pages. 1991.

Vol. 1476: M. Bekkali, Topics in Set Theory. VII, 120 pages. 1991.

Vol. 1477: R. Jajte, Strong Limit Theorems in Noncommutative L_2-Spaces. X, 113 pages. 1991.

Vol. 1478: M.-P. Malliavin (Ed.), Topics in Invariant Theory. Seminar 1989-1990. VI, 272 pages. 1991.

Vol. 1479: S. Bloch, I. Dolgachev, W. Fulton (Eds.), Algebraic Geometry. Proceedings, 1989. VII, 300 pages. 1991.

Vol. 1480: F. Dumortier, R. Roussarie, J. Sotomayor, H. Żoładek, Bifurcations of Planar Vector Fields: Nilpotent Singularities and Abelian Integrals. VIII, 226 pages. 1991.

Vol. 1481: D. Ferus, U. Pinkall, U. Simon, B. Wegner (Eds.), Global Differential Geometry and Global Analysis. Proceedings, 1991. VIII, 283 pages. 1991.

Vol. 1482: J. Chabrowski, The Dirichlet Problem with L^2-Boundary Data for Elliptic Linear Equations. VI, 173 pages. 1991.

Vol. 1483: E. Reithmeier, Periodic Solutions of Nonlinear Dynamical Systems. VI, 171 pages. 1991.

Vol. 1484: H. Delfs, Homology of Locally Semialgebraic Spaces. IX, 136 pages. 1991.

Vol. 1485: J. Azéma, P. A. Meyer, M. Yor (Eds.), Séminaire de Probabilités XXV. VIII, 440 pages. 1991.

Vol. 1486: L. Arnold, H. Crauel, J.-P. Eckmann (Eds.), Lyapunov Exponents. Proceedings, 1990. VIII, 365 pages. 1991.

Vol. 1487: E. Freitag, Singular Modular Forms and Theta Relations. VI, 172 pages. 1991.

Vol. 1488: A. Carboni, M. C. Pedicchio, G. Rosolini (Eds.), Category Theory. Proceedings, 1990. VII, 494 pages. 1991.

Vol. 1489: A. Mielke, Hamiltonian and Lagrangian Flows on Center Manifolds. X, 140 pages. 1991.

Vol. 1490: K. Metsch, Linear Spaces with Few Lines. XIII, 196 pages. 1991.

Vol. 1491: E. Lluis-Puebla, J.-L. Loday, H. Gillet, C. Soulé, V. Snaith, Higher Algebraic K-Theory: an overview. IX, 164 pages. 1992.

Vol. 1492: K. R. Wicks, Fractals and Hyperspaces. VIII, 168 pages. 1991.

Vol. 1493: E. Benoît (Ed.), Dynamic Bifurcations. Proceedings, Luminy 1990. VII, 219 pages. 1991.

Vol. 1494: M.-T. Cheng, X.-W. Zhou, D.-G. Deng (Eds.), Harmonic Analysis. Proceedings, 1988. IX, 226 pages. 1991.

Vol. 1495: J. M. Bony, G. Grubb, L. Hörmander, H. Komatsu, J. Sjöstrand, Microlocal Analysis and Applications. Montecatini Terme, 1989. Editors: L. Cattabriga, L. Rodino. VII, 349 pages. 1991.

Vol. 1496: C. Foias, B. Francis, J. W. Helton, H. Kwakernaak, J. B. Pearson, H_∞-Control Theory. Como, 1990. Editors: E. Mosca, L. Pandolfi. VII, 336 pages. 1991.

Vol. 1497: G. T. Herman, A. K. Louis, F. Natterer (Eds.), Mathematical Methods in Tomography. Proceedings 1990. X, 268 pages. 1991.

Vol. 1498: R. Lang, Spectral Theory of Random Schrödinger Operators. X, 125 pages. 1991.

Vol. 1499: K. Taira, Boundary Value Problems and Markov Processes. IX, 132 pages. 1991.

Vol. 1500: J.-P. Serre, Lie Algebras and Lie Groups. VII, 168 pages. 1992.

Vol. 1501: A. De Masi, E. Presutti, Mathematical Methods for Hydrodynamic Limits. IX, 196 pages. 1991.

Vol. 1502: C. Simpson, Asymptotic Behavior of Monodromy. V, 139 pages. 1991.

Vol. 1503: S. Shokranian, The Selberg-Arthur Trace Formula (Lectures by J. Arthur). VII, 97 pages. 1991.

Vol. 1504: J. Cheeger, M. Gromov, C. Okonek, P. Pansu, Geometric Topology: Recent Developments. Editors: P. de Bartolomeis, F. Tricerri. VII, 197 pages. 1991.

Vol. 1505: K. Kajitani, T. Nishitani, The Hyperbolic Cauchy Problem. VII, 168 pages. 1991.

Vol. 1506: A. Buium, Differential Algebraic Groups of Finite Dimension. XV, 145 pages. 1992.

Vol. 1507: K. Hulek, T. Peternell, M. Schneider, F.-O. Schreyer (Eds.), Complex Algebraic Varieties. Proceedings, 1990. VII, 179 pages. 1992.

Vol. 1508: M. Vuorinen (Ed.), Quasiconformal Space Mappings. A Collection of Surveys 1960-1990. IX, 148 pages. 1992.

Vol. 1509: J. Aguadé, M. Castellet, F. R. Cohen (Eds.), Algebraic Topology - Homotopy and Group Cohomology. Proceedings, 1990. X, 330 pages. 1992.

Vol. 1510: P. P. Kulish (Ed.), Quantum Groups. Proceedings, 1990. XII, 398 pages. 1992.

Vol. 1511: B. S. Yadav, D. Singh (Eds.), Functional Analysis and Operator Theory. Proceedings, 1990. VIII, 223 pages. 1992.

Vol. 1512: L. M. Adleman, M.-D. A. Huang, Primality Testing and Abelian Varieties Over Finite Fields. VII, 142 pages. 1992.

Vol. 1513: L. S. Block, W. A. Coppel, Dynamics in One Dimension. VIII, 249 pages. 1992.

Vol. 1514: U. Krengel, K. Richter, V. Warstat (Eds.), Ergodic Theory and Related Topics III, Proceedings, 1990. VIII, 236 pages. 1992.

Vol. 1515: E. Ballico, F. Catanese, C. Ciliberto (Eds.), Classification of Irregular Varieties. Proceedings, 1990. VII, 149 pages. 1992.

Vol. 1516: R. A. Lorentz, Multivariate Birkhoff Interpolation. IX, 192 pages. 1992.

Vol. 1517: K. Keimel, W. Roth, Ordered Cones and Approximation. VI, 134 pages. 1992.

Vol. 1518: H. Stichtenoth, M. A. Tsfasman (Eds.), Coding Theory and Algebraic Geometry. Proceedings, 1991. VIII, 223 pages. 1992.

Vol. 1519: M. W. Short, The Primitive Soluble Permutation Groups of Degree less than 256. IX, 145 pages. 1992.

Vol. 1520: Yu. G. Borisovich, Yu. E. Gliklikh (Eds.), Global Analysis – Studies and Applications V. VII, 284 pages. 1992.

Vol. 1521: S. Busenberg, B. Forte, H. K. Kuiken, Mathematical Modelling of Industrial Process. Bari, 1990. Editors: V. Capasso, A. Fasano. VII, 162 pages. 1992.

Vol. 1522: J.-M. Delort, F. B. I. Transformation. VII, 101 pages. 1992.

Vol. 1523: W. Xue, Rings with Morita Duality. X, 168 pages. 1992.

Vol. 1524: M. Coste, L. Mahé, M.-F. Roy (Eds.), Real Algebraic Geometry. Proceedings, 1991. VIII, 418 pages. 1992.

Vol. 1525: C. Casacuberta, M. Castellet (Eds.), Mathematical Research Today and Tomorrow. VII, 112 pages. 1992.

Vol. 1526: J. Azéma, P. A. Meyer, M. Yor (Eds.), Séminaire de Probabilités XXVI. X, 633 pages. 1992.

Vol. 1527: M. I. Freidlin, J.-F. Le Gall, Ecole d'Eté de Probabilités de Saint-Flour XX – 1990. Editor: P. L. Hennequin. VIII, 244 pages. 1992.

Vol. 1528: G. Isac, Complementarity Problems. VI, 297 pages. 1992.

Vol. 1529: J. van Neerven, The Adjoint of a Semigroup of Linear Operators. X, 195 pages. 1992.

Vol. 1530: J. G. Heywood, K. Masuda, R. Rautmann, S. A. Solonnikov (Eds.), The Navier-Stokes Equations II – Theory and Numerical Methods. IX, 322 pages. 1992.

Vol. 1531: M. Stoer, Design of Survivable Networks. IV, 206 pages. 1992.

Vol. 1532: J. F. Colombeau, Multiplication of Distributions. X, 184 pages. 1992.

Vol. 1533: P. Jipsen, H. Rose, Varieties of Lattices. X, 162 pages. 1992.

Vol. 1534: C. Greither, Cyclic Galois Extensions of Commutative Rings. X, 145 pages. 1992.

Vol. 1535: A. B. Evans, Orthomorphism Graphs of Groups. VIII, 114 pages. 1992.

Vol. 1536: M. K. Kwong, A. Zettl, Norm Inequalities for Derivatives and Differences. VII, 150 pages. 1992.

Vol. 1537: P. Fitzpatrick, M. Martelli, J. Mawhin, R. Nussbaum, Topological Methods for Ordinary Differential Equations. Montecatini Terme, 1991. Editors: M. Furi, P. Zecca. VII, 218 pages. 1993.

Vol. 1538: P.-A. Meyer, Quantum Probability for Probabilists. X, 287 pages. 1993.

Vol. 1539: M. Coornaert, A. Papadopoulos, Symbolic Dynamics and Hyperbolic Groups. VIII, 138 pages. 1993.

Vol. 1540: H. Komatsu (Ed.), Functional Analysis and Related Topics, 1991. Proceedings. XXI, 413 pages. 1993.

Vol. 1541: D. A. Dawson, B. Maisonneuve, J. Spencer, Ecole d´ Eté de Probabilités de Saint-Flour XXI - 1991. Editor: P. L. Hennequin. VIII, 356 pages. 1993.

Vol. 1542: J.Fröhlich, Th.Kerler, Quantum Groups, Quantum Categories and Quantum Field Theory. VII, 431 pages. 1993.

Vol. 1543: A. L. Dontchev, T. Zolezzi, Well-Posed Optimization Problems. XII, 421 pages. 1993.

Vol. 1544: M.Schürmann, White Noise on Bialgebras. VII, 146 pages. 1993.

Vol. 1545: J. Morgan, K. O'Grady, Differential Topology of Complex Surfaces. VIII, 224 pages. 1993.

Vol. 1546: V. V. Kalashnikov, V. M. Zolotarev (Eds.), Stability Problems for Stochastic Models. Proceedings, 1991. VIII, 229 pages. 1993.

Vol. 1547: P. Harmand, D. Werner, W. Werner, M-ideals in Banach Spaces and Banach Algebras. VIII, 387 pages. 1993.

Vol. 1548: T. Urabe, Dynkin Graphs and Quadrilateral Singularities. VI, 233 pages. 1993.

Vol. 1549: G. Vainikko, Multidimensional Weakly Singular Integral Equations. XI, 159 pages. 1993.

Vol. 1550: A. A. Gonchar, E. B. Saff (Eds.), Methods of Approximation Theory in Complex Analysis and Mathematical Physics IV, 222 pages, 1993.

Vol. 1551: L. Arkeryd, P. L. Lions, P.A. Markowich, S.R. S. Varadhan. Nonequilibrium Problems in Many-Particle Systems. Montecatini, 1992. Editors: C. Cercignani, M. Pulvirenti. VII, 158 pages 1993.

Vol. 1552: J. Hilgert, K.-H. Neeb, Lie Semigroups and their Applications. XII, 315 pages. 1993.

Vol. 1553: J.-L- Colliot-Thélène, J. Kato, P. Vojta. Arithmetic Algebraic Geometry. Trento, 1991. Editor: E. Ballico. VII, 223 pages. 1993.

Vol. 1554: A. K. Lenstra, H. W. Lenstra, Jr. (Eds.), The Development of the Number Field Sieve. VIII, 131 pages. 1993.

Vol. 1555: O. Liess, Conical Refraction and Higher Microlocalization. X, 389 pages. 1993.

Vol. 1556: S. B. Kuksin, Nearly Integrable Infinite-Dimensional Hamiltonian Systems. XXVII, 101 pages. 1993.

Vol. 1557: J. Azéma, P. A. Meyer, M. Yor (Eds.), Séminaire de Probabilités XXVII. VI, 327 pages. 1993.

Vol. 1558: T. J. Bridges, J. E. Furter, Singularity Theory and Equivariant Symplectic Maps. VI, 226 pages. 1993.

Vol. 1559: V. G. Sprindžuk, Classical Diophantine Equations. XII, 228 pages. 1993.

Vol. 1560: T. Bartsch, Topological Methods for Variational Problems with Symmetries. X, 152 pages. 1993.

Vol. 1561: I. S. Molchanov, Limit Theorems for Unions of Random Closed Sets. X, 157 pages. 1993.

Vol. 1562: G. Harder, Eisensteinkohomologie und die Konstruktion gemischter Motive. XX, 184 pages. 1993.

Vol. 1563: E. Fabes, M. Fukushima, L. Gross, C. Kenig, M. Röckner, D. W. Stroock, Dirichlet Forms. Varenna, 1992. Editors: G. Dell'Antonio, U. Mosco. VII, 245 pages. 1993.

Vol. 1564: J. Jorgenson, S. Lang, Basic Analysis of Regularized Series and Products. IX, 122 pages. 1993.

Vol. 1565: L. Boutet de Monvel, C. De Concini, C. Procesi, P. Schapira, M. Vergne. D-modules, Representation Theory, and Quantum Groups. Venezia, 1992. Editors: G. Zampieri, A. D'Agnolo. VII, 217 pages. 1993.

Vol. 1566: B. Edixhoven, J.-H. Evertse (Eds.), Diophantine Approximation and Abelian Varieties. XIII, 127 pages. 1993.

Vol. 1567: R. L. Dobrushin, S. Kusuoka, Statistical Mechanics and Fractals. VII, 98 pages. 1993.

Vol. 1568: F. Weisz, Martingale Hardy Spaces and their Application in Fourier Analysis. VIII, 217 pages. 1994.

Vol. 1569: V. Totik, Weighted Approximation with Varying Weight. VI, 117 pages. 1994.

Vol. 1570: R. deLaubenfels, Existence Families, Functional Calculi and Evolution Equations. XV, 234 pages. 1994.

Vol. 1571: S. Yu. Pilyugin, The Space of Dynamical Systems with the C^0-Topology. X, 188 pages. 1994.

Vol. 1572: L. Göttsche, Hilbert Schemes of Zero-Dimensional Subschemes of Smooth Varieties. IX, 196 pages. 1994.

Vol. 1573: V. P. Havin, N. K. Nikolski (Eds.), Linear and Complex Analysis – Problem Book 3 – Part I. XXII, 489 pages. 1994.

Vol. 1574: V. P. Havin, N. K. Nikolski (Eds.), Linear and Complex Analysis – Problem Book 3 – Part II. XXII, 507 pages. 1994.

Vol. 1575: M. Mitrea, Clifford Wavelets, Singular Integrals, and Hardy Spaces. XI, 116 pages. 1994.